Springer Geography

The Springer Geography series seeks to publish a broad portfolio of scientific books, aiming at researchers, students, and everyone interested in geographical research. The series includes peer-reviewed monographs, edited volumes, textbooks, and conference proceedings. It covers the entire research area of geography including, but not limited to, Economic Geography, Physical Geography, Quantitative Geography, and Regional/Urban Planning.

More information about this series at http://www.springer.com/series/10180

Piyushimita (Vonu) Thakuriah • Nebiyou Tilahun
Moira Zellner
Editors

Seeing Cities Through Big Data

Research, Methods and Applications
in Urban Informatics

 Springer

Editors
Piyushimita (Vonu) Thakuriah
Urban Studies and Urban Big
　Data Centre
University of Glasgow
Glasgow, UK

Nebiyou Tilahun
Department of Urban Planning
　and Policy
University of Illinois at Chicago
Chicago, IL, USA

Moira Zellner
Department of Urban Planning
　and Policy
University of Illinois at Chicago
Chicago, IL, USA

ISSN 2194-315X 　　　　　　　ISSN 2194-3168　(electronic)
Springer Geography
ISBN 978-3-319-82213-6　　　　ISBN 978-3-319-40902-3　(eBook)
DOI 10.1007/978-3-319-40902-3

© Springer International Publishing Switzerland 2017
Softcover reprint of the hardcover 1st edition 2016
This work is subject to copyright. All rights are reserved by the Publisher, whether the whole or part of the material is concerned, specifically the rights of translation, reprinting, reuse of illustrations, recitation, broadcasting, reproduction on microfilms or in any other physical way, and transmission or information storage and retrieval, electronic adaptation, computer software, or by similar or dissimilar methodology now known or hereafter developed.
The use of general descriptive names, registered names, trademarks, service marks, etc. in this publication does not imply, even in the absence of a specific statement, that such names are exempt from the relevant protective laws and regulations and therefore free for general use.
The publisher, the authors and the editors are safe to assume that the advice and information in this book are believed to be true and accurate at the date of publication. Neither the publisher nor the authors or the editors give a warranty, express or implied, with respect to the material contained herein or for any errors or omissions that may have been made.

Printed on acid-free paper

This Springer imprint is published by Springer Nature
The registered company is Springer International Publishing AG
The registered company address is: Gewerbestrasse 11, 6330 Cham, Switzerland

Preface

Big Data is spawning new areas of research, new methods and tools, and new insights into Urban Informatics. This edited volume presents several papers highlighting the opportunities and challenges of using Big Data for understanding urban patterns and dynamics. The volume is intended for researchers, educators, and students who are working in this relatively new area and outlines many of the considerations that are likely to rise in research, applications, and education. The papers tackle a myriad of issues—while some empirical papers showcase insights that Big Data can provide on urban issues, others consider methodological issues or case studies which highlight how Big Data can enrich our understanding of urban systems in a variety of contexts.

The chapters in this book are peer-reviewed papers selected among those originally presented in a 2-day workshop on Big Data and Urban Informatics sponsored by the National Science Foundation and held at the University of Illinois at Chicago in 2014. The workshop brought together researchers, educators, practitioners, and students representing a variety of academic disciplines including Urban Planning, Computer Science, Civil Engineering, Economics, Statistics, and Geography. It was a unique opportunity for urban social scientists and data scientists to exchange ideas in how Big Data can or is being used to address a variety of urban challenges. This edited volume draws from these various disciplines and seeks to address the numerous important issues emerging from these areas.

This volume is intended to introduce and familiarize the reader with how Big Data is being used as well as to highlight different technical and methodological issues that need to be addressed to ensure urban Big Data can answer critical urban questions. The issues explored in this volume cover eight broad categories and span several urban sectors including energy, the environment, transportation, housing, and emergency and crisis management. Authors have also considered the complexities and institutional factors involved in the use of Big Data, from meeting educational needs to changing organizational and social equity perspectives regarding data innovations and entrepreneurship. Others consider the methodological and technical issues that arise in collecting, managing, and analyzing unstructured user-generated content and other sensed urban data. We have aimed to make the volume comprehensive by incorporating papers that show both the immense potential Big Data holds for Urban Informatics and the challenges it poses.

We would like to acknowledge the support of the National Science Foundation which funded the Big Data and Urban Informatics workshop, without which this volume would not have been possible. We would also like to thank the Department of Urban Planning and Policy at the University of Illinois at Chicago which provided additional support for the workshop. A number of people helped us in preparing this edited volume and in the events that led up to the workshop. A special thank you to Alison Macgregor of the University of Glasgow who helped us organize and manage the review process and to Keith Maynard for providing editing support. We are immensely grateful to Ms. Nina Savar whose efforts ensured a successful workshop. We are also indebted to all of the anonymous reviewers who took their time to provide useful feedback to the authors in this volume.

Glasgow, UK   Piyushimita (Vonu) Thakuriah
Chicago, IL   Nebiyou Tilahun
                                                                 Moira Zellner

Contents

Introduction to Seeing Cities Through Big Data: Research, Methods and Applications in Urban Informatics 1
Piyushimita (Vonu) Thakuriah, Nebiyou Y. Tilahun, and Moira Zellner

Big Data and Urban Informatics: Innovations and Challenges to Urban Planning and Knowledge Discovery 11
Piyushimita (Vonu) Thakuriah, Nebiyou Y. Tilahun, and Moira Zellner

Part I Analytics of User-Generated Content

Using User-Generated Content to Understand Cities 49
Dan Tasse and Jason I. Hong

Developing an Interactive Mobile Volunteered Geographic Information Platform to Integrate Environmental Big Data and Citizen Science in Urban Management 65
Zhenghong Tang, Yanfu Zhou, Hongfeng Yu, Yue Gu, and Tiantian Liu

CyberGIS-Enabled Urban Sensing from Volunteered Citizen Participation Using Mobile Devices 83
Junjun Yin, Yizhao Gao, and Shaowen Wang

Part II Challenges and Opportunities of Urban Big Data

The Potential for Big Data to Improve Neighborhood-Level Census Data 99
Seth E. Spielman

Big Data and Survey Research: Supplement or Substitute? 113
Timothy P. Johnson and Tom W. Smith

**Big Spatio-Temporal Network Data Analytics
for Smart Cities: Research Needs** 127
Venkata M.V. Gunturi and Shashi Shekhar

**A Review of Heteroscedasticity Treatment with Gaussian
Processes and Quantile Regression Meta-models** 141
Francisco Antunes, Aidan O'Sullivan, Filipe Rodrigues,
and Francisco Pereira

**Part III Changing Organizational and Educational Perspectives
with Urban Big Data**

**Urban Informatics: Critical Data and Technology
Considerations** 163
Rashmi Krishnamurthy, Kendra L. Smith,
and Kevin C. Desouza

**Digital Infomediaries and Civic Hacking in Emerging
Urban Data Initiatives** 189
Piyushimita (Vonu) Thakuriah, Lise Dirks,
and Yaye Mallon Keita

How Should Urban Planners Be Trained to Handle Big Data? 209
Steven P. French, Camille Barchers, and Wenwen Zhang

**Energy Planning in Big Data Era: A Theme Study
of the Residential Sector** 219
Hossein Estiri

Part IV Urban Data Management

**Using an Online Spatial Analytics Workbench
for Understanding Housing Affordability in Sydney** 233
Christopher Pettit, Andrew Tice, and Bill Randolph

**A Big Data Mashing Tool for Measuring Transit
System Performance** 257
Gregory D. Erhardt, Oliver Lock, Elsa Arcaute,
and Michael Batty

**Developing a Comprehensive U.S. Transit
Accessibility Database** 279
Andrew Owen and David M. Levinson

Seeing Chinese Cities Through Big Data and Statistics 291
Jeremy S. Wu and Rui Zhang

Part V Urban Knowledge Discovery Applied to Different Urban Contexts

Planning for the Change: Mapping Sea Level Rise and Storm Inundation in Sherman Island Using 3Di Hydrodynamic Model and LiDAR 313
Yang Ju, Wei-Chen Hsu, John D. Radke, William Fourt, Wei Lang, Olivier Hoes, Howard Foster, Gregory S. Biging, Martine Schmidt-Poolman, Rosanna Neuhausler, Amna Alruheli, and William Maier

The Impact of Land-Use Variables on Free-Floating Carsharing Vehicle Rental Choice and Parking Duration 331
Mubassira Khan and Randy Machemehl

Dynamic Agent Based Simulation of an Urban Disaster Using Synthetic Big Data 349
A. Yair Grinberger, Michal Lichter, and Daniel Felsenstein

Estimation of Urban Transport Accessibility at the Spatial Resolution of an Individual Traveler 383
Itzhak Benenson, Eran Ben-Elia, Yodan Rofe, and Amit Rosental

Modeling Taxi Demand and Supply in New York City Using Large-Scale Taxi GPS Data 405
Ci Yang and Eric J. Gonzales

Detecting Stop Episodes from GPS Trajectories with Gaps 427
Sungsoon Hwang, Christian Evans, and Timothy Hanke

Part VI Emergencies and Crisis

Using Social Media and Satellite Data for Damage Assessment in Urban Areas During Emergencies 443
Guido Cervone, Emily Schnebele, Nigel Waters, Martina Moccaldi, and Rosa Sicignano

Part VII Health and Well-Being

'Big Data': Pedestrian Volume Using Google Street View Images 461
Li Yin, Qimin Cheng, Zhenfeng Shao, Zhenxin Wang, and Laiyun Wu

**Learning from Outdoor Webcams: Surveillance
of Physical Activity Across Environments**.......................... 471
J. Aaron Hipp, Deepti Adlakha, Amy A. Eyler,
Rebecca Gernes, Agata Kargol, Abigail H. Stylianou,
and Robert Pless

Mapping Urban Soundscapes via Citygram....................... 491
Tae Hong Park

Part VIII Social Equity and Data Democracy

Big Data and Smart (Equitable) Cities........................... 517
Mai Thi Nguyen and Emma Boundy

**Big Data, Small Apps: Premises and Products
of the Civic Hackathon**... 543
Sara Jensen Carr and Allison Lassiter

Erratum... E1

Contributors

Deepti Adlakha, M.U.D. Brown School, Washington University in St. Louis, St. Louis, MO, USA

Prevention Research Center, Washington University in St. Louis, St. Louis, MO, USA

Amna Alruheli Department of Landscape Architecture and Environmental Planning, University of California, Berkeley, CA, USA

Francisco Antunes Center of Informatics and Systems of the University of Coimbra, Coimbra, Portugal

Elsa Arcaute Centre for Advanced Spatial Analysis (CASA), University College London, London, UK

Camille Barchers School of City and Regional Planning, Georgia Institute of Technology, Atlanta, GA, USA

Michael Batty Centre for Advanced Spatial Analysis (CASA), University College London, London, UK

Itzhak Benenson Department of Geography and Human Environment, Tel Aviv University, Tel Aviv, Israel

Eran Ben-Elia Department of Geography and Environmental Development, Ben-Gurion University of the Negev, Beersheba, Israel

Gregory S. Biging Department of Environmental Science, Policy, and Management, University of California, Berkeley, CA, USA

Emma Boundy, M.A. Department of City and Regional Planning, University of North Carolina at Chapel Hill, Chapel Hill, NC, USA

Sara Jensen Carr School of Architecture/Office of Public Health Studies, University of Hawaii Manoa, Honolulu, HI, USA

Guido Cervone GeoInformatics and Earth Observation Laboratory, Department of Geography and Institute for CyberScience, The Pennsylvania State University, University Park, PA, USA

Research Application Laboratory, National Center for Atmospheric Research, Boulder, CO, USA

Qimin Cheng Department of Electronics and Information Engineering, Huazhong University of Science and Technology, Wuhan, China

Kevin C. Desouza School of Public Affairs, Arizona State University, Phoenix, AZ, USA

Lise Dirks Urban Transportation Center, University of Illinois at Chicago, Chicago, IL, USA

Gregory D. Erhardt Department of Civil Engineering, University of Kentucky, Lexington, KY, USA

Hossein Estiri, Ph.D. Institute of Translational Health Sciences, University of Washington, Seattle, WA, USA

Christian Evans Physical Therapy, Midwestern University, Glendale, AZ, USA

Amy A. Eyler, Ph.D. Brown School, Washington University in St. Louis, St. Louis, MO, USA

Prevention Research Center, Washington University in St. Louis, St. Louis, MO, USA

Daniel Felsenstein Department of Geography, Hebrew University of Jerusalem, Jerusalem, Israel

Howard Foster The Center for Catastrophic Risk Management, University of California, Berkeley, CA, USA

William L. Fourt Department of Landscape Architecture and Environmental Planning, University of California, Berkeley, CA, USA

Steven P. French, Ph.D., F.A.I.C.P College of Architecture, Georgia Institute of Technology, Atlanta, GA, USA

Yizhao Gao CyberGIS Center for Advanced Digital and Spatial Studies, University of Illinois at Urbana—Champaign, Urbana, IL, USA

CyberInfrastructure and Geospatial Information Laboratory, University of Illinois at Urbana—Champaign, Urbana, IL, USA

Department of Geography and Geographic Information Science, University of Illinois at Urbana—Champaign, Urbana, IL, USA

Rebecca Gernes, M.S.W., M.P.H. Brown School, Washington University in St. Louis, St. Louis, MO, USA

Prevention Research Center, Washington University in St. Louis, St. Louis, MO, USA

Eric J. Gonzales, Ph.D. Department of Civil and Environmental Engineering, University of Massachusetts Amherst, Amherst, MA, USA

A. Yair Grinberger Department of Geography, Hebrew University of Jerusalem, Jerusalem, Israel

Yue Gu Community and Regional Planning Program, University of Nebraska—Lincoln, Lincoln, NE, USA

Venkata M. V. Gunturi Department of Computer Science, IIIT-Delhi, New Delhi, India

Timothy Hanke Physical Therapy, Midwestern University, Glendale, AZ, USA

J. Aaron Hipp, Ph.D. Department of Parks, Recreation, and Tourism Management, Center for Geospatial Analytics, North Carolina State University, Raleigh, NC, USA

Olivier Hoes Faculty of Civil Engineering and Geosciences, Delft University of Technology, Delft, The Netherlands

Jason I. Hong Human-Computer Interaction Institute, Carnegie Mellon University, Pittsburgh, PA, USA

Wei-Chen Hsu Department of Landscape Architecture and Environmental Planning, University of California, Berkeley, CA, USA

Sungsoon Hwang Geography, DePaul University, Chicago, IL, USA

Timothy P. Johnson Survey Research Laboratory, University of Illinois at Chicago, Chicago, IL, USA

Yang Ju Department of Landscape Architecture and Environmental Planning, University of California, Berkeley, CA, USA

Agata Kargol Computer Science and Engineering, Washington University in St. Louis, St. Louis, MO, USA

Yaye Mallon Keita Department of Urban Planning and Policy, University of Illinois at Chicago, Chicago, IL, USA

Mubassira Khan HDRAustin, TX, USA

Department of Civil, Architectural and Environmental Engineering, The University of Texas at Austin, Austin, TX, USA

Rashmi Krishnamurthy School of Public Affairs, AArizona State University, Phoenix, AZ, USA

Wei Lang Department of Building and Real Estate, The Hong Kong Polytechnic University, Hung Hom, Hong Kong

Allison Lassiter Department of Economics, Monash University, Clayton, VIC, Australia

David M. Levinson Department of Civil, Environmental, and Geo-Engineering, University of Minnesota, Minneapolis, MN, USA

Michal Lichter Department of Geography, Hebrew University of Jerusalem, Jerusalem, Israel

Tiantian Liu Community and Regional Planning Program, University of Nebraska—Lincoln, Lincoln, NE, USA

Oliver Lock Arup, 1 Nicholson St, East Melbourne, VIC, Australia

Randy Machemehl Department of Civil, Architectural and Environmental Engineering, The University of Texas at Austin, Austin, TX, USA

William Maier Department of Environmental Science, Policy, and Management, University of California, Berkeley, CA, USA

Martina Moccaldi GeoInformatics and Earth Observation Laboratory, Department of Geography and Institute for CyberScience, The Pennsylvania State University, University Park, PA, USA

Rosanna Neuhausler Department of City and Regional Planning, University of California, Berkeley, CA, USA

Department of Environmental Science, Policy, and Management, University of California, Berkeley, CA, USA

Mai Thi Nguyen, Ph.D. Department of City and Regional Planning, University of North Carolina at Chapel Hill, Chapel Hill, NC, USA

Aidan O'Sullivan Singapore-MIT Alliance for Research and Technology, Singapore, Singapore

Andrew Owen Department of Civil, Environmental, and Geo-Engineering, University of Minnesota, Minneapolis, MN, USA

Tae Hong Park Department of Music and Performing Arts Professions, Steinhardt School, New York University, New York, NY, USA

Francisco Pereira Technical University of Denmark, Lyngby, Denmark

Christopher Pettit Faculty of Built Environment, City Futures Research Centre, University of New South Wales, Kensington, NSW, Australia

Robert Pless, Ph.D. Computer Science and Engineering, Washington University in St. Louis, St. Louis, MO, USA

John D. Radke Department of Landscape Architecture and Environmental Planning, University of California, Berkeley, CA, USA

Department of City and Regional Planning, University of California, Berkeley, CA, USA

Bill Randolph Faculty of Built Environment, City Futures Research Centre, University of New South Wales, Kensington, NSW, Australia

Filipe Rodrigues Technical University of Denmark, Lyngby, Denmark

Yodan Rofe Switzerland Institute for Drylands Environmental and Energy Research, Ben-Gurion University of the Negev, Beersheba, Israel

Amit Rosental Department of Geography and Human Environment, Tel Aviv University, Tel Aviv, Israel

Martine Schmidt-Poolman The Center for Catastrophic Risk Management, University of California, Berkeley, CA, USA

Emily Schnebele GeoInformatics and Earth Observation Laboratory, Department of Geography and Institute for CyberScience, The Pennsylvania State University, University Park, PA, USA

Zhenfeng Shao The State Key Laboratory of Information Engineering on Surveying Mapping and Remote Sensing, Wuhan University, Wuhan, China

Shashi Shekhar Department of Computer Science and Engineering, University of Minnesota, Minneapolis, MN, USA

Rosa Sicignano GeoInformatics and Earth Observation Laboratory, Department of Geography and Institute for CyberScience, The Pennsylvania State University, University Park, PA, USA

Kendra L. Smith Morrison Institute for Public Policy, Arizona State University, Phoenix, AZ, USA

Tom W. Smith General Social Survey, NORC at the University of Chicago, Chicago, IL, USA

Seth E. Spielman University of Colorado, Boulder, CO, USA

Abigail H. Stylianou Computer Science and Engineering, Washington University in St. Louis, St. Louis, MO, USA

Zhenghong Tang Community and Regional Planning Program, University of Nebraska—Lincoln, Lincoln, NE, USA

Dan Tasse Human-Computer Interaction Institute, Carnegie Mellon University, Pittsburgh, PA, USA

Piyushimita (Vonu) Thakuriah Urban Studies and Urban Big Data Centre, University of Glasgow, Glasgow, UK

Andrew Tice Faculty of Built Environment, City Futures Research Centre, University of New South Wales, Kensington, NSW, Australia

Nebiyou Tilahun Department or Urban Planning and Policy, College or Urban Planning and Public Affairs, University of Illinois at Chicago, Chicago, IL, USA

Shaowen Wang CyberGIS Center for Advanced Digital and Spatial Studies, University of Illinois at Urbana—Champaign, Urbana, IL, USA

CyberInfrastructure and Geospatial Information Laboratory, University of Illinois at Urbana—Champaign, Urbana, IL, USA

Department of Geography and Geographic Information Science, University of Illinois at Urbana—Champaign, Urbana, IL, USA

Department of Computer Science, University of Illinois at Urbana—Champaign, Urbana, IL, USA

Department of Urban and Regional Planning, University of Illinois at Urbana—Champaign, Urbana, IL, USA

National Center for Supercomputing Applications, University of Illinois at Urbana—Champaign, Urbana, IL, USA

Zhenxin Wang Center for Human-Engaged Computing, Kochi University of Technology, Kochi, Japan

Department of Urban and Regional Planning, University at Buffalo, The State University of New York, Buffalo, NY, USA

Nigel Waters GeoInformatics and Earth Observation Laboratory, Department of Geography and Institute for CyberScience, The Pennsylvania State University, University Park, PA, USA

Laiyun Wu Department of Urban and Regional Planning, University at Buffalo, The State University of New York, Buffalo, NY, USA

Jeremy S. Wu, Ph.D. Retired, Census Bureau, Suitland, Maryland, and Department of Statistics, George Washington University, Washington, DC, USA

Ci Yang, Ph.D. Senior Transportation Data Scientist, DIGITALiBiz, Inc., Cambridge, MA, USA

Junjun Yin CyberGIS Center for Advanced Digital and Spatial Studies, University of Illinois at Urbana—Champaign, Urbana, IL, USA

CyberInfrastructure and Geospatial Information Laboratory, University of Illinois at Urbana—Champaign, Urbana, IL, USA

Department of Geography and Geographic Information Science, University of Illinois at Urbana—Champaign, Urbana, IL, USA

Li Yin Department of Urban and Regional Planning, University at Buffalo, The State University of New York, Buffalo, NY, USA

Hongfeng Yu Department of Computer Science and Engineering, University of Nebraska—Lincoln, Lincoln, NE, USA

Moira Zellner Department or Urban Planning and Policy, College or Urban Planning and Public Affairs, University of Illinois at Chicago, Chicago, IL, USA

Rui Zhang Public Relations and Communications Department, Digital China Holdings Limited, Beijing, China

Wenwen Zhang School of City and Regional Planning, Georgia Institute of Technology, Atlanta, GA, USA

Yanfu Zhou Community and Regional Planning Program, University of Nebraska—Lincoln, Lincoln, NE, USA

Introduction to Seeing Cities Through Big Data: Research, Methods and Applications in Urban Informatics

Piyushimita (Vonu) Thakuriah, Nebiyou Y. Tilahun, and Moira Zellner

1 Scope of Workshop and the Book

The chapters in this book were first presented in a 2-day workshop on Big Data and Urban Informatics held at the University of Illinois at Chicago in 2014. The workshop, sponsored by the National Science Foundation, brought together approximately 150 educators, practitioners and students from 91 different institutions in 11 countries. Participants represented a variety of academic disciplines including Urban Planning, Computer Science, Civil Engineering, Economics, Statistics, and Geography and provided a unique opportunity for discussions by urban social scientists and data scientists interested in the use of Big Data to address urban challenges. The papers in this volume are a selected subset of those presented at the workshop and have gone through a peer-review process.

Our main motivation for the workshop was to convene researchers and professionals working on the emerging interdisciplinary research area around urban Big Data. We sought to organize a community with interests in theoretical developments and applications demonstrating the use of urban Big Data, and the next-generation of Big Data services, tools and technologies for Urban Informatics. We were interested in research results as well as idea pieces and works in progress that highlighted research needs and data limitations. We sought papers that clearly create or use novel, emerging sources of Big Data for urban and regional analysis in transportation, environment, public health, land-use, housing, economic

---

P. Thakuriah (✉)
Urban Studies and Urban Big Data Centre, University of Glasgow, Glasgow, UK
e-mail: piyushimita.thakuriah@glasgow.ac.uk

N.Y. Tilahun • M. Zellner
Department or Urban Planning and Policy, College or Urban Planning and Public Affairs, University of Illinois at Chicago, Chicago, IL, USA
e-mail: ntilahun@uic.edu; mzellner@uic.edu

development, labor markets, criminal justice, population demographics, urban ecology, energy, community development, and public participation. A background paper titled *Big Data and Urban Informatics: Innovations and Challenges to Urban Planning and Knowledge Discovery* (Thakuriah et al. 2016b) documenting the major motivations for the workshop is a chapter in this book.

2 Topics on Big Data and Urban Informatics

The chapters in this book are organized around eight broad categories: (1) Analytics of user-generated content; (2) Challenges and opportunities of urban Big Data; (3) Changing organizational and educational perspectives with urban Big Data; (4) Urban data management; (5) Urban knowledge discovery applied to a variety of urban contexts; (6) Emergencies and Crisis; (7) Health and well-being; and (8) Social equity and data democracy.

2.1 Analytics of User-Generated Content

The first set focuses on how to analyze user-generated content. Understanding urban dynamics or urban environmental problems is challenged by the paucity of public data. The ability to collect and analyze geo-tagged social media is emerging as a way to address this shortage or to supplement existing data, for use by planners, businesses and citizens. New platforms to integrate these forms of data are proposed (Tasse and Hong 2016) but are not without their limitations. In particular, GIS platforms have been evaluated (Tang et al. 2016) that hint at the critical role of committed users in ensuring the successful and reliable use of these tools, and the consequent need for integration of online and off-line activities and for the effective transfer of information to individuals' mobile devices.

Other GIS-enabled frameworks are proposed (Yin et al. 2016a) to support citizen sensing of urban environmental pollution like noise. Such participatory computing architecture supports scalable user participation and data-intensive processing, analysis and visualization.

2.2 Challenges and Opportunities of Urban Big Data

The second set of papers considers the challenges and opportunities of urban Big Data, particularly as an auxiliary data source that can be combined with more traditional survey data, or even as a substitute for large survey-based public datasets. Big Data exists within a broader data economy that has changed in recent years (e.g., the American Community Survey (ACS) data quality). Spielman (2016) argues that carefully considered Big Data sources hold potential to increase confidence in the estimates provided by data sources such as the ACS. Recognizing that Big Data appears as an attractive alternative to design-based survey data, Johnson and Smith (2016) caution the potential of serious methodological costs and call on efforts to find ways of integrating these data sources, which have different qualities that make them valuable to understand cities (Johnson and Smith 2016).

In addition to the cost savings, the potential for data fusion strategies lies in the integration of a rich diversity of data sources shedding light on complex urban phenomena from different angles, and covering different gaps. There are, however, major barriers to doing so, stemming from the difficulty in controlling the quality and quantity of the data, and privacy issues (Spielman 2016). The proliferation of Big Data sources also demand new approaches to computation and analysis. Gunturi and Shekhar (2016), explore the computational challenges posed by spatio-temporal Big Data generated from location-aware sensors and how these may be addressed by use of scalable analytics. In another application, Antunes et al. (2016) discuss how explicitly addressing heteroscedasticity greatly improves the quality of model predictions and the confidence associated with those predictions in regression analysis using Big Data.

2.3 Changing Organizational and Educational Perspectives with Urban Big Data

The third set of papers focus on the organizational and educational perspectives that change with Big and Open Urban Data. Cities are investing on technologies to enhance human and automated decision-making. For smarter cities, however, urban systems and subsystems require connectivity through data and information management. Conceptualizing cities as platforms, Krishnamurthy et al. (2016) discuss the importance of how data and technology management are critical for cities to become agile, adaptable and scalable while also raising critical considerations to ensure such goals are achieved. Thakuriah et al. (2016a) review organizations in the urban data sector with the aim of understanding their role in the production of data and service delivery using data. They identify nine organizational types in this dynamic and rapidly evolving sector, which they align along five dimensions to account for their mission, major interest, products and activities: techno-managerial, scientific, business and commercial, urban engagement, and openness and transparency.

Despite the rapid emergence of this data rich world, French et al. (2016) ask if the urban planners of tomorrow are being trained to leverage these emerging resources for creating better urban spaces. They argue that urban planners are still being educated to work in a data poor environment, taking courses in statistics, survey research and projection and estimation that are designed to fill in the gaps in this environment. With the advent of Big Data, visualization, simulation, data mining and machine learning may become the appropriate tools planners can use, and planning education and practice need to reflect this new reality (French et al. 2016). In line with this argument, Estiri (2016) proposes new planning frameworks for planning for urban energy demand, based on improvements that non-linear modeling approaches provide over mainstream traditional linear modeling.

2.4 Urban Data Management

The book also includes examples of online platforms and software tools that allow for urban data management and applications that use such urban data for measurement of urban indicators. The AURIN (Australian Urban Research Infrastructure Network) workbench (Pettit et al. 2016), for example, provides a machine-to-machine online access to large scale distributed and heterogeneous data resources from across Australia, which can be used to understand, among other things, housing affordability in Australia. AURIN allows users to systematically access existing data and run spatial-statistical analysis, but a number of additional software tools are required to undertake data extraction and manipulation. In another application to measure the performance of transit systems in San Francisco, researchers have developed software tools to support the fusion and analysis of large, passively collected data sources like automated vehicle location (AVL) and automated passenger counts (APC) (Erhardt et al. 2016). The tools include methods to expand the data from a sample of buses, and is able to report and track performance in several key metrics and over several years. Queries and comparisons support the analysis of change over time.

Owen and Levinson (2016) also showcase a national public transit job accessibility evaluation at the Census block level. This involved assembling and processing a comprehensive national database of public transit network topology and travel times, allowing users to calculate accessibility continuously for every minute within a departure time window of interest. The increased computational complexity is offset by the robust representation of the interaction between transit service frequency and accessibility at multiple departure times.

Yet, the data infrastructure needed to support Urban Informatics does not materialize overnight. Wu and Zhang (2016) demonstrate how resources at the scale of an entire country is needed to establish basic processes required to develop comprehensive citizen-oriented services. By focusing on China's emerging smart cities program, they demonstrate the need for a proactive data-driven approach to meet challenges posed by China's urbanization. The approach needs not only a number of technological and data-oriented solutions, but also a change in culture towards statistical thinking, quality management, and data integration. New investments in smart cities have the potential to design systems such that the data can lead to much-needed governmental innovations towards impact.

2.5 Urban Knowledge Discovery

Big Data is playing a major role in urban knowledge discovery and planning support. For example, a high-resolution digital surface model (DSM) from Light Detection and Ranging (LiDAR) have supported the dynamic simulation of flooding due to sea level rise in California (Ju et al. 2016). This study provides more detailed information than static mapping, and serves as a fine database for better planning, management, and governance to understand future scenarios. In another example, Khan and Machemehl (2016) study how land use and different social and policy variables affect free-floating carsharing vehicle choice and parking duration, for which there is very little empirical data. The authors use two approaches; logistic regression and a duration model and find that land-use level socio-demographic attributes are important factors in explaining usage patterns of carsharing services. This has implications for carsharing parking policy and the availability of transit around intermodal transportation. Another example by Grinberger et al. (2016) shows that synthetic big data can also be generated from standard administrative small data for applications in urban disaster scenarios. The data decomposition process involves moving from a database describing only hundreds or thousands of spatial units to one containing records of millions of buildings and individuals (agents) over time, that then populate an agent-based simulation of responses to a hypothetical earthquake in downtown Jerusalem. Simulations show that temporary shocks to movement and traffic patterns can generate longer term lock-in effects, which reduce commercial activity. The issue arising here is the ability to identify when this fossilization takes place and when a temporary shock has passed the point of no return. A large level of household turnover and 'churning' through the built fabric of the city in the aftermath of an earthquake was also observed, which points to a waste of resources, material, human and emotional. Less vulnerable socio-economic groups 'weather the storm' by dispersing and then re-clustering over time.

A suite of studies focuses on new methods to apply Big Data to transportation planning and management, particularly with the help of GIS tools. Benenson et al. (2016) use big urban GIS data that is already available to measure accessibility from the viewpoint of an individual traveler going door-to-door. In their work, a computational application that is based on the intensive querying of relational database management systems was developed to construct high-resolution accessibility maps for an entire metropolitan area, to evaluate new infrastructure projects. High-resolution representations of trips enabled unbiased accessibility estimates, providing more realistic assessments of such infrastructure investments, and a platform for transportation planning. Similarly, Yang and Gonzales (2016) show that Big Data derived from taxicabs' Global Positioning Systems (GPS) can be used to refine travel demand and supply models and street network assessments, by processing and integrating with GIS. Such evaluations can help identify service mismatch, and support fleet regulation and management. Hwang et al. (2016) demonstrate a case where GPS trajectory data is used to study travel behavior and to estimate carbon emission from vehicles. They propose a reliable method for partitioning GPS trajectories into meaningful elements for detecting a stay point (where an individual stays for a while) using a density-based spatial clustering algorithm.

2.6 Emergencies and Crisis

Big Data has particular potential in helping to deal with emergencies and urban crises in real time. Cervone et al. (2016) propose a new method to use real-time social media data (e.g., Twitter, photos) to augment remote sensing observations of transportation infrastructure conditions in response to emergencies. Challenges remain, however, associated with producer anonymity and geolocation accuracy, as well as differing levels in data confidence.

2.7 Health and Well-Being

Health and well-being is another major area where Big Data is making significant contributions. Data on pedestrian movement has however proven difficult and costly to collect and analyze. Yin et al. (2016b) propose and test a new image-based machine learning method which processes panoramic street images from Google Street View to detect pedestrians. Initial results with this method resemble the pedestrian field counts, and thus can be used for planning and design. Another paper, by Hipp et al. (2016) using the Archive of Many Outdoor Scenes (AMOS) project aims to geolocate, annotate, archive, and visualize outdoor cameras and images to serve as a resource for a wide variety of scientific applications. The AMOS image dataset, crowdsourcing, and eventually machine learning can be used to develop reliable, real-time, non-labor intensive and valid tools to improve physical activity assessment via online, outdoor webcam capture of global physical activity patterns and urban built environment characteristics.

A third paper (Park 2016) describes research conducted under the Citygram project umbrella and illustrates how a cost-effective prototype sensor network, remote sensing hardware and software, database interaction APIs, soundscape analysis software, and visualization formats can help characterize and address urban noise pollution in New York City. This work embraces the idea of time-variant, poly-sensory cartography, and reports on how scalable infrastructural technologies can capture urban soundscapes to create dynamic soundmaps.

2.8 Social Equity and Data Democracy

Last, but not least, Nguyen and Boundy (2016) discuss issues surrounding Big Data and social equity by focusing on three dimensions: data democratization, digital access and literacy, and promoting equitable outcomes. The authors examine how Big Data has changed local government decision-making, and how Big Data is being used to address social equity in New York, Chicago, Boston, Philadelphia, and Louisville. Big Data is changing decision-making by supplying more data sources, integrating cross agency data, and using predictive rather than reactive analytics. Still, no study has examined the cost-effectiveness of these programs to determine the return on investment. Moreover, local governments have largely focused on tame problems and gains in efficiency. Technologies remain largely accessible to groups that are already advantaged, and may exacerbate social inequalities and inhibit democratic processes. Carr and Lassiter (2016) question the effectiveness of civic apps as an interface between urban data and urban residents, and ask who is represented by and who participates in the solutions offered by apps. They determine that the transparency, collaboration and innovation that hackathons aim to achieve are not yet fully realized, and suggest that a first step to improving the outcomes of civic hackathons is to subject these processes to the same types of scrutiny as any other urban practice.

3 Conclusions

The urban data landscape is changing rapidly. There has been a tremendous amount of interest in the use of emerging forms of data to address complex urban problems. It is therefore an opportune time for an interdisciplinary research community to have a discussion on the range of issues relating to the objectives of Urban Informatics, the research approaches used, the research applications that are emerging, and finally, the many challenges involved in using Big Data for Urban Informatics.

We hope this volume familiarizes the reader to both the potential and the technological and methodological challenges of Big Data, the complexities and institutional factors involved, as well as the educational needs for adopting these emerging data sources into practice, and for adapting to the new world of urban Big Data. We have also sought to incorporate papers that highlight the challenges that need to be addressed so the promise of Big Data is fulfilled. The challenges of representativeness and of equity in the production of such data and in applications that use Big Data are also areas needing continued attention. We have aimed for making the volume comprehensive but we also recognize that a single volume cannot completely cover the broad range of applications using Big Data in urban contexts. We hope this collection proves an important starting point.

References

Antunes F, O'Sullivan A, Rodrigues F, Pereira F (2016) A review of heteroscedasticity treatment with Gaussian Processes and Quantile Regression meta-models. In: Thakuriah P, Tilahun N, Zellner M (eds) Seeing cities through big data: research, methods and applications in urban informatics. Springer, New York

Benenson I, Elia EB, Rofe Y, Rosental A (2016) Estimation of urban transport accessibility at the spatial resolution of an individual traveler. In: Thakuriah P, Tilahun N, Zellner M (eds) Seeing cities through big data: research, methods and applications in urban informatics. Springer, New York

Carr SJ, Lassiter A (2016) Big data, small apps: premises and products of the civic hackathon. In: Thakuriah P, Tilahun N, Zellner M (eds) Seeing cities through big data: research, methods and applications in urban informatics. Springer, New York

Cervone G, Schnebele E, Waters N, Moccaldi M, Sicignano R (2016) Using social media and satellite data for damage assessment in urban areas during emergencies. In: Thakuriah P, Tilahun N, Zellner M (eds) Seeing cities through big data: research, methods and applications in urban informatics. Springer, New York

Erhardt GD, Lock O, Arcaute E, Batty M (2016) A big data mashing tool for measuring transit system performance. In: Thakuriah P, Tilahun N, Zellner M (eds) Seeing cities through big data: research, methods and applications in urban informatics. Springer, New York

Estiri H (2016) Energy planning in big data era: a theme study of the residential sector. In: Thakuriah P, Tilahun N, Zellner M (eds) Seeing cities through big data: research, methods and applications in urban informatics. Springer, New York

French SP, Barchers C, Zhang W (2016) How should urban planners be trained to handle big data? In: Thakuriah P, Tilahun N, Zellner M (eds) Seeing cities through big data: research, methods and applications in urban informatics. Springer, New York

Grinberger AY, Lichter M, Felsenstein D (2016) Dynamic agent based simulation of an urban disaster using synthetic big data. In: Thakuriah P, Tilahun N, Zellner M (eds) Seeing cities through big data: research, methods and applications in urban informatics. Springer, New York

Gunturi VMV, Shekhar S (2016) Big spatio-temporal network data analytics for smart cities: research needs. In: Thakuriah P, Tilahun N, Zellner M (eds) Seeing cities through big data: research, methods and applications in urban informatics. Springer, New York

Hipp JA, Adlakha D, Eyler AA, Gernes R, Kargol A, Stylianou AH, Pless R (2016) Learning from outdoor webcams: surveillance of physical activity across environments. In: Thakuriah P, Tilahun N, Zellner M (eds) Seeing cities through big data: research, methods and applications in urban informatics. Springer, New York

Hwang S, Evans C, Hanke T (2016) Detecting stop episodes from GPS trajectories with gaps. In: Thakuriah P, Tilahun N, Zellner M (eds) Seeing cities through big data: research, methods and applications in urban informatics. Springer, New York

Johnson TP, Smith TW (2016) Big data and survey research: supplement or substitute? In: Thakuriah P, Tilahun N, Zellner M (eds) Seeing cities through big data: research, methods and applications in urban informatics. Springer, New York

Ju Y, Hsu W, Radke JD, Fourt WL, Lang W, Hoes O, Foster H, Biging G, Schmidt-Poolman M, Neuhausler R, Alruheli A, Maier WF (2016) Planning for the change: mapping sea level rise and storm inundation in Sherman Island using 3Di hydrodynamic model and LiDAR. In: Thakuriah P, Tilahun N, Zellner M (eds) Seeing cities through big data: research, methods and applications in urban informatics. Springer, New York

Khan M, Machemehl R (2016) The impact of land-use variables on free-floating carsharing vehicle rental choice and parking duration. In: Thakuriah P, Tilahun N, Zellner M (eds) Seeing cities through big data: research, methods and applications in urban informatics. Springer, New York

Krishnamurthy R, Smith KL, Desouza KC (2016) Urban informatics: critical data and technology considerations. In: Thakuriah P, Tilahun N, Zellner M (eds) Seeing cities through big data: research, methods and applications in urban informatics. Springer, New York

Nguyen MT, Boundy E (2016) Big data and smart (equitable) cities. In: Thakuriah P, Tilahun N, Zellner M (eds) Seeing cities through big data: research, methods and applications in urban informatics. Springer, New York

Owen A, Levinson DM (2016) Developing a comprehensive U.S. transit accessibility database. In: Thakuriah P, Tilahun N, Zellner M (eds) Seeing cities through big data: research, methods and applications in urban informatics. Springer, New York

Park TH (2016) Mapping Urban Soundscapes via Citygram. In: Thakuriah P, Tilahun N, Zellner M (eds) Seeing cities through big data: research, methods and applications in urban informatics. Springer, New York

Pettit C, Tice A, Randolph B (2016) Using an online spatial analytics workbench for understanding housing affordability in Sydney. In: Thakuriah P, Tilahun N, Zellner M (eds) Seeing cities through big data: research, methods and applications in urban informatics. Springer, New York

Spielman SE (2016) The potential for big data to improve neighborhood-level census data. In: Thakuriah P, Tilahun N, Zellner M (eds) Seeing cities through big data: research, methods and applications in urban informatics. Springer, New York

Tang Z, Zhou Y, Yu H, Gu Y, Liu T (2016) Developing an interactive mobile volunteered geographic information platform to integrate environmental big data and citizen science in urban management. In: Thakuriah P, Tilahun N, Zellner M (eds) Seeing cities through big data: research, methods and applications in urban informatics. Springer, New York

Tasse D, Hong JI (2016) Using user-generated content to understand cities. In: Thakuriah P, Tilahun N, Zellner M (eds) Seeing cities through big data: research, methods and applications in urban informatics. Springer, New York

Thakuriah P, Dirks L, Mallon-Keita K (2016a) Digital Infomediaries and Civic Hacking in Emerging Urban Data Initiatives. In: Thakuriah P, Tilahun N, Zellner M (eds) Seeing cities through big data: research, methods and applications in urban informatics. Springer, New York

Thakuriah P, Tilahun N, Zellner M (2016b) Big data and urban informatics: innovations and challenges to urban planning and knowledge discovery. In: Thakuriah P, Tilahun N, Zellner M (eds) Seeing cities through big data: research, methods and applications in urban informatics. Springer, New York

Wu J, Zhang R (2016) Seeing Chinese cities through big data and statistics. In: Thakuriah P, Tilahun N, Zellner M (eds) Seeing cities through big data: research, methods and applications in urban informatics. Springer, New York

Yang C, Gonzales EJ (2016) Modeling taxi demand and supply in New York City using large-scale taxi GPS data. In: Thakuriah P, Tilahun N, Zellner M (eds) Seeing cities through big data: research, methods and applications in urban informatics. Springer, New York

Yin J, Gao Y, Wang S (2016a) CyberGIS-enabled urban sensing from volunteered citizen participation using mobile devices. In: Thakuriah P, Tilahun N, Zellner M (eds) Seeing cities through big data: research, methods and applications in urban informatics. Springer, New York

Yin L, Cheng Q, Shao Z, Wang Z, Wu L (2016b) 'Big Data': pedestrian volume using Google street view images. In: Thakuriah P, Tilahun N, Zellner M (eds) Seeing cities through big data: research, methods and applications in urban informatics. Springer, New York

Big Data and Urban Informatics: Innovations and Challenges to Urban Planning and Knowledge Discovery

Piyushimita (Vonu) Thakuriah, Nebiyou Y. Tilahun, and Moira Zellner

Abstract Big Data is the term being used to describe a wide spectrum of observational or "naturally-occurring" data generated through transactional, operational, planning and social activities that are not specifically designed for research. Due to the structure and access conditions associated with such data, their use for research and analysis becomes significantly complicated. New sources of Big Data are rapidly emerging as a result of technological, institutional, social, and business innovations. The objective of this background paper is to describe emerging sources of Big Data, their use in urban research, and the challenges that arise with their use. To a certain extent, Big Data in the urban context has become narrowly associated with sensor (e.g., Internet of Things) or socially generated (e.g., social media or citizen science) data. However, there are many other sources of observational data that are meaningful to different groups of urban researchers and user communities. Examples include privately held transactions data, confidential administrative micro-data, data from arts and humanities collections, and hybrid data consisting of synthetic or linked data.

The emerging area of Urban Informatics focuses on the exploration and understanding of urban systems by leveraging novel sources of data. The major potential of Urban Informatics research and applications is in four areas: (1) improved strategies for dynamic urban resource management, (2) theoretical insights and knowledge discovery of urban patterns and processes, (3) strategies for urban engagement and civic participation, and (4) innovations in urban management, and planning and policy analysis. Urban Informatics utilizes Big Data in innovative ways by retrofitting or repurposing existing urban models and simulations that are underpinned by a wide range of theoretical traditions, as well as through data-driven modeling approaches that are largely theory agnostic, although these divergent research approaches are starting to converge in some ways. The paper surveys

---

P. Thakuriah (✉)
Urban Studies and Urban Big Data Centre, University of Glasgow, Glasgow, UK
e-mail: piyushimita.thakuriah@glasgow.ac.uk

N.Y. Tilahun • M. Zellner
Department or Urban Planning and Policy, College or Urban Planning and Public Affairs, University of Illinois at Chicago, Chicago, IL, USA
e-mail: ntilahun@uic.edu; mzellner@uic.edu

the kinds of urban problems being considered by going from a data-poor environment to a data-rich world and the ways in which such enquiries have the potential to enhance our understanding, not only of urban systems and processes overall, but also contextual peculiarities and local experiences. The paper concludes by commenting on challenges that are likely to arise in varying degrees when using Big Data for Urban Informatics: technological, methodological, theoretical/epistemological, and the emerging political economy of Big Data.

Keywords Big Data • Urban Informatics • Knowledge discovery • Dynamic resource management • User generated content

1 Introduction

Urban and regional analysis involves the use of a wide range of approaches to understand and manage complex sectors, such as transportation, environment, health, housing, the built environment, and the economy. The goals of urban research are many, and include theoretical understanding of infrastructural, physical and socioeconomic systems; developing approaches to improve urban operations and management; long-range plan making; and impact assessments of urban policy.

Globally, more people live in urban areas than in rural areas, with 54 % of the world's population estimated to be residing in urban areas in 2014 (United Nations 2014), levying unprecedented demand for resources and leading to significant concerns for urban management. Decision-makers face a myriad of questions as a result, including: What strategies are needed to operate cities effectively and efficiently? How can we evaluate potential consequences of complex social policy change? What makes the economy resilient and strong, and how do we develop shockproof cities? How do different cities recover from man-made or natural disasters? What are the technological, social and policy mechanisms needed to develop interventions for healthy and sustainable behavior? What strategies are needed for lifelong learning, civic engagement, and community participation, adaptation and innovation? How can we generate hypothesis about the historical evolution of social exclusion and the role of agents, policies and practices?

The Big Data tsunami has hit the urban research disciplines just like many other disciplines. It has also stimulated the interest of practitioners and decision-makers seeking solutions for governance, planning and operations of multiple urban sectors. The objective of this background paper is to survey the use of Big Data in the urban context across different academic and professional communities, with a particular focus on Urban Informatics. Urban Informatics is the exploration and understanding of urban systems for resource management, knowledge discovery of patterns and dynamics, urban engagement and civic participation, and planning and policy analysis. Urban Informatics research approaches involve both a theory-driven as well as an empirical data-driven perspective centered on emerging Big Data sources. New sources of such data are arising as a result of technological, institutional, social and business innovations, dramatically increasing possibilities for urban researchers. Of equal importance are innovations in accessing hard-to-access data sources and in data linkage, which are leading to new connected data systems. We identify major research questions that may be possible to investigate with the data, as well as existing questions that can be revisited with improved data, in an attempt to identify broad themes for the use of Big Data in Urban Informatics.

While the main research agenda is about knowledge discovery and the better understanding of urban systems, there are equally important questions relating to technical challenges in managing the data and in addressing the methodological and measurement questions that arise. The use of Big Data in Urban Informatics pose significant epistemological challenges regarding the overall modes of research inquiry, and about institutions and the overall political economy regarding the access and use.

This chapter is organized as follows: in Sect. 2, we review background information and different types of Big Data being used for urban research. This is followed in Sect. 3 by a discussion of research approaches and applications in Urban Informatics that involve the use of Big Data. Challenges that arise with the use of such data are discussed in Sect. 4 and conclusions are drawn in Sect. 5.

2 Big Data: Complexities and Types

For many, Big Data is just a buzzword and to a certain extent, the ambiguity in its meaning reflects the different ways in which it is used in different disciplines and user communities. The ambiguity is further perpetuated by the multiple concepts that have become associated with the topic. However, the vagueness and well-worn clichés surrounding the subject have overshadowed potentially strong benefits in well-considered cases of use.

Based on a review of 1437 conference papers and articles that contained the full term "Big Data" in either the title or within the author-provided keywords, De Mauro et al. (2015) arrived at four groups of definitions of Big Data. These definitions focus on: (1) the characteristics of Big Data (massive, rapid, complex, unstructured and so on), with the 3-Vs—Volume, Variety and Velocity—referring to the pure amount of information and challenges it poses (Laney 2001) being a particularly over-hyped example; (2) the technological needs behind the processing of large amounts of data (e.g., as needing serious computing power, or, scalable architecture for efficient storage, manipulation, and analysis); (3) as Big Data being associated with crossing of some sort of threshold (e.g., exceeding the processing capacity of conventional database systems); and (4) as highlighting the impact of Big Data advancement on society (e.g., shifts in the way we analyze information that transform how we understand and organize society).

Moreover, the term Big Data has also come to be associated with not just the data itself, but with curiosity and goal-driven approaches to extract information out of the data (Davenport and Patil 2012), with a focus on the automation of the entire scientific process, from data capture to processing to modeling (Pietsch 2013). This is partly an outcome of the close association between Big Data and data science, which emphasizes data-driven modeling, hypothesis generation and data description in a visually appealing manner. These are elements of what has become known as the Fourth Paradigm of scientific discovery (Gray 2007 as given in Hey et al. 2009), which focuses on exploratory, data-intensive research, in contrast to earlier research paradigms focusing on describing, theory-building and computationally simulating observed phenomena.

Quantitative urban research has historically relied on data from censuses, surveys, and specialized sensor systems. While these sources of data will continue to play a vital role in urban analysis, declining response rates to traditional surveys, and increasing costs of administering the decennial census and maintaining and replacing sensor systems have led to significant challenges to having high-quality data for urban research, planning and operations. These challenges have led to increasing interest in looking at alternative ways of supplementing the urban data infrastructure.

For our purposes, Big Data refers to structured and unstructured data generated naturally as a part of transactional, operational, planning and social activities, or the linkage of such data to purposefully designed data. The use of such data gives rise to technological and methodological challenges and complexities regarding the scientific paradigm and political economy supporting inquiry. Established and emerging sources of urban Big Data are summarized in Table 1: sensor systems, user-generated content, administrative data (open and confidential micro-data), private sector transactions data, data from arts and humanities collections, and hybrid data sources, including linked data and synthetic data. While there are many ways to organize Big Data for urban research and applications, the grouping here is primarily informed by the user community typically associated with each type of data; other factors such as methods of generation, and issues of ownership and access are also considered. The grouping is not mutually exclusive; for example, sensor systems might be owned by public agencies for administrative and operational purposes as well as by private companies to assist with transactions.

2.1 Sensor Systems: Infrastructure and Moving Object Sensors and Internet of Things

Sensors in urban infrastructure (transportation, health, energy, water, waste, weather systems, structural health monitoring systems, environmental management, buildings and so on) result in vast amounts of data on urban systems.

Table 1 Types of Urban Big Data and illustrative user communities

Urban Big Data	Examples	Illustrative user communities
Sensor systems (infrastructure-based or moving object sensors)	Environmental, water, transportation, building management sensor systems; connected systems; Internet of Things	Public and private urban operations and management organizations, independent ICT developers, researchers in the engineering sciences
User-Generated Content ("social" or "human" sensors)	Participatory sensing systems, citizen science projects, social media, web use, GPS, online social networks and other socially generated data	Private businesses, customer/client-focused public organizations, independent developers, researchers in data sciences and urban social sciences
Administrative (governmental) data (open and confidential micro-data)	Open administrative data on transactions, taxes and revenue, payments and registrations; confidential person-level micro-data on employment, health, welfare payments, education records	Open data: innovators, civic hackers, researchers
		Confidential data: government data agencies, urban social scientists involved in economic and social policy research, public health and medical researchers
Private Sector Data (customer and transactions records)	Customer transactions data from store cards and business records; fleet management systems; customer profile data from application forms; usage data from utilities and financial institutions; product purchases and terms of service agreements	Private businesses, public agencies, independent developers, researchers in data sciences and urban social sciences
Arts and Humanities Data	Repositories of text, images, sound recordings, linguistic data, film, art and material culture, and digital objects, and other media	Urban design community, historical, art, architecture and digital humanities organizations, community organizations, data scientists and developers, private organizations
Hybrid data (linked and synthetic data)	Linked data including survey-sensor, census-administrative records	Urban planning and social policy community, government data organizations, private businesses and consultants

Novel patterns of demand and usage can be extracted from these data. The sensors detect activity and changes in a wide variety of urban phenomena involving inanimate objects (infrastructure, building structure), physical aspects of urban areas (land cover, water, tree cover, atmospheric conditions), movement (of cars, people, animals), and activity (use patterns, locations).

As noted earlier, sensor systems might be government or privately owned, with very different access and data governance conditions, and some have been operational for a long time. Typical user communities are public and private urban operations management organizations, independent ICT developers, and researchers in the engineering sciences. However, sensor data, if linked to other sources and archived over long periods of time, can be used by urban social scientists studying long-term social, economic and environmental changes, and their effects on neighborhoods and communities. The emerging smart cities community has become increasingly involved with urban sensor systems, particularly with their integration and performance enhancement through ICT solutions. Many urban sensor systems are now likely to be wirelessly connected, mobile, and significantly more embedded and distributed. Examples from a vast range of operational and planned applications include cooperative or connected vehicle systems, Vehicle-to-Grid systems, Smart Grid systems, and a wide range of indoor and outdoor assistive technologies for seniors and persons with disabilities. The diverse field of remote sensing has been undergoing rapid developments as well, with massive amounts of high-resolution temporal and spatial data being collected more rapidly than ever before with sensors that are mounted on satellites, planes, and lately, drones.

Potentially the "next phase" in this ever-transforming technology landscape is the creation of tiny, intelligent devices that are embedded into everyday objects such as houses, cars, furniture, and clothes, and which can "listen in" and produce recommendations and interventions as needed. The concept of "Internet of Things" (IOT), attributed to Ashton in 1999 (Ashton, 2009), is still primarily a vision at this stage, although there are many individual IoT technologies and systems that are operational. A future with Machine-to-Machine (M2M) communications is envisioned by some, where "billions to trillions of everyday objects and the surrounding environment are connected and managed through a range of devices, communication networks, and cloud-based servers" (Wu et al. 2011). Needless to say, the number and variety of data streams available to study cities will greatly increase.

2.2 User-Generated Content: Social and Human Sensing Systems

Transformative changes have taken place in the last decade regarding ways in which citizens are being involved in co-creating information, and much has been written about crowd-sourcing, Volunteered Geographic Information, and, generally, User-Generated Content (UGC). Citizens, through the use of sensors or social media, and other socially generated information resulting from their participation in social, economic or civic activities, are going from being passive subjects of survey and research studies to being active generators of information. Citizen-based approaches can be categorized as contributory, collaborative, or co-created (Bonney et al. 2009). UGC can generally occur: (1) proactively when users voluntarily generate data on ideas, solve problems, and report on events, disruptions or activities that are of social and civic interest, or (2) retroactively, when analysts process secondary sources of user-submitted data published through the web, social media and other tools (Thakuriah and Geers 2013).

UGC can be proactively generated through idea generation, feedback and problem solving. Developments in Information and Communications Technology (ICT) have expanded the range and diversity of ways in which citizens provide input into urban planning. It has enabled sharing ideas and voting on urban projects, and providing feedback regarding plans and proposals with the potential to affect life in cities. These range from specialized focus groups where citizens provide input to "hackathons" where individuals passionate about ICT and cities get together to generate solutions to civic problems using data. Citizens also solve problems; for example, through human computation (described further in Sect. 3.4) to assess livability or the quality of urban spaces where objective metrics from sensors and machines are not accurate. These activities produce large volumes of structured and unstructured data that can be analyzed to obtain insights into preferences, behaviors and so on.

There has been an explosive growth in the wealth of data proactively generated through different sensing systems. Depending on the level of decision-making needed on the part of users generating information, proactive sensing modes can be disaggregated into participatory (active) and opportunistic (passive) sensing modes. In participatory sensing, users voluntarily opt into systems that are specifically designed to collect information of interest (e.g., through apps which capture information on quality of local social, retail and commercial services, or websites that consolidate information for local ride-sharing), and actively report or upload information on objects of interest. In opportunistic sensing, users enable their wearable or in-vehicle location-aware devices to automatically track and passively transmit their physical sensations, or activities and movements (e.g., real-time automotive tracking applications which measure vehicle movements yielding data on speeds, congestion, incidents and the like, as well as biometric sensors, life loggers and a wide variety of other devices for personal informatics relating to health and well-being). The result of these sensing programs are streams of content including text, images, video, sound, GPS trajectories, physiological signals and others, which are available to researchers at varying levels of granularity depending on, among other factors, the need to protect personally identifiable information.

In terms of retroactive UCG, massive volumes of content are also created every second of every day as a result of users providing information online about their lives and their experiences. The key difference from the proactive modes is that users are not voluntarily opting into specific systems for the purpose of sharing information on particular topics and issues. There are many different types of retroactive UGC that could be used for urban research including Internet search terms, customer ratings, web usage data, and trends data. Data from social networks, micro-blogs or social media streams have generated a lot of interest among researchers, with the dominant services at the current time being online social networks such as Twitter, Facebook, LinkedIn and Foursquare, the latter being a location-based social network. Additionally, there are general question-and-answer databases from which data relevant to urban researchers could be retrieved, as well as a wide variety of multimedia online social sharing platforms such as YouTube and Flickr, and user-created online content sites such Wikipedia, TripAdvisor and Yelp. Such naturally occurring UGC provide a rich source of secondary data on the social fabric of cities, albeit through the lens of their user communities, raising questions regarding bias and lack of generalizability. However, provided appropriate information retrieval and analytics techniques are used, such data can allow detection and monitoring of events and patterns of interest, as well as the ability to identify concerns, emotions and preferences among citizens. Measuring responses to new alerts, service disruptions and policy changes are of particular use for real-time understanding of urban dynamics.

2.3 Administrative Data

Governments collect micro-data on citizens as a part of everyday processes on registrations, transactions and record keeping which typically occur during the delivery of a service. Tax and revenue agencies record data on citizens and taxes paid, revenues generated, licenses issued and real estate or vehicle transactions made. Employment and benefits agencies collect information on income, earnings and disability or retirement benefits. Administrative micro-data in particular contain a wealth of information that is relevant to urban policy evaluation. The advantages often cited regarding the use of administrative data in research include being relatively cheap and potentially less intrusive and yet comprehensive (Gowans et al. 2015), as well as having larger sample sizes, and fewer problems with attrition, non-response, and measurement error compared to traditional survey data sources (Card et al. n.d.).

One particular development with administrative data is its increasing availability through "Open Data" initiatives. These initiatives have largely been driven by open government strategies, generally thought of as reflecting transparent government, collaborative government, and innovative government, with some degree of confusion and ideological tensions about what these terms mean in practice (Shkabatur 2013; Pyrozhenko 2011). Open data initiatives are based on the idea that governmental data should be accessible for everyone to use and to republish without copyright or other restrictions in order to create a knowledgeable, engaged, creative citizenry, while also bringing about accountability and transparency. Open Data initiatives have the potential to lead to innovations (Thorhildur et al. 2013) and to address the needs of the disadvantaged (Gurstein 2011).

National and local governments around the world have now supported open data policies. This has led to a proliferation of open data portals where government agencies upload administrative data that are aggregated or anonymized by removing personal identifiers, and are license-free and in non-proprietary formats. Although they present many opportunities, open data initiatives can face challenges due to a number of reasons including closed government culture in some localities, privacy legislation, limitations in data quality that prohibit publication, and limited user-friendliness (Huijboom and van den Broek 2011).

Many valuable uses of administrative data require access to personally identifiable information, typically micro-data at the level of individual persons, which are usually strictly protected by data protection laws or other governance mechanisms. Personally-identifiable information are those that can be used on its own or together with other information to identify a specific individual, and the benefits of accessing and sharing identifiable administrative data for research purposes have to be balanced against the requirements for data security to ensure the protection of individuals' personal information. Confidential administrative micro-data are of great interest to urban social scientists involved in economic and social policy research, as well as to public health and medical researchers.

There are several current activities involving confidential data that are likely to be of interest to urban researchers. The UK Economic and Social Research Council recently funded four large centers on administrative data research, including running data services to support confidential administrative data linkage, in a manner similar to that offered in other countries such as Denmark, Finland, Norway and Sweden. In the US, the Longitudinal Employment Household Dynamics (LEHD) program of the Census Bureau is an example of an ambitious nationwide program combining federal, state and Census Bureau data on employers and employees from unemployment insurance records, data on employment and wages, additional administrative data and data from censuses and surveys (Abowd et al. 2005), to create detailed estimates of workforce and labor market dynamics.

Administrative data in some cases can be linked both longitudinally for the same person over time and between registers of different types, e.g. linking employment data of parents to children's test scores, or linking medical records to a person's historical location data and other environmental data. The latter, for example, could potentially allow research to investigate questions relating to epigenetics and disease heritability (following Aguilera et al. 2010). Such linkages are also likely to allow in-depth exploration of spatial and temporal variations in health and social exclusion.

2.4 Private Sector Transaction Data

Like government agencies, businesses collect data as a part of their everyday transactions with customers. They also develop detailed customer profiles from different sources. Such privately held data may be contrasted with the aforementioned privately owned sensor systems data as those that continuously track customer activity and use patterns. In a report titled "New Data for Understanding the Human Condition: International Perspectives" (OECD Global Science Forum 2013), customer transactions was identified as a major data category, within which the following were noted as useful data sources: store cards such as supermarket loyalty cards, customer accounts on utilities, financial institutions, and other customer records such as product purchases and service agreements.

Companies have historically used such data to improve business process management, market forecasts and to improve customer relations. Many of these data sources can provide key insights into challenges facing cities and have been increasingly of interest to urban researchers. For example, utility companies have records on energy consumption and transactions, which can help to understand variations in energy demand and impact for sustainable development policy, while also understanding implications for fuel poverty where households spend over an acceptable threshold to maintain adequate heating (NAREC 2013).

2.5 Arts and Humanities Collections and Historical Urban Data

There are vast arts and humanities collections that depict life in the city in the form of text, image, sound recording, and linguistic collections, as well as media repositories such as film, art, material culture, and digital objects. These highly unstructured sources of data allow the representation of the ocular, acoustic and other patterns and transformations in cities to be mapped and visualized, to potentially shed light on social, cultural and built environment patterns in cities. For example, a recent project seeks to digitize a treasure trove of everyday objects such as "advertisements, handbills, pamphlets, menus, invitations, medals, pins, buttons, badges, three-dimensional souvenirs and printed textiles, such as ribbons and sashes" to provide "visual and material insight into New Yorkers' engagement with the social, creative, civic, political, and physical dynamics of the city, from the Colonial era to the present day" (Museum of the City of New York 2014). This collection will contain detailed metadata making it searchable through geographic querying.

Inferring knowledge from such data involves digitization, exploratory media analysis, text and cultural landscape mapping, 3-D mapping, electronic literary analysis, and advanced visualization techniques. With online publishing and virtual archives, content creators and users have the potential to interact with source materials to create new findings, while also facilitating civic engagement, community building and information sharing. Recent focus has been on humanities to foster civic engagement; for example, the American Academy of Arts and Sciences (2013), while making a case for federal funding for the public humanities, emphasized the need to encourage "civic vigor" and to prepare citizens to be "voters, jurors, and consumers". There is potential for this line of work in improving the well-being of cities by going beyond civic engagement, for example, to lifelong learning (Hoadley and Bell 1996; CERI/OECD 1992). Stakeholders engaged in this area are typically organizations involved in cultural heritage and digital culture, such as museums, galleries, memory institutions, libraries, archives and institutions of learning. Typical user communities for this type of data are history, urban design, art and architecture, and digital humanities organizations, as well as community and civic organizations, data scientists, and private organizations. The use of such data in quantitative urban modeling opens up a whole new direction of urban research.

2.6 Hybrid Data and Linked Data Systems

Data combinations can occur in two ways: combination through study design to collect structured and unstructured data during the same data collection effort (e.g., obtaining GPS data from social survey participants, so that detailed movement data are collected from the persons for whom survey responses are available), or through a combination of different data sources brought together data by data linkage or multi-sensor data fusion under the overall banner of what has recently been called "broad data" (Hendler 2014).

There are now several examples where data streams have been linked by design such as household travel surveys and activity diaries administered using both a questionnaire-based survey instrument and a GPS element. One of many examples is the 2007/2008 Travel Tracker data collection by the Chicago Metropolitan Agency for Planning (CMAP), which included travel diaries collected via computer assisted telephone interviews (CATI) and GPS data collected from a subset of participants over 7 days. Recent efforts have expanded the number of sensing devices used and the types of contextual data collected during the survey period. For example, the Integrated Multimedia City Data (iMCD) (Urban Big Data Centre n.d., Thakuriah et al, forthcoming) involves a questionnaire-based survey covering travel, ICT use, education and literacy, civic and community engagement, and sustainable behavior of a random sample of households in Glasgow, UK. Respondents undertake a sensing survey using GPS and life logging sensors leading to location and mobility data and rapid still images of the world as the survey respondent sees it. In the survey background is a significant Information Retrieval effort from numerous social media and multimedia web sources, as well as retrieval of information from transport, weather, crime-monitoring CCTV and other urban sectors. Alongside these data streams are Very High Resolution satellite data and LiDAR allowing digital surface modeling creating 3D urban representations.

The census is the backbone for many types of urban analysis; however, its escalating costs has been noted to be unsustainable, with the cost of the 2010 US Census being almost $94 per housing unit, representing a 34 % increase in the cost per housing unit over Census 2000 costs, which in turn represents a 76 % increase over the costs of the 1990 Census (Reist and Ciango 2013). There was an estimated net undercount of 2.07 % for Blacks, 1.54 % for Hispanics, and 4.88 % for American Indians and Alaska Natives, while non-Hispanic whites had a net over-count of 0.83 % (Williams 2012). Vitrano and Chapin (2014) estimated that without significant intervention, the 2020 Census would cost about $151 per household. This has led the US Census Bureau to actively consider innovative solutions designed to reduce costs while maintaining a high quality census in 2020. Some of the strategies being considered include leveraging the Internet and new methods of communications to improve self-response by driving respondents to the Internet and taking advantage of Internet response processes. Another census hybridization step being considered is the use of administrative records to reduce or eliminate some interviews of households that do not respond to the census and related field contacts.

Similar concerns in the UK led to the Beyond 2011 program where different approaches to produce population statistics were considered. The program recommended several potential approaches such as the use of an online survey for the decennial census and a census using existing government data and annual compulsory surveys (Office for National Statistics 2015). The ONS Big Data project also evaluated the possibility of using web scraping through a series of pilot projects including Internet price data for the Consumer Price Index (CPI) and the Retail Price Index (RPI) and Twitter data to infer student movement, which is a population that has been historically been difficult to capture through traditional surveys (Naylor et al. 2015). Other Big Data sources being studied as a part of the pilots are smart meter data to identify household size/structure and the likelihood of occupancy during the day, and mobile phone positioning data to infer travel patterns of workers.

Another situation is where data on survey respondents are linked to routine administrative records; one approach involved the use of an informed consent process where respondents who agree to participate in a survey are explicitly asked if the information they provide can be linked to their administrative records. One example of this approach is the UK Biobank Survey (Lightfoot and Dibben 2013). Having survey responses linked to administrative data enables important urban policy questions to be evaluated; the key issue here is that participants understand and agree to such linkage.

From urban an operations point of view, connected systems allow a degree of sophistication and efficiency not possible with data from individual data systems. This was touched upon briefly in Sect. 2.1; clearly weather-responsive traffic management systems (Thakuriah and Tilahun 2013) and emergency response systems (Salasznyk et al. 2006) require extensive integration of very different streams of data, often in real time. This can be computationally challenging, but also perhaps equally challenging to get data owners to share information. These types of linked data are likely to accrue a diverse user community including urban planning and operations management researchers, as well as the economic and social policy community, in addition to public and private data organizations.

3 Urban Informatics

Overall, developments with urban Big Data have opened up several opportunities for urban analysis. Building on previous definitions (Foth et al. 2011; Bays and Callanan 2012; Batty 2013; Zheng et al. 2014), we view Urban Informatics as the

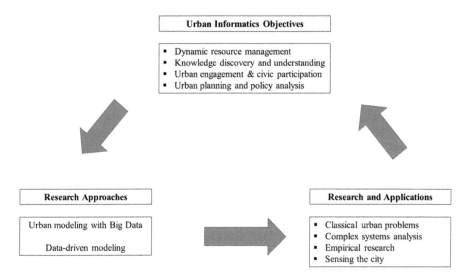

Fig. 1 Relationships among Urban Informatics objectives, research approaches and applications

exploration and understanding of urban patterns and processes, and that it involves analyzing, visualizing, understanding, and interpreting structured and unstructured urban Big Data for four primary objectives:

1. Dynamic resource management: developing strategies for managing scarce urban resources effectively and efficiently and often making decisions in real-time regarding competitive use of resources;
2. Knowledge discovery and understanding: discovering patterns in, and relationships among urban processes, and developing explanations for such trends;
3. Urban engagement and civic participation: developing practices, technologies and other processes needed for an informed citizenry and for their effective involvement in social and civic life of cities;
4. Urban planning and policy analysis: developing robust approaches for urban planning, service delivery, policy evaluation and reform, and also for the infrastructure and urban design decisions.

The overall framework used here, in terms of the objectives, research approach and applications, and their interdependencies, is shown in Fig. 1.

3.1 Research Approaches in Urban Informatics

The analysis of urban systems is theoretically underpinned by a myriad of economic, social, behavioral, biological and physical principles, allowing the simulation of complex interactions, movements, transactions, trading, diffusion and other urban dynamics and diffusion patterns. While some urban models aim to improve long-range economic and infrastructural planning and program evaluation, others attempt to generate empirical understanding of urban dynamics and verification of theoretical urban concepts, and to provide input into shorter-term operations and management of urban sectors. However, Big Data has become closely associated with data-driven science and modeling, which is typically an empirical approach without the social, psychological, economic and regional planning theory which frame urban research. Data-driven modeling brings novel new methodological approaches for using some of the highly unstructured and voluminous types of Big Data, as well as a bottom-up approach to understanding urban systems, particularly for improved dynamic resource management, knowledge discovery and citizen engagement.

The research approaches utilized in Urban Informatics are:

1. **Urban modeling and analysis with Big Data**: The use of Big Data within existing urban modeling and simulation frameworks, and in practical empirical approaches, are grounded in theoretical urban research paradigms, by: (a) reconfiguring/restructuring emerging Big Data through specialized data preparation techniques so that it meets the input requirements of existing urban modeling approaches; or (b) retrofitting or repurposing existing methods through the integration of data-driven approaches (e.g., machine learning, data mining) in the overall analysis scheme, so that urban models are able to use emerging forms of data.
2. **Data-driven models towards "bottom-up" sensing of the city**: Empirical data-driven methods that are derived primarily from the data science and statistical learning communities which focus on retrieval and extraction of information from unstructured or very voluminous streams of data that are not easily accessible to non-specialists. These methods consider pattern detection, knowledge discovery, empirical explanation and hypothesis generation regarding urban phenomena, events and trends.

3.2 Urban Informatics Applications with Big Data

We organize the discussion on Urban Informatics applications using Big Data through urban models and data-driven models in the following ways: (1) reconsidering classical urban problems with new forms of data, (2) use of Big Data for complex systems analysis, (3) applications to address complex urban challenges through empirical research, and (4) through methods to collaboratively sense the city. The applications in turn, help to fine-tune the objectives of Urban Informatics for more comprehensive knowledge discovery, urban planning and operations.

3.2.1 Reconsidering Classical Urban Problems with Big Data

Classical approaches to urban systems analysis include mathematical models of human spatial interaction to measure flows of travelers and services between pairs of points in urban areas (Wilson 1971; Erlander 1980; Sen and Smith 1995), models of urban development, and study of urban structure, and the interaction between transportation and land-use systems (Burgess 1925; Alonso 1960; Lowry 1964; Fujita and Ogawa 1982; Fujita 1988). Other areas are transportation network dynamics and travel mode choice analysis (Beckman et al. 1956; Sheffi 1985; Ben-Akiva and Lerman 1985), models of housing dynamics and residential location theory (Ellis 1967; Muth 1969; Beckmann 1973; Richardson 1977); and models of regional and local economies, labor markets and industry location and agglomeration (Marshall 1920; Isard 1956; Krugman 1991; Fujita et al. 1999). These methods are typically equation-based and draw from operations research and statistical techniques. These models and their numerous variants have contributed to a voluminous and diverse literature on which several decades of planning, policy and operational decisions have been based.

These approaches typically use data from traditional sources such as the census or surveys, and to a lesser degree aggregated administrative or sensor data. Use of the many other urban Big Data sources would require significant modifications to such models "to see around the corners". Perhaps new model development or integration with data science approaches, in addition to specialist curation and processing of the data itself would also be necessary. However, there are now an increasing number of examples where emerging forms of data have been used within the framework of these classical urban models. Recent examples are in the areas of travel demand models, e.g., the use of GPS data to estimate flows of travelers from travel origins to destinations traditionally achieved using census journey-to-work data (Zhang et al. 2010), and the use of detailed freeway and arterial street sensor data along with the synthetic LEHD and other data to measure job accessibility (Levinson et al. 2010). Other examples include studies of labor market dynamics using administrative data (e.g., Bijwaard et al. 2011), use of social media data to measure labor market flows and indexes of job loss, job search, and job posting (Antenucci et al. 2014) and the use of online housing searches to study housing market dynamics in terms of area definition, submarket geography and search pressure locations (Rae 2014).

3.3 Complex Systems Analysis

Large-scale urban modeling practices also use complex systems approaches utilizing Agent-based Models (ABM) and myriad forms of specialized survey, administrative, synthetic and other data sources, to study outcomes that are emergent from individual agent action in interaction with other agents and the environment while also incorporating agent heterogeneity. Well-known early implementations of ABMs include Schelling's segregation model (Schelling 1971) and Conway's Game of Life (Conway 1970). ABMs have found widespread application in diverse areas of urban research. Examples include urban and environmental planning (e.g. Zellner et al. 2009; Zellner and Reeves 2012), transportation (e.g. Tilahun and Levinson 2013; Zellner et al. forthcoming), environmental studies (e.g. Evans and Kelley 2004), large-scale agent based micro-simulation models such as ILUTE (Salvini and Miller 2005), and integrated land, transportation and environment modeling system such as MATSim (Balmer et al. 2009), which provides agent-based mobility simulations. Related developments in computational network perspectives to study a variety of phenomena have also entered modeling practice, including studies of community structure (Girvan and Newman 2002) and the susceptibility of power grids to failure (Kinney et al. 2005).

ABMs have recently used unstructured sources of data, one example of which is the use of GPS trajectories to obtain a better understanding of human mobility patterns within an ABM framework (Jia et al. 2012). Some researchers have also focused on the use of social network data (Kowald and Axhausen 2015 gives examples for the case of transportation planning), while others have utilized social networks to examine peer effects, and processes to exchange opinions, preferences and to share experiences, as well as to see how individuals' participation in social networks lead to outcomes of interest (e.g., Christakis and Fowler 2007, demonstrated the spread of obesity via social relationships in a social network while Tilahun et al. 2011 examined the role of social networks in location choice). The use of online social networks in ABMs has been an interesting development in this respect allowing the flexibility of ABMs to incorporate detailed representation and analysis of the effects of social networks that underlie complex decision problems. One example of this nascent literature is the use of Twitter data within an ABM framework to model diffusion of crisis information (Rand et al. 2015).

3.4 Empirical Urban Research

A vast body of empirical work embedded in the urban disciplines is among the most active consumers of urban data, for better understanding, hypothesis testing and inference regarding urban phenomenon. Among these, one vast research area with requirements for specialized data sources, models and tools is that of environmental sustainability and issues relating to clean air, non-renewable energy dependence and climate change. While difficult to generalize, a recent OECD report identified gaps in quantitative urban and regional modeling tools to systematically assess the impacts of urban systems on climate change and sustainability (OECD 2011). Significant developments in sensor technology have led to the emergence of smart commodities ranging from household appliances to smart buildings. This has in turn led to cost-efficiencies and energy savings, and to the design of Vehicle-to-Grid (V2G) systems (Kempton et al. 2005), and to personal carbon trading

(Bottrill 2006) and vehicular cap-and-trade systems (Lundquist 2011), with data-analytic research around technology adoption, and behavioral and consumption patterns.

Urban models that detect disparities relating to social justice and distributional aspects of transportation, housing, land-use, environment and public health are other consumers of such data. These approaches provide an empirical understanding of the social inclusion and livability aspects of cities, and operational decisions and policy strategies needed to address disparities. This line of work has focused on social exclusion and connections to work and social services (Kain and Persky 1969; Wilson 1987; Krivo and Peterson 1996; Thakuriah et al. 2013), issues of importance to an aging society (Federal Interagency Forum on Aging-Related Statistics 2010), health and aging in place (Black 2008; Thakuriah et al. 2011) and needs of persons with disabilities (Reinhardt et al. 2011). Administrative data has played a significant role in this type of research leading to knowledge discovery about urban processes as well as in evaluation of governmental actions such as welfare reform and post-recession austerity measures. Linked and longitudinal administrative data can support understanding of complex aspects of social justice and changes in urban outcomes over time. For example, Evans et al. (2010) highlighted the importance of using longitudinal administrative data to understand the long-term interplay of multiple events associated with substance abuse over time, while Bottoms and Costello (2009) discuss the role that longitudinal police crime records can play in studying repeat victimization of crime.

New ICT-based solutions to track and monitor activities allow urban quality and well-being to be assessed at more fine-grained levels. Personalized data generated by assistive technologies, ambient assisted living (Abascal et al. 2008) and related ICT applications can contribute to urban quality of life as well as to design solutions supporting urban wellness (e.g., hybrid qualitative-GPS data as described by Huang et al. 2012 to understand barriers to accessing food by midlife and older adults with mobility disability). Mobile heath and awareness technologies (Consolvo et al. 2006) particularly those embedded within serious medical pervasive gaming environments (e.g., DiabetesCity—Collect Your Data, Knoll 2008) and numerous mobile, wearable and other sensor-based physical health recommender systems, one example of which is Lin et al. (2011), open up possibilities for urban researchers to tap into a wealth of data to understand overall built environment and activity-based conditions fostering health and well-being.

3.5 *Approaches to Collaboratively Sense the City*

The discussion above shows that urban information generation and strategies to analyze the data increasingly involve ICT solutions and the active participation of users. Strategies such as focus groups, SWOT, Strategic Approach, Future Workshops and other approaches have been extensively used in the past as a part of urban participatory practice to generate ideas and find solutions to problems. However, advances in ICT solutions have led to the emergence of new models of citizens input into problem solving, plan and design sourcing, voting on projects, and sharing of ideas on projects. Examples range from civic hackers analyzing data from Open Data portals to generate ideas about fixing urban problems to using serious games and participatory simulations for the ideation process (Poplin 2014; Zellner et al. 2012).

As noted earlier, citizens may also engage by generating content through human computation, or by performing tasks that are natural for humans but difficult for machines to automatically carry out (von Ahn et al. 2008). Human computation approaches provide structured ways for citizens to engage in play, to provide input and to interact with, and learn about the urban environment. For example, citizens may be able to judge different proposed urban design, or they may be used to assess the quality of urban spaces where objective metrics from data derived through machine vision algorithms are not accurate. Celino et al. (2012) gives the example of UrbanMatch, a location-based Game with a Purpose (GWAP), which is aimed at exploiting the experience that players have of the urban environment to make judgments towards correctly linking points of interests in the city with most representative photos retrieved from the Internet. There are multiple variants of human computation including social annotations (where users tag or annotate photos or real-world objects), information extraction (e.g., where users are asked to recognize objects in photos), and others.

By "sensing" the city and its different behavioral and use patterns, data-driven models have stimulated research into a broad range of social issues relevant to understanding cities, including building participatory sensing systems for urban engagement, location-based social networks, active travel and health and wellness applications, and mobility and traffic analytics. Other objectives include dynamic resource management of urban assets and infrastructure, assisted living and social inclusion in mobility, and community and crisis informatics. For example, one of the major cited benefits of social media analysis has been the ability to instantaneously and organically sense sentiments, opinions and moods to an extent not previously possible, and ways in which these diffuse over space and time, thereby enabling the policy community to monitor public opinion, and predict social trends. A part of this trend is being stimulated by major governmental agencies which are increasingly realizing the power of social media in understanding where needs are, and how the public are reacting to major policy changes and political events and people's political preferences (Golbeck and Hansen 2013).

A data-driven focus is also being seen in learning analytics (e.g., Picciano 2012), location-based social networks (Zheng and Xie 2011), recommender systems based on collaborative filtering for travel information (Ludwig et al. 2009) and approaches to detect disruptions from social media (Sasaki et al. 2012). Presumably if these information streams are collected over time and linked to other sociodemographic data, it would be possible to examine variations in the outcomes currently measured by the socially generated data to capture urban dynamics to a greater degree.

Overall, Big Data is being increasingly utilized for a range of Urban Informatics research and applications. By using existing urban models with new forms of data, or through data-driven modeling, urban processes and behaviors can be studied in a timely manner and contextual peculiarities of urban processes and local experiences can examined in greater detail. Yet significant challenges arise in their use, which are addressed next.

4 Challenges in Using Big Data for Urban Informatics

The challenges associated with the use of Big Data for Urban Informatics are: (1) technological, (2) methodological, (3) theoretical and epistemological, and (4) due to political economy that arises from accessing and using the data. These challenges are given in Table 2 along with the characteristics of the challenges and examples of the complexities involved with different types of Big Data.

4.1 Technological Challenges

Technological challenges arise due to the need to generate, capture, manage, process, disseminate and discover urban information. The challenges to managing large volumes of structured and unstructured information have been extensively documented elsewhere. Some of the major information management challenges are those relating to building a data infrastructure, cloud stores and multi-cloud architectures, as well as resource discovery mechanisms, and language and execution environments. Other considerations include hardware, software, and the need for well-defined Application Programming Interfaces (API) to capture, integrate, organize, search and query and analyze the data. Of equal importance are scalability, fault-tolerance, and efficiency, and platforms for scalable execution. Various Big Data solutions have emerged in the market such as Hadoop, MapReduce and others, some of which are open source.

One of the biggest challenges with using Big Data for Urban Informatics is not that the data are necessarily huge as in the case of finance, genomics, high-energy physics or other data (although this may change with the incoming deluge of data from the connected vehicles and the IoT worlds). Rather, it is that urban Big Data tends to be fragmented, messy and sometimes unstructured. Particularly for data linkage, when one goes beyond structured, rectangular databases to streaming data through APIs leading to text, image and other unstructured data formats, the diversity and fragmentation can pose significant problems.

Data privacy also becomes all-important with many sources of Big Data, whether they are administrative micro-data or user-generated image or GPS data, and is often a major roadblock to data acquisition for research, particularly for research that requires potentially personally identifiable data. There are many

Table 2 Challenges in using Big Data for Urban Informatics and illustrative topics

Challenges	Characteristics	Challenges by type of data
Technological	Urban information management challenges: 1. Information generation and capture 2. Management 3. Processing 4. Archiving, curation and storage 5. Dissemination and discovery	Information management challenges likely to be very high with real-time, high-volume sensor and UGC data which require specific IT infrastructure development and information management solutions
Methodological	1. Data Preparation Challenges 　(a) Information retrieval and extraction 　(b) Data linkage/information integration 　(c) Data cleaning, anonymization and quality assessment 2. Urban Analysis Challenges 　(a) Developing methods for data-rich urban modeling and data-driven modeling 　(b) Ascertaining uncertainty, biases and error propagation	Data preparation challenges likely to be very high with unstructured or semi-structured sensor, UCG and arts and humanities data, and data from real-time private-sector and administrative transactional systems All types of observational Big Data pose significant methodological challenges in deriving generalizable knowledge requiring specialist knowledge to assess and address measurement issues and error structures
Theoretical and epistemological	1. Understanding metrics, definitions, concepts and changing ideologies and methods to understanding "urban" 2. Determining validity of approaches and limits to knowledge 3. Deriving visions of future cities and the links to sustainability and social justice	All types of observational Big Data pose limitations in deriving theoretical insights and in hypothesis generation without adequate cross-fertilization of knowledge between the data sciences and the urban disciplines, but the challenges are greater with certain forms of UGC and sensor data which yield high-value descriptions but are less amenable to explanations and explorations of causality
Political economy	1. Data entrepreneurship, innovation networks and power structures 2. Value propositions and economic issues 3. Data access, governance framework and provenance 4. Data confidentiality, security and trust management 5. Responsible innovation and emergent ethics	Data confidentiality and power structures pose significant challenges to use of administrative data in open government and program evaluation, while access to private sector transactions data, and privately-controlled sensor and UGC are potentially susceptible to changing innovation and profitability motivations; challenges to ethics and responsible innovation are significantly high for certain sensor-based (e.g., IOT) applications

approaches to data privacy, and these range from technological encryption and anonymization solutions to design, access and rights management solutions. A vast range of Privacy Enhancing Technologies (PETs) (Beresford and Stajano 2003; Gruteser and Grunwald 2003) that are relevant to urban Big Data focus on anonymization of GPS data, images and so on. In the case of administrative micro-data, many approaches to ensure confidentiality are used, including de-identified data, simulated micro-data (called synthetic data) that is constructed to mimic some features of the actual data using micro-simulation methods (Beckman et al. 1996; Harland et al. 2012) and utilization of Trusted Third Party (TTP) mechanisms to minimize the risks of the disclosure of an individual's identity or loss of the data (Gowans et al. 2012).

One major capability needed to progress from data-poor to data-rich urban models is that data should be archived over time, enabling storage of very high-resolution and longitudinal spatio-temporal data. The linkage to other socio-economic, land-use and other longitudinal data opens up additional avenues for in-depth exploration of changes in urban structure and dynamics. Although this was previously a challenge, the decrease in storage costs and increase in linkage capacity has made this possible.

Another important determinant in data access is having access to high-quality resource discovery tools for urban researchers to find and understand data, ontologies for knowledge representation, and a data governance framework that includes harmonization of standards, key terms and operational aspects. Given the vast and dispersed sources of urban Big Data, resource discovery mechanisms to explore and understand data are critical. This includes metadata, or data about the data, containing basic to advanced information describing the data and the management rights to it, including archiving and preservation, in a consistent, standardized manner so that it is understandable and usable by others. Other issues are data lifecycle management (the strategic and operational principles underpinning long-term publication and archiving), access to necessary para-data, (i.e., data about the processes used to collect data), and social annotations (i.e., social bookmarking that allows users to annotate and share metadata about various information sources). These issues not only have significant technical requirements in terms of integrating urban data from different sources; they also have legal (e.g., licensing, terms of service, non-disclosure), ethical (e.g., regarding lack of informed consent in some cases, or use by secondary organizations which did not seek consent), and research culture implications (e.g., establishing a culture of reanalysis of evidence, reproduction and verification of results, minimizing duplication of effort, and building on the work of others, as in Thanos and Rauber 2015).

The above technology issues are not likely to be directly relevant to urban researchers in many cases. However, methodological aspects of Big Data such as information retrieval, linkage and curation or the political economy of Big Data including data access, governance, privacy, and trust management may have IT requirements that could limit the availability of data for urban research.

4.2 Methodological Challenges

We consider two types of methodological challenges: data preparation methods (such as cleaning, retrieving, linking, and other actions needed to prepare data for the end-user) and empirical urban analysis methods (data analytics for knowledge discovery and empirical applications). Sensor and co-created data require special processing and analysis methods to manage very large volumes of unstructured data, from which to retrieve and extract information. With traditional sources of urban data, the specific aspects of the workflow from data collection/generation to analysis are clearly demarcated among professionals from different backgrounds (e.g., data collection is typically done by census takers or surveyors who create a clean data file along with the necessary data documentation, which is then used by urban researchers for further analysis). In contrast to this model, in the case of certain forms of unstructured data (e.g., social media data such as Twitter), the analytics of the data (e.g., using machine learning for topic detection and classification algorithms) happens alongside with, or as a part of, information retrieval or the "gathering" of information from the raw data streams. Thus the "data gathering" and the "data analytics" aspects of the workflow are much more tightly coupled, requiring new skills to be learned by urban researchers wishing to use such data or to have close collaboration with data scientists who have this type of skill.

Observational Big Data involves having to address several methodological challenges for inference. Administrative data, for example, may pose challenges due to causality, endogeneity, and other issues that can bias inference. Socially generated data obtained from participatory sensing and crowd-sourcing are likely to be non-representative in the sense that participants probably do not resemble random samples of the population. Those who are easiest to recruit may also have strong opinions about what the data should show and can provide biased information. Social media users are typically not representative of the overall population since they are more likely to be younger and more digitally savvy (Mislove et al. 2011), and they are also more likely to be concentrated in certain areas or generate differing content depending on where they live (Ghosh and Guha 2013). These patterns may however change over time as the technology becomes more widely used.

In addition, technology changes rapidly and there would always be the issue of the first adopters with specific, non-representative demographics and use patterns. Aside from this, there is dominance by frequent issues and lack of data generation by passive users, and the proliferation of fake accounts which does not add a true representation of moods, opinions and needs, and are sometimes maliciously created to swell sentiments in one direction or the other. Other challenges include the lack of independence or herding effects, which is the effect of prior crowd decisions on subsequent actions. Samples may need to be routinely re-weighted, again on the fly, with the weights depending on the purpose of the analysis. Recent work by Dekel and Shamir (2009), Raykar et al. (2010), and Wauthier and Jordan (2011) consider issues on sampling and sampling bias in crowd-sourcing or citizen science while others have considered sampling issues relating to social networks (Gjoka et al. 2010) and social media (Culotta 2014). However, this work is in its infancy, and further developments are necessary in order to use highly unstructured forms of data for urban inference.

Using Big Data for Urban Informatics requires methods for information retrieval, information extraction, GIS technologies, and multidisciplinary modeling and simulation methods from urban research as well as the data sciences (e.g., machine learning and tools used to analyze text, image and other unstructured sources of data). Methods of visualization and approaches to understanding uncertainty, error propagation and biases in naturally occurring forms of data are essential in order to use and interpret Big Data for urban policy and planning.

4.3 Theoretical and Epistemological Challenges

The theoretical and epistemological challenges pertain to the potential for insights and hypothesis generation about urban dynamics and processes, as well as validity of the approaches used, and the limits to knowledge discovery about urban systems derived from a data focus. As noted earlier, Big Data for Urban Informatics has two distinct roots: quantitative urban research and data science. Although the walls surrounding what may be considered "urban models and simulations" are pervious, these are typically analytical, simulation-based or empirical approaches that are derived from diverse conceptual approaches (e.g., queuing theory, multi-agent systems) and involve strong traditions of using specialist data to calibrate. These models support the understanding of urban systems using theory-driven forecasting of urban resources, simulation of alternative investment scenarios, strategies for engagement with different communities, and evaluation of planning and policy, as well as efficient operation of transportation and environmental systems. Such models are now using Big Data in varying degrees.

At the same time, exploratory data-driven research is largely devoid of theoretical considerations but is necessary to fully utilize emerging data sources to better discover and explore interesting aspects of various urban phenomena. Social data streams and the methods that are rapidly building around them to extract, analyze and interpret information are active research areas, as are analytics around data-driven geography that may be emerging in response to the wealth of geo-referenced data flowing from sensors and people in the environment (e.g., Miller and Goodchild 2014). The timely discovery and continuous detection of interesting urban patterns possible with Big Data and the adoption of innovative data-driven urban management is an important step forward and serves useful operational purposes. The knowledge discovery aspects of data-driven models are important to attract the attention of citizens and decision-makers on urban problems and to stimulate new hypotheses about urban phenomena, which could potentially be rigorously tested using inferential urban models.

The limitation of the data-driven research stream is that there is less of a focus on the "why" or "how" of urban processes and on complex cause-and-effect type relationships. In general, data-driven methods have been the subject of interesting debates regarding the scope, limitations and possibility of such approaches to provide solutions to complex problems beyond pattern detection, associations, and correlations. The current focus on data-driven science and the advocacy for it has in some cases led to rather extreme proclamations to the effect that the data deluge means the "end of theory" and that it will render the scientific process of hypothesizing, modeling, testing, and determining causation obsolete (Anderson 2008). Quantitative empirical research has always been a mainstay for many urban researchers but there is inevitably some conceptual underpinning or theoretical framing which drives such research. Long before Big Data and data science became options to derive knowledge, the well-known statistician, Tukey (1980), noted in an article titled "We Need Both Exploratory and Confirmatory" that "to try to replace one by the other is madness", while also noting that "ideas come from previous exploration more often than from lightning strikes".

A part of the debate is being stimulated by the fact that data-driven models have tended to focus on the use of emerging sources of sensor or socially co-created data, and is closely connected to the data science community. At the time of writing this paper, entering "GPS data" into the Association for Computing Machinery (ACM) Digital Library, a major computer science paper repository returns about 11,750 papers, while entering the same term in IEEE XPlore Digital Library, another such source, returns another 6727 papers; these numbers are in fact higher than the counts obtained when the first author was writing her book "Transportation and Information: Trends in Technology and Policy" (Thakuriah and Geers 2013), indicating not just a voluminous literature on these topics in the data sciences but one that continues to grow very fast. Such data sources have become most closely associated with the term Big Data in the urban context, to the exclusion of administrative data, hybrids of social survey and sensing data, humanities repositories and other novel data sources, which play an important role in substantive, theoretically-informed urban inquiry, beyond detection, correlations and association.

Sensor and ICT-based UGC has also become closely associated with smart cities, or the use of ICT-based intelligence as a development strategy mostly championed and driven by large technology companies for efficient and cost-effective city management, service delivery and economic development in cities. There are numerous other definitions of smart cities, as noted by Hollands (2008). The smart cities movement has been noted to have several limitations, including having "a one-size fits all, top-down strategic approach to sustainability, citizen well-being and economic development" (Haque 2012) and for being "largely ignorant of this (existing and new) science, of urbanism in general, and of how computers have been used to think about cities since their deployment in the mid-20th century" (Batty 2013), a point also made by others such as Townsend (2013). It needs to be pointed out that the scope of smart cities has expanded over time to include optimal delivery of public services to citizens and on processes for citizen engagement and civic participation, as encapsulated by the idea of "future cities".

Urban Big Data is also now being strongly associated with Open Data; Open Data is now being increasingly linked to smart cities, along with efforts to grow data entrepreneurship involving independent developers and civic hackers to stimulate innovations and promote social change. Nevertheless, at least in the European Union, the European Innovation Partnership (EIP) on Smart Cities and Communities has received some 370 commitments to fund and develop smart solutions in the areas of energy, ICT and transport. These commitments involve more than 3000 partners from across Europe, offering "a huge potential for making our cities more attractive, and create business opportunities" (European Commission 2015). It is little wonder that the term Big Data for cities is being referred to in some circles almost exclusively in the context of smart cities, to the exclusion of a long-standing urban operations literature and research that includes contributions on using sensor and at least some types of user-generated data.

Notwithstanding these tensions, some of the benefits of using sensor and socially generated forms of Big Data are in identifying contextual particularities and local experiences that are very often smoothed over by the systems-oriented view of quantitative urban research; the latter often emphasizes generalizability, sometimes masking elements of complex urban challenges. Such "local" focus lends the hope that observations of unique local features from data will stimulate interest in exploring previously unknown hypothesis of urban processes and that the unique local urban problems identified potentially lends itself to context-dependent urban policy and plan-making. The "new geographies of theorizing the urban" (Robinson 2014; Roy 2009) is oriented to skepticism regarding authoritative and universalizing claims to knowledge about urban experiences and is committed to giving attention to contextual particularities and local experiences within places (Brenner and Schmid 2015). Although epistemological links between data-driven urban modeling and critical urban theory is virtually non-existent at the current time, and may never be explicitly articulated, novel sources of Big Data have the potential to allow the capture of data on social, behavioral and economic aspects of urban phenomena that have either not been previously measured or have been measured at resolutions that are too aggregated to be meaningful. However, such localized observations are far from being a substitute for qualitative social science research, as noted by Smith (2014), who advocates a continued need for ethnographic approaches and qualitative methods and cautions against the continued separation of method from methodology and discipline.

Further, causality assessment is challenging with many forms of Big Data and it does not lend itself easily to the derivation of counterfactuals and to forming an etiologic basis for complex urban processes. Instead of moving from using urban models to a completely different data-driven era, as noted earlier, the focus may be to shift to using administrative, sensing or socially generated urban Big Data as input into estimating and testing traditional models. Urban Big Data analysis would also benefit from being linked to behavioral models needed to build alternative scenarios to understand the effects of unobserved assumptions and factors, or to derive explanations for parts of the urban environment not measured by data. The linked data hybrids suggested previously potentially offers a way to address these limitations.

4.4 Challenges Due to the Political Economy of Big Data

The political economy of Big Data arises due to the agendas and actions of the institutions, stakeholders and processes involved with the data. Many of the challenges facing urban researchers in using Big Data stem from complexities with data access, data confidentiality and security, and responsible innovation and emergent ethics. Access and use conditions are in turn affected by new types of data entrepreneurship and innovation networks, which make access easier in some cases, through advocacy for Open Data, or more difficult through conditions imposed as a result of commercialization. Such conditions are generally underpinned by power structures and value propositions arising from Big Data.

The economic, legal and procedural issues that relate to data access and governance are non-trivial and despite the current rhetoric around the open data movement, vast collections of data that are useful for urban analysis are locked away in a mix of legacy and siloed systems owned and operated by individual agencies and private organizations, with their own internal data systems, metadata, semantics and so on. Retrieving information from social media and other online content databases, and the analytics of the resulting retroactive UGC either in real-time or from historical archives have mushroomed into a significant specialized data industry, but the data availability itself is dictated by the terms of service agreements required by the private companies which own the system or which provide access, giving rise to a new political economy of Big Data. User access is provided in some cases using an API, but often there are limits on how much data can be accessed at any one time by the researcher and the linkage of a company's data to other data. Others may mandate user access under highly confidential and secure access conditions requiring users to navigate a complex legal landscape of data confidentiality, and special end-user licensing and terms of service and non-disclosure agreements. Data users may also be subject to potentially changing company policy regarding data access and use. There are also specific restrictions on use including data storage in some cases, requiring analytics in real-time.

Undoubtedly, a part of the difficulty in access stems from data confidentiality and the need to manage trust with citizens, clients and the like. Privacy, trust and security are concepts that are essential to societal interactions. Privacy is a fundamental human right and strategies to address privacy involve privacy-enhancing technology, the legal framework for data protection, as well as consumer awareness of the privacy implications of their activities (Thakuriah and Geers 2013), especially as users leave a digital exhaust with their everyday activities. However, privacy is also not a static, immutable constant. People are likely to trade off some privacy protection in return for utility gained from information, benefits received, or risks minimized (Cottrill and Thakuriah 2015). Aside from technological solutions to maintain privacy, a process of user engagement is necessary to raise consumer awareness, in addition to having the legal and ethical processes in place in order to be able to offer reassurance about confidential use of data. Further, many private data owners may not release data due to being able to reserve competitive advantage through data analytics. However, lack of knowledge about the fast-moving legal landscape with regards to data confidentiality, copyright violations and other unintended consequences of releasing data are central elements of the political economy of Big Data.

The social arguments for and against Big Data, connected systems and IoT are similar to other technology waves that have been previously witnessed, and these considerations also generate multiple avenues for research. Increasingly pervasive sensing and connectivity associated with IoT, and the emphasis on large-scale highly coupled systems that favor removing human input and intervention has been seen to increase exposure to hacking and major system crashes (BCS 2013). Aside from security, the risks for privacy are greatly enhanced as the digital trail left by human activities may be masked under layers of connected systems. Even those systems that explicitly utilize privacy by design are potentially susceptible to various vulnerabilities and unanticipated consequences since the technological landscape is changing very rapidly and the full implications cannot be thought through in their entirety. This has prompted the idea of "responsible innovation", which "seeks to promote creativity and opportunities for science and innovation that are socially desirable and undertaken in the public interest" and which makes clear that "innovation can raise questions and dilemmas, is often ambiguous in terms of purposes and motivations and unpredictable in terms of impacts, beneficial or otherwise. Responsible Innovation creates spaces and processes to explore these aspects of innovation in an open, inclusive and timely way" (Engineering and Physical Sciences Research Council n.d.).

Against this backdrop of complex data protection and governance challenges, and the lure of a mix of objectives such as creating value and generating profit as well as public good, a significant mix of private, public, non-profit and informal infomediaries, ranging from very large organizations to independent developers that are leveraging urban Big Data have emerged. Using a mixed-methods approach, Thakuriah et al. (2015) identified four major groups of organizations within this dynamic and diverse sector: general-purpose ICT companies, urban information service providers, open and civic data infomediaries, and independent and open source developer infomediaries. The political economy implication of these developments is that publicly available data may become private as value is added to such data, and the publicly-funded data infrastructure, due to its complexity and technical demands, may increasingly be managed by private companies that in turn, potentially restricts access and use.

5 Conclusions

In this paper, we discussed the major sources of urban Big Data, their benefits and shortcomings, and ways in which they are enriching Urban Informatics research. The use of Big Data in urban research is not a distinct phase of a technology but rather a continuous process of seeking novel sources of information to address concerns emerging from high cost or design or operational limitations. Although Big Data has often been used quite narrowly to include sensor or socially generated data, there are many other forms that are meaningful to different types of urban researchers and user communities, and we include administrative data and other data sources to capture these lines of scholarship. But even more importantly, it is necessary to bring together (through data linkage or otherwise), data that have existed in fragmented ways in different domains, for a holistic approach to urban analysis.

We note that both theory-driven as well as data-driven approaches are important for Urban Informatics but that retrofitting urban models to reflect developments in a data-rich world is a major requirement for comprehensive understanding of urban processes. Urban Informatics in our view is the study of urban patterns using novel sources of urban Big Data that is undertaken from both a theory-driven empirical perspective as well as a data-driven perspective for the purpose of: urban resource management, knowledge discovery and understanding, urban engagement and civic participation, and planning and policy implementation. The research approaches utilized to progress these objectives are a mix of enriched urban models underpinned by theoretical principles and retrofitted to accommodate emerging forms of data, or data-driven modeling that are largely theory-agnostic and emerge bottom-up from the data. The resulting Urban Informatics research applications have focused on revisiting classical urban problems using urban modeling frameworks but with new forms of data; evaluation of behavioral and structural interactions within enriched complex systems approach; empirical research on sustainable, socially-just and engaged cities; and applications to engage and collaboratively sense cities.

The use of Big Data poses a considerable challenge for Urban Informatics research. This includes technology-related challenges putting requirements for special information management approaches, methodological challenges to retrieve, curate and draw knowledge from the data; theoretical or epistemological challenges to frame modes of inquiry to derive knowledge and understand the limits of Urban Informatics research; and finally, an issue that is likely to play an increasingly critical role for urban research—the emerging political economy of urban Big Data, arising from complexities associated with data governance and ownership, privacy and information security, and new modes of data entrepreneurship and power structures emerging from the economic and political value of data. From the perspective of urban analysts, the use of sensor data, socially generated data, and certain forms of arts and humanities and private sector data may pose significant technical and methodological challenges. With other sources such as administrative micro-data, the data access challenges and issues relating to political economy and data confidentiality might be non-trivial. Issues such as sustainability of the data infrastructure, dealing with data quality, and having access to the skills and knowledge to make inferences, apply to all forms of naturally occurring data.

While many types of urban Big Data such as administrative data and specific sensor systems have been used for a long time, there are many novelties as well, such as new, connected sensor systems, and socially generated or hybrid linked data that result in data in new formats or structure. There is a need for a wide variety of skills due to the tight coupling of preparing unstructured data and data analysis, but also due to the wide variety of technological, methodological and political economy issues involved. Additionally, data and analytics are only one part of the data-focused approach to urban operations, planning and policy-making; having the mechanisms to interpret the results and to highlight the value derived is critical for adoption of data-driven strategies by decision-making, and for its eventual impact on society.

References

Abascal J, Bonail B, Marco Á, Casas R, JL Sevillano (2008). AmbienNet: an intelligent environment to support people with disabilities and elderly people. In: Proceedings of tenth international ACM SIGACCESS conference on computers and accessibility (Assets '08), pp 293–294

Abowd JM, Stephens BE, Vilhuber L, Andersson F, McKinney KL, Roemer M, Woodcock S (2005) The LEHD infrastructure files and the creation of the quarterly workforce indicators. In: Producer dynamics: new evidence from micro data. Published 2009 by University of Chicago Press, pp 149–230. http://www.nber.org/chapters/c0485.pdf. Accessed 1 March 2014

Federal Interagency Forum on Aging-Related Statistics (2010) Older American 2010: key indicators of well-being. http://www.agingstats.gov/agingstatsdotnet/Main_Site/Data/2010_Documents/Docs/OA_2010.pdf. Accessed 31 July 2010

Aguilera O, Fernández AF, Muñoz A, Fraga MF (2010) Epigenetics and environment: a complex relationship. J Appl Physiol 109(1):243–251

Alonso W (1960) A theory of the urban land market. Pap Reg Sci 6(1):149–157

American Academy of the Arts and Sciences (2013) The heart of the matter. http://www.amacad.org. Accessed 1 April 2015

Anderson C (2008) The end of theory: the data deluge makes the scientific method obsolete. Wired Magazine, 23 June 2008. http://www.wired.com/science/discoveries/magazine/16-07/pb_theory. Accessed 10 Feb 2012

Antenucci D, Cafarella M, Levenstein MC, Ré C, Shapiro MD (2014) Using social media to measure labor market flows. Report of the University of Michigan node of the NSF-Census Research Network (NCRN) supported by the National Science Foundation under Grant No. SES 1131500

Ashton K (2009) That "Internet of Things" Thing. RFID Journal, May/June 2009

Auto-id Labs. http://www.autoidlabs.org

Balmer M, Rieser M, Meister K, Charypar D, Lefebvre N, Nagel K, Axhausen K (2009) MATSim-T: architecture and simulation times. In: Multi-agent systems for traffic and transportation engineering. IGI Global, Hershey, pp 57–78

Batty M (2013) Urban informatics and Big Data: a report to the ESRC Cities Expert Group. http://www.smartcitiesappg.com/wp-content/uploads/2014/10/Urban-Informatics-and-Big-Data.pdf. Accessed 15 Dec 2014

Bays J, Callanan L (2012) 'Urban informatics' can help cities run more efficiently. McKinsey on Society. http://mckinseyonsociety.com/emerging-trends-in-urban-informatics/. Accessed 1 July 2014

BCS, The Chartered Institute for IT (2013) The societal impact of the internet of things. www.bcs.org/upload/pdf/societal-impact-report-feb13.pdf. Accessed 10 April 2015

Beckman MJ, McGuire CB, Winston CB (1956) Studies in the economics of transportation. Yale University Press, Connecticut

Beckman R, Baggerly KA, McKay MD (1996) Creating synthetic baseline populations. Transp Res A Policy Pract 30(6):415–429

Beckmann MJ (1973) Equilibrium models of residential location. Reg Urban Econ 3:361–368

Ben-Akiva M, Lerman SR (1985) Discrete choice analysis: theory and application to travel demand. MIT Press, Cambridge

Beresford AR, Stajano F (2003) Location privacy in pervasive computing. IEEE Pervasive Comput 2(1):46–55

Bijwaard GE, Schluter C, Wahba J (2011) The impact of labour market dynamics on the return—migration of immigrants. CReAM Discussion Paper No. 27/12

Black K (2008) Health and aging-in-place: implications for community practice. J Commun Pract 16(1):79–95

Bonney R, Ballard H, Jordan R, McCallie E, Phillips T, Shirk J, Wilderman CC (2009) Public participation in scientific research: defining the field and assessing its potential for informal science education. Technical report, Center for Advancement of Informal Science Education

Bottoms AE, Costello A (2009) Crime prevention and the understanding of repeat victimization: a longitudinal study. In: Knepper P, Doak J, Shapland J (eds) Urban crime prevention, surveillance, and restorative justice: effects of social technologies. CRC, Boca Raton, pp 23–54

Bottrill C (2006) Understanding DTQs and PCAs. Technical report, Environmental Change. Institute/UKERC, October

Burgess EW (1925) The growth of the city: an introduction to a research project. In: Park RE, Burgess EW, Mackenzie RD (eds) The city. University of Chicago Press, Chicago, pp 47–62

Card D, Chetty R, Feldstein M, Saez E (n.d.) Expanding access to administrative data for research in the United States. NSF-SBE 2020 White Paper. http://www.nsf.gov/sbe/sbe_2020/all.cfm. Accessed 10 April 2015

Celino I, Contessa S, Della Valle E, Krüger T, Corubolo M, Fumeo S (2012) UrbanMatch—linking and improving Smart Cities Data. LDOW2012, Lyon, France

U.S. Census Bureau (2012) Press release, 22 May 2012. https://www.census.gov/2010census/news/releases/operations/cb12-95.html

CERI/OECD (1992) City strategies for lifelong learning. In: A CERI/OECD study No. 3 in a series of publications from the Second congress on educating cities, Gothenburg, November

Christakis NA, Fowler JH (2007) The spread of obesity in a large social network over 32 years. N Engl J Med 357(4):370–379

Consolvo S, Everitt K, Smith I, Landay JA (2006) Design requirements for technologies that encourage physical activity. In: Proceedings of SIGCHI conference on human factors in computing systems (CHI '06), pp 457–466

Conway J (1970) The game of life. Sci Am 223(4):4

Cottrill CD, Thakuriah P (2015) Location privacy preferences: a survey-based analysis of consumer awareness, trade-off and decision-making. Transp Res C Emerg Technol 56:132–148

Culotta A (2014) Reducing sampling bias in social media data for county health inference. In: JSM proceedings

Davenport TH, Patil DJ (2012) Data scientist: the sexiest job of the 21st century. Harvard Business Review, October, pp 70–76

Dekel O, Shamir O (2009) Vox Populi: collecting high-quality labels from a crowd. In: Proceedings of the 22nd annual conference on learning theory (COLT), pp 377–386. http://www.cs.mcgill.ca/~colt2009/proceedings.html

De Mauro A, Greco M, Grimaldi M (2015) What is big data? A consensual definition and a review of key research topics. In: AIP conference proceedings, 1644, pp 97–104

NAREC Distributed Energy (2013) ERDF social housing energy management project—final project report. UK National Renewable Energy Centre. https://ore.catapult.org.uk/documents/10619/127231/Social%20Housing%20final%20report/6ca05e01-49cc-43ca-a78c-27fe0e2dd239. Accessed 1 April 2015

Drake JS, Schofer JL, May A, May AD (1965) Chicago area expressway surveillance project, and Expressway Surveillance Project (Ill.). A statistical analysis of speed-density hypotheses: a summary. Report (Expressway Surveillance Project (Ill.)). Expressway Surveillance Project

Ellis RH (1967) Modelling of household location: a statistical approach. Highw Res Rec 207:42–51

Engineering and Physical Sciences Research Council (n.d) Framework for responsible innovation. https://www.epsrc.ac.uk/research/framework/. Accessed 10 April 2015

Erlander S (1980) Optimal spatial interaction and the gravity model. Lecture notes in economics and mathematical systems, vol 173. Springer, Berlin

European Commission (2015) Digital agenda for Europe: a Europe 2020 initiative: European Innovation Partnership (EIP) on Smart Cities and Communities. https://ec.europa.eu/digital-agenda/en/smart-cities. Accessed 1 Aug 2015

Evans TP, Kelley H (2004) Multi-scale analysis of a household level agent-based model of landcover change. J Environ Manage 72(1):57–72

Evans E, Grella CE, Murphy DA, Hser Y-I (2010) Using administrative data for longitudinal substance abuse research. J Behav Health Serv Res 37(2):252–271

Foth M, Choi JH, Satchell C (2011) Urban informatics. In: Proceedings of the ACM 2011 conference on computer supported cooperative work (CSCW '11). ACM, New York, pp 1–8

Fujita M (1988) A monopolistic competition model of spatial agglomeration: differentiated product approach. Reg Sci Urban Econ 18(1):87–124

Fujita M, Ogawa H (1982) Multiple equilibria and structural transition of non-monocentric urban configurations. Reg Sci Urban Econ 12(2):161–196

Fujita M, Krugman P, Venables AJ (1999) The spatial economy: cities, regions, and international trade. MIT Press, Cambridge

Ghosh D, Guha R (2013) What are we tweeting about obesity? Mapping tweets with topic modeling and geographic information system. Cartogr Geogr Inf Sci 40(2):90–102

Girvan M, Newman ME (2002) Community structure in social and biological networks. Proc Natl Acad Sci 99(12):7821–7826

Gjoka M, Kurant M, Butts CT, Markopoulou A (2010) Walking in Facebook: a case study of unbiased sampling of OSNs. In: Proceedings of IEEE 2010 INFOCOM 2010

OECD Global Science Forum (2013) New data for understanding the human condition: international perspectives. Report on data and research infrastructure for the social sciences. http://www.oecd.org/sti/sci-tech/new-data-for-understanding-the-human-condition.pdf. Accessed 1 April 2015

Golbeck J, Hansen D (2013) A method for computing political preference among Twitter followers. Soc Netw 36:177–184

Gowans H, Elliot M, Dibben C, Lightfoot D (2012) Accessing and sharing administrative data and the need for data security, Administrative Data Liaison Service

Gruteser M, Grunwald D (2003) Anonymous usage of location-based services through spatial and temporal cloaking. In: Proceedings of first international conference on mobile systems, applications and services, MobiSys '03, pp 31–42

Gurstein M (2011) Open data: empowering the empowered or effective data use for everyone? First Monday, vol 16, no 2, 7 Feb 2011. http://firstmonday.org/ojs/index.php/fm/article/view/3316/2764. Accessed 1 July 2013

Haque U (2012) Surely there's a smarter approach to smart cities? Wired Magazine. 17 April 2012. http://www.wired.co.uk/news/archive/2012-04/17/potential-of-smarter-cities-beyond-ibm-and-cisco. Accessed 10 April 2012

Harland K, Heppenstall A, Smith D, Birkin M (2012) Creating realistic synthetic populations at varying spatial scales: a comparative critique of population synthesis techniques. J Artif Soc Soc Simul, vol 15(1):1. http://jasss.soc.surrey.ac.uk/15/1/1.html. Accessed 19 April 2015

Hendler J (2014) Data integration for heterogeneous datasets. Big Data 2(4):205–215

Hey T, Tansley S, Tolle K (2009) The fourth paradigm: data-intensive scientific discovery. Microsoft Research, Redmond

Hoadley CM, Bell P (1996) Web for your head: the design of digital resources to enhance lifelong learning. D-Lib Magazine, September. http://www.dlib.org/dlib/september96/kie/09hoadley.html. Accessed 15 Jan 2015

Hollands RG (2008) Will the real smart city please stand up? Intelligent, progressive or entrepreneurial? City: analysis of urban trends, culture, theory, policy, action, vol 12(3), pp 303–320

Huang DL, Rosenberg DE, Simonovich SD, Belza B (2012) Food access patterns and barriers among midlife and older adults with mobility disabilities. J Aging Res 2012:1–8

Huijboom N, van den Broek T (2011). Open data: an international comparison of strategies. Eur J ePractice, vol 12, March/April 2011

Isard W (1956) Location and space-economy. MIT Press, Cambridge

Jia T, Jiang B, Carling K, Bolin M, Ban YF (2012) An empirical study on human mobility and its agent-based modeling. J Stat Mech Theory Exp P11024

Kain JF, Persky JJ (1969) Alternatives to the "Gilded Ghetto". In The Public Interest, Winter, pp 77–91

Kempton W, Letendre SE (1997) Electric vehicles as a new power source for electric utilities. Transp Res D Transp Environ 2(3):157–175

Kempton W, Tomic J (2005) Vehicle-To-Grid power fundamentals: calculating capacity and net revenue. J Power Sources 144(1):268–279

Kinney R, Crucitti P, Albert R, Latora V (2005) Modeling cascading failures in the North American power grid. Eur Phys J B 46(1):101–107

Kinsley S (n.d.) A political economy of Twitter data? Conducting research with proprietary data is neither easy nor free. http://blogs.lse.ac.uk/impactofsocialsciences/2014/12/30/a-political-economy-of-twitter-data/. Accessed 1 April 2015

Knoll M (2008) Diabetes city: how urban game design strategies can help diabetics. In: eHealth'08, pp 200–204

Kowald M, Axhausen KW (2015) Social networks and travel behaviour. Ashgate, Burlington

Krivo LJ, Peterson RD (1996) Extremely disadvantaged neighborhoods and urban crime. Soc Forces 75:619–648

Krugman P (1991) Geography and trade. MIT Press, Cambridge

Laney D (2001) 3-D data management: controlling data volume, velocity and variety. Application delivery strategies by META Group Inc., February, p 949

Lee DB Jr (1973) Requiem for large-scale models. J Am Inst Plann 39:163–178

Levinson D, Marion B, Iacono M (2010) Access to destinations, phase 3: measuring accessibility by automobile. http://www.cts.umn.edu/Research/ProjectDetail.html?id=2009012

Lightfoot D, Dibben C (2013) Approaches to linking administrative records to studies and surveys—a review. Administrative Data Liaison Service, University of St Andrews. https://www.google.co.uk/search?q=Approaches+to+linking+administrative+records+to+studies+and+surveys+-+a+review&ie=utf-8&oe=utf-8&gws_rd=cr&ei=rfxFVZWeM4KUaumTgdAC#. Accessed 15 March 2015

Lin Y, Jessurun J, de Vries B, Timmermans H (2011) Motivate: towards context-aware recommendation mobile system for healthy living. In: 2011 fifth international conference on pervasive computing technologies for healthcare (PervasiveHealth), pp 250–253

Liu B (2007) Web data mining: data-centric systems and applications. Springer, Berlin

Lowry IS (1964) A model of metropolis. The Rand Corporation, Santa Monica

Ludwig Y, Zenker B, Schrader J (2009) Recommendation of personalized routes with public transport connections. In: Tavangarian D, Kirste T, Timmermann D, Lucke U, Versick D (eds) Intelligent interactive assistance and mobile multimedia computing of communications in computer and information science, vol 53. Springer, Berlin, pp 97–107

Lundquist D (2011) Pollution credit trading in Vehicular Ad Hoc Networks. http://connected.vehicle.challenge.gov/submissions/2926-pollution-credit-trading-in-vehicular-ad-hocnetworks

Marshall A (1920) Principles of economics. MacMillan, London

Miller H, Goodchild MF (2014) Data-driven geography. GeoJournal, October, pp 1–13

Mislove A, Lehmann S, Ahn Y, Onnela J, Rosenquist JN (2011) Understanding the demographics of twitter users. In: Proceedings of the fifth international AAAI conference on weblogs and social media (ICWSM'11), Barcelona, Spain

Moor JH (2005) Why we need better ethics for emerging technologies. Ethics Inf Technol 7:111–119

Museum of the City of New York (2014) Museum of the City of New York receives grant from the National Endowment for the Humanities to process, catalog, digitize and rehouse the Ephemera Collections. http://www.mcny.org/sites/default/files/Press_Release_NEH_Grant_FINAL.pdf. Accessed 5 April 2015

Muth RF (1969) Cities and housing: the spatial pattern of urban residential land use. University of Chicago Press, Chicago

Naylor J, Swier N, Williams S, Gask K, Breton R (2015) ONS Big Data Project—progress report: Qtr 4 October to Dec 2014. ONS Big Data Project Qtr 4 Report. http://www.ons.gov.uk/ons/about-ons/who-ons-are/programmes-and-projects/the-ons-big-data-project/index.html. Accessed 15 April 2015

Office for National Statistics (2015) Beyond 2011 research strategy and plan—2015 to 2017. http://www.ons.gov.uk/ons/about-ons/who-ons-are/programmes-and-projects/beyond-2011/reports-and-publications/research-strategy-and-plan---2015-2017.pdf. Accessed 1 March 2015

Organisation for Economic Co-operation and Development (OECD) Global Science Forum (2011) Effective modelling of urban systems to address the challenges of climate change and sustainability, October 2011. www.oecd.org/sti/sci-tech/49352636.pdf. Accessed 13 April 2013

Ortega F, Gonzalez-Barahona J, Robles G (2008) On the inequality of contributions to Wikipedia. In: HICSS '08 Proceedings of 41st Annual Hawaii International Conference on System Sciences, p 304

Picciano AG (2012) The evolution of Big Data and learning analytics in American Higher Education. J Asynchronous Learn Netw 16(3):9–20

Pietsch W (2013) Big Data—the new science of complexity. In: Sixth Munich-Sydney-Tilburg conference on models and decisions, Munich, 10–12 April 2013. http://philsci-archive.pitt.edu/9944/. Accessed 1 April 2015

Poplin A (2014) Digital serious game for urban planning: B3—design your marketplace! Environ Plann B Plann Design 41(3):493–511

Pyrozhenko V (2011) Implementing open government: exploring the ideological links between open government and the free and open source software movement. Prepared for 11th Annual public management meeting

Quinn AJ, Bederson BB (2011) Human computation: a survey and taxonomy of a growing field. In: Proceedings of annual conference on human factors in computing systems (CHI '11), pp 1403–1412

Rae A (2014) Online housing search and the geography of submarkets. Hous Stud 30(3):453–472

Rand W, Herrmann J, Schein B, Vodopivec N (2015) An agent-based model of urgent diffusion in social media. J Artif Soc Soc Simul 18(2):1

Raykar VC, Yu S, Zhao LH, Valadez GH, Florin C, Bogoni L, Moy L (2010) Learning from crowds. J Mac Learn Res 11:1297–1322

Reinhardt J, Miller J, Stucki G, Sykes C, Gray D (2011) Measuring impact of environmental factors on human functioning and disability: a review of various scientific approaches. Disabil Rehabil 33(23-24):2151–2165

Reist BH, Ciango A (2013) Innovations in census taking for the United States in 2020. In: Proceedings of 59th ISI World Statistics Congress, Hong Kong. http://www.statistics.gov.hk/wsc/STS017-P2-S.pdf. Accessed 15 March 2015

Richardson HW (1977) A generalization of residential location theory. Reg Sci Urban Econ 7:251–266

Robinson J (2014) New geographies of theorizing the urban: putting comparison to work for global urban studies. In: Parnell S, Oldfield S (eds) The Routledge handbook on cities of the global south. Routledge, New York, pp 57–70

Roy A (2009) The 21st century metropolis: new geographies of theory. Reg Stud 43(6):819–830

Salasznyk PP, Lee EE, List GF, Wallace WA (2006) A systems view of data integration for emergency response. Int J Emerg Manage 3(4):313–331

Salvini P, Miller EJ (2005) ILUTE: an operational prototype of a comprehensive microsimulation model of urban systems. Netw Spat Econ 5(2):217–234

Sasaki K, Nagano S, Ueno K, Cho K (2012) Feasibility study on detection of transportation information exploiting Twitter as a sensor. In: Sixth international AAAI conference on weblogs and social media. Workshop on when the city meets the citizen, AAAI technical report WS-12-0

Schelling TC (1971) Dynamic models of segregation. J Math Sociol 1(2):143–186

Sen A, Smith TE (1995) Gravity models of spatial interaction behavior, Advances in spatial and network economics series. Springer, Berlin

Sheffi Y (1985) Urban transportation networks: equilibrium analysis with mathematical programming methods. Prentice-Hall, Englewood Cliffs

Shkabatur J (2013) Transparency with(out) accountability: open government in the United States. Yale Law & Policy Review, vol 31, 25 March 2012

Smith RJ (2014) Missed miracles and mystical connections: qualitative research and digital social sciences and big data. In: Hand M, Hillyard S (eds) Big Data?: Qualitative approaches to digital research. Edward Group, pp 181–204. http://www.eblib.com. Accessed 1 Aug 2015

Tang KP, Lin J, Hong JI, Siewiorek DP, Sadeh N (2010) Rethinking location sharing: exploring the implications of social-driven vs. purpose-driven location sharing. In: Proceedings of 12th ACM international conference on ubiquitous computing (Ubicomp '10), pp 85–94

Thakuriah P, Geers G (2013) Transportation and information: trends in technology and policy. Springer, New York

Thakuriah P, Mallon-Keita Y (2014) An analysis of household transportation spending during the 2007-2009 US economic recession. In: Transportation Research Board 93rd annual meeting, Washington, DC, 12–16 Jan 2014

Thakuriah P, Tilahun N (2013) Incorporating weather information into real-time speed estimates: comparison of alternative models. J Transp Eng 139(4):379–389

Thakuriah P, Soot S, Cottrill C, Tilahun N, Blaise T, Vassilakis W (2011) Integrated and continuing transportation services for seniors: case studies of new freedom program. Transp Res Rec 2265:161–169

Thakuriah P, Persky J, Soot S, Sriraj P (2013) Costs and benefits of employment transportation for low-wage workers: an assessment of job access public transportation services. Eval Program Plann 37:31–42

Thakuriah P, Dirks L, Mallon-Keita Y (2016) Digital Infomediaries and Civic Hacking in Emerging Urban Data Initiatives. In Thakuriah P, Tilahun N, Zellner M (eds) Seeing cities through big data: research, methods and applications in urban informatics. Springer, New York

Thakuriah P, Sila-Nowicka K, Gonzalez Paule J (forthcoming) Sensing Spatiotemporal Patterns in Urban Areas: Analytics and Visualizations Using the Integrated Multimedia City Data Platform. In Big Data and the City, a special issue of Built Environment.

Thanos C, Rauber A (2015) Scientific data sharing and re-use. ERCIM News, no. 100, 13 Jan 2015

Thorhildur J, Avital M, BjÃrn-Andersen N (2013) The generative mechanisms of open government data. In: ECIS 2013 proceedings, Paper 179

Tilahun N, Levinson D (2013) An agent-based model of origin destination estimation (ABODE). J Transp Land Use 6(1):73–88

Townsend A (2013) Smart cities: Big Data, civic hackers and the quest for a New Utopia. W. W. Norton, New York

Tukey JW (1980) We need both exploratory and confirmatory. Am Stat 34(1):23–25

United Nations, Department of Economic and Social Affairs, Population Division (2014) World Urbanization prospects: the 2014 revision, highlights (ST/ESA/SER.A/352)

Urban Big Data Centre (n.d.) Integrated Multimedia City Data (iMCD). http://ubdc.ac.uk/research/research-projects/methods-research/integrated-multimedia-city-data-imcd/. Accessed 18 September 2016

Vitrano FA, Chapin MM (2010) Possible 2020 census designs and the use of administrative records: what is the impact on cost and quality? U.S. Census Bureau, Suitland. https://fcsm.sites.usa.gov/files/2014/05/Chapin_2012FCSM_III-A.pdf

Vitrano FA, Chapin MM (2014) Possible 2020 census designs and the use of administrative records: what is the impact on cost and quality? https://fcsm.sites.usa.gov/files/2014/05/Chapin_2012FCSM_III-A.pdf. Accessed 1 March 2015

von Ahn L, Blum M, Hopper NJ, Langford J (2003) CAPTCHA: using hard AI problems for security. Technical Report 136

von Ahn L, Maurer B, McMillen C, Abraham D, Blum M (2008) reCAPTCHA: human-based character recognition via web security measures. Science 321(5895):1465–1468

Wauthier FL, Jordan MI (2011) Bayesian bias mitigation for crowdsourcing. In: Proceedings of the conference on neural information processing system, no 24, pp 1800–1808. http://machinelearning.wustl.edu/mlpapers/papers/NIPS2011_1021

Wegener M (1994) Operational urban models state of the art. J Am Plann Assoc 60(1):17–29

Weil R, Wootton J, Garca-Ortiz A (1998) Traffic incident detection: sensors and algorithms. Math Comput Model 27(911):257–291

Williams JD (2012) The 2010 decennial census: background and issues. Congressional Research Service R40551. http://www.fas.org/sgp/crs/misc/R40551.pdf. Accessed 1 March 2015

Wilson AG (1971) A family of spatial interaction models, and associated developments. Environ Plann 3(1):1–32

Wilson WJ (1987) The truly disadvantaged: the inner city, the underclass and public policy Chicago. University of Chicago Press, Chicago

Wilson AG (2013) A modeller's utopia: combinatorial evolution. Commentary. Environ Plann A 45:1262–1265

Wu G, Talwar S, Johnsson K, Himayat N, Johnson KD (2011) M2m: from mobile to embedded internet. IEEE Communications Magazine 49(4):36–43, April

Zellner ML, Reeves HW (2012) Examining the contradiction in 'sustainable urban growth': an example of groundwater sustainability. J Environ Plann Manage 55(5):545–562

Zellner ML, Page SE, Rand W, Brown DG, Robinson DT, Nassauer J, Low B (2009) The emergence of zoning games in exurban jurisdictions. Land Use Policy 26(2009):356–367

Zellner ML, Lyons L, Hoch CJ, Weizeorick J, Kunda C, Milz D (2012) Modeling, learning and planning together: an application of participatory agent-based modeling to environmental planning. URISA J (GIS in Spatial Planning Issue) 24(1):77–92

Zellner M, Massey D, Shiftan Y, Levine J, Arquero M (2016) Overcoming the last-mile problem with transportation and land-use improvements: an agent-based approach. Int J Transport 4 (1):1–26

Zhang X, Qin S, Dong B, Ran B (2010) Daily OD matrix estimation using cellular probe data. In: Proceedings of ninth annual meeting Transportation Research Board

Zheng Y, Xie X (2011) Location-based social networks: locations. In: Zheng Y, Zhou X (eds) Computing with spatial trajectories. Springer, New York, pp 277–308

Part I
Analytics of User-Generated Content

Using User-Generated Content to Understand Cities

Dan Tasse and Jason I. Hong

Abstract Understanding urban dynamics is crucial for a number of domains, but it can be expensive and time consuming to gather necessary data. The rapid rise of social media has given us a new and massive source of geotagged data that can be transformative in terms of how we understand our cities. In this position paper, we describe three opportunities in using geotagged social media data: to help city planners, to help small businesses, and to help individuals adapt to their city better. We also sketch some possible research projects to help map out the design space, as well as discuss some limitations and challenges in using this kind of data.

Keywords Social media • User generated content • Geotagged data • Data visualization • Urban analytics

1 Introduction

Over half of the world's population now lives in cities (Martine et al. 2007), and understanding the cities we live in has never been more important. Urban planners need to plan future developments, transit authorities need to optimize routes, and people need to effectively integrate into their communities.

Currently, a number of methods are used to collect data about people, but these methods tend to be slow, labor-intensive, expensive, and lead to relatively sparse data. For example, the US census cost $13 billion in 2010 (Costing the Count 2011), and is only collected once every 10 years. The American Community Survey is collected annually, and cost about $170 million in 2012, but only samples around 1 % of households in any given year (Griffin and Hughes 2013). While data like this can benefit planners, policy makers, researchers, and businesses in understanding changes over time and how to allocate resources, today's methods for understanding people and cities are slow, expensive, labor-intensive, and do not scale well.

Researchers have looked at using proprietary call detail records (CDRs) from telecoms to model mobility patterns (Becker et al. 2011; González et al. 2008;

D. Tasse (✉) • J.I. Hong
Human-Computer Interaction Institute, Carnegie Mellon University, Pittsburgh, PA, USA
e-mail: dantasse@cmu.edu; jasonh@cs.cmu.edu

Isaacman et al. 2012) and other social patterns, such as the size of one's social network and one's relationship with others (Palchykov et al. 2012). These studies leverage millions of data points; however, these approaches also have coarse location granularity (up to 1 sq mile), are somewhat sparse (CDRs are recorded only when a call or SMS is made), have minimal context (location, date, caller, callee), and use data not generally available to others. Similarly, researchers have also looked at having participants install custom apps. However, this approach has challenges in scaling up to cities, given the large number of app users needed to get useful data. Corporations have also surreptitiously installed software on people's smartphones (such as CallerIQ (McCullagh 2011) and Verizon's Precision Market Insights (McCullagh 2012)), though this has led to widespread outcry due to privacy concerns.

We argue that there is an exciting opportunity for creating new ways to conceptualize and visualize the dynamics, structure, and character of a city by analyzing the social media its residents already generate. Millions of people already use Twitter, Instagram, Foursquare, and other social media services to update their friends about where they are, communicate with friends and strangers, and record their actions. The sheer quantity of data is also tantalizing: Twitter claims that its users send over 500 million tweets daily, and Instagram claims its users share about 60 million photos per day (Instagram 2014). Some of this media is geotagged with GPS data, making it possible to start inferring people's behaviors over time. In contrast to CDRs from telcos, we can get fine-grained location data, and at times beyond when people make phone calls. In contrast to having people install custom apps (which is hard to persuade people to do), we can leverage social media data that millions of people are already creating every day.

We believe that this kind of geotagged social media data, combined with new kinds of analytics tools, will let urban planners, policy analysts, social scientists, and computer scientists explore how people actually use a city, in a manner that is cheap, highly scalable, and insightful. These tools can shed light onto the factors that come together to shape the urban landscape and the social texture of city life, including municipal borders, demographics, economic development, resources, geography, and planning.

As such, our main question here is, how can we use this kind of publicly visible, geotagged social media data to help us understand cities better? In this position paper, we sketch out several opportunities for new kinds of analytics tools based on geotagged social media data. We also discuss some longer-term challenges in using this kind of data, including biases in this kind of data, issues of privacy, and fostering a sustainable ecosystem where the value of this kind of data is shared with more people.

2 Opportunities

In this section, we sketch out some design and research opportunities, looking at three specific application areas. Many of the ideas we discuss below are speculative. We use these ideas as a way of describing the potential of geotagged social media data, as well as offering possible new directions for the research community.

2.1 For City Planners

First, and perhaps most promisingly, we believe that geotagged social media data can offer city planners and developers better information that can be used to improve planning and quality of life in cities. This might include new kinds of metrics for understanding people's interactions in different parts of a city, new methods of pinpointing problems that people are facing, and new ways of identifying potential opportunities for improving things.

2.1.1 Mapping Socioeconomic Status

It is important for governments to know the socioeconomic status of different sections of their jurisdiction in order to properly allocate resources. In England, for example, the government uses the Index of Multiple Deprivation (IMD) to measure where the problems of poverty are the most severe, and to therefore mitigate those effects. The IMD is based on surveys and other statistics collected by different areas of government. However, even in developed countries, surveys and statistics can be difficult and expensive to collect.

Past work with cell phone call logs suggests that it is possible to find valuable demographic information using communication records. For example, Eagle et al found that network diversity in phone calls correlated with the IMD (Eagle et al. 2010). A recent project by Smith-Clarke et al. (2014) explored the use of call logs to map poverty. Especially in developing countries, maps of poverty are often very coarse and out of date. This is not a simple problem of improving an already-good metric; data at different granularities can tell very different stories. For example, Fig. 1 shows two different maps of socioeconomic data in the UK, one very coarse-grained and one fine-grained.

However, call log data, while more complete than surveys, still presents the limitations mentioned earlier: it is proprietary, coarse-grained, and lacking context and transparency. Much social media data, on the other hand, are publicly visible and accessible to researchers. Different forms of social media data also offer their own advantages. For example, Twitter users follow other users, and Foursquare check-ins often have "likes" or comments attached.

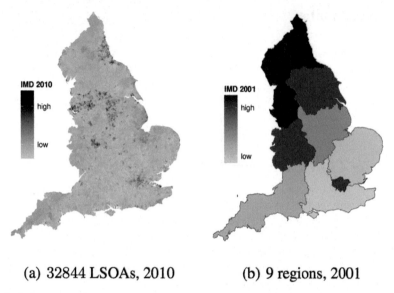

(a) 32844 LSOAs, 2010   (b) 9 regions, 2001

Fig. 1 *Left*: accurate data about socio-economic deprivation in England; *darker* indicates more deprivation. *Right*: much coarser, out-of-date information about the same index. Figure from (Smith-Clarke et al. 2014)

2.1.2 Mapping Quality of Life

Socioeconomic status, however, is not the only metric that matters. A community can be poor but flourishing, or rich but suffering. Other metrics like violence, pollution, location efficiency, and even community coherence are important for cities to track. Some of these are even more difficult to track than socioeconomic status.

We believe some aspects of quality of life can be modeled using geotagged social media data. For example, approximating violence may be possible by analyzing the content of posts. Choudhury et al. (2014) showed that psychological features associated with desensitization appeared over time in tweets by people affected by the Mexican Drug War. Other work has found that sentiments in tweets are correlated with general socio-economic wellbeing (Quercia et al. 2011). Measuring and mapping posts that contain these emotional words may help us find high crime areas and measure the change over time.

As another example, location efficiency, or the total cost of transportation for someone living in a certain location, can be approximated by sites like Walkscore.com. However, Walkscore currently only relies on the *spaces*, that is, where services are on the map. It does not take into account the *places*, the ways that people use these services. A small market may be classified as a "grocery store", but if nobody goes to it for groceries, maybe it actually fails to meet people's grocery needs. We believe geotagged social media data can be used as a new way of understanding how people actually use places, and thereby offer a better measure of location efficiency.

2.1.3 Mapping Mobility

One more analysis that could be useful for city planners is in understanding the mobility patterns of people in different parts of different cities. This kind of information can help, for example, in planning transportation networks (Kitamura et al. 2000). Mobility can help planners with social information as well, such as how public or private people feel a place is (Toch et al. 2010). Previously, mobility information has been gathered from many sources, but they all lack the granularity and ease of collection of social media data. Cell tower data has been used to estimate the daily ranges of cell phone users (Becker et al. 2013). At a larger scale, data from moving dollar bills has been used to understand the range of human travel (Brockmann et al. 2006). Among other applications, this data could be used for economic purposes, such as understanding the value of centralized business districts like the Garment District in New York (Williams and Currid-Halkett 2014). It seems plausible that pre-existing social media data could help us find the similar information without needing people to enter dollar bill serial numbers or phone companies to grant access to expensive and sensitive call logs. Geotagged social media data is also more fine-grained, allowing us to pinpoint specific venues that people are going to.

2.1.4 "Design Patterns" for Cities

Originating in the field of architecture, *design patterns* are good and reusable solutions to common design problems. Geotagged social media data offers new ways of analyzing physical spaces and understanding how the design of those spaces influences people's behaviors.

For example, in his book A Pattern Language (1977), Alexander and colleagues present several kinds of patterns characterizing communities and neighborhoods. These patterns include Activity Nodes (community facilities should not be scattered individually through a city, but rather clustered together), Promenades (a center for its public life, a place to see people and to be seen), Shopping Streets (shopping centers should be located near major traffic arteries, but should be quiet and comfortable for pedestrians), and Night Life (places that are open late at night should be clustered together).

By analyzing geotagged social media data, we believe it is possible to extract known design patterns. One possible scenario is letting people search for design patterns in a given city, e.g. "where is Night Life in this city?" or "show the major Promenades". Another possibility is to compare the relationship of different patterns in different cities, as a way of analyzing why certain designs work well and others do not. For example, one might find that areas that serve both as Shopping Streets and as Night Life are well correlated with vibrant communities and general well-being.

2.2 For Small Businesses

Understanding one's customers is crucial for owners of small businesses, like restaurants, bars, and coffee shops. We envision two possible scenarios for how geotagged social media data can help small business owners.

2.2.1 Knowing Demographics of Existing Customers

Small businesses cannot easily compete with big-box stores in terms of data and analytics about existing customers. This makes it difficult for small businesses to tailor their services and advertisements effectively.

Businesses can already check their reviews on Yelp or Foursquare. We believe that geotagged social media data can offer different kinds of insights about the behaviors and demographics of customers. One example would be knowing what people do before and after visiting a given venue. For example, if a coffee shop owner finds that many people go to a sandwich shop after the coffee shop, they may want to partner with those kinds of stores or offer sandwiches themselves. This same analysis could be done with classes of venues, for example, cafés or donut shops.

As another example, an owner may want to do retail trade analysis (Huff 1963), which is a kind of marketing research for understanding where a store's customers are coming from, how many potential customers are in a given area, and where one can look for more potential customers. Some examples include quantifying and visualizing the flow and movement of customers in the area around a given store. Using this kind of analysis, a business can select potential store locations, identify likely competitors, and pinpoint ideal places for advertisements.

Currently, retail trade analysis is labor intensive, consisting of numerous observations by field workers (e.g. watching where customers come from and where they go, or shadowing customers) or surveys given to customers. Publicly visible social media data offers a way of scaling up this kind of process, and extending the kind of analysis beyond just the immediate area. For example, one could analyze more general kinds of patterns. For example, what are the most popular stores in this area, and how does the store in question stack up? How does the store in question compare against competitors in the same city?

Knowing more about the people themselves would be useful as well. For example, in general, what kinds of venues are most popular for people who come to this store? If a business finds that all of its patrons come from neighborhoods where live music is popular, they may want to consider hosting musicians themselves. All of the information offered by a service like this would have to be rather coarse, but it could provide new kinds of insights for small businesses.

2.2.2 Knowing Where to Locate a New Business

New businesses often have many different potential locations, and evaluating them can be difficult. Public social media data could give these business owners more insight into advantages and disadvantages of their potential sites. For example, if they find that a certain neighborhood has many people who visit Thai restaurants in other parts of the city, they could locate a new Thai restaurant there.

2.3 For Individuals

There are also many opportunities for using geotagged social media to benefit individuals as well. Below, we sketch out a few themes.

2.3.1 Feeling at Home in New Cities

Moving to a new city can make it hard for people to be part of a community. The formally defined boundaries of neighborhoods may help people understand the *spaces* where they live, but not so much the socially constructed *places* (Harrison and Dourish 1996). Currently it is difficult for non-locals to know the social constructs of a city as well as locals do. This is particularly important when someone changes places, either as a tourist or a new resident.

Imagine a new person arriving in a diverse city like San Francisco, with multiple neighborhoods and sub-neighborhoods. It would be useful for that person to know the types of people who live in the city, and where each group goes: students go to this neighborhood in the evening, members of the Italian community like to spend time in this area, families often live in this neighborhood but spend time in that neighborhood on weekends.

Some work has been done in this area, but we believe it could be extended. Komninos et al. collected data from Foursquare to examine patterns over times of day and days of the week (Komninos et al. 2013), showing the daily variations of people's activity in a city in Greece. They showed when people are checking in, and at what kind of venue, but not who was checking in. Cheng et al, too, showed general patterns of mobility and times of checkins (Cheng et al. 2011), but these statistics remain difficult for individuals to interpret.

Related, the Livehoods project (Cranshaw et al. 2012) and Hoodsquare (Zhang et al. 2013) both look at helping people understand their cities by clustering nearby places into neighborhoods. Livehoods used Foursquare checkins, clustering nearby places where the same people often checked in. Hoodsquare considered not only checkins but also other factors including time, location category, and whether tourists or locals attended the place. Both of these projects would be helpful for people to find their way in a new city, but even their output could be more informative. Instead of simply knowing that a neighborhood has certain boundaries, knowing why those boundaries are drawn or what people do inside those boundaries would be helpful. Andrienko et al. (2011) also describe a visual analytic approach to finding important places based on mobility data that could help newcomers understand which places were more popular or important.

Another approach to helping people get to know the city is to help them get to know the people in the city, rather than the places. We look to the work of Joseph et al. (2012) who used topic models to assign people to clusters such as "sports enthusiast" or "art enthusiast". We could imagine this information being useful for individuals to find other like-minded people.

2.3.2 Discovering New Places

In the previous section, we described applications to help people become accustomed to a new city. However, sometimes the opposite problem may arise: people become too comfortable in their routines and they want to discover new places. Some recent projects have helped people to discover new places based on actions that can be performed there (Dearman et al. 2011) or aspects that people love about the place (Cranshaw et al. 2014). However, we believe that this idea can be pushed further. Perhaps combining a person's current mobility patterns with visualizations of other people's mobility patterns would help a person to put their own actions in context. They may realize that there are entire social flows in the city that they did not even know existed.

2.3.3 Understanding People, Not Just Places

Tools like Yelp and Urbanspoon already exist to help people find places they would like to go or discover new places that they didn't know about. Previous work like Livehoods (Cranshaw et al. 2012) also worked to enable a richer understanding of the places there. The benefit of incorporating social media data, though, is that users can start to understand the people who go to places, not just the places themselves.

3 Potential Research in This Design Space

In this section we sketch out some potentially interesting research projects in using geotagged social media data. Our goal here is to map out different points in the overall design space, which can be useful in understanding the range of applications as well as the pros and cons of various techniques.

3.1 Who Goes Here?

The approach is simple: select a geographic region (which may be as small as an individual store) and retrieve all the tweets of the people who have ever tweeted there. Then, compute a heat map or other geographic visualization of the other places that they tweet. This kind of visualization could help elucidate all the places associated with a given place, and could be useful to small business owners or managers of a larger organization like a university.

3.2 Groceryshed

Watershed maps show where water drains off to lakes and oceans. Researchers have extended this metaphor to map "laborsheds" and "paradesheds" (Becker et al. 2013) to describe where people who work in a certain area come from, or people who attend a certain parade. We could extend this metaphor even further to describe "sheds" of smaller categories of business, such as Thai restaurants.

More interestingly, we could map change over time in various "sheds". This could be particularly important for grocery stores. Places that are outside any "groceryshed" could be candidate areas for a new store, and showing the change in people's behavior after a new store goes in could help measure the impact of that store. The content of Tweets or other social data could also show how people's behavior changed after a new grocery store was put in.

3.3 How Is This Place Relevant To Me?

We envision a system that can convey not only what a place *is*, but also what it *means*. Imagine a user looking up a particular coffee shop. Currently, they can look up the coffee shop's web site, find basic information like store hours, and find reviews on sites like Yelp. Using geotagged social media data, however, we could surface information like:

- Your friends (or people you follow on Twitter) go here five times per week.
- Friends of your friends go here much more than nearby coffee shops.
- People who are music enthusiasts like you (using topic modeling as in Joseph et al. (2012)) often go to this coffee shop.
- You've been to three other coffee shops that are very similar to this one.
- People who tweet here show the same profiles of emotions as your tweets.

These could help people form a deeper relationship with a place than one based on locality or business type alone. In addition, we could pre-compute measures of relevance for a particular user, giving them a map of places that they might enjoy.

3.4 Human Network Visualizations

We can go beyond assigning people to groups by topics, also showing where they tweet over time. This could help people understand the dynamics of neighborhoods where, for example, one group of more affluent people are pricing out a group of previous residents. One interesting work in this area is the Yelp word maps (2013), which show where people write certain words, like "hipster", in reviews of businesses. However, this still describes the places; using social media data, we could show maps that describe the people. Instead of a map of locations tagged as "hipster", we could identify groups of people based on their check-in patterns and tag where they go during the day. Perhaps the hipsters frequent certain coffee shops in the morning and certain bars at night, but during the day hang out in parks where they do not check in.

3.5 Cheaper, Easier, and Richer Demographics

For all of our groups and stakeholders, it is important to understand demographic information of city regions. We could improve the process in two main ways. First, we could make it cheaper and easier to infer existing demographic information. We plan to investigate whether Twitter volume, or volume of certain topics of discussion, correlates with deprivation or other measures of socioeconomic status or quality of life. If so, then we can use the social media measures as proxies for real measures, and thereby collect that information cheaply and in real time.

Second, we plan to create more descriptive demographics. In each neighborhood, we can calculate each person's average distance traveled, radius of gyration, and other measures of mobility. We could either simply output statistics or we could create interactive visualizations that show the daily movements of people in each neighborhood.

4 Data Sources and Limitations

For our research, we are currently using data from Twitter, due to its richness and volume. Twitter claims over 500 million tweets are posted per day (Krikorian 2013). Furthermore, this data is publicly available. While only a small fraction of these tweets are geotagged, even a small fraction of tweets from any given day forms a large and rich data set. Furthermore, past work suggests that the sampling bias from only selecting these tweets is limited (Priedhorsky et al. 2014).

4.1 Biases in the Data

Of course, neither Twitter nor Foursquare provides an exactly accurate view of people's mobility. Both are communicative media, not purely representative. For example, Rost et al. (2013) report that the Museum of Modern Art in New York has more check-ins than Atlanta's airport, even though the airport had almost three times as many visitors in the time period that was studied. In some cases this will not matter; if we are clustering people, for example, grouping people who communicate that they go to the same places will be nearly as successful as grouping people who actually go to the same places. In other cases, we hope to minimize this bias by primarily comparing similar businesses, but we must remain aware of it.

Second, these media can be performative as well. People check in not because the check-in represents their location the most accurately, but because they want to show off that they have performed the check-in (Cramer et al. 2011). Sometimes people may avoid checking in for the same reason; they do not want it to be known that they checked in at a certain venue, like a fast food restaurant (Lindqvist et al. 2011).

Third, there are currently several demographic biases in these data sets. For example, Twitter, Flickr, and Foursquare are all more active per capita in cities than outside them (Hecht and Stephens 2014). Furthermore, these social media sites are all used by predominantly young, male, technology-savvy people.

One effect of this bias is shown in Fig. 2. This screenshot shows the results of some of our clusters in Livehoods (Cranshaw et al. 2012), with each dot representing a venue, and different colors representing different clusters. Note the lack of data in the center of the figure. This area is Pittsburgh's Hill District, a historic area which was the center of Pittsburgh's jazz scene in the early twentieth century. Currently, the median income for residents in the Hill District is far lower than in other parts of Pittsburgh. This neighborhood has also seen some revitalization with new senior housing, a library, a YMCA, several small office buildings, and a grocery store. However, there is still a notable lack of geotagged social media data in this area.

In short, while geotagged social media data has great potential, we also need to be careful because this data may not necessarily be representative of all people that live in a city. It is possible that this demographic bias may solve itself over time. Currently, about 58 % of Americans have a smartphone (Pew Internet 2014), and the number is rapidly growing. However, it may still be many years before demographics are more representative, and there is still no guarantee that the demographics of geotagged social media data will follow. For now, one approach is to look for ways of accounting for these kinds of biases in models. Another approach is to make clearer what the models do and do not represent.

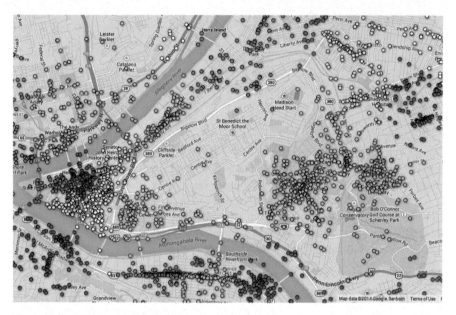

Fig. 2 Several livehoods in Pittsburgh. Each *dot* represents one Foursquare location. Venues with the *same color* have been clustered into the same livehood (the *colors* are arbitrary). The center of the map is a residential area where people with relatively low socioeconomic status live. There is also a notable lack of foursquare data in this area

4.2 Privacy Implications

Privacy is also a clear concern in using geotagged social media to understand cities. From an Institutional Review Board (IRB) perspective, much of social media data is considered exempt, because the researchers do not directly interact with participants, the data already exists, and the data is often publicly visible. However, as researchers, we need to go beyond IRB and offer stronger privacy protections, especially if we make our analyses available as interactive tools.

Here, there are at least two major privacy concerns. The first is making it easy to access detailed information about specific individuals. Even if a person's social media data is public data, interactive tools could make a person's history and inferences on that history more conspicuously available. Some trivial examples include algorithms for determining a user's home and work locations based on their tweets (Komninos et al. 2013). More involved examples might include other aspects of their behaviors, such as their activities, preferences, and mobility patterns. In the Livehoods project, we mitigated this aspect of user privacy by only presenting information about locations, not people. We also removed all venues labeled as private homes.

Second, we need to be more careful and more thoughtful about the kinds of inferences that algorithms can make about people, as these inferences can have far-reaching effects, regardless of whether they are accurate or not. There are numerous examples of inferences outside of geotagged social media that might be viewed as intrusive, embarrassing, or even harmful. For example, Jernigan and Mistree (2009) found that, given a social network with men who did not self-report their sexuality, they could identify gay men simply by analyzing the self-reported sexuality of an individual's friends. As another example, the New York Times reported on how Target had developed algorithms that could infer if a customer was pregnant (Duhigg 2012). A separate New York Times article reported on how people were assessed for credit risks based on what they purchased as well as where they shopped (Duhigg 2009).

It is important to note that these risks are not just hypothetical. At least one person had his credit card limit lowered, with the explanation that "other customers who have used their card at establishments where you recently shopped have a poor repayment history with American Express" (Cuomo et al. 2009).

It is not yet clear what the full range and extent of inferences is with geotagged social media. A significant concern is that inferences like the ones above and ones using social media data can become a proxy for socioeconomic status, gender, or race, inadvertently or even intentionally skirting around charged issues under the guise of an "objective" algorithm. It is also not clear if there is anything that can be done about these kinds of inferences, given that these inferences would be done on private servers. It is unlikely that there is a technical solution to this problem. It may very well be the case that society will require new kinds of laws governing how these inferences are used, rather than trying to control the inferences themselves.

4.3 Creating a Sustainable Ecosystem

We hope to find a way to co-create value both to social media users and to people in our work. As it exists now, value flows only from users to marketers and analysts. To create a more sustainable tool, and to avoid impinging on users' freedoms, it is important that the users gain some benefit from any system we create as well. Some of our projects point in this direction, especially the ones aimed at individual users. People may be more amenable to a tool that offers businesses insights based on their public tweets if they can have access to those insights as well.

A successful example of co-creation of value is Tiramisu (Zimmerman et al. 2011). This app aimed to provide real-time bus timing information by asking people to send messages to a server when they were on a bus. Users were allowed to get more information if they shared more information. In contrast, OneBusAway (Ferris et al. 2010) provides real-time bus information using sensors that are installed on buses. Using a collaborative approach, it may not be necessary to implement a costly instrumentation project. In addition, people may feel more ownership of a system if they contribute to it.

5 Conclusion

Despite the challenges, social media remains a potentially transformative, yet underused, source of geographic data. The works we have cited here represent useful early attempts, but we hope to inspire more. Analytics tools based on geotagged social media data can help city planners plan future developments, businesses understand the pulse of their customers, and individuals fit into new cities much more seamlessly. As our cities grow quickly and more of the world moves into heavily urbanized areas, instead of using costly methods to understand our cities, researchers of all kinds will be able to mine existing data to understand social patterns that are already there.

References

Alexander C et al (1977) A pattern language: towns, buildings, construction. Oxford University Press, New York

Andrienko G, Andrienko N, Hurter C, Rinzivillo S, Wrobel S (2011) From movement tracks through events to places: extracting and characterizing significant places from mobility data. IEEE VAST, pp 159–168

Becker RA, Cáceres R, Hanson K, Varshavsky A (2011) A tale of one city: using cellular network data for urban planning. IEEE Pervasive Comput 10(4):18–26

Becker R, Volinsky C, Cáceres R et al (2013) Human mobility characterization from cellular network data. Commun ACM 56(1):74–82

Brockmann D, Hufnagel L, Geisel T (2006) The scaling laws of human travel. Nature 439 (7075):462–465

Cheng Z, Caverlee J, Lee K, Sui D (2011) Exploring millions of footprints in location sharing services. In: Proceedings of ICWSM

Choudhury MD, Monroy-Hernández A, Mark G (2014) Narco emotions: affect and desensitization in social media during the Mexican drug war. In: Proceedings of CHI

Cramer H, Rost M, Holmquist LE (2011) Performing a check-in: emerging practices, norms and 'conflicts' in location-sharing using foursquare. In: Proceedings of MobileHCI

Cranshaw J, Schwartz R, Hong JI, Sadeh N (2012) The livehoods project: utilizing social media to understand the dynamics of a city. In: Proceedings of ICWSM

Cranshaw J, Luther K, Kelley PG, Sadeh N (2014) Curated city: capturing individual city guides through social curation. In: Proceedings of CHI

(2011) Costing the count. The Economist, June 2, 2011

Cuomo C, Shaylor J, Mcguirt M, Francescani C (2009) 'GMA' gets answers: some credit card companies financially profiling customers. ABC News

Dearman D, Sohn T, Truong KN (2011) Opportunities exist: continuous discovery of places to perform activities. In: Proceedings of CHI, pp 2429–2438

Duhigg C (2009) What does your credit-card company know about you? The New York Times. http://www.nytimes.com/2009/05/17/magazine/17credit-t.html

Duhigg C (2012) How companies learn your secrets. New York Times, Feb 16, 2012. Retrieved from http://www.nytimes.com/2012/02/19/magazine/shopping-habits.html

Eagle N, Macy M, Claxton R (2010) Network diversity and economic development. Science 328:1029–1031

Ferris B, Watkins K, Borning A (2010) OneBusAway: results from providing real-time arrival information for public transit. In: Proceedings of CHI, pp 1–10

González MC, Hidalgo CA, Barabási A-L (2008) Understanding individual human mobility patterns. Nature 453(7196):779–782

Griffin D, Hughes T (2013) Projected 2013 costs of a voluntary American Community Survey. United States Census Bureau

Harrison S, Dourish P (1996) Re-place-ing space: the roles of place and space in collaborative systems. In: Proceedings of CSCW, pp 67–76

Hecht B, Stephens M (2014) A tale of cities: urban biases in volunteered geographic information. ICWSM

Huff DL (1963) A probabilistic analysis of shopping center trade areas. Land Econ 39(1):81–90

(2014) Instagram: our story. http://www.instagram.com/press. Retrieved 24 May 2014

Isaacman S, Becker R, Cáceres R et al (2012) Human mobility modeling at metropolitan scales. In: Proceedings of MobiSys

Jernigan C, Mistree BFT (2009) Gaydar: Facebook friendships expose sexual orientation. First Monday 14:10

Joseph K, Tan CH, Carley KM (2012) Beyond "Local", "Categories" and "Friends": clustering foursquare users with latent "Topics." In: Proceedings of Ubicomp

Kitamura R, Chen C, Pendyala RAMM, Narayanan R (2000) Micro-simulation of daily activity-travel patterns for travel. Transportation 27:25–51

Komninos A, Stefanis V, Plessas A, Besharat J (2013) Capturing urban dynamics with scarce check-in data. IEEE Pervasive Comput 12(4):20–28

Krikorian R (2013) New tweets per second record, and how! Twitter Engineering Blog. 16 Aug 2013. Retrieved from https://blog.twitter.com/2013/new- tweets-per-second-record-and-how

Lindqvist J, Cranshaw J, Wiese J, Hong J, Zimmerman J (2011) I'm the Mayor of my house: examining why people use foursquare - a social-driven location sharing application. In: Proceedings of CHI

Lowenthal TA (2006) American Community Survey: evaluating accuracy

Mahmud J, Nichols J, Drews C (2013) Home location identification of Twitter users. ACM Trans Intell Syst Technol. doi:10.1145/2528548

Martine G et al (2007) The state of world population 2007: unleashing the potential of urban growth. United Nations Population Fund, New York, Retrieved from https://www.unfpa.org/public/home/publications/pid/408

Mccullagh D (2011) Carrier IQ: more privacy alarms, more confusion | Privacy Inc. - CNET News. C|Net. http://www.cnet.com/news/carrier-iq-more-privacy-alarms-more-confusion/

Mccullagh D (2012) Verizon draws fire for monitoring app usage, browsing habits. C|Net. http://www.cnet.com/news/verizon-draws-fire-for-monitoring-app-usage-browsing-habits/

Palchykov V, Kaski K, Kertész J, Barabási A-L, Dunbar RIM (2012) Sex differences in intimate relationships. Sci Rep 2:370

Pew Internet (2014) Mobile technology fact sheet. http://www.pewinternet.org/fact-sheets/mobile-technology-fact-sheet/. January 2014

Priedhorsky R, Culotta A, Del Valle SY (2014) Inferring the origin locations of tweets with quantitative confidence. In: Proceedings of CSCW, pp 1523–1536

Quercia D, Ellis J, Capra L et al (2011) Tracking "Gross Community Happiness" from Tweets. CSCW

Rost M, Barkhuus L, Cramer H, Brown B (2013) Representation and communication: challenges in interpreting large social media datasets. In: Proceedings of CSCW

Smith-Clarke C, Mashhadi A, Capra L (2014) Poverty on the cheap: estimating poverty maps using aggregated mobile communication networks. In: Proceedings of CHI

Toch E, Cranshaw J, Drielsma PH et al (2010) Empirical models of privacy in location sharing. In: Proceedings of Ubicomp, pp 129–138

Williams S, Currid-Halkett E (2014) Industry in motion: using smart phones to explore the spatial network of the garment industry in New York City. PLoS One 9(2):1–11

Yelp Word Maps (2013) http://www.yelp.com/wordmap/sf/

Zhang AX, Noulas A, Scellato S, Mascolo C (2013) Hoodsquare: modeling and recommending neighborhoods in location-based social networks. In: Proceedings of SocialCom, pp 1–15

Zimmerman J et al (2011) Field trial of tiramisu: crowd-sourcing bus arrival times to spur co-design. In: Proceedings of CHI

Developing an Interactive Mobile Volunteered Geographic Information Platform to Integrate Environmental Big Data and Citizen Science in Urban Management

Zhenghong Tang, Yanfu Zhou, Hongfeng Yu, Yue Gu, and Tiantian Liu

Abstract A significant technical gap exists between the large amount of complex scientific environmental big data and the limited accessibility to these datasets. Mobile platforms are increasingly becoming important channels through which citizens can receive and report information. Mobile devices can be used to report Volunteered Geographic Information (VGI), which can be useful data in environmental management. This paper evaluates the strengths, weaknesses, opportunities, and threats for the selected real cases: "Field Photo," "CoCoRaHS," "OakMapper," "What's Invasive!," "Leafsnap," "U.S. Green Infrastructure Reporter", and "Nebraska Wetlands". Based on these case studies, the results indicate that active, loyal and committed users are key to ensuring the success of citizen science projects. Online and off-line activities should be integrated to promote the effectiveness of public engagement in environmental management. It is an urgent need to transfer complex environmental big data to citizens' daily mobile devices which will then allow them to participate in urban environmental management. A technology framework is provided to improve existing mobile-based environmental engagement initiatives.

Keywords Environmental big data • Citizen science • Urban environmental management • Mobile • Volunteered geographic information

---

Z. Tang (✉) • Y. Zhou • Y. Gu • T. Liu
Community and Regional Planning Program, University of Nebraska-Lincoln, Lincoln, NE, USA
e-mail: ztang2@unl.edu

H. Yu
Department of Computer Science and Engineering, University of Nebraska-Lincoln, Lincoln, NE, USA

© Springer International Publishing Switzerland 2017
P. Thakuriah et al. (eds.), *Seeing Cities Through Big Data*, Springer Geography,
DOI 10.1007/978-3-319-40902-3_4

1 Introduction

Professional environmental datasets such as the Soil Survey Geographic database (SSURGO), Flood Insurance Rate Map (FIRM), National Wetland Inventory (NWI), and the water quality dataset (STORET) provide centralized environmental information on a nationwide basis. Although more accurate, detailed, geo-referenced, real-time information is being collected on a daily basis, these datasets are increasingly becoming much larger and more technical. At the same time, technical barriers still exist for the general public to be able to access nearby environmental information. Specific software (e.g., ArcGIS) and certain technical skills are needed to read these datasets. In fact, professional educators, researchers, planners, and urban managers often have difficulty in accessing some professional datasets (e.g., SSURGO data, atrazine pollution datasets) in the field, which may require using additional expensive equipment (e.g., a hand-held GPS unit). A limited number of user-friendly systems are available on mobile platforms in spite of the fact that mobile devices are rapidly becoming a predominant information channel, particularly for younger generations including college students. Most existing big environmental datasets are only accessible through websites or even hard copies. The major information channel used by our younger generations is dramatically shifting to GPS-enabled mobile devices, this critical time also represents a great opportunity to transfer big environmental datasets to mobile platforms. A mobile information platform can not only improve public awareness of environmental conditions in a community, but it can also improve the efficiency and effectiveness of how environmental data are used in urban management.

More importantly, most of these environmental datasets are still read-only in nature, whereby one can only view the data. There is no real-time reporting functionality to enable citizen science. In addition, many of the professional environmental datasets (e.g., NWI, FIRM, and SSURGO) have inaccurate or out-of-date information. Today, citizen science is recognized as a critical resource in verifying and updating environmental data (Conrad and Hilchey 2011; Tang and Liu 2016). However, citizens have inadequate tools or channels with which to share their observed information with stakeholders such as educators, researchers, and managers. From a crowdsourcing perspective, citizen scientists have not been fully empowered to participate in the traditional urban environmental management framework. However, mobile mapping tools, geo-tagged social networks, and GPS-enabled mobile devices provide robust tools for collecting real-time environmental data. With these rapidly developing technologies, citizen science is well positioned to make even more important contributions to environmental management, including air quality monitoring, water quality monitoring, and biodiversity conservation.

This paper uses a case study methodology to review pioneering crowdsourcing environmental information platforms. A technical framework, with learned experiences and lessons from two federally-funded projects, and case studies are provided to further integrate big environmental data with citizen science in urban environmental management. The SWOT (Strengths, Weaknesses, Opportunities, Threats) analysis methodology is used to qualitatively evaluate user information, and cite the use of these cases. A technical framework is provided to guide the future development of citizen science projects. This paper aims to advance current mobile information systems to provide better services for citizen science in environmental management.

2 VGI in Environmental Management

Volunteered Geographic Information (VGI) is a special case of the more general web phenomenon of user-generated content. The term VGI was first coined by Michael F. Goodchild (2007). Volunteered behavior, whether performed by an individual or an anonymous group, has existed in human history for a long time and played a significant role when humans first sought to control nature. Although it remains unclear why people volunteer themselves to generate content, the concept of VGI and citizen science has been applied to many research fields and business areas, and has resulted in countless positive effects in society (Goodchild and Glennon 2010; Jordan et al. 2012; Devictor et al. 2010; Dragicevic and Balram 2006; Elwood 2008, 2010; McCall 2003; Werts et al. 2012). For example, in 1854, Dr. John Snow discovered the reason behind the fact that cholera had caused massive deaths in London, when he showed that a central water source was responsible for the outbreak of cholera on his map (Johnson 2006). However, because there were no modern technologies available at that time, it took a long time to spread Dr. Snow's idea to the general public. With modern technologies, it is possible for an individual to share ideas over the world quickly through social media such as Facebook, Wikipedia, Twitter, and Flickr.

From the nineteenth century through to the Internet era, there have been many examples of the successful use of volunteered information. We note that the spatial property enabled by today's Web 2.0 technology can be regarded as a special enhancement for volunteered information, which is under the umbrella of VGI. VGI methods are a type of crowdsourcing, a practice that is expected to improve the theory and methodology of citizen science (Brabham 2009; Tang and Liu 2016). This theory could also be a guide for using VGI methods. In addition, the crowdsourcing data generated by VGI users can provide invaluable results for decision-making.

The integration of VGI with environmental monitoring and management allows users, whether they are citizens or scientists, to create new databases or to maintain and update existing databases. However, existing environmental databases are typically not user-friendly and are difficult for inexperienced users to access. Because many of the environmental databases were built in the early stage of the Internet, their web interfaces are not compatible with modern devices such as smart phones or tablets. Also, some existing databases are massive and if users want to access the data, they must follow specific instructions on the websites, download the data and then open it using professional tools.

Authoritative information and tools, such as remote sensing data from NASA (NASA, National Aeronautics and Space Administration) or NWI data from USWFS (NWI, a product of the U.S. Fish & Wildlife Service), are reliable and accurate to a high resolution. However, they are expensive to use, and time-consuming, inaccessible, and geographically limited. In contrast, using VGI to aid environmental monitoring and management does not require expensive updating and maintenance for high-resolution remote sensing data. This is because users, instead of agencies and corporations, update the maps. VGI also does not involve the obvious costs of building large databases, because once a VGI system is built, it functions as a data-driven website (data-driven website, one of the types of dynamic web pages). However, there are still certain costs associated with maintenance and monitoring, security, privacy or trust management, obtaining open source licenses, updating underlying databases and linking to new applications. Other forms of expense are general start-up costs and the cost of forming strategies for growing the user community. In addition, VGI can also bridge the gap between citizens, scientists, and governments. The application of VGI in environmental monitoring emphasizes the importance of participants in knowledge production and reduction of existing gaps between the public, researchers, and policymakers (Peluso 1995; Bailey et al. 2006; Mason and Dragicevic 2006; Parker 2006; Walker et al. 2007). Applying VGI in environmental monitoring enables access to potential public knowledge (Connors et al. 2012). It is only limited by a user's spatial location, and therefore it is more flexible than traditional methods in certain cases such as wetland locations and water quality management.

While some users may just want to browse rather than modify data, the learning process required for accessing existing databases may be very time-consuming and therefore off-putting. For example, if a user wants to check whether his or her property is in a national wetland inventory area, the user doesn't need to do extensive research on federal website and open data from ESRI's ArcMap program (ESRI, Environmental Systems Research Institute). Instead, the user can simply access wetland data on a smartphone anytime, anywhere. The enabled Web 2.0 technologies and VGI methods can resolve these previously-mentioned issues, incorporate websites to be compatible with smart devices, and transfer databases to crowdsourcing clients, which would significantly benefit environmental monitoring and management.

3 Criteria for Case Selection

In order to ensure the objectivity of the case study methodology, this paper adopts six criteria for case selection: information platform, issue addressed, data collection method, data presentation, service provider, and coverage. The information platform of the target cases should have interactive mobile-accessible platforms to allow citizens to view the spatial information and engage their participation through mobile devices. The selected cases should be environmentally-related topics. The data collection should rely on the public's submissions as primary data sources. The data presentation should have geospatial maps. The service providers should be research institutes and the projects chosen should have been designed for research purposes. The coverage indicates each case should address a different topic to avoid repetition and ensure the diversity of selected cases. Based on the above criteria, seven real cases were selected: "Field Photo", "CoCoRaHS," "OakMapper," "What's Invasive!", "Leafsnap," "U.S. Green Infrastructure Reporter", and "Nebraska Wetlands".

4 Case Studies

"*Field Photo*" was developed by the Earth Observation Modeling Facility at the University of Oklahoma. It has a mobile system to enable citizens to share their field photos, show footprints of travel, support monitoring of earth conditions, and verify satellite image data. Field Photo can document citizen observations of landscape conditions (e.g., land use types, natural disasters, and wildlife). Citizen-reported photos are included in the Geo-Referenced Field Photo Library which is an open-sourcing data archive. Researchers, land managers, and citizens can share, visualize, edit, query, and download the field photos. More importantly, these datasets provide crowdsourcing geospatial datasets for research on land use and changes to land coverage, the impacts of extreme weather events, and environmental conservation. Registered users have more accessibility to the photo library than guest users do. Both the iOS and Android versions of "Field Photo" applications have been available since 2014 for public download. Potential users include researchers, students and citizens. They can use GPS-enabled cameras, smartphones, or mobile devices to take photos to document their observations of landscape conditions and events.

"*CoCoRaHS*" represents the Community Collaborative Rain, Hail and Snow Network that was set up and launched by three high school students with local funding. CoCoRaHS is a non-profit, community-based network of volunteers who collaboratively measure and map precipitation (rain, hail and snow) (Cifelli et al. 2005). Beginning with several dozens of enthusiastic volunteers in 1998, the number of participants has increased every year. Besides active volunteers, there are some people who have participated in this program for a few weeks but have not remained active over the long-term. In 2000, CoCoRaHS received funding from the National Science Foundation's Geoscience Education program and was operated by the Colorado Climate Center at Colorado State University. Based on real-time statistical data, around 8000 daily precipitation reports were received in 2013 across the United States and Canada. Mobile applications of CoCoRaHS Observer for iOS and Andriod systems were provided by Steve Woodruff and Appcay Software (not CoCoRaHS) to allow registered volunteers to submit their daily precipitation reports via their mobile devices. The potential users are general public who are interested in measuring precipitation issues.

"*OakMapper*" was developed by the University of California-Berkeley in 2001 (Kelly and Tuxen 2003). Sudden oak death is a serious problem in California and Oregon forests. Because there are so many people who walk or hike in these forests, "OakMapper" extended its original site, which further allowed communities to monitor sudden oak death. "OakMapper" has successfully explored the potential synergy of citizen science and expert science efforts for environmental monitoring in order to provide timely detection of large-scale phenomena (Connors et al. 2012). By 2014, it had collected 3246 reports, most of which came from California. However, "OakMapper" is not a full real-time reporting system. Submitted data can only be displayed if it contains a specific address. As of 2014, OakMapper only had an iOS-based mobile application. Users can view the data, but they need to register in order to submit volunteered reports. The potential users include either general citizens or professional stakeholders (e.g. forest managers, biologists, etc.). This study provides a unique data source for examining sudden oak death issues in California.

"**What's Invasive!**" is a project that attempts to get volunteered citizens to locate invasive species anywhere in the United States by making geo-tagged observations and taking photos that provide alerts of habitat-destroying invasive plants and animals. This project is hosted and supported by the Center for Embedded Networked Sensing (CENS) at the University of California, the Santa Monica Mountains National Recreation Area, and Invasive Species Mapping Made Easy, a web-based mapping system for documenting invasive species distribution developed by the University of Georgia's Center for invasive Species and Ecosystem Health. Any user who registers as a project participant must provide an accurate email address. Users can self-identify as a beginner, having had some species identification training, or an expert. This project only tracks statistical data such as the frequency with which users log in for research use. No personal background data on users are collected. Both the iOS and Android versions of mobile applications have been available since 2013 for citizen download. The potential users are general citizens who are interested in invasive species.

"**Leafsnap**" is a pioneer in a series of electronic field guides being developed by Columbia University, the University of Maryland, and the Smithsonian Institution. It uses visual recognition software to help identify tree species from photographs of their leaves. Leafsnap provides high-resolution images of leaves, flowers, fruit, petiole, seed, and bark in locations that span the entire continental United States. Leafsnap allows users to share images, species identifications, and geo-coded stamps of species locations and map and monitor the ebb and flow of flora nationwide. Both the iOS and Android versions of mobile applications are available for citizen use. Citizens with interests in trees and flowers are the potential users for this application.

"**U.S. Green Infrastructure Reporter**" was developed in the Volunteered Geographic Information Lab at the University of Nebraska-Lincoln in 2012. Its main purpose is to allow stakeholders and citizens to report green infrastructure sites and activities through their mobile devices. This mobile information system has a GPS-synchronous real-time reporting function with its own geospatial database that can be used for analysis. It provides both iOS and Android versions of mobile applications. More than 6700 reports were collected across the United States by 2013. Both professional stakeholders and general citizens are the potential users.

"**Nebraska Wetlands**" was developed in the Volunteered Geographic Information Lab at the University of Nebraska-Lincoln in 2014. This application translates the wetland datasets such as NWI and SSURGO data to mobile devices. It can allow citizens to easily access the complex environmental datasets. In addition, this application also incorporates a GPS-synchronous real-time reporting system to allow citizens to upload their observations. This system has both iOS and Android versions of mobile applications. More than 600 reports were collected in Nebraska in 2005. Both the professional stakeholders and general citizens are the potential users.

5 SWOT Analysis

This study adopts a qualitative analysis method to analyze the selected cases. SWOT (Strengths, Weaknesses, Opportunities, and Threats) analysis is a structured method used to evaluate these cases. The strengths indicate the advanced characteristics of one case over other cases. The weaknesses indicate the disadvantages of a case in this usage. Opportunities mean the elements in an environment that can be exploited to their advantage. Threats indicate the elements in an environment that could cause trouble or uncertainty for development. SWOT analysis results can provide an overview of these cases in terms of their existing problems, barriers, and future directions. Through identifying the strengths, weaknesses, opportunities, and threats from the seven real cases; "Field Photo", "CoCoRaHS," "OakMapper," "What's Invasive!", "Leafsnap", "U.S. Green Infrastructure Reporter", and "Nebraska Wetlands", this paper proposes a technical framework to guide future mobile application development.

6 Results

6.1 Strengths

The mobile-based information platform has more unique strengths than other web-based platforms. All the cases were built on mobile platforms. Citizens can use their portable mobile devices to participate in these projects and do not need to rely on special devices. The projects, such as "Field Photo", "CoCoRaHS", or "OakMapper", can be incorporated into their daily activities. Another example is the identification of invasive species, a user-friendly participatory tool of "What's Invasive!" is helpful in solving this large-scale and long-term environmental problems across the national parks. The significant strengths of these selected cases clearly address the specific tasks. All of these cases address very well-defined topics (e.g. "Leafsnap" for leaf identification) that are easily understood by citizens. When citizens find topics that fit their interests and have the opportunity to contribute to a problem-solving process, they tend to convert their interests into real action in order to participate in these projects. For example, citizens can participate in recording the green infrastructure sites through "U.S. Green Infrastructure Reporter" and wetland conditions through "Nebraska Wetlands". Compared with web-based technologies, mobile information platforms can transcend the limitations of time and space and thus attract grassroots participation. For example, "Field Photo" allows citizens to use GPS-enabled mobile devices to report geo-referenced field information anytime, anywhere. Compared to traditional methods, VGI has many advantages for environmental management including reduced costs and adaptability. The crowdsourcing data generated by volunteers can be an alternative data source for environmental management; there is only a slight difference in quality between the data collected by experts and the data collected by non-experts (Engel and Voshell 2002; Harvey et al. 2001). Additionally, VGI can be used as an effective means to promote public awareness and engagement, compensating for the weaknesses of traditional methods (Tang and Liu 2016).

Although the VGI method will not replace traditional methods, it will augment traditional methods for environmental management. VGI can also be considered as a new way to rapidly collect information from third parties, nonprofit organizations, and municipal departments. The databases, such as "What's Invasive!" and "U.S. Green Infrastructure Reporter", created by VGI users can provide multiple benefits for environmental management in future studies such as invasive species research and green infrastructure management. When designing an application for a VGI method, planners also need to think about people's urgent and daily needs (e.g. "Field Photo"), which could be very helpful for planners to reach the public and improve citizen engagement. In addition, using VGI methods could accelerate the planning process and save costs, since residents can contribute to part of the planning process instead of the government.

6.2 Weaknesses

The weaknesses of these projects include limited availability of mobile devices, age gaps, data quality issues, and data verification issues. Some citizens, particularly the minority groups, still cannot afford GPS-enabled mobile devices with expensive data plans (Abascal et al. 2015). From the current usage of mobile devices, aging population may have relatively lower levels of knowledge and/or interest in using mobile devices. The quality of citizen-reported data is a fundamental challenge for any citizen science project, including our seven cases. For example, data verification for citizen-reported invasive species needs a significant amount of time and resources. In addition, experiences with the "U.S. Green Infrastructure Reporter" also bear out that the number of users does not equal the number of active users. Most of the green infrastructure were submitted by a small number of active users. Thus, the total number of users only provides an overview of project coverage, but the number of registered users is a better indicator of those who contribute to citizen science projects. More importantly, a high level of participation can bring a large amount of citizen-submitted data, but this does not mean that citizens provide a large amount of useful data, such as the precipitation, snow, hails data from the "CoCoRaHS" project. Data quality comes from better users' understanding, judgment and operational experiences. The quality verification depends on an expert's knowledge and experience. For example, a manual verification procedure for green infrastructure sites, on the other hand, can reduce the probability of error, but needs more time and money to implement. Compared to the authoritative data from agencies or corporations, the quality of data collected from VGI systems is typically a concern. However, it has been proven that there is no significant difference between data collected by scientists and volunteers and, in fact, volunteers can made valuable contributions to the data collection process (Engel and Voshell 2002; Harvey et al. 2001).

The VGI method has similar problems to other crowdsourcing practices. The VGI method still has occasional data quality problems, especially when volunteers are not motivated to contribute their knowledge to the VGI system. For example, in the "Field Photo" and "CoCoRaHS," projects, some volunteers may not be willing to finish all the inventory questions when the results of the VGI methods cannot be seen immediately. In addition to data quality issues, crowdsourcing also has the following problem: There is no time constraint for the data collection process with the VGI method. When using the VGI method, it is hard for planners to define or identify what citizens' motivations and interests are, and whether their motivation and interests will have a positive or a negative impact to the planning process. For example, in the "Field Photo" and ""Leafsnap"" projects, how to prohibit intellectual property leakage for sensitive topics is another concern. There are currently no standards for designing a VGI system, so there is not much control over the development or the ultimate products, such as "U.S. Green Infrastructure Reporter" and "Nebraska Wetlands". Thus, when using the VGI method, planners also need to think about what things can be crowd sourced and what cannot, such as content that has copyright issues.

6.3 Opportunity

Mobile-based projects provide a new avenue to participate in environmental management. Mobile devices are currently very popular as a part of people's daily lives. Mobile information platforms are increasingly becoming the predominant information channel for people to receive and report information. Many of these projects, such as "Field Photo", "Leafsnap,", "OakMapper," and "What's Invasive!", have successfully combined online activities and off-line engagement activities to attract new participants and retain older users. The case study for "CoCoRaHS" project also found that informal communication is a valuable portion of mobile engagement. From the experiences in the "CoCoRaHS" project, the integration of nonverbal communication and personal touches can improve the effectiveness of mobile engagement. In addition, volunteered coordinators are helpful in promoting volunteer activities. From the "U.S. Green Infrastructure Reporter", and "Nebraska Wetlands", we found that social media and social networks are promising channels through which to attract people in the virtual community. These mobile VGI projects are still not empowered by social media and social networks. In general, applying VGI methods to planning means not only building a VGI system. It is both a planning process and a crowdsourcing practice. Crowdsourcing can be considered as a new and upcoming planning issue in the future, especially when e-government practices are expected to become more popular in the future.

6.4 Threats

The first threat is limited motivation from citizens (Agostinho and Paço 2012). Projects without adequate motivation are the greatest threat to long-term success. Most citizens have no interest in scientific research projects that cannot bring any direct economic or social benefits to them (Tang and Liu 2016). A successful engagement should build a shared vision and devote responsibility and commitment in specific tasks, such as invasive species observation and green infrastructure site report. From the seven cases, we found that a two-way communication platform cannot be fully engaged due to lack of timely feedback from the organizers. None of the cases has provided timely online feedbacks for the users. The reported data can only be directly used by the project organizers for analysis. Current databases (e.g. "Field Photo", "U.S. Green Infrastructure Reporter", and "Nebraska Wetlands") do not have efficient approaches to manage citizen science data such as photos and videos. The analysis of the collected data is still very superficial and does not have any in-depth data mining. The collected data types mainly only include text and photos and do not include more scientific data. These projects still lack strategic management and sustainability. Most of these projects depend on specific funding to support their daily operations. Both the "CoCoRaHS" and "What's Invasive!" projects seek for alternative funding resources to reduce the threats and to ensure effective implementation.

Although technology has improved greatly on mobile devices, there are still two technology problems that have an impact on adopting VGI methods: signal coverage range and signal quality for mobile devices, and battery life. Although most mobile devices can get good signals in urban areas, in rural areas mobile devices can disconnect suddenly due to inadequate signal range or poor signal quality, which can result in missing or erroneous data. For example, the "Field Photo" and "Nebraska Wetlands" were not accessible in the rural areas. In addition, information service providers have different signal coverage. Even in the same city, the signal quality and strength varies greatly and VGI users may lose patience if their phone service isn't available in a particular area. Signal quality and strength not only has an impact on the adoption of VGI methods, it also has an impact on the volunteers themselves. When we used the "Nebraska Wetlands", the speed of mobile maps was very slow in some areas. Adopting VGI methods also requires users to have better mobile devices, although VGI developers and planners will try to cover as many popular devices as they can and reduce memory requirements for mobile applicants.

There are still some hardware limits that cannot be solved, however, such as battery life. If the VGI users are a group of professionals, they may use VGI apps on their mobile devices to collect data for an entire 24 h, and the screen will be constantly refreshing, reducing the battery life. Another problem with adopting VGI methods is covering the entire mobile phone market. It is in the VGI developers' best interest to have their apps deployed on both Android and iOS phones. This is not a simple task because Android and iOS systems have different policies for reviewing a developer's app. Google is open source and allows more freedom to access phone abilities, whereas Apple is closed and is sensitive to accessing phone abilities such as sensors. Trying to keep VGI apps working both on Google and Apple phones is a challenge for developers and planners. In general, Google and Apple both provide a good development space for promoting VGI methods, and each has its own advantages and disadvantages.

The invasion of hackers is still a threat, and the same concerns extend to the mobile era. From a technological viewpoint, it is true that some smartphone systems have safety issues, and hacking software targets to smart phones is common. Utilizing VGI methods can also become a new opportunity for hackers to invade public privacy. With these risks in minds, users question the necessity of installing software on their smartphones that they don't need to rely on or use every day.

7 Discussion

Based on the findings from the selected cases, this paper suggests a technical development framework that can be used to improve the existing mobile-based environmental information system. The below sections explain the technical development framework that can be used for future VGI system development. The proposed technical development framework aims to continue the strengths, overcome the weaknesses, expand the opportunities, and reduce the threats.

8 Technical Development Framework

From the seven selected cases, we analyzed the strengths, weaknesses, opportunities, and threats, and thus propose a new technical development framework that can maximize the VGI applications in environmental management. Understanding users' needs is essential for using VGI methods in the environmental management field. When designing mobile information platforms, designers also need to think about what kinds of needs the public has. System deployment includes two tasks: front-end tasks and back-end tasks. Front ends represent the client side of the system, such as desktop users and mobile users. Back ends are the server side of the system. It includes two web servers and a GIS server. There are two types of front ends: web front ends and mobile front ends. Web front ends are web applications that allow Internet clients or users to request back-end services through a URL (Uniform Resource Locator) via web browsers. Mobile front-ends are mobile applications that allow mobile clients to request back-end services through smartphones or tablets. Mobile applications can be downloaded and installed via Apple Store or Google Play. Typically, multi-requests occur when there are too many front-end users accessing the server simultaneously. The number of multi-requests from mobile clients is usually higher than the number of multi-requests from desktop clients, which means that the server back-ends should be optimized to account for these differences. Besides the REST (REST, Representational state transfer) service hosted on a GIS server, there is also educational information hosted on the back end. Educational information is typically placed on static web pages which does not require communication with a GIS server to access. By adding an additional web server, the system can filter the client into two groups: those requiring use of GIS services and those who do not. Using this deployment method, the total number of multi-requests for a GIS server can be reduced significantly if a certain number of clients only browse the education web page and, at the same time, those who search on the map or report data can still have a smooth experience. In short, this deployment method can reduce the peak number of multi-requests from clients, especially when there are many mobile users who are only viewing static information through the front end. A workstation was also added to the back end in order to give data access to professional users such as GIS analysts (Fig. 1). In addition, by publishing web maps through ArcGIS online.com for related research, data can also be shared between other professionals such as environmental scientists and planners (Fig. 1).

The key point of system architecture is using embedded Javascripts in Html (Html, HyperText Markup Language) (Fig. 2). JavaScripts is a client-side scripting language. Using embedded JavaScripts is an affordable way for deploying small business applications through clients, because most current browsers, such as Firefox, Safari or Chrome, support JavaScripts very well. Thus, it is possible to develop only one application and make it executable on different platforms. The PhoneGap (PhoneGap, a mobile development framework produced by Nitobi) has realized this possibility; developers do not need to update their code frequently or

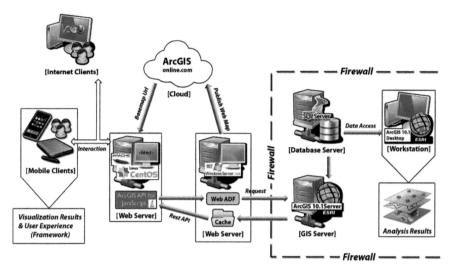

Fig. 1 Deployment of front ends and back ends

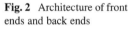

Fig. 2 Architecture of front ends and back ends

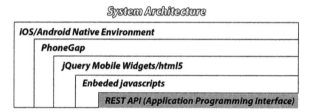

develop applications for each different platform, which can significantly reduce the total cost of the system. The Html code with embedded JavaScripts can be wrapped into the native environment on the client side, because PhoneGap provides a bridge between the native environment and web environment (Fig. 2).

8.1 Mobile Front-End Design Framework

The design framework for mobile front ends, including visualization and user interfaces, includes five key features (Fig. 3): (1) GPS reporting and mapping features will enable users to browse maps, send geocoded photos or data, and query attributes of geometric objects. (2) Publication and education features will be designed for novice technicians or students to study green infrastructure strategies. Third party publications are also posted as a link in this feature. (3) With the linkage to social media, users, experts and advocates can share their ideas through

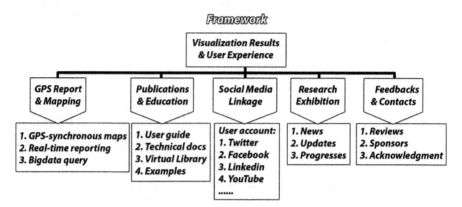

Fig. 3 Design framework for a mobile front end

social networks. Several popular social networks will be included, such as Twitter and Facebook. (4) News and research progress will be posted through a research exhibition feature for environmental experts. (5) Users can also find reviews and contact information through feedback and contacts features.

8.2 Native App vs. Hybrid App vs. Web App

A mobile application, the front end of the VGI system, has a great impact on attracting volunteers and promoting VGI concepts to the public. Issues to consider include whether it appears user-friendly, and whether the key functionality enhances the user experience. There are three different types of mobile applications: native apps, hybrid apps, and web apps. All of these applications have their advantages and disadvantages when building a VGI system. Developers and planners also need to choose a suitable application type for their projects. It is difficult to assess which type is the perfect option for building a VGI system; planners and developers need to balance development costs and time, as well as key functional features.

Native applications are those that are developed by native programming languages, which are specified by mobile Operating Systems (OS). A native application can only run on the mobile OS that supports it. One immediate advantage of a native application is that all the documentation can be done with or without Internet connectivity. A native application can also run smoothly on the specified OS, and it usually has fewer bugs. However, a native application also has disadvantages. Since the native programming language is used to develop native apps, it is hard for a native app to be cross-platform. If planners or developers choose to develop a native application, they need to program on every OS platform by using different native

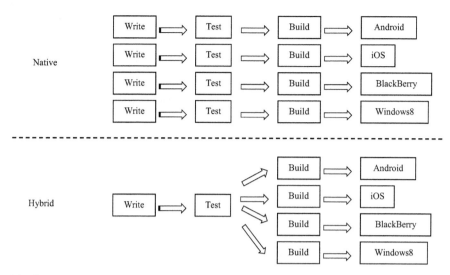

Fig. 4 Comparison of the development workflows between a native app and a hybrid app

OS coding languages, such as Objective-C and Java (Fig. 4). In addition, updating the native app is also a problem since it requires knowledge of different programming languages. It can be expensive and time-consuming to develop a native app for the VGI concept, but in some cases, such as local data analysis, choosing a native application is a wise solution.

Web applications actually are websites in mobile form, which rely on browser and Internet connectivity. Web applications don't have a native look, but they are much cheaper to develop than native apps and hybrid apps. Web apps are not popular, because they require users to remember website links. In general, web apps are not suitable for promoting VGI concepts because they don't look user-friendly, and they often has a blurred user experience. Hybrid apps are part native apps, and part web apps. Hybrid app development requires web experience, such as Html5 and JavaScript. Although hybrid apps are not developed using native programming languages, they can have a native look and can access the functionalities of the smart phone. The native appearance of a hybrid app relies on some open sourced or third party user interface framework, such as jQuery Mobile and Sencha Touch. Thus, its performance may not be as smooth as a native app, but the cost and maintenance of the hybrid app is usually cheaper and easier than the native app workflow. In addition, choosing a hybrid app doesn't require planners and developers to have native development experience; it only requires web development experience, making it easier for a hybrid app to be cross-platform. Most social media apps are hybrid apps, because they are easily distributed on different mobile OS platforms. In general, choosing the right type of app for building a VGI system is very important, because it has a direct impact on the volunteers. It is hard to say which type of app workflow is better than others. All have their own advantages and disadvantages.

9 Conclusions

This paper is a pilot study to summarize the strengths, weaknesses, opportunities, and threats of current mobile applications from the seven cases—"Field Photo", "CoCoRaHS", "OakMapper", "What's Invasive!", "Leafsnap", "U.S. Green Infrastructure Reporter", and "Nebraska Wetlands". The findings provide insightful information for environmental managers and researchers to further integrate mobile information system in environmental monitoring and management. The technical framework can help development of new mobile information system in environmental management. At the same time, however, we also have to recognize the major limitations of mobile information system. The limitations may include the lower motivation on scientific applications, difficulty to verify data quality, and uncertainty to maintain the sustainability of mobile information system. This paper provides suggestions for better use of mobile information tool in the environmental management field. In general, the VGI method is not expected to replace traditional methods (e.g. governmental data collection methods), but rather be an alternative solution for environmental monitoring and management. The data collected by non-professionals from a VGI system is reliable, and can be an additional data source for environmental monitoring and management.

Acknowledgements This paper has been funded wholly or partially by the United States Environmental Protection Agency (EPA) under an assistance agreement (CD-977422010; UW-97735101). The contents do not necessarily reflect the views and policies of the EPA, nor does the mention of trade names or commercial products constitute endorsement or recommendation for use.

References

Abascal J, Barbosa SDJ, Nicolle C, Zaphiris P (2015) Rethinking universal accessibility: a broader approach considering the digital gap. Univ Access Inf Soc. doi:10.1007/s10209-015-0416-1

Agostinho D, Paço A (2012) Analysis of the motivations, generativity and demographics of the food bank volunteer. Int J Nonprofit Volunt Sect Mark 17:249–261

Bailey C, Convery I, Mort M, Baxter J (2006) Different public health geographies of the 2001 foot and mouth disease epidemic: "Citizen" versus "professional" epidemiology. Health Place 12:157–166

Brabham DC (2009) Crowdsourcing the public participation process for planning projects. Plan Theory 8(3):242–262

Cifelli R, Doesken N, Kennedy P, Carey LD, Rutledge SA, Gimmestad C, Depue T (2005) The community collaborative rain, hail, and snow network. Am Meteorol Soc 86(8):1069–1077

Connors JP, Lei S, Kelly M (2012) Citizen science in the age of neogeography: utilizing volunteered geographic information for environmental monitoring. Ann Assoc Am Geogr 102(X):1–23

Conrad CC, Hilchey KG (2011) A review of citizen science and community-based environmental monitoring: issues and opportunities. Environ Monit Assess 176(1-4):273–291

Devictor V, Whittaker RJ, Beltrame C (2010) Beyond scarcity: citizen science programmes as useful tools for conservation biogeography. Divers Distrib 16:354–362

Dragicevic S, Balram S (2006) Collaborative geographic information systems and science: a transdisciplinary evolution. In: Dragicevic S, Balram S (eds) Collaborative geographic information systems. Idea Group, Hershey, PA, pp 341–350

Elwood S (2008) Grassroots groups as stakeholders in spatial data infrastructures: challenges and opportunities for local data development and sharing. Int J Geogr Inf Sci 22:71–90

Elwood S (2010) Geographic information science: emerging research on the societal implications of the geospatial web. Prog Hum Geogr 34(3):349–357

Engel S, Voshell JR (2002) Volunteer biological monitoring: can it accurately assess the ecological condition of streams? Am Entomol 48(3):164–177

Goodchild MF (2007) Citizens as sensors: the world of volunteered geography. GeoJournal 69:211–221

Goodchild MF, Glennon J (2010) Crowdsourcing geographic information for disaster response: a research frontier. Int J Digit Earth 3(3):231–241

Harvey E, Fletcher D, Shortis M (2001) A comparison of the precision and accuracy of estimates of reef-fish lengths determined visually by divers with estimates produced by a stereo-video system. Fish Bull 99:63–71

Johnson S (2006) The Ghost Map: the story of London's most terrifying epidemic, and how it changed science, cities, and the modern world. Riverhead, New York

Jordan RC, Brooks WR, Howe DV, Ehrenfeld JG (2012) Evaluating the performance of volunteers in mapping invasive plants in public conservation lands. Environ Manag 49:425–434

Kelly NM, Tuxen K (2003) WebGIS for monitoring "Sudden Oak Death" in coastal California. Comput Environ Urban Syst 27:527–547

Mason BC, Dragicevic S (2006) Web GIS and knowledge management systems: an integrated design for collaborative community planning. In: Balram S, Dragicevic S (eds) Collaborative geographic information systems. Idea Group Publishing, Hershey, PA, pp 263–284

McCall MK (2003) Seeking good governance in participatory-GIS: a review of processes and governance dimensions in applying GIS to participatory spatial planning. Habitat Int 27:549–573

Parker B (2006) Constructing community through maps? Power and praxis in community mapping. Prof Geogr 58(4):470–484

Peluso NL (1995) Whose woods are these? Counter-mapping forest territories in Kalimantan, Indonesia. Antipode 27(4):383–406

Tang Z, Liu T (2016) Evaluating Internet-based public participation GIS (PPGIS) and volunteered geographic information (VGI) in environmental planning and management. J Environ Plan Manage 59(6):1073–1090

Walker D, Jones JP, Roberts SM, Fröhling OR (2007) When participation meets empowerment: the WWF and the politics of invitation in the Chimalapas, Mexico. Ann Assoc Am Geogr 97:423–444

Werts JD, Mikhailova EA, Post CJ, Sharp JL (2012) An integrated WebGIS framework for volunteered geographic information and social media in soil and water conservation. Environ Manag 49:816–832

CyberGIS-Enabled Urban Sensing from Volunteered Citizen Participation Using Mobile Devices

Junjun Yin, Yizhao Gao, and Shaowen Wang

Abstract Environmental pollution has significant impact on citizens' health and wellbeing in urban settings. While a variety of sensors have been integrated into today's urban environments for measuring various pollution factors such as air quality and noise, to set up sensor networks or employ surveyors to collect urban pollution datasets remains costly and may involve legal implications. An alternative approach is based on the notion of volunteered citizens as sensors for collecting, updating and disseminating urban environmental measurements using mobile devices. A Big Data scenario emerges as large-scale crowdsourcing activities tend to generate sizable and unstructured datasets with near real-time updates. Conventional computational infrastructures are inadequate for handling such Big Data, for example, designing a "one-fits-all" database schema to accommodate

---

J. Yin • Y. Gao
CyberGIS Center for Advanced Digital and Spatial Studies, University of Illinois at Urbana-Champaign, Urbana, IL 61801, USA

CyberInfrastructure and Geospatial Information Laboratory, University of Illinois at Urbana-Champaign, Urbana, IL 61801, USA

Department of Geography and Geographic Information Science, University of Illinois at Urbana-Champaign, Urbana, IL 61801, USA
e-mail: jyn@illinois.edu

S. Wang (✉)
CyberGIS Center for Advanced Digital and Spatial Studies, University of Illinois at Urbana-Champaign, Urbana, IL 61801, USA

CyberInfrastructure and Geospatial Information Laboratory, University of Illinois at Urbana-Champaign, Urbana, IL 61801, USA

Department of Geography and Geographic Information Science, University of Illinois at Urbana-Champaign, Urbana, IL 61801, USA

Department of Computer Science, University of Illinois at Urbana-Champaign, Urbana, IL 61801, USA

Department of Urban and Regional Planning, University of Illinois at Urbana-Champaign, Urbana, IL 61801, USA

National Center for Supercomputing Applications, University of Illinois at Urbana-Champaign, Urbana, IL 61801, USA
e-mail: shaowen@illinois.edu

diverse measurements, or dynamically generating pollution maps based on visual analytical workflows.

This paper describes a CyberGIS-enabled urban sensing framework to facilitate the volunteered participation of citizens in sensing environmental pollutions using mobile devices. Since CyberGIS is based on advanced cyberinfrastructure and characterized as high performance, distributed, and collaborative GIS, the framework enables interactive visual analytics for big urban data. Specifically, this framework integrates a MongoDB cluster for data management (without requiring a predefined schema), a MapReduce approach to extracting and aggregating sensor measurements, and a scalable kernel smoothing algorithm using a graphics processing unit (GPU) for rapid pollution map generation. We demonstrate the functionality of this framework though a use case scenario of mapping noise levels, where an implemented mobile application is used for capturing geo-tagged and time-stamped noise level measurements as engaged users move around in urban settings.

Keywords Volunteered Geographic Information • Urban sensing • Noise mapping • CyberGIS • Mobile devices

1 Introduction

In today's urban environments, various pollution problems have become significant concerns to people's health and well-being. Being able to monitor and measure the status of environmental pollution with high spatiotemporal resolution for producing accurate and informative pollution maps is crucial for citizens and urban planners to effectively contribute to decision making for improving living quality of urban environments. Traditionally, government agencies are responsible for measuring and collecting urban pollution data, which is done either by employing surveyors with specialized equipment or by setting up monitoring networks. For example, under the EU environmental noise directive (2002/49/EC) (Directive 2002), some cities commenced the installation of permanent ambient sound-monitoring networks. This approach is, however, subject to several limitations. For instance, it is often costly to build such sensor networks and hire surveyors. Furthermore, such sensors are statically placed and each can only cover an area or space of certain size. The sensor measurements themselves are usually sampled and aggregated for a period of time resulting in low update frequency.

Due to these limitations, alternative approaches have been investigated including the utilization of citizens as sensors to contribute to collecting, updating and disseminating information of urban environments, also known as crowdsourcing (Howe 2006; Goodchild 2007). In particular, some previous studies have explored the idea of encouraging participatory noise monitoring using mobile devices. For example, the NoiseTube mobile application utilizes the combination of microphone and embedded GPS receiver to monitor noise pollution at various sites of a city

(Maisonneuve et al. 2009, 2010). This effort also showed some promising results regarding the effectiveness of participatory noise mapping. Compared to the traditional noise monitoring approach that relies on centralized sensor networks, the mobile approach is less costly; and with collective efforts, this approach using humans as sensors can potentially reach a significantly larger coverage of the city.

With integrated environmental sensors,[1] the new generation mobile devices can instrument comprehensive environmental properties, such as ambient temperature, air pressure, humidity, and sound pressure level (i.e., noise level). However, when the involvement of a large number of participants engaging in crowdsourcing activities becomes a realization, a large volume of, near real-time updated, unstructured datasets are produced. Conventional end-to-end computational infrastructures will have difficulties in coping with managing, processing, and analyzing such datasets (Bryant 2009), requiring support from more advanced cyberinfrastructure regarding data storage and computational capabilities.

This paper describes a CyberGIS-enabled urban sensing framework to facilitate the participation of volunteered citizens in monitoring urban environmental pollution using mobile devices. CyberGIS represents a new-generation GIS (Geographic Information System) based on the synthesis of advanced cyberinfrastructure, GIS and spatial analysis (Wang 2010). It provides abundant cyberinfrastructure resources and toolkits to facilitate the development of applications that require access to, for example, high performance and distributed computing resources and massive data storage. This framework enables scalable data management, analysis, and visualization intended for massive spatial data collected by mobile devices. To demonstrate its functionality, we focus on the case of noise mapping. In general, this framework integrates a MongoDB[2] cluster for data storage, a MapReduce approach (Dean and Ghemawat 2008) to extracting and aggregating noise records collected and uploaded by mobile devices, and a parallel kernel smoothing algorithm using graphics processing unit (GPU) for efficiently creating noise pollution maps from massive collection of records. This framework also implements a mobile application for capturing geo-tagged and time-stamped noise level measurements as users move around in urban settings.

The remainder of this paper is organized as follows: Section "Participatory Urban Sensing and CyberGISParticipatory Urban Sensing and CyberGIS" describes the related work in the context of volunteered participation of citizens in sensing urban environment. We focus on the research challenges in terms of data management, processing, analysis, and visualization. In particular, CyberGIS is argued to be suitable for addressing these challenges. Section "System Design and Implementation" illustrates the details of the design and implementation of the CyberGIS-enabled urban sensing framework. Section "User Case Scenario" details a user case scenario for noise mapping using mobile devices. Section "Conclusions and Future Work" concludes the paper and discusses future work.

---

[1] http://developer.android.com/guide/topics/sensors/index.html
[2] http://www.mongodb.org/

2 Participatory Urban Sensing and CyberGIS

To monitor and study urban environmental pollution, data collection and processing are two major steps in our framework. In terms of data collection from citizens engaged in reporting noise levels around a city, researchers found a low cost solution of using the microphone of mobile device to record and calculate the sound levels, such as the SPL android application.[3] Combining the embedded GPS receiver on mobile devices, the noise-level measurements are geo-tagged with geographic locations, which allow researchers to generate heatmap like noise maps (Maisonneuve et al. 2009; Stevens and DHondt 2010). In addition to appending the geo-location as a tag to the measurement, other applications, such as NoiseTube also encourages the appending of environmental tags, such as the type of noise (e.g., cars and aircraft) as additional attributes to the records. To encourage participants to contribute to the sensing activity as much as possible, such measurement can even take place whenever a user posts a social media message using their mobile device. However, since the availability of sensors varies in different devices, the collection of users' measurements can seem to be "unstructured", which makes it difficult to design a "one-fits-all" database schema to accommodate all the user inputs. Furthermore, when a large number of citizens participate in sensing urban environments using mobile devices simultaneously, it poses challenges for efficient data management, processing and visualizations. A Big Data scenario emerges in large-scale crowdsourcing activities, which requires an innovative system to support scalable data handling, such as data storage with flexible data schema and efficient database querying. Many applications, such as NoiseTube, use a relational database for data storage and processing. Relational databases, with a rigidly defined, schema-based approach, make it difficult to incorporate new types of data (Stonebraker et al. 2007) and achieve dynamic scalability while maintaining the performance users demand (Han et al. 2011).

The large volume and dynamic nature of the datasets also causes visualization problems for noise map generation. Existing GIS libraries, such as heatmap.js[4] and map servers (e.g., GeoServer[5]) provide inadequate support for this type of data. In particular, the ability to perform visualization based on customized queries regarding, e.g., a specified time window or an individual user (or a particular group of users) from the accumulated large volume of data, is limited in the existing applications. To embrace the characteristics of Big Data from large-scale crowdsourcing activities and accommodate the geographic attributes of the user generated content, CyberGIS integrates high performance computing resources and scalable computing architecture to support data intensive processing, analysis and visualization (Ghosh et al. 2012; Wang et al. 2012). CyberGIS represents a new-generation of GIS based on the synthesis of advanced cyberinfrastructure

---

[3] http://play.google.com/store/apps/details?id=com.julian.apps.SPLMeter&hl=en
[4] http://www.patrick-wied.at/static/heatmapjs/
[5] http://geoserver.org/

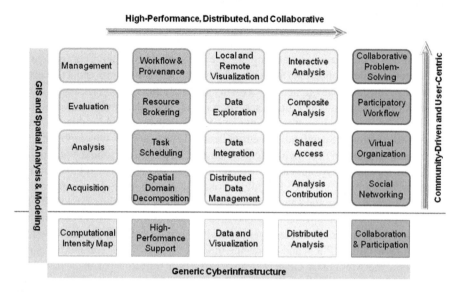

Fig. 1 An overview of the CyberGIS architecture. *Source*: Wang et al. (2013a)

GIS and spatial analysis (Wang 2010). As illustrated in Fig. 1 for the overview of the CyberGIS architecture, CyberGIS provides a range of capabilities for tackling the data and computation-intensive challenges, where the embedded middleware can link different components to form a holistic platform tailored to specific requirements.

In particular, our framework utilizes several components within this architecture. In the "distributed data management" component, we deploy a MongoDB cluster over multiple computing nodes for monitoring data intake and storage, which is scalable to the growth of collected data volume. Compared to a relational database, the NoSQL database supports more flexible data models with easy scale-out ability and high performance advantages (Han et al. 2011; Wang et al. 2013b). In the "high performance support" layer, we rely on the MapReduce functionality of the MongoDB cluster for data processing, such as individual user trajectory extraction, which is used to visualize the pollution exposure to a particular participant; and aggregation of data provided by all participants to a 1-h (this value is defined for the ease of implementation and can be changed according to user specifications) time window. This is then used to dynamically produce noise maps for the monitored environment. And finally, in the "data and visualization" layer, we apply a parallel kernel smoothing algorithm for rapid noise map generation using GPUs. Specific design and implementation details will be discussed in the following section.

3 System Design and Implementation

The framework is designed and implemented to include two main components: a dedicated mobile application (for Android devices) for participants and a CyberGIS workflow for data management, processing and pollution map generation. A diagram for the overall architecture is shown in Fig. 2. For this framework, we employ a service-oriented architecture for the integration between mobile devices and a CyberGIS platform. Specifically, the mobile application utilizes the combination of GPS receivers and environmental sensors on mobile devices to produce geo-tagged and time-stamped environmental measurements. In addition, this application provides a background service that allows user to choose to store or append the measurement to other apps a user is interacting with in the mobile device. It is up to participants to decide when to upload their records to the CyberGIS platform via the implemented RESTful (Representational state transfer) web service interface. CyberGIS workflow filters and parses the input data (into JSON[6] format) and stores them into the MongoDB cluster. It also extracts a trajectory of each individual participant to visualize the pollution exposure along the trajectory. For pollution map generation from the measurements that are uploaded by all of the participants, the data aggregation process is carried out using a specified time window. A pollution map is dynamically generated as a GeoTIFF[7] image via a parallel kernel smoothing method using GPU, which will be displayed as a map overlay on top of the ESRI world street map.[8]

3.1 CyberGIS Workflow

The workflow first filters out invalid data records (e.g., records without valid coordinates) and then parses each record as a JSON object before saved to the MongoDB cluster. The MongoDB cluster is chained in a master-slave style in order to achieve scalability as datasets are accumulated into significant size, which is one of the significant advantages over the existing relational databases. Another advantage brought by the MongoDB cluster is the embedded mechanism for performing MapReduce tasks. Since there is no predefined data schema and the input data are simply raw documents with the only structure of <key, value> pairs, the MapReduce function can efficiently sort the "unstructured" records based on the specified keys, e.g. timestamp, unique user id or even geographical coordinates (or a combination of these). More importantly, the data are stored in a distributed fashion, meaning multiple instances of computing nodes can perform such tasks simultaneously, which is otherwise nearly impossible for conventional database

---

[6] http://json.org/
[7] http://en.wikipedia.org/wiki/GeoTIFF
[8] http://www.esri.com/software/arcgis/arcgisonline/maps/maps-and-map-layers

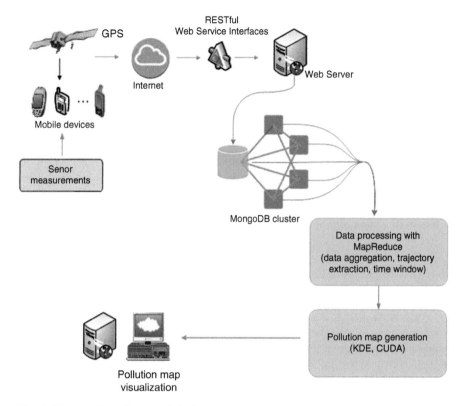

Fig. 2 The overall architecture of the framework

queries. To visualize the pollution exposure to each individual user, we utilize MapReduce to simply use the device ID as the key to extract the trajectory of a specific user from the database.

In producing pollution maps for the measured environmental properties, many existing applications provide the visualization based on individual points. However, such an approach is subject to two major challenges. (1) When dealing with a massive collection of measurements in the form of geographical points, the visualization process will experience longer processing time, which may not be able to provide an effective response to dynamic user requests. (2) Since the framework is intended for a large group of people's collected measurements, visualizing the measurements (even calibrated) collected at the same location (or locations nearby) at different times will provide confusing results. In this regard, we aggregate all users inputs based on a predefined time window and kernel band-width and calibrate according to factors such as the sound decay distance. To simplify the process, we define a 1-h time window and 50-m kernel bandwidth. In other words, we assumed that each measurement will last for 1 h and covers an area of 50-m radius. The value of this assumption needs to be more carefully determined based on the real-world measurements once there are enough data collected by multiple users. The aggregation is implemented also using the MapReduce method, where the device ID is treated as the map key and the reduction process is based on the timestamps that fall in a specified 1-h time window.

The pollution map is dynamically generated by using a kernel smoothing method. Kernel smoothing is used to estimate a continuous surface of environmental measures (e.g. noise level) from point observations. The estimated measurement at each location (target location) is calculated as a weighted average of the observations within a search window (or bandwidth). The weight of each observation is decided by applying a kernel function to the distance between the target location and that observation. The kernel function is typically a distance decay function with a maximum value when the distance is zero and with a zero value when the distance exceeds the bandwidth. The formula of kernel smoothing is shown below, where $K(\ )$ is the kernel function, h is the bandwidth, (X_i, Y_i) is the location of observation i, and Z_i is the environmental measures of observation i.

$$\frac{\sum_{i=1}^{n} K\left(\frac{x-X_i}{h}, \frac{y-Y_i}{h}\right) Z_i}{\sum_{i=1}^{n} K\left(\frac{x-X_i}{h}, \frac{y-Y_i}{h}\right)}$$

Performing kernel smoothing with a massive number of observations from multiple users is extremely computationally intensive. Hence, a parallel kernel smoothing algorithm is implemented based on CUDA[9] (Compute Unified Device Architecture) to exploit the computational power of GPUs. Multiple parallel threads are launched simultaneously, each of which estimates the measurement at one location (one cell for the output raster). Each thread searches through each of the observations, calculates the weight of this observation to its cell, and outputs the weighted average of these observations as an estimated measurement of its cell. In this case, the 50-m kernel bandwidth distance is also incorporated as the bandwidth of the kernel smoothing method, and the output is a GeoTIFF image, which is overlaid on top of ESRI world street map for visualization purposes.

4 User Case Scenario

A noise mapping user case is investigated by collecting data of sound pressure using a mobile application. The application utilizes the microphone of a mobile device to measure sound pressure with the noise level calculated in decibels (dB) using the following equation (Bies and Hansen 2009; Maisonneuve et al. 2009):

---

[9] http://www.nvidia.com/object/cudahome new.html

$$L_p = 10\log_{10}\left(\frac{p_{rms}^2}{p_{ref}^2}\right) = 20\log_{10}\left(\frac{p_{rms}}{p_{ref}}\right) dB$$

where p_{ref} is the reference sound pressure level with a value of 0.0002 dynes/cm^2 and p_{rms} is the measured sound pressure level. According to the World Health Organization Night Noise Guidelines (NNGL) for Europe,[10] the annual average noise level of 40 dB is considered as equivalent to the lowest observed adverse effect level (LOAEL) for night noise, whereas a noise level above 55 dB can become a major public health concern and over 70 dB can cause severe health problems. This calculated value is also calibrated by users according to physical environment conditions and the type of mobile device.

The mobile application assigns a pair of geographic coordinates (in the format of latitude and longitude) to each measured value together with a timestamp. The update time interval for each recording is set to every 5 s. The recorded measurements are saved directly on the mobile device and we let users decide when to upload their data to the server, whether immediately after taking the measurements or at a later time. An example of the data format of the measurements is shown in Fig. 3. Note that the measurements of other sensors on a mobile device can be included. Given the diversity of sensors on different devices, we use a flexible data management approach based on MongoDB.

In this user case scenario, we choose the campus of University of Illinois at Urbana—Champaign and its surroundings as the study area and asked the participants to go around the campus to collect the noise level measurements. The user interface of the mobile application is shown in Fig. 4, where users have the options to record, upload and interact with noise maps. The mobile application is implemented as a background service on the device and therefore participants are free to engage in other activities.

From a generated noise map, we can identify those spots at which the noise level exceeds such ranges. In Fig. 5, we can examine the visualization of the noise exposure to an individual participant along their trajectory. At the current stage, we have not quantitatively estimated accumulated noise exposure, which will be taken into account in our future work. Figure 6 shows the noise map of a specified hour using a 50-m kernel bandwidth, which is generated from the measurements uploaded by all of the participants during this period. From the visualized results, we can identify the spots where the noise pollution occurs (shown in red) within the specified hour. A new feature to be evaluated for providing in-depth information about what causes such noise pollution is to allow users to append descriptive text when they carry out monitoring using their mobile devices (Maisonneuve et al. 2009). Figure 7 is the noise map of the same hour but using 100-m kernel bandwidth, which demonstrates the effects of choosing different sound decay distance since the value can be changed in framework.

---

[10] http://www.euro.who.int/data/assets/pdf file/0017/43316/E92845.pdf

	A	B	C	D	E
1	DeviceID	LAT	LNG	NoiseLevel	TimeStamp
2	5beceacccOffdfd	40.112101	-88.230728	41.62454071	09-July-2014 01:21:09
3	5beceacccOffdfd	40.112101	-88.230728	59.85949238	09-July-2014 01:21:14
4	5beceacccOffdfd	40.112101	-88.230728	42.88586133	09-July-2014 01:21:19
5	5beceacccOffdfd	40.112101	-88.230728	63.21920352	09-July-2014 01:21:24
6	5beceacccOffdfd	40.112101	-88.230728	45.48609774	09-July-2014 01:21:29
7	5beceacccOffdfd	40.112101	-88.230728	42.81638778	09-July-2014 01:21:34
8	5beceacccOffdfd	40.1127555	-88.2302514	71.98051497	09-July-2014 01:46:43
9	5beceacccOffdfd	40.1127555	-88.2302514	74.86597743	09-July-2014 01:46:48

Fig. 3 An example of recorded noise measurements saved on a mobile device

Fig. 4 The user interface of the mobile application

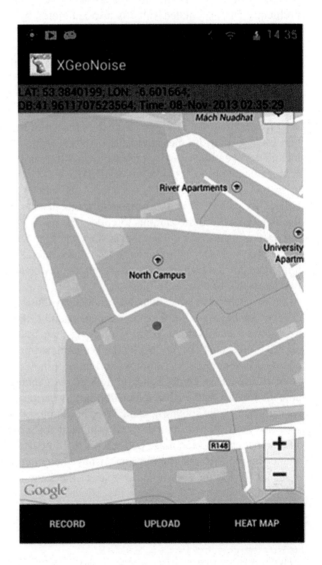

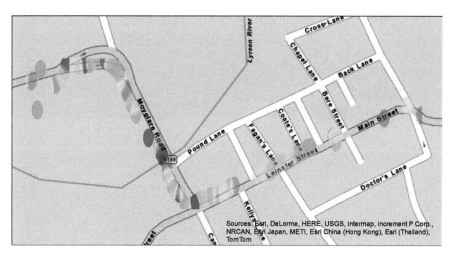

Fig. 5 Noise mapping along the trajectory of an individual participant

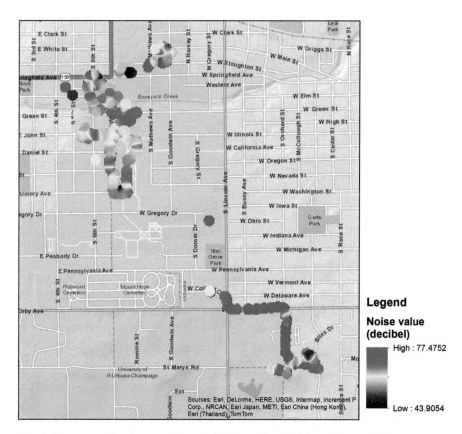

Fig. 6 The generated noise map using a 100-m kernel bandwidth during a specified hour

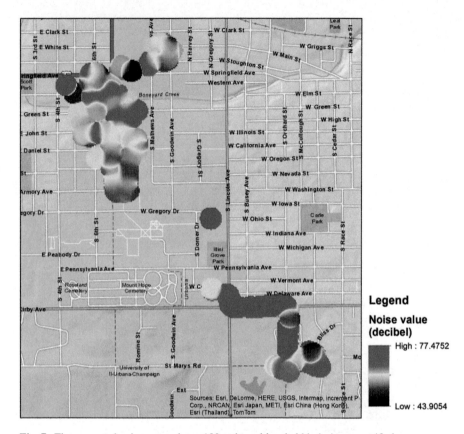

Fig. 7 The generated noise map using a 100-m kernel bandwidth during a specific hour

5 Conclusions and Future Work

The availability of a variety of affordable mobile sensors is fostering volunteered participation of citizens in sensing urban environments using mobile devices. By utilizing embedded GPS receivers to append geographic coordinates to sensor measurements, the collective efforts from participatory urban sensing activities can provide high-resolution spatiotemporal data for creating pollution maps of large cities. In relation to the big data collected from such crowdsourcing activities, CyberGIS provides high performance computing and participatory computing architecture to support scalable user participation and data-intensive processing, analysis and visualization.

In this paper, we present a framework that utilizes several components of the CyberGIS platform to facilitate citizens in engaging with environmental monitoring using mobile devices. This framework is intended to incorporate readings from the environmental sensors on the mobile device. As the availability of sensors varies on different devices, this framework chooses a MongoDB (without the requirement for a predefined schema) cluster for data storage. A MapReduce approach is used to filter and extract trajectories of each individual participant to visualize the pollution exposure. It is also used for dynamically generating pollution maps by aggregating the collected sensor measurements using a time window. The pollution maps are rapidly generated by using kernel method via paralleled GPU. In this study, we only demonstrate the functionality of the framework using the case for dealing with the geo-tagged and timestamped noise level measurements, which is collected from our dedicated prototype mobile application using the combination of an integrated GPS receiver and a microphone on a mobile device.

At the current stage, there are still some limitations regarding the implementation of the framework. For example, the selection of the kernel method assumes the measured values stay the same within the kernel bandwidth, which may not be the case in real-world scenarios. Also, the kernel method may not be suitable for generating other pollution maps, for example, air pressure. Therefore, some domain knowledge is required for future improvement of the framework. In relation to trajectory extraction for visualizing pollution exposure to individual participants, quantitative methods for estimating actual exposure need to be explored. Furthermore, we plan to acquire environmental measurements from pertinent government agencies to validate the results that are produced based on data from volunteered participants. Finally, the current MapReduce method relies on the MongoDB cluster, where Apache Hadoop is being explored to improve computational performance.

Acknowledgements This material is based in part upon work supported by the U.S. National Science Foundation under grant numbers: 0846655, 1047916, and 1354329. Any opinions, findings, and conclusions or recommendations expressed in this material are those of the authors and do not necessarily reflect the views of the National Science Foundation. The authors are grateful for insightful comments on the earlier drafts received from Yan Liu, Anand Padmanabhan.

References

Bies DA, Hansen CH (2009) Engineering noise control: theory and practice. CRC press, Boca Raton, FL
Bryant RE (2009) Data-intensive scalable computing harnessing the power of cloud computing (Tech. Rep.). CMU technical report. Retrieved from http://www.cs.cmu.edu/bryant/pubdir/disc-overview09.pdf
Dean J, Ghemawat S (2008) MapReduce: simplified data processing on large clusters. Commun ACM 51(1):107–113
Directive E (2002) Directive 2002/49/ec of the European parliament and the council of 25 June 2002 relating to the assessment and management of environmental noise. Off J Eur Communities 189(12):12–26
Ghosh S, Raju P.P, Saibaba J, Varadan G (2012) Cybergis and crowdsourcing–a new approach in e-governance. In Geospatial Communication Network. Retrieved from http://www.geospatialworld.net/article/cybergis-and-crowdsourcing-a-new-approach-in-e-governance/
Goodchild MF (2007) Citizens as sensors: the world of volunteered geography. GeoJournal 69 (4):211–221

Han J, Haihong E, Le G, Du J (2011) Survey on NoSQL database. In Proceeding of 6th international conference on pervasive computing and applications (ICPCA), pp 363–366
Howe J (2006) Crowdsourcing: a definition. In: Crowdsourcing: tracking the rise of the amateur
Maisonneuve N, Stevens M, Niessen ME, Steels L (2009) Noisetube: measuring and mapping noise pollution with mobile phones. In: Information technologies in environmental engineering. Springer, New York, pp 215–228
Maisonneuve N, Stevens M, Ochab B (2010) Participatory noise pollution monitoring using mobile phones. Inform Polity 15(1):51–71
Stevens M, DHondt E (2010) Crowdsourcing of pollution data using smartphones. In: Workshop on ubiquitous crowdsourcing, Ubicomp'10, September 26–29, 2010, Copenhagen, Denmark
Stonebraker M, Madden S, Abadi DJ, Harizopoulos S, Hachem N, Helland P (2007) The end of an architectural era: (it's time for a complete rewrite). In Proceedings of the 33rd international conference on very large data bases, pp 1150–1160
Wang S (2010) A cybergis framework for the synthesis of cyberinfrastructure, GIS, and spatial analysis. Ann Assoc Am Geogr 100(3):535–557
Wang S, Wilkins-Diehr NR, Nyerges TL (2012) Cybergis-toward synergistic advancement of cyberinfrastructure and giscience: a workshop summary. J Spat Inform Sci 4:125–148
Wang S, Anselin L, Bhaduri B, Crosby C, Goodchild MF, Liu Y, Nyerges TL (2013a) Cybergis software: a synthetic review and integration roadmap. Int J Geogr Inf Sci 27(11):2122–2145
Wang S, Cao G, Zhang Z, Zhao Y, Padmanabhan A, Wu K (2013b) A CyberGIS environment for analysis of location-based social media data. In: Hassan AK, Amin H (eds) Location-based computing and services, 2nd edn. CRC Press, Boca Raton, FL

Part II
Challenges and Opportunities of Urban Big Data

Part II

The Potential for Big Data to Improve Neighborhood-Level Census Data

Seth E. Spielman

Abstract The promise of "big data" for those who study cities is that it offers new ways of understanding urban environments and processes. Big data exists within broader national data economies, these data economies have changed in ways that are both poorly understood by the average data consumer and of significant consequence for the application of data to urban problems. For example, high resolution demographic and economic data from the United States Census Bureau since 2010 has declined by some key measures of data quality. For some policy-relevant variables, like the number of children under 5 in poverty, the estimates are almost unusable. Of the 56,204 census tracts for which a childhood poverty estimate was available 40,941 had a margin of error greater than the estimate in the 2007–2011 American Community Survey (ACS) (72.8 % of tracts). For example, the ACS indicates that Census Tract 196 in Brooklyn, NY has 169 children under 5 in poverty ±174 children, suggesting somewhere between 0 and 343 children in the area live in poverty. While big data is exciting and novel, basic questions about American Cities are all but unanswerable in the current data economy. Here we highlight the potential for data fusion strategies, leveraging novel forms of big data and traditional federal surveys, to develop useable data that allows effective understanding of intra urban demographic and economic patterns. This paper outlines the methods used to construct neighborhood-level census data and suggests key points of technical intervention where "big" data might be used to improve the quality of neighborhood-level statistics.

Keywords Census • American Community Survey • Neighborhood data • Uncertainty • Data fusion

---

S.E. Spielman (✉)
University of Colorado, Boulder, CO, USA
e-mail: seth.spielman@colorado.edu

1 Introduction

The promise of "big data[1]" for those who study cities is that it offers new ways of understanding urban environments and their affect on human behavior. Big data lets one see urban dynamics at much higher spatial and temporal resolutions than more traditional sources of data, such as survey data collected by national statistical agencies. Some see the rise of big data as a revolutionary of mode of understanding cities, this "revolution" holds particular promise for academics because, as argued by Kitchin (2014), revolutions in science are often preceded by revolutions in measurement. That is, big data could give rise to something even bigger, a new science of cities. Others, such as Greenfield (2014) argue that real urban problems cannot be solved by data and are deeply skeptical of the potential for information technologies to have meaningful impacts on urban life. Here, we aim to contextualize the enthusiasm about urban big data within broader national data economies, particularly focusing on the US case. This paper argues that changes to national data infrastructures, particularly in the US, have led to decline of important sources of neighborhood-level demographic and economic data and that these changes complicate planning and policymaking, even in a big data economy. We argue that in spite of the shortcomings of big data, such as uncertainty over who or what is being measured (or not measured), it is possible to leverage these new forms of data to improve traditional survey based data from national statistical agencies.

The utility of data is contingent upon the context within which the data exist. For example, one might want to know the median income in an area and the crime rate in isolation these data are far less useful then they are in combination. Knowing that place is high crime might help guide policy. However, policy might be much more effectively targeted in one knew context within which crime was occurring. In a general sense we might refer to the context within which data exist as the "data economy." For those who work on urban problems, the data economy has changed in ways that are both poorly understood by the average data consumer and of consequence to the application of big data to urban problems. Traditional sources of information about cities in the US have recently changed in profound ways. We argue that these changes create potential, and problems, for the application of data to urban questions.

In particular, the data collected by the US Census Bureau has recently undergone a series of dramatic changes, some of these changes are a result of the gradual accrual of broader social changes and some have been abrupt, the result of changes to federal policy. The National Research Council (2013) document a gradual long term national trend of increases in the number of people who refuse to respond to public (and private) surveys. Geographic and demographic patterns in survey

---

[1] Defining big data is difficult, most existing definitions, include some multiple of V's (see Laney 2001). All are satisfactory for our purposes here. We use the term to distinguish between census/survey data which we see as "designed" measurement instruments and big data which we see as "accidental" measurement instruments.

non-response make it difficult for surveys to accurately describe populations and create the need for complex statistical adjustments to ensure that the estimates produced by the survey are representative of the target population. If for example, low income immigrants do not respond to official surveys they would be invisible to the data-centric urban analyst. More realistically, if they respond with a much lower frequency than the average person then they would appear much less prevalent than they actually are unless one accounts for their differential response rate when producing estimates. However, efforts to reduce bias due to non-response can add uncertainty to the final estimate creating large margins of error and complicating data use.

In fact, high levels of uncertainty now plague almost all fine resolution[2] urban data produced by the United States Census Bureau (USCB). Neighborhood-level data from the Census Bureau are terribly imprecise, for some policy-relevant variables, like the number of children in poverty, the estimates are almost unusable—of the 56,204 tracts for which a poverty estimate for children under 5 was available 40,941 had a margin of error greater than the estimate in the 2007–2011 ACS (72.8 % of tracts). For example, the ACS indicates that Census Tract 196 in Brooklyn, NY has 169 children under 5 in poverty ±174 children, suggesting somewhere between 0 and 343 children in the area live in poverty. Users of survey data often face the situation in Table 1, which shows the ACS median income estimates for African-American households for a contiguous group of census tracts in Denver, Colorado. Income estimates range from around $21,000 to $60,000 (American Factfinder website accessed 7/15/2013). Without taking account of the margin of error, it would seem that Tract 41.06 had the highest income, however, when one accounts for the margin of error, the situation is much less clear—Tract 41.06 may be either the wealthiest or the poorest tract in the group.

The uncertainty in Table 1 is all but ignored by practicing planners, a voluntary online survey of 180 urban planners that we conducted during 2013 found that most planners (67 %) simply delete or ignore information about the quality of estimates, like the margin of error, when preparing maps and reports. This practice, according to planners is driven by the "demands" of their "consumers." That is, the audience for their maps and reports would have difficulty incorporating the margins of error into decision-making processes. This practice is further reinforced by federal agencies, which use only the tract level estimates to determine eligibility for certain programs (for example, see the eligibility guidelines for the Treasury's New Markets Tax Credit program). The problem with the margins of error is especially pronounced for the Census Transportation Planning Package (CTTP), a key input for transportation planning and travel demand models.

---

[2] We use the terms "fine" and "high" resolution to refer to census tract or smaller geographies, these data are commonly conceived of as "neighborhood-scale" data. We conceive of resolution in the spatial sense, higher/finer resolution means a smaller census tabulation unit. However, the geographic scale high resolution of census units is a function of population density.

Table 1 2006–2010 ACS estimates of African-American median household income in a selected group of proximal tracts in Denver County, Colorado

Tract number	African-American median household income	Margin of error
Census Tract 41.01	$28,864	$8650
Census Tract 41.02	$21,021	$4458
Census Tract 41.03	$43,021	$14,612
Census Tract 41.04	$36,092	$3685
Census Tract 41.06	$60,592	$68,846

The decline in the quality of neighborhood scale data in the United States began in 2010, the year the American Community Survey (ACS) replaced the long form of the United States decennial census as the principal source of high-resolution geographic information about the U.S. population. The ACS fundamentally changed the way data about American communities are collected and produced. The long form of the decennial census was a large-sample, low frequency national survey; the ACS is a high-frequency survey, constantly measuring the American population using small monthly samples. One of the primary challenges for users of the ACS is that the margins of error are on average 75 % larger than those of the corresponding 2000 long-form estimate (Alexander 2002; Starsinic 2005). This loss in precision was justified by the increase in timeliness of ACS estimates, which are released annually (compared to the once a decade long form). This tradeoff prompted Macdonald (2006) to call the ACS a "warm" (current) but "fuzzy" (imprecise) source of data. While there are clear advantages to working with "fresh" data, the ACS margins of error are so large that for many variables at the census tract and block group scales the estimates fail to meet even the loosest standards of data quality.

Many of the problems of the American Community Survey are rooted in data limitations. That is at critical stages in the creation of neighborhood-level estimates the census bureau lacks sufficient information and has to make assumptions and/or use data from a coarser level of aggregation (municipality or county). We argue that one of the major potential impacts of big data for the study of cities is the reduction of variance in more traditional forms demographic and economic information. To support this claim, we describe the construction of the ACS in some detail, with the hope that these details illuminate the potential for big data to improve federal and/or state statistical programs.

2 Understanding the American Community Survey

Like the decennial long form before it, the ACS is a sample survey. Unlike complete enumerations, sample surveys do not perfectly measure the characteristics of the population—two samples from the same population will yield different estimates. In the ACS, the margin of error for a given variable expresses a range of values around the estimate within which the true value is expected to lie. The margin of error reflects the variability that could be expected if the survey were repeated with a different random sample of the same population. The statistic used to describe the magnitude of this variability is referred to as *standard error* (SE). Calculating standard errors for a complex survey like the ACS is not a trivial task, the USCB uses a procedure called Successive Differences Replication to produce variance estimates (Fay and Train 1995). The margins of error reported by the USCB with the ACS estimates are simply 1.645 times the standard errors.

One easy way to understand the ACS Margin of Error is to consider the simple case, in which errors are simply a function of the random nature of the sampling procedure. Such sampling error has two main causes, the first is the sample size— the larger the sample the smaller the standard error, intuitively more data about a population leads to less uncertainty about its true characteristics. The second main cause of sampling error is heterogeneity in the population being measured (Rao 2003). Consider two jars of U.S. coins, one contains U.S. pennies and the other contains a variety of coins from all over the world. If one randomly selected five coins from each jar, and used the average of these five to estimate the average value of the coins in each jar, then there would be more uncertainty about the average value in the jar that contained a diverse mixture of coins. If one took repeated random samples of five coins from each jar the result would always be the same for the jar of pennies but it would vary substantially in the diverse jar, this variation would create uncertainty about the true average value.[3] In addition, a larger handful of coins would reduce uncertainty about the value of coins in the jar. In the extreme case of a 100 % sample the uncertainty around the average value would be zero. What is important to realize is that in sample surveys the absolute number of samples is much more important than the relative proportion of people sampled, a 5 % sample of an area with a large population will provide a much better estimate than a 5 % sample of a small population. While the ACS is much more complicated than pulling coins from a jar, this analogy helps to understand the standard error of ACS estimates. Census Tracts (and other geographies) are like jars of coins. If a tract is like the jar of pennies, then the estimates will be more precise, whereas if a tract is like the jar of diverse coins or has a small population, then the estimate will be less precise.

While the simple example is illustrative of important concepts it overlooks the central challenge in conducting surveys; many people included in a sample will choose not to respond to the survey. While a group's odds of being included in the

---

[3] The Census Bureau generally is not actually estimating the "average" value, they are estimating the "total" value of coins in the jar. Repeatedly grabbing five coins and computing the average will over many samples get you a very precise estimate of the average value, but it will give you no information on the total value. To get the total value, you need a good estimate of the average AND a good estimate of the total number of coins in the jar. The loss of cotemporaneous population controls caused by decoupling the ACS from the Decennial enumeration means that the census does not have information about the number of coins in the jar. This is discussed in more details later.

ACS sample are proportional to its population size, different groups of people have different probabilities of responding. Only 65 % of the people contacted by the ACS actually complete the survey (in 2011, 2.13 million responses were collected from 3.27 million samples). Some groups are more likely to respond than others, this means that a response collected from a person in a hard to count group is worth more than a response from an easy to count group. Weighting each response controls for these differential response rates. In the ACS each completed survey is assigned a single weight through a complex procedure involving dozens of steps. The important point, as far as this paper is concerned, is that these weights are *estimated* and uncertainty about the appropriate weight to give each response is an important source of uncertainty in the published data.

3 Sampling

Before 1940, the census was a complete enumeration; each and every housing unit (HU) received the same questionnaire. By 1940 the census forms had become a long, complicated set of demographic and economic questions. In response, the questionnaire was split in 1940 into a short set of questions asked of 100 % of the population and an additional "long form" administered to a subset of the population. Originally, this long form was administered to a 5 % random sample, but in later years it was sent to one HU in six (Anderson et al. 2011). Before 1940 any error in the data could be attributed either to missing or double counting a HU, to incorrect transcription of a respondent's answer, or to intentional/unintentional errors by the respondent. After 1940, however, the adoption of statistical sampling introduced new sources of uncertainty for those questions on the long form.

Up until 2010 the sample based (long form) and the complete enumeration (short form) of the census were administered at the same time. In 2010 the ACS replaced the sample based long form. The American Community Survey constantly measures the population; it does not co-occur with a complete census. The lack of concurrent complete count population data from the short form is a key source of uncertainty in the ACS. Prior to the rise of the ACS, short form population counts could serve as controls for long-form based estimates. The decoupling of the sample from the complete enumeration accounts for 15–25 % of the difference in margin of error between the ACS and the decennial long form (Navarro 2012). Population controls are essential to the ACS sample weighting process, now population controls are only available for relatively large geographic areas such as municipalities and counties. This is a key data gap which as discussed later might be addressed with big data.

4 Spatial and Temporal Resolution of Census Estimates

Prior to the advent of sampling, the complete count census data could, in principle, be tabulated using any sort of geographic zone. Tract based census data has become a cornerstone of social science and policy making. the decennial census by the late twentieth century. However, users of the once a decade census were increasingly concerned about the timeliness of the data (Alexander 2002). A solution to this problem was developed by Leslie Kish, a statistician who developed the theory and methods for "rolling" surveys (Kish 1990).

Kish's basic idea was that a population could be divided into a series of non-overlapping annual or monthly groups called subframes. Each subframe would then be enumerated or sampled on a rolling basis. If each subframe were carefully constructed so as to be representative of the larger population, then the annual estimates would also be representative, and eventually, the entire population would be sampled. The strength of this rolling framework is its efficient use of surveys. The decennial census long form had to sample at a rate appropriate to make reasonable estimates for small geographic areas such as census tracts, which contain on average 4000 people. Therefore, citywide data released for a municipality of, say, one million people would be based on considerably more samples than necessary. Spreading the samples over time lets larger areas receive reasonable estimates annually, while smaller areas wait for more surveys to be collected. The rolling sample therefore increases the frequency of data on larger areas. The primary cost comes in the temporal blurring of data for smaller areas. The advent of sampling made census data for small geographic areas less precise. Since there are a finite number of samples in any geographic area, as tabulation zones become smaller sample sizes decline, making estimates more uncertain. The rise uncertainty is greater for small populations; for instance the effects of reducing a sample size from 200 to 100 is much greater than the effect of reducing a sample size from 20,000 to 10,000. The USCB counteracts this decline in sample size by pooling surveys in a given area over multiple years, thus diluting the temporal resolution of the estimates.

Rolling sampling is straightforward in the abstract. For example, suppose that there are $K=5$ annual subframes, that the population in a tract is known ($N=1000$), that the sampling rate is $r=1/6$, and that the response rate is 100%; then one would sample $n=N/(K*1/r)$ people per year. Over a 5-year period 1/6 of the population would be sampled and each returned survey would represent $w=(N/n)/K$ people, where w is the weight used to scale survey responses up to a population estimate. In this simple case, the weight assigned to each survey would be the same. For any individual attribute y, the tract level estimate would be $y_t = \Sigma w_i y_i$ (equation 1), a weighted summation of all i surveys collected in tract t. If the weights are further adjusted by ancillary population controls X, then the variance of the estimate is $\Sigma w_i^2 \text{VAR}[y_i|X]$ (equation 2; Fuller 2011, assuming independence.). If the rolling sample consisting of long-form-type questions were administered simultaneously with a short form census, then all the parameters in our simple example (N,K, X) would be known.

However, in the ACS good population controls are not available for small areas (N and X are unknown) because, unlike the long form, the survey is not contemporaneous with the complete enumeration decennial census. Thus weights (w) for each response must be estimated and this is an important source of uncertainty in the ACS.

5 Weighting

In the ACS each completed survey is assigned a weight (w) that quantifies the number of persons in the total population that are represented by a sampled household/individual. For example, a survey completed by an Asian male earning $45,000 per year and assigned a weight of 50 would in the final tract-level estimates represent 50 Asian men and $2.25 million in aggregate income. The lack of demographically detailed population controls, and variations in response rate all necessitate a complex method to estimate w. The construction of ACS weights is described in the ACS technical manual (which runs hundreds of pages, U.S. Census Bureau 2009a). Individually these steps make sense but they are so numerous and technically complex that in the aggregate they make the ACS estimation process nearly impenetrable for even the most sophisticated data users. The cost of extensive tweaking of weights is more than just lack of transparency and complexity. Reducing bias by adjusting weights carries a cost. Any procedure that increases the variability in the survey weights also increases the uncertainty in tract-level estimates (Kish 2002). Embedded in this process is a trade-off between estimate accuracy (bias) and precision (variance/margin of error), refining the survey weights reduces bias in the ACS but it also leads to variance in the sample weights.

6 Big Data and Public Statistics

Without traditional survey data from national statistical agencies, like the USCB, it is difficult to contextualize big data, its hard to know who is (and who is not) represented in big data. It is difficult to know if there are demographic, geographic, and or economic biases in the coverage of big data without traditional census data as a baseline. Ironically, as this baseline data declines in quality, many of the populations most in need of urban services are least well served by the traditional census data and quite possibly the big data as well—consider the example of young children in poverty discussed in the introduction.

In the preceding sections we identified several key data gaps and methodological decisions that might be addressed with big data:

1. Sampling is constrained by a lack of detailed high geographic and demographic resolution population data.
2. Small area geographies are not "designed" and this leads to degradation in the quality of estimates and the utility of the published data.
3. Weights are complex and difficult to accurately estimate without additional data.

In this section we outline how big data might be used to address these issues. This section is by no means exhaustive, the aim more to draw attention to the potential for new forms of data to mitigate emerging problems with neighborhood statistics. It is also important to note that, for reasons discussed in the conclusion, this section is largely speculative, that is, very few of the ideas we propose have seen implementation.

So far this paper has emphasized the mechanics of the construction of the ACS—sampling, the provision of small area estimates, the provision of annual estimates, and the estimate of survey weights. The prior discussion had a fair amount of technical detail because such detail is necessary in order to understand how novel forms of "big" data might be integrated into the production process. Directly integrating big data into the production of estimates is not the only way to use new forms of data concurrently with traditional national statistics, but in this paper the emphasis is on such an approach.

It should be apparent that the data gaps and methodological choices we have identified thus far are intertwined. For example, the use of sampling necessitates the estimation of survey weights which are complicated to estimate when very little is known about the target population in the areas under investigation. Spatial and temporal resolution are related because the reliability of the estimate depends on the number of surveys, which accrue over time, and the size (population) and composition of the area under investigation.

The lack of detailed small area population controls is makes it very difficult to estimate the weight for each survey. Since the US Census Bureau does not know how many low income Caucasian males live in each census tract it is difficult to know if the number of surveys returned by low income Caucasian males higher or lower than expected—this affects the weight assigned to a response. For example, imagine a hypothetical census tract with 2000 housing units and a population of 4000 people. 10 % of the population is low-income white males and this tract was sampled at a 5 % rate, one would expect 10 % of the completed surveys to be filled in by low-income white males. However, if this group is less likely than others to respond perhaps the only 2 % of the completed surveys would be completed by low-white males. If the number of low-income white males was known in advance one could "up-weight" these responses to make sure that in the final data low income-white males represented 10 % of the population. However, the census bureau has no idea how many low-income white males are in each census tract. This is where big data might help.

If, for example, the number of low-income white males could be estimated by using credit reporting data, social media profiles, or administrative records from other government agencies, then a lot of the guesswork in deciding how to weight survey responses could be eliminated. It's important to realize that these forms of "big" data might not be of the highest quality. However, they could be used to establish meaningful benchmarks for sub-populations making simple comparisons of "big" and traditional data possible. While it would be difficult to say which data was "correct" it is reasonable to suggest that large discrepancies would warrant closer inspection and would highlight key differences in the coverage of the various data sets. These coverage differences are not especially well understood at the time of writing.

A more sophisticated strategy would be to employ what is called are called "model assisted estimation" strategies (see Särndal 1992). Model assisted estimation is a set of strategies for using ancillary data and regression models to estimate survey weights. Currently, the ACS uses a model assisted strategy called "Generalized Regression Estimator" (GREG). In the ACS GREG takes advantage of person-level administrative data on age, race, and gender of residents from auxiliary sources such as the Social Security Administration, the Internal Revenue Service, and previous decennial census tabulations. The procedure builds two parallel datasets for each census tract: one using the administrative data on all people in the tract, and the second using administrative data for only the surveyed housing units. The second dataset can be viewed, and tested, as an estimate of the demographic attributes of the first—e.g., proportions of males aged 30–44, non-Hispanic blacks, etc. A weighted least squares regression is then run on the second dataset, in which the dependent variable is weighted HU counts and the independent variables are the various weighted attribute counts.

The strength of model assisted estimation procedure depends entirely on the quality of the regression. A well-fit regression should reduce overall uncertainty in the final ACS estimates by reducing the variance of the weights, while a poorly fit regression can actually increase the margin of error. The data used in models assisted estimation in the ACS is terrible for its intended purpose, that is age, sex, and race are only loosely correlated with many of the economic and demographic characteristics of most interest to urban planners and policy makers. In spite of these weaknesses Age, Sex, and Race data are used because they are available to the USCB from other Federal agencies, more sensitive data, like income, is not incorporated into estimates.

However, data on homeownership, home values, spending patterns, employment, education and many other attributes may be obtainable through big data sets and this could be used to improve the quality of estimates through model assisted estimation. For example, housing data from cadastral records and home sales could be (spatially) incorporated into the ACS weighting strategy. The exact home value of each house is unknown, so they are unusable as hard benchmarks. But, it is possible to approximate the value of each house based upon location, characteristics, and nearby sales. Even if it was not possible to directly match survey respondent to records in other datasets, it might be possible to geospatially impute such characteristics. For example, recent nearby home sales might be used to estimate the value of a respondents' home. This approximation is used to great effect by the mortgage industry and by local governments for property tax assessments. Since these models are approximations, the data may enter the weighting phase as "soft" benchmarks (i.e. implemented a mixed effects models). It is not appropriate for the weights to exactly duplicate the estimated home value, but it is appropriate for the weights to approximate the estimated home value. For example, Porter et al. (2013) use the prevalence of Spanish language Google queries to improve census estimates of the Hispanic population. Carefully chosen controls have the potential to dramatically reduce the bias and margin of error in ACS estimate for certain variables. The estimates most likely to be impacted are socioeconomic variables, which are poorly correlated with the currently available demographic benchmarks, and thus have a disproportionately large margin of error.

A second mechanism for using big data to improve estimates is through zone design. Census geographies are designed to be stable over time, that is, local committees at some point designed them in the past (often 3o years ago) and they have only evolved through splits and merges with other census tracts. Splits and merges can only occur when the tract population crosses some critical threshold. The size and shape of census fundamentally affects the quality of estimates. Larger population census tracts, because they generally have more surveys supporting estimates have higher quality data. However, as geographies grow in size there is potential to loose information on intra urban variation. However, information loss does not necessarily occur as a result of changes in zone size. Consider two adjacent census tracts that are very similar to each other in terms of ethnic composition, housing stock, and economic characteristics. The cost of combining these two census tracts into a single area is very small. That is, on a thematic map these two adjacent areas would likely appear as a single unit (because they would be the same legend color because they would likely have the same value). Combining similar places together boosts the number of completed surveys and thus reduces the margin of error. The challenge is how does one tell if adjacent places are similar (or not) when the margins of error on key variables are very large? Again, big data, if it provides a reasonable approximation of the characteristics of the places at high spatial resolutions it maybe possible to combine lower level census geographies into units large enough to provide high quality estimates. For example, Spielman and Folch (2015) develop an algorithm to combine existing lower-level census geographies, like tracts and block groups, into larger geographies while producing new estimates for census variables such that the new estimates leverage the larger population size and have smaller margins of error. For example, they demonstrate that even for variables like childhood poverty, it is possible to produce usable estimates for the city of Chicago by intelligently combining census geographies into new "regions". This strategy results in some loss of geographic detail, but the loss is minimized by ensuring that only similar and proximal geographies are merged together

7 Conclusion

Little (2012) argues that a fundamental philosophical shift is necessary within both federal statistical agencies and among data users, "we should see the traditional survey as one of an array of data sources, including administrative records, and other information gleaned from cyberspace. Tying this information together to yield cost-effective and reliable estimates..." However, Little also notes that for the Census "combining information from a variety of data sources is attractive in principle, but difficult in practice" (Little 2012, p. 309). By understanding the causes of uncertainty in the ACS the implications of Little's statement become clear, there is enormous potential to mash-up multiple forms of information to provide a more detailed picture of US cities.

However, there are major barriers to incorporating non-traditional forms of data into official neighborhood statistics. The reasons for this range from organizational to technical. Institutionally, there is a resistance to barriers to the adoption of non-standard forms of data in public statistics. This resistance stems from the fact such data sources are outside of the control of the agencies producing the estimates are relying on such data, that may be subject to changes in quality and availability, poses a problem for the tight production schedules faced by national statistical agencies. Technically, it is often unclear how to best leverage such information, while we have outlined some possibilities they are difficult to test given the sensitive and protected nature of census/survey data itself. Very few people have access to this protected data, it is protected by statute, and thus must be handled in very cumbersome secure computing environments. This makes it difficult to "prove" or "test" concepts. In the US and UK there are some efforts underway to publish synthetic data to allow research on/with highly detailed micro data without releasing the data itself. The barriers to innovative data fusion are unlikely to be resolved and until clear and compelling examples are developed that push national statistical agencies away from their current practices.

To summarize, the growing enthusiasm over big data makes it easy to disregard the decline of traditional forms of public statistics. As these data decline in quality it becomes difficult to plan, provide services, or understand changes in cities. The enthusiasm over big data should be tempered by a holistic view of the current data economy. While it is true that many new data systems have come online in the last 10 years, it is also true that many critical public data sources are withering. Is big data a substitute for the carefully constructed, nationally representative, high resolution census data that many practicing planners and policymakers rely upon? I think not, and while federal budgets are unlikely to change enough to yield a change to the quality of federal statistical programs, the use of new forms of data to improve old forms of data is a promising avenue for investigation.

References

Alexander CH (2002) Still rolling: Leslie Kish's "rolling samples" and the American Community Survey. Surv Methodol 28(1):35–42

Anderson MJ, Citro C, Salvo JJ (2011) Encyclopedia of the US Census: from the Constitution to the American Community Survey. CQ Press, Washington, DC

Fay RE, Train GF (1995) Aspects of survey and model-based postcensal estimation of income and poverty characteristics for states and counties. In Proceedings of the Government Statistics Section, American Statistical Association, pp 154–159

Fuller WA (2011) Sampling statistics. Wiley, Hoboken, NJ

Against the smart city (The city is here for you to use) by Adam Greenfield Kindle Edition, 152 pages, 2013

National Research Council (2013) Nonresponse in social science surveys: a research agenda. In: Tourangeau R, Plewes TJ (eds) Panel on a research agenda for the future of social science data collection, Committee on National Statistics. Division of Behavioral and Social Sciences and Education. The National Academies Press, Washington, DC

Kish L (1990) Rolling samples and censuses. Surv Methodol 16(1):63–79

Kish L (2002) Combining multipopulation statistics. J Stat Plan Inference 102(1):109–118

Kitchin (2014) Big Data & Society 1(1)2053951714528481; DOI: 10.1177/2053951714528481

Little RJ (2012) Calibrated Bayes: an alternative inferential paradigm for official statistics. J Off Stat 28(3):309–372

MacDonald H (2006) The American community survey: warmer (more current), but fuzzier (less precise) than the decennial census. J Am Plan Assoc 72(4):491–503

Navarro F (2012) An introduction to ACS statistical methods and lessons learned. Measuring people in place conference, Boulder, CO. http://www.colorado.edu/ibs/cupc/workshops/measuring_people_in_place/themes/theme1/asiala.pdf. Accessed 30 Dec 2012

Porter AT, Holan SH, Wikle CK, Cressie N (2014) Spatial Fay-Herriot models for small area estimation with functional covariates. Spat Stat 10:27–42

Rao JNK (2003) Small area estimation, vol 327. Wiley-Interscience, New York

Särndal C-E (1992) Model assisted survey sampling. Springer Science & Business Media, New York

Spielman SE, Folch DC (2015) Reducing uncertainty in the American Community Survey through data-driven regionalization. PLoS One 10(2):e0115626

Starsinic M (2005) American Community Survey: improving reliability for small area estimates. In Proceedings of the 2005 Joint Statistical Meetings on CD-ROM, pp 3592–3599

Starsinic M, Tersine A (2007) Analysis of variance estimates from American Community Survey multiyear estimates. In: Proceedings of the section on survey research methods. American Statistical Association, Alexandria, VA, pp 3011–3017

U.S. Census Bureau (2009a) Design and methodology. American Community Survey. U.S. Government Printing Office, Washington, DC

Big Data and Survey Research: Supplement or Substitute?

Timothy P. Johnson and Tom W. Smith

Abstract The increasing availability of organic Big Data has prompted questions regarding its usefulness as an auxiliary data source that can enhance the value of design-based survey data, or possibly serve as a replacement for it. Big Data's potential value as a substitute for survey data is largely driven by recognition of the potential cost savings associated with a transition from reliance on expensive and often slow-to-complete survey data collection to reliance on far less-costly and readily available Big Data sources. There may be, of course, serious methodological costs of doing so. We review and compare the advantages and disadvantages of survey-based vs. Big Data-based methodologies, concluding that each data source has unique qualities and that future efforts to find ways of integrating data obtained from varying sources, including Big Data and survey research, are most likely to be fruitful.

Keywords Survey research • Big Data • Data quality • Design-based data • Organic data

1 Introduction

As response rates and survey participation continue to decline, and as costs of data collection continue to grow, researchers are increasingly looking for alternatives to traditional survey research methods for the collection of social science information. One approach has involved modifying scientific survey research methods through the abandonment of probability sampling techniques in favor of less expensive non-probability sampling methodologies (c.f. Cohn 2014). This strategy has

---

T.P. Johnson (✉)
Survey Research Laboratory, University of Illinois at Chicago, 412 S. Peoria St., Chicago, IL 60607, USA
e-mail: timj@uic.edu

T.W. Smith
General Social Survey, NORC at the University of Chicago, 1155 E 60th St., Chicago, IL 60637, USA
e-mail: Smith-tom@norc.org

become popular enough that the American Association for Public Opinion Research (AAPOR) recently felt it necessary to appoint a Task Force to investigate the issue and release a formal report (Baker et al. 2013). Others have explored the usefulness of supplementing, or replacing completely, surveys with information captured efficiently and inexpensively via "Big Data" electronic information systems. In this paper, we explore the advantages and disadvantages of using survey data versus Big Data for purposes of social monitoring and address the degree to which Big Data can become a supplement to survey research or a complete alternative or replacement for it.

Survey research originally evolved out of social and political needs for better understandings of human populations and social conditions (Converse 1987). Its genesis predates considerably the pre-electronic era to a time when there were few alternative sources of systematically collected information. Over the past 80 years, survey research has grown and diversified, and complex modern societies have come to increasingly rely on survey statistics for a variety of public and private purposes, including public administration and urban planning, consumer and market research, and academic investigations, to name a few. In contrast, Big Data became possible only recently with the advent of reliable, high speed and relatively inexpensive electronic systems capable of prospectively capturing vast amounts of seemingly mundane process information. In a very short period of time, Big Data has demonstrated its potential value as an alternative method of social analysis (Goel et al. 2010; Mayer-Schönberger and Cukier 2013).

Before proceeding further, however, it is important to define what we mean exactly by survey research and "Big Data." Vogt (1999: 286) defines a survey as "a research design in which a sample of subjects is drawn from a population and studied (often interviewed) to make inferences about the population." Groves (2011) classifies surveys as forms of inquiry that are "design-based," as the specific methodology implemented for any given study is tailored (or designed) specifically to address research questions or problems of interest. In contrast, Webopedia (2014) defines Big Data as "a buzzword...used to describe a massive volume of both structured and unstructured data that is so large that it's difficult to process using traditional database and software techniques." Thakuriah et al. (2016), more carefully define Big Data as "structured and unstructured data generated naturally as a part of transactional, operational, planning and social activities, or the linkage of such data to purposefully designed data." In addition to these attributes, Couper (2013) observes that Big Data is produced at a rapid pace. In contrast to design-based data, Groves classifies Big Data as being organic in nature. Although similar to survey data in the systematic manner in which it is collected, organic data is not typically designed to address specific research questions. Rather, such data, referred to by Harford (2014) as "digital exhaust," is a by-product of automated processes that can be quantified and reused for other purposes. There are, of course, exceptions, such as the National Weather Service's measurements, which are design-based and otherwise fit the definition of Big Data.

Although they do not fit today's electronic-based definitions of Big Data, there are several examples of survey-based data sets that are uncharacteristically "big" by any reasonable standards. Examples of Big Surveys include national censuses, which routinely attempt to collect information from millions of citizens. The U.S. micro decennial Census is an example of this. Also included here is the infamous *Literary Digest* Poll, which attempted, and failed badly, to predict the outcome of the 1936 Presidential election, based on more than two million postcard responses collected from individuals sampled from published telephone directories and automobile registration lists (Squire 1988). The *Literary Digest* had been conducting similar straw polls since 1908, but did not run into trouble with a failed election prediction until 1936. The *Literary Digest* experience taught the still young survey research community of the 1930s that big does not necessarily mean better. Subsequent to that experience, survey statisticians worked to develop sampling theory, which enabled them to rely on much smaller, but more carefully selected, random samples to represent populations of interest.

2 What Distinguishes Surveys from Big Data?

While censuses and the *Literary Digest* examples share with today's Big Data large observation-to-variable ratios, they do not have Big Data's electronic-based longitudinal velocity, or rate of data accumulation. Rather, even Big Surveys are only snapshots that represent at best a brief moment in time. Perhaps even more importantly, the structures of these design-based data sources are carefully constructed, unlike many sources of Big Data, which are known for their "messy" nature (Couper 2013). Hence, there are several important differences between design-based survey data, and the organic data sources that represent Big Data. These include differences in volume, data structures, the velocity and chronicity with which data are accumulated, and the intended purposes for which the data are collected.

2.1 Volume

Big Data is big by definition. As Webopedia (2014) suggests, Big Data represents "a massive volume of both structured and unstructured data that is so large that it's difficult to process using traditional database and software techniques." Most of the information generated in the history of our planet has probably been produced in the past several years by automated Big Data collection systems. Google's search database alone collects literally billions of records on a daily basis and will presumably continue to do so into the foreseeable future, accumulating an almost impossibly large amount of organic information. Prewitt (2013: 229) refers to this as a "digital data tsunami." Survey data, by contrast, is many orders of magnitude more modest in volume, and as mentioned earlier, is becoming more expensive and difficult to collect.

2.2 Data Structures

By data structures, we mean the ratio of observations to variables. Big Data commonly have higher ratios (i.e., vastly more observation points than variables), and surveys have much lower ratios (i.e., many more variables but for vastly fewer observations). Prewitt (2013) describes survey data as case-poor-and-variable-rich, and Big Data as case-rich-and-variable-poor.

2.3 Velocity

Data velocity is the speed with which data is accumulated. Big Data's velocity, of course, means that it can be acquired very quickly. Not so with surveys, which require greater planning and effort, depending on mode. Well-done telephone surveys can take weeks to complete, and well-done face-to-face and mail surveys can require months of effort. Even online surveys require at least several days of effort to complete all "field" work. Where government and business decisions must be made quickly, Big Data may increasingly become the most viable option for instant analysis. Indeed, many complex organizations now employ real-time "dashboards" that display up-to-the-minute sets of indicators of organizational functioning and activity to be used for this purpose, and one of the stated advantages of Google's Flu Index (to be discussed below) and similar efforts has been the almost real-time speed with which the underlying data become available, vastly outperforming surveys, as well as most other forms of data collection. Big Data is collected so quickly, without much in the way of human intervention or maintenance, that its velocity is sometimes compared to that of water emitting from a fire hose. Survey research will continue to have difficulty competing in this arena.

2.4 Data Chronicity

Data chronicity refers to time dimensions. The chronicity of Big Data is much more continuous (or longitudinal) than that of most common cross-sectional surveys. With few exceptions, survey data are almost invariably collected over relatively short time intervals, typically over a matter of days, weeks or months. Some data collection systems for Big Data, in contrast, are now systematically collecting information on an ongoing, more or less, permanent basis. There is an often incorrect assumption that the methods, coverage and content of Big Data remains static or unchanging over time. In fact, Big Data systems are often quite changeable and hence there is a danger that time series measurements may not always be comparable.

2.5 Intended Purpose

Design-based survey data are collected to address specific research questions. There are few examples of Big Data being intentionally constructed for research purposes, mostly by governmental agencies interested in taking, for example, continuous weather or other environmental or economic measurements. Most Big Data initiatives, rather, seem driven by commercial interests. Typically, researchers have a good deal of control over the survey data they collect, whereas most analysts of Big Data are dependent on the cooperative spirit and benevolence of large corporate enterprises who collect and control the data that the researchers seek to analyze.

3 Relative Advantages of Big Data

The main advantages of Big Data over survey data collection systems are costs, timeliness and data completeness.

3.1 Costs of Data Collection

As mentioned earlier, Big Data has an important advantage in terms of data collection costs. Surveys, particularly those using an interviewer-assisted mode, continue to become increasingly expensive, whereas the costs of using available Big Data collected for other purposes may be less expensive. The cost of original collection of Big Data, though, is often very high. As research funding becomes more difficult to obtain, the economic attractiveness of Big Data make it difficult to not seriously consider it as an alternative data source.

3.2 Timeliness

As discussed earlier, the velocity of Big Data greatly exceeds that of traditional survey research. As such, it theoretically provides greater opportunities for the real-time monitoring of social, economic and environmental processes. It has been noted, however, that the processing of Big Data can in some cases be a lengthy and time-consuming process (Japec et al. 2015). In addition, being granted real-time access by the original collectors of this information is not always allowed.

3.3 Data Completeness

Missing data at both the item and unit levels is a difficult problem in survey research and the errors associated with it preoccupy many researchers. Big Data sets do not typically share this problem. Because most Big Data sets are based on varied data collection systems that do not rely directly on the participation of volunteers, and subjects are typically not even aware that they are contributing information to Big Data systems (on this point, see the section on *Ethical Oversight* below), non-observations due to failure to contact individuals, or to their unwillingness or inability to answer certain questions, or to participate at all, is not a problem. But Big Data is also not perfect, as we would expect for example that monitors and other recording devices will occasionally malfunction, rendering data streams incomplete. As with surveys, the information missing from Big Data sets may also be biased in multiple ways.

4 Relative Advantages of Survey Research

Advantages of survey research data over Big Data include its emphasis on theory, the ease of analysis, error assessment, population coverage, ethical oversight and transparency.

4.1 The Role of Theory

Some have argued that the we are facing "the end of theory," as the advent of Big Data will make "the scientific method obsolete" (Anderson 2008). Although some of the survey research reported in the popular news media is descriptive only, much of the research conducted using survey methods is theory-driven. Survey data are routinely employed to test increasingly sophisticated and elaborate theories of the workings of our social world. Rather than allowing theory to direct their analyses, Big Data users tend to be repeating some earlier criticisms of empirical survey research by inductively searching for patterns in the data, behaviors that left earlier generations of survey researchers vulnerable to accusations of using early high-speed computers for "fishing expeditions." Fung (2014) criticizes Big Data as being observational (without design) and lacking in the controls that design-based data typically collect and employ to rule-out competing hypotheses.

4.2 Ease of Analysis

The sheer size of many Big Data sets and their often unstructured nature make them much more difficult to analyze, compared to typical survey data files. There are numerous packaged data management and statistical analysis systems readily available to accommodate virtually any survey data set. Big Data, in contrast, typically requires large, difficult-to-access computer systems to process, and there is a shortage of experts with the knowledge and experience to manage and analyze Big Data (Ovide 2013). The time necessary to organize and clean Big Data sets may offset, to some extent, the speed advantage with which Big Data is accumulated.

4.3 Measurement Error

The error sources associated with survey data are reasonably well understood and have been the subject of robust, ongoing research initiatives for many decades (Groves et al. 2009; Schuman and Presser 1981; Sudman and Bradburn 1974). We know that the Literary Digest poll was discredited by several error sources, including coverage and nonresponse errors that have been well documented (Lusinchi 2012; Squire 1988). Errors associated with Big Data, however, are currently not well understood and efforts to systematically investigate them are only now beginning. Prewitt (2013: 230) observes that "there is no generally accepted understanding of what constitutes errors when it is machines collecting data from other machines." Measurement error is an important example. Survey measures are typically the subject of considerable research and refinement, with sophisticated methodologies readily available for the design, testing, and assessment of measurement instruments (Madans et al. 2011; Presser et al. 2004). Big Data shares many of the challenges of secondary analyses of survey data in which specific indicators of the construct(s) of interest may not always be available, challenging the analyst's creativity and cleverness to sometimes "weave a silk purse from a sow's ear." Indeed, those analyzing Big Data must work with what is available to them and there is seldom an opportunity to allow theory to drive the design of Big Data collection systems. There is also concern that those who generate Big Data are sometimes unwilling to share details of how their data are collected, to provide definitions of the terms and measures being used, and to allow replication of measurements and/or analyses based on their measurements.

One interesting example is the Google Flu Index. In 2009, a team from Google Inc. and the Centers for Disease Control and Prevention (CDC) published a paper in *Nature* that described the development of a methodology for examining billions of Google search queries in order to monitor influenza in the general population (Ginsberg et al. 2009).[1] They described a non-theoretical procedure that involved

---

[1] In 2008, a team of academic investigators and Yahoo! Employees published a similar paper (Polgreen et al. 2008).) That team, however, had not continued to report on this topic.

identifying those Google search queries that were most strongly correlated with influenza data from the CDC; a large number of models were fit during the development of the flu index. They reported the ability to accurately estimate weekly influenza within each region of the U.S. and to do so with only a very short time lag. Shortly thereafter, the flu index *underestimated* a non-seasonal outbreak, and researchers speculated that changes in the public's online search behaviors, possibly due to seasonality, might be responsible (Cook et al. 2011). Despite an ongoing effort to revise, update and improve the predictive power of Google Flu Trends, it also greatly *overestimated* influenza at the height of the flu season in 2011–2012 (Lazer et al. 2014a) and especially in 2012–2013 (Butler 2013). Lazer et al. (2014a) also demonstrated that Google Flu Trends had essentially overestimated flu prevalence during 100 of 108 weeks (starting with August 2011). A preliminary analysis of the 2013–2014 season suggests some improvement, although it is still *overestimating* flu prevalence (Lazer et al. 2014b).

Couper (2013) has made the interesting point that many users of social media, such as Facebook, are to some extent motivated by impression management, and we can thus not be certain of the extent to which information derived from these sources accurately represents the individuals who post information there. Social desirability bias would thus appear to be a threat to the quality of Big Data as well as survey data. The fact that a significant proportion of all Facebook accounts, for example, are believed to represent fictitious individuals is another cause for concern. One estimate from 2012 suggests the number of fake Facebook accounts may be as many as 83 million (Kelly 2012). Hence, concerns with data falsification also extend to Big Data.

4.4 Population Coverage

The *Literary Digest* Poll was big, but many believe it did not provide adequate coverage of the population to which it was attempting to make inferences. Rather, it likely over-represented upper income households with political orientations decidedly unrepresentative of the Depression Era U.S. citizenry. Clearly, volume could not compensate for or fix coverage error. Big Data faces similar problems. For Big Data that captures online activities, it is important to be reminded that not everyone is linked to the internet, not everyone on the web uses Google search engines, Twitter and Facebook, and everyone who does certainly does not do so in a similar manner. Among those who do interact with the web, the manners in which they do are very diverse. The elderly, who are less likely to engage the internet, are particularly vulnerable to influenza, yet none of the Google Flu Index papers referenced here address this issue. A related concern is the problem of selection bias. As Couper (2013) has observed, Big Data tends to focus on society's "haves" and less so on the "have-nots." In addition, in Big Data there can be a problem with potential violations of the "one-person-one-vote" rule. As Smith (2013) has commented, a large preponderance of some social media activities, such as Twitter and Facebook, are the products of the activities of relatively small concentrations of individuals, further calling in to question the adequacy of their coverage. Indeed, many Big Data systems have what Tufekci (2014) refers to as a denominator problem "created by vague, unclear or unrepresentative sampling." Others have expressed concerns regarding the danger that Big Data "can be easily gamed" (Marcus and Davis 2014). Campbell (1979) wrote more than 40 years ago about the corruptibility of social data as it becomes more relevant to resource allocation decisions. Marcus and Davis (2014) discuss several Big Data examples of this. Design-based, "small data" surveys, in comparison, go to great lengths to insure that their samples adequately cover the population of interest.

4.5 Ethical Oversight

Unlike survey researcher's insistence on obtaining informed consent from respondents prior to data collection, and emphasis on the distribution of de-identified data only, many Big Data operations routinely collect identifying information without the consent, or even the knowledge, of those being monitored. In comparison to the careful ethical reviews and oversight academic and government-based survey research routinely receives, the ethical issues surrounding Big Data are not yet well understood or recognized. There is little transparency or oversight in Big Data research, much of it being conducted by private groups using proprietary data.

Unfortunately, recent events, such as Facebook's mood experiments (Albergott and Dwoskin 2014), are reminiscent of some of the ethical transgressions of past generations that led to ethical review requirements for federally funded research (Humphreys 1970; Milgram 1974). For example, in 2014, Kramer et al. (2014) published in the *Proceedings of the National Academy of Sciences (PNAS)* findings from a field experiment that examined the extent to which emotional states could be manipulated by altering the content that Facebook users were exposed to. They demonstrated that reductions in displays of emotionally negative postings from others resulted in both reductions in the amount of positive posting and increases in negative emotional postings among the Facebook users being monitored. The paper's authors reported that 689,003 individuals and more than three million Facebook postings were studied as part of the experiment. Shortly after the paper's publication, *PNAS* published an "Editorial Expression of Concern and Correction," acknowledging potential contradictions between established ethical principles of research conduct—specifically the degree to which the Facebook study subjects had sufficient opportunity to provide informed consent and/or to opt out of the research—and the Facebook data user policies, under which users agree to corporate use of their data at the time they establish their personal account. In response to ambiguities such as these, some have called for a new Big Data Code of Ethical Practices (Rayport 2011). The National Science Foundation recognized this need and launched a Council for Big Data, Ethics, and Society in early 2014 "to provide critical social and cultural perspectives on big data initiatives" (see: http://www.

datasociety.net/initiatives/council-for-big-data-ethics-and-society/). There is no consensus, however, regarding the ethical issues surrounding cases such as the Facebook experiments (Puschmann and Bozdag 2014).

4.6 Transparency

Transparency of methods is central to the ability to replicate research findings. There is a need for greater and more general understanding of how Big Data sets are constructed (Mayer-Schönberger and Cukier 2013). Big Data is not yet transparent, and most Big Data is proprietary and commercially controlled, and the methods employed to analyze these data are seldom described in a manner that would facilitate replication. In fact, commercial interests often dictate against transparency. The Google Flu Index, for example, has never revealed the 45 or so search terms it uses to make its prevalence estimates. Lazer et al. (2014b) have accused Google of reporting misleading information regarding the search terms they employ. While survey research is far from perfect when it comes to transparency of methods, there is general recognition of its importance. Most high-quality professional journals demand disclosure of survey methods. In 2010, AAPOR launched a Transparency Initiative, designed "to promote methodological disclosure through a proactive, educational approach that assists survey organizations in developing simple and efficient means for routinely disclosing the research methods associated with their publicly-released studies" (see: http://www.aapor.org/). In addition, codebooks, methodological reports, and other forms of documentation are considered to be standard products of any reputable survey, and have been so for many decades. The documentation requirements of social science data archives, such as the Inter-University Consortium of Social and Political Research (ICPSR; see http://www.icpsr.umich.edu/icpsrweb/content/deposit/guide/chapter3docs.html) are very stringent. Documentation of internet data, by comparison, is extremely limited (Smith 2013).

5 Supplement or Substitute?

Lazer and colleagues (2014a: 1203) have coined the term "Big Data Hubris" to refer to "the often implicit assumption that big data are a substitute for, rather than a supplement to, traditional data collection and analysis." Others share this sentiment. The British sociologists Savage and Burrows (2007: 890) have considered the historicity of survey research and suggest that its "glory years" were between 1950 and 1990. Taking the long view, one has to wonder as to whether or not surveys might merely represent one of the first generations of social research methods, destined to be replaced by more efficient methodologies in an increasingly digital world? Just as the horse-drawn carriage was replaced by more advanced forms of transportation, might we be now witnessing the passing of a traditional methodology?

Only time will tell. Big Data, in its current stage of evolution, though, does not appear capable of serving as a wholesale replacement or substitute for survey research. Even Savage and Burrows (2007: 890) acknowledge that there are some niches "in which the sample survey will continue to be a central research tool because of the limits of transactional data" (i.e., Big Data). They cite crime victimization surveys, which consistently demonstrate victimization rates well in excess of estimates derived from administrative records. There are no doubt many other examples. But, Big Data is an important new and highly valuable source of information about our social world, one with the potential to help us examine and better understand social problems, including many of those being addressed in this book. So how do we reconcile small surveys with Big Data?

Several observers, including another AAPOR Task Force concerned specifically with the rise of Big Data (Japec et al. 2015), see important opportunities for surveys and Big Data to be supplements or adjuncts to one another (Butler 2013; Couper 2013; Marcus and Davis 2014; Smith 2011; 2013); for Big Data to contribute rich context to surveys, and for surveys to help make sense of patterns uncovered, but not well understood, in Big Data. Combining multiple data sources to take advantage of the strengths of each and to help compensate for the limits of each approach, seems to be what the future holds for these largely unique data resources. Smith and Kim (2014) have proposed a multi-level, multi-source (ML-MS) approach to reducing survey-related errors through a coordinated effort to more systematically link survey data with information from multiple auxiliary sources, including Big Data. These linkages would take place at each possible level of analysis, from high levels of geographies through unique paradata sources that are themselves by-products of survey data collection activities, such as contact attempts and even computer key-stroke data from interviewers and/or respondents (c.f., Kreuter 2013). In addition to private Big Data, administrative data files from governmental sources would also be linked to develop better understandings of social phenomena and the strengths and limitations of the various data sources themselves. As the former U.S. Census Bureau Director Robert Groves (2011: 869) has commented: "combining data sources to produce new information not contained in any single sources is the future."

References

Albergott R, Dwoskin E (2014) Facebook study sparks soul-searching and ethical questions. Wall Street J. [Online] 30th June. http://online.wsj.com/articles/facebook-study-sparks-ethical-questions-1404172292. Accessed 3 Aug 2014

Anderson C (2008) The end of theory: the data deluge makes the scientific method obsolete. Wired Magazine, 23rd June. http://archive.wired.com/science/discoveries/magazine/16-07/pb_theory. Accessed 29 Jul 2014

Baker R, Brick JM, Bates NA, Battaglia M, Couper MP, Dever JA, Gile KJ, Tourangeau R (2013) Report of the AAPOR task force on non-probability sampling. http://www.aapor.org/AM/Template.cfm?Section=Reports1&Template=/CM/ContentDisplay.cfm&ContentID=5963. Accessed 1 Aug 2014

Butler D (2013) When Google got flu wrong. Nature 494:155–156

Campbell DT (1979) Assessing the impact of planned social change. Eval Program Plann 2:67–90

Cohn N (2014) Explaining online panels and the 2014 midterms. New York Times. [Online] 27 July. http://www.nytimes.com/2014/07/28/upshot/explaining-online-panels-and-the-2014-midterms.html?_r=0. Accessed 1 Aug 2014

Converse JM (1987) Survey research in the United States: roots and emergence 1890-1960. University of California Press, Berkeley

Cook S, Conrad C, Fowlkes AL, Mohebbi MH (2011) Assessing Google flu trends performance in the United States during the 2009 influenza virus A (H1N1) pandemic. PLoS One 6:e23610

Couper MP (2013) Is the sky falling? New technology, changing media, and the future of surveys. Surv Res Methods 7:145–156

Fung K (2014) Google flu trends' failure shows good data > big data. Harvard Business Review/HBR Blog Network. [Online] 25 March. http://blogs.hbr.org/2014/03/google-flu-trends-failure-shows-good-data-big-data/. Accessed 17 Jun 2014

Ginsberg J, Mohebbi MH, Patel RS, Brammer L, Smolinsji MS, Brilliant L (2009) Detecting influenza epidemics using search engine query data. Nature 457:1012–1014

Goel S, Hofman JM, Lahaie S, Pennock DM, Watts DJ (2010) Predicting consumer behavior with web search. PNAS 107:17486–17490, http://www.pnas.org/content/107/41/17486. Accessed 15 Aug 2015

Groves RM (2011) Three eras of survey research. Public Opin Quart 75:861–871

Groves RM, Fowler FJ, Couper MP, Lepkowski JM, Singer E, Tourangeau R (2009) Survey methodology, 2nd edn. Wiley, New York

Harford T (2014) Big data: are we making a big mistake? Financial Times. [Online] 26 March. http://www.ft.com/intl/cms/s/2/21a6e7d8-b479-11e3-a09a-00144feabdc0.html#axzz39NlqxnU8. Accessed 17 Jun 2014

Humphreys L (1970) Tearoom trade: impersonal sex in public places. Duckworth, London

Japec L, Kreuter F, Berg M, Biemer P, Decker P, Lampe C, Lane J, O'neil C, Usher A (2015) AAPOR report on big data. http://www.aapor.org/AAPORKentico/AAPOR_Main/media/Task-Force-Reports/BigDataTaskForceReport_FINAL_2_12_15.pdf. Accessed: 27 Jul 2015

Kelly H (2012) 83 million Facebook accounts are fakes and dupes. CNN Tech. [Online] 2 August. http://www.cnn.com/2012/08/02/tech/social-media/facebook-fake-accounts/. Accessed 3 Aug 2014

Kramer ADI, Guillory JE, Hancock JT (2014) Experimental evidence of massive-scale emotional contagion through social networks. PNAS 111:8788–8790, http://www.pnas.org/content/111/24/8788.full. Accessed 2 Aug 2014

Kreuter F (2013) Improving surveys with paradata: analytic uses of process information. Wiley, New York

Lazer D, Kennedy R, King G, Vespignani A (2014a) The parable of Google flu: traps in big data analysis. Science 343:1203–1205

Lazer D, Kennedy R, King G, Vespignani A (2014b) Google flu trends still appears sick: an evaluation of the 2013-2014 flu season. [Online] http://ssrn.com/abstract=2408560. Accessed 26 Jul 2014

Lusinchi D (2012) "President" Landon and the 1936 *Literary Digest* Poll: were automobile and telephone owners to blame? Soc Sci Hist 36:23–54

Madans J, Miller K, Maitland A, Willis G (2011) Question evaluation methods: contributing to the science of data quality. Wiley, New York

Marcus G, Davis E (2014) Eight (no, nine!) problems with big data. New York Times. [Online] 7 April. http://www.nytimes.com/2014/04/07/opinion/eight-no-nine-problems-with-big-data.html?_r=0. Accessed 7 Aug 2014

Mayer-Schönberger V, Cukier K (2013) Big data: a revolution that will transform how we live, work, and think. Houghton Mifflin Harcourt, New York

Milgram S (1974) Obedience to authority: an experimental view. Harper, New York

Ovide S (2013) Big data, big blunders. Wall Street Journal. [Online] 10 March. http://online.wsj.com/news/articles/SB10001424127887324196204578298381588348290. Accessed: 30 Jul 2014

Polgreen PM, Chen Y, Pennock DM, Nelson FD (2008) Using internet searches for influenza surveillance. Clin Infect Dis 47:1443–1448

Presser S, Rothgeb JM, Couper MP, Lessler JT, Martin E, Martin J, Singer E (2004) Methods for testing and evaluating survey questionnaires. Wiley, New York

Prewitt K (2013) The 2012 Morris Hansen lecture: Thank you Morris, et al., for Westat, et al. J Off Stat 29:223–231

Puschmann C, Bozdag E (2014) Staking out the unclear ethical terrain of online social experiments. Internet Policy Review 3(4). http://policyreview.info/articles/analysis/staking-out-unclear-ethical-terrain-online-social-experiments. Accessed: 5 Aug 2016

Rayport JF (2011) What big data needs: a code of ethnical practices. MIT Technology Review, May 26. http://www.technologyreview.com/news/424104/what-big-data-needs-a-code-of-ethical-practices/. Accessed 2 Aug 2014

Savage M, Burrows R (2007) The coming crisis of empirical sociology. Sociology 41:885–899

Schuman H, Presser S (1981) Questions and Answers in Attitude Surveys. Wiley, New York

Smith TW (2011) The report of the international workshop on using multi-level data from sample frames, auxiliary databases, paradata and related sources to detect and adjust for nonresponse bias in surveys. Int J Public Opin Res 23:389–402

Smith TW (2013) Survey-research paradigms old and new. Int J Public Opin Res 25:218–229

Smith TW, Kim J (2014) The multi-level, multi-source (ML-MS) approach to improving survey research. GSS Methodological Report 121. NORC at the University of Chicago, Chicago

Squire P (1988) Why the 1936 *Literary Digest* poll failed. Public Opin Quart 52:125–133

Sudman S, Bradburn NM (1974) Response effects in surveys: a review and synthesis. Aldine Press, Chicago

Thakuriah P, Tilahun N, Zellner M (2016) Big data and urban informatics: innovations and challenges to urban planning and knowledge discovery. In: Thakuriah P, Tilahun N, Zellner M (eds) Seeing cities through big data: research methods and applications in urban informatics. Springer, New York

Tufekci Z (2014) Big questions for social media big data: representativeness, validity and other methodological pitfalls. In ICWSM'14: Proceedings of the 8th international AAAI conference on weblogs and social media, forthcoming. http://arxiv.org/ftp/arxiv/papers/1403/1403.7400.pdf. Accessed 26 Jul 2014

Vogt WP (1999) Dictionary of statistics & methodology, 2nd edn. Sage, Thousand Oaks, CA

Webopedia (2014) Big data. http://www.webopedia.com/TERM/B/big_data.html. Accessed 29 Jul 2014

Big Spatio-Temporal Network Data Analytics for Smart Cities: Research Needs

Venkata M.V. Gunturi and Shashi Shekhar

Abstract Increasingly, location-aware sensors in urban transportation networks are generating a wide variety of data which has spatio-temporal network semantics. Examples include temporally detailed roadmaps, GPS tracks, traffic signal timings, and vehicle measurements. These datasets, which we collectively call Big Spatio-Temporal Network (BSTN) Data, have value addition potential for several smart-city use-cases including navigation services which recommend eco-friendly routes. However, BSTN data pose significant computational challenges regarding the assumptions of the current state-of-the-art analytic-techniques used in these services. This article attempts to put forth some potential research directions towards addressing the challenges of scalable analytics on BSTN data. Two kinds of BSTN data are considered here, viz., the vehicle measurement big data and the travel-time big data.

Keywords Spatial databases • Spatial data mining • Road networks • Routing • Spatio-temporal networks

1 Introduction

Smart cities can be considered as an "interconnected eco-system" thriving through a diverse set of networks for collective sustenance, communication, distribution and delivery, for its entities. Examples of these networks include, transportation networks (for distribution and delivery); water, electricity, gas and sewage-disposal networks (for sustenance); internet and tele-communication networks (for communication) (Kelmelis and Loomer 2003). A common characteristic among all these networks is that they are either physically embedded in space (e.g. transportation,

---

V.M.V. Gunturi (✉)
Department of Computer Science, IIIT-Delhi, New Delhi, India
e-mail: gunturi@iiitd.ac.in

S. Shekhar
Department of Computer Science and Engineering, University of Minnesota, Minneapolis, MN, USA
e-mail: shekhar@cs.umn.edu

© Springer International Publishing Switzerland 2017
P. Thakuriah et al. (eds.), *Seeing Cities Through Big Data*, Springer Geography,
DOI 10.1007/978-3-319-40902-3_8

water, sewage-disposal networks) or, use space for increasing their reach (e.g. transmission towers and repeaters for electricity and tele-communication networks).

Increasingly, a large number of urban sensors in major cities (Boyle et al. 2013) have started to produce a variety of datasets representing both historic and evolving characteristics of these smart-city networks. Each of these datasets record a certain property or a phenomena on the smart-city network spread over space and time. A collection of such datasets is referred to as Big Spatio-Temporal Network (BSTN) data. On transportation network alone we have emerging datasets such as temporally detailed (TD) roadmaps (Shekhar et al. 2012) that provide travel-time for every road-segment at 5 min granularity, traffic signal timings and coordination (Liu and Hu 2013), GPS tracks (Shekhar et al. 2012; Yuan et al. 2010), vehicle measurements (Ali et al. 2015) and traffic information from vehicle-to-vehicle communications (Lee et al. 2009; Wolfson and Xu 2014). Other sample BSTN datasets include, meter readings from water (Yang et al. 2011) and energy distribution (electricity and gas) (Boyle et al. 2013) networks, traffic information from communication networks (internet and telecom). To maintain coherence of the paper, we will focus only on BSTN data generated from transportation networks. The goal of this paper is to highlight some of the open computer science research questions for a potential realization of analytics on BSTN data.

Realization of BSTN analytics is important due to its value addition potential in several smart-city application use-cases e.g., urban navigation, transportation infrastructure management and planning. For instance, BTSN can make significant extensions to the traditional routing query "what is the shortest route between University of Minnesota and the airport?" through novel preference metrics such as "which route has the least amount of waiting at traffic signals?", "Which route typically produces the least amount of greenhouse gas emissions?", "Which route do commuters typically prefer?" Also, BTSN data can allow for exploratory analysis through queries like "Which parts of the transportation networks typically tend to have more greenhouse gas emissions than others?" etc.

This paper categorizes BSTN into the following two types: (a) vehicle measurement big data and, (b) Travel-time big data. We define vehicle measurement big data (VMBD) only to contain data coming from sensors inside a vehicle. VMBD allows us to study how vehicles typically perform in a real world transportation network. We define travel-time big data as to contain information on typical travel-times and traffic signal delays observed by the traffic. Note that this segregation was made for communication purpose only. Both VMBD and travel-time big data have spatio-temporal network semantics and can be referred to as BSTN itself. Also, we take a broad definition of the word "data analytics" which includes both knowledge discovery and query processing.

This chapter is organized as follows: Section "Vehicle Measurement Big Data (VMBD)" describes VMBD in detail and presents some open research questions for realizing statistically significant pattern mining on it. In section "Travel-Time Big Data" we present the travel-time big data and discuss some open research questions which could pave the way towards scalable query processing this kind of big data. We conclude in section "Conclusion".

2 Vehicle Measurement Big Data (VMBD)

Rich instrumentation (e.g., GPS receivers and engine sensors) in modern fleet vehicles allow us to periodically measure vehicle sub-system properties (Kargupta et al. 2006, 2010; Ali et al. 2015) of the vehicle. These datasets, which we refer to as vehicle measurement big data (VMBD), contains a collection of trips on a transportation network. Here, each trip is a time-series of attributes such as vehicle location, fuel consumption, vehicle speed, odometer values, engine speed in revolutions per minute (RPM), engine load, and emissions of greenhouse gases (e.g., CO_2 and NOx). Figure 1 illustrates a sample VMBD from a trip using a plot (Fig. 1a) and tabular representations (Fig. 1b). The geographic route associated with the trip is shown as a map in lower part of Fig. 1a, where color indicates the value of a vehicle measurement, e.g., NOx emission.

Computationally, VMBD may be modeled as a spatio-temporal network (STN) (George and Shekhar 2008; George et al. 2007), which is a generalization of location-aware adjacency graph representation for roadmaps (Fig. 2), where road intersections are modeled as vertices and the road segments connecting adjacent intersections are represented as edges. For example, the intersection of SE 5th Ave and SE University Ave is modeled as node N1 and the segment between SE University Ave and SE 4th Street is represented by the edge N1-N4 with the direction information on the segment. Note that location-aware adjacency graphs are richer than graphs in graph theory (Trudeau 2013; Agnarsson and Greenlaw 2006; Ahuja et al. 1993), since its vertices and edges are associated with their geographic locations (e.g., Map-matched edge-id and Location colums in Fig. 1b) to facilitate modeling spatial relationships (e.g., left-turn, right-turn) that are not explicitly *modeled via edges*. Location-aware adjacency graph representations are generalized to spatio-temporal networks (STN) representations (Fig. 2b, c), where nodes or edges or their properties vary over time. Figure 2b shows the snapshot model of STNs across four different time points (t = 1,2,3 and 4) to represent the time-dependence of edge weights. Figure 2c shows a time-expanded graph (TEG)

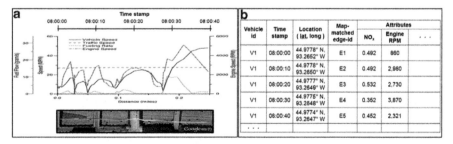

Fig. 1 Vehicle measurement data and its tabular representation. (**a**) Vehicle measurement data with route map. (**b**) Tabular representation

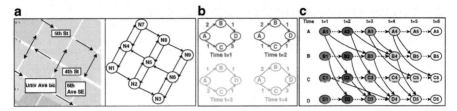

Fig. 2 Spatio-temporal networks (STN): snapshop and time expanded. (**a**) Example roadmap and its adjacency representation. (**b**) Snapshot model. (**c**) Time-expanded graph

(Köhler et al. 2002) representation, which stitches all snapshots together via edges spanning multiple snapshots. For example, the edge (A1–B3) in Fig. 2c represents a trip starting at node A at time 1 and reaching node B at time 3. Even though not explicit in Fig. 2c, properties of edge (A1–B3) may record fuel-consumption, NOx emissions and other vehicle measurements made during the trip.

2.1 Statistically Significant Hot Route Discovery Using VMBD

Given the spatio temporal network (STN) representation of vehicle measurement big data (VMBD), the goal of statistically significant hot route discovery is to identify simple STN paths in the underlying spatio-temporal graph which have a significantly higher instances of discrepancies (e.g., deviation from US EPA standards (US Environmental Protection Agency 2015)) in a given target engine variable (e.g., NOx or CO_2) than the other parts of the STG.

Finding such specialized hotspots of discrepancies could spur the next level of research questions such as: What is the scientific explanation behind a certain high emission observation? Was it due to a particular acceleration-breaking pattern? Can this pattern be coded into the engine control unit for better fuel economy? Figure 3 illustrates NOx hotspots (red portions of trajectories) on a transportation network for a conventional diesel engine using measured engine data from buses in Minneapolis-St Paul area.

Figure 4 illustrates a sample scenario for hot route discovery. Here, a target engine variable (e.g., NOx) is recorded as a bus travels along two different paths between node A and node D in the transportation network containing nodes A, B, C and D (Fig. 4a). The path A-B-D involves a left turn at B which takes one unit of time for times $t = 1, 2$ and 3 (mimicking rush hours), whereas the A-C-D path has a right turn which does not involve any waiting. Further assume that the bus takes one time unit to travel on each of the edges in Fig. 4a. Figure 4b illustrates seven different journeys between A and D on a STN representation of the transportation network. For the sake of simplicity, the figure illustrates only those ST edges that

Fig. 3 Sample VMBD showing instances of high NOx at a left turn

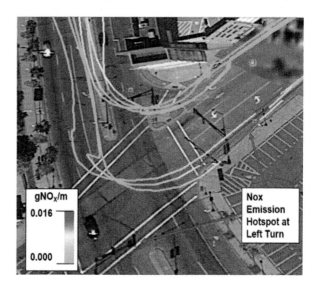

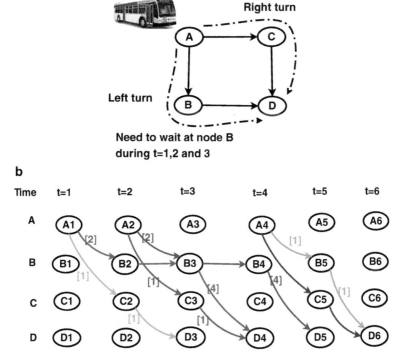

Fig. 4 (a) Two routes between A and D for the bus. (b) STN based representation of the two bus routes between A and D

were traveled in some journey. Each journey is shown using a different color. The number mentioned in "[]" on the edges denote the number of discrepancies in the target variable observed on that edge. The journeys colored red and purple (both over the route with left turn during "rush hours") can be seen to have the highest number (six each, compared to others with only two) of discrepancies, making them a potential output for the hot route discovery problem.

Mining statistically significant hot routes is challenging due to the large data volume and the STN data semantics, which violates critical assumptions underlying traditional techniques for identifying statistically significant hotspots (Kulldorff 1997; Neill and Moore 2004; Anselin and Getis 2010) that data samples are embedded in isotropic Euclidean spaces and spatial footprints of hotspots are simple geometric shapes (e.g., circles or rectangle with sides parallel to coordinate axes). This is illustrated in Fig. 5 using a transportation safety dataset about pedestrian fatalities. SatScan (Kulldorff 1997, 1999), a popular hotspot detection technique, finds a few circular hotspots (Fig. 5a) with statisfical significance of around 0.1, even though many highway stretches (shown in color in Fig. 5b) have higher statististical significance (0.035–0.052) and can potentially be detected from non-geometric methods such as (Oliver et al. 2014; Ernst et al. 2011).

For our proposed hot route discovery problem, one needs to generalize spatial statistics beyond iso-tropic Eucliean space to model the STN semantics such as linear routes. Computationally, the size of the set of candidate patterns (all possible directed paths in the given spatio-temporal network) is potentially exponential. In addition, statistical interest measures (e.g., likelihood ration, p-value) may lack monotonicity. In other words, a longer directed path may have a higher likelihood ratio. We now describe some potential research questions for the problem of hot route discovery.

2.1.1 Potential Next Steps for Research

Step-1: *Design interest measure for statistically significant hot routes*: The goal here would be to develop interest measures for identifying statistically significant hot routes. This would require balancing conflicting requirements of statistical interpretation and support for computationally efficient algorithms. A statistically interpretable measure is one whose values would typically conform to a known distribution. On the other hand, computational properties which are known to facilitate efficient algorithms include properties such as monotonicity. For this task, one could start by exploring a ratio based interest measure containing the densities of non-compliance inside the path to that of outside the path. We could investigate the merit of this measure through the following research questions: What are the characteristics of this interest measure? Does it have monotonicity? If not then can we design an upper bound which has monotonicity? Do the values of this measure (or its upper bound) follow a known distribution? Does it belong to one of the general classes of distributions? If not then then how would an appropriate null hypothesis be designed for monte-carlo simulations? What distribution

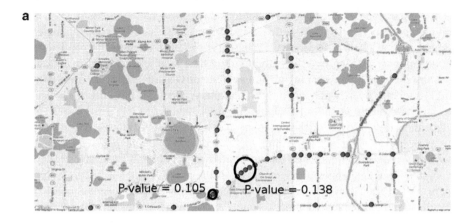

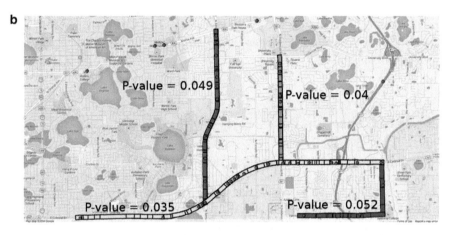

Fig. 5 Significant patterns discovered by geometric-shape based approach and transportation network aware approach. (**a**) Geometric-shape based technique (*circles*). (**b**) Transportation network aware technique (*routes*)

does the underlying VMBD normally tends to follow (needed for testing null hypothesis)? Can we derive an interest measure based on the underlying distribution of the data (similar to log likelihood ratio (Kulldorff et al. 1998; Kulldorff 1997))?

Step-2: *Design a Computational Structure of exploring the space of STN shortest paths*: As a first step towards addressing the challenge of the large number of candidate paths, one would have to investigate scalable techniques for computing all-pair STN shortest paths. This would require addressing the challenge of non-stationary ranking among alternative paths between two points in an STN. This non-stationary ranking precludes the possibility of a dynamic programming

(DP) based approach. A potential approach for this challenge could involve a divide-and-conquer strategy called critical-time-point (CTP) based approaches (Gunturi et al. 2011, 2015). A CTP based approach addresses the challenge of non-stationary ranking by efficiently dividing the given time interval (over which we observe non-stationary ranking among alternative paths) into a set of disjointed time intervals over which ranking is stationary. One can now use a DP based technique over these sub-intervals. Using the concept of CTPs, one could adapt algorithms (e.g., Floyd Warshall's, Johnson's) for the all-pair shortest path problem.

Step-3: Algorithm for exploring the space of all simple paths in a spatio-temporal graph: The third step for hot route discovery would be to investigate the computational challenges of enumerating all the possible paths in spatio-temporal network (STN). Clearly, in any real STN, there would be an exponential number of candidate paths. A naïve approach could involve an incremental spatio-temporal join based strategy, which starts off with singleton edges is the STN and iteratively joins them (based on spatio-temporal neighborhood constraints) to create longer paths. This naïve algorithm raises several unanswered questions: Would paths enumerated this way still be simple, i.e., with no repeated nodes? How can we avoid creating loops?

3 Travel-Time Big Data

Figure 6 shows a sample transportation network in the Minneapolis area (to the left). On the right is its simplified version, where arrows represent road segments and labels (in circles) represent an intersection between two road segments. Locations of traffic signals are also annotated in the figure.

We consider an instance of travel-time big data on our sample networked system shown in Fig. 6 by considering following three datasets: (a) temporally detailed (TD) roadmaps (Shekhar et al. 2012), (b) traffic signal data (Liu and Hu 2013) and (c) map matched GPS traces (Yuan et al. 2010; Shekhar et al. 2012). Each of these measure either historical or evolving aspects of certain travel related phenomena on our transportation network. TD roadmaps store historical travel-time on road segments for several departure-times (at 5 min intervals) in a typical week. The essence of TD roadmaps and traffic signal data is illustrated in Fig. 7. For simplicity, TD roadmaps are illustrated by highlighting the morning (7:00 am to 11:00 am) travel time only on segments A-F, F-D and S-A (7 min, 11 min and 9 min respectively). The travel-times of other road segments in the figure (shown next to arrows representing roads) are assumed to remain constant. The figure also shows the traffic signal delays during the 7:00 am to 11:00 am period. Additionally, the traffic signals SG1, SG2, SG3 are coordinated such that in a journey towards D (from S within certain speed-limits), one would typically wait only at SG1. Similarly, a traveler starting on segment B-C (after SG1) would have to wait only at SG2.

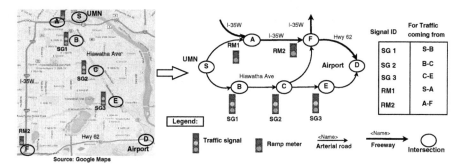

Fig. 6 Sample transportation network for illustrating travel-time big data (best in color)

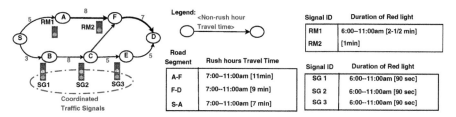

Fig. 7 Sample travel-time big data on transportation network shown in Fig. 6

Map matched and pre-processed (Zheng and Zhou 2011) GPS tracks, another component of our travel-time big data, consists of a sequence of road-segments traversed in the journey along with its schedule denoting the exact time when the traversal of a particular segment began (via map-matching the GPS points). GPS traces can potentially capture the evolving aspect of our system. For instance, if segment E-D in Fig. 7 is congested due to an event (a non-equilibrium phenomena), travel-time denoted by TD roadmaps may no longer be accurate. In such a case, one may prefer to follow another route (say C-F-D) which other commuters may be taking to reach D.

3.1 Moving Towards a Navigation System Using Travel-Time Big Data

As a first step towards building a navigation system which can harness travel-time big data for recommending routes, we need a unified querying framework across the previous described TD roadmaps, traffic signal data and GPS traces. This querying framework would integrate all the datasets into one model such that routing algorithms can access and compare information from multiple datasets. This problem is formally defined below:

Given TD roadmaps, traffic signal data and annotated GPS traces, and a set of use-case queries, the goal is to build a unified logical data-model across these datasets which can express travel related concepts explicitly while supporting efficient algorithms for given use-cases. The objective here would be to balance the trade-off between expressiveness and computational efficiency.

Challenges: Designing a logical data-model for travel-time big data is non-trivial. For instance, the model should be able to conveniently express all the properties of the n-ary relations in the data. For example, consider a typical journey along S-B-C-E-D through a series of coordinated traffic signals SG1, SG2 and SG3 in Fig. 7. Here, the red-light durations and phase gaps between the traffic signals SG1, SG2 and SG3 are set in a way that a traveler starting at S and going towards D (within certain speed-limits) would typically wait only at SG1, before being smoothly transferred through intersections C and E with no waiting at SG2 or SG3. In other words, in this journey, initial waiting at SG1 would render SG2 and SG3 wait-free. If the effects of the immediate spatial neighborhood are referred to as local-interactions, e.g. waiting at SG1 delaying entry into segment B-C, then this would be referred to as a non-local interaction as SG1 is not in the immediate neighborhood of SG2.

Typical signal delays (and travel-time) measured under signal coordination cannot be decomposed to get correct, reliable information. For instance, if we decompose the above mentioned journey through S-B-C-E-D into experiences at individual road-segments and traffic signals, we would note that we typically never wait at SG2 or SG3. However, this is not true as any typical journey starting at intersection B (after SG1) would have to wait at SG2. This kind of behavior where properties measured over larger instances lose their semantic meaning when decomposed into smaller instances is called *holism*. And we refer to such properties as *holistic properties*. Travel-time observed in GPS traces also show such holism as time spent on a road segment depends on the initial speed attained before entering the segment (Gunturi and Shekhar 2014).

Limitations of Current Logical Data-models: Current approaches for modeling travel-time data such as time aggregated graphs (George and Shekhar 2006, 2007), time expanded graphs (Köhler et al. 2002; Kaufman and Smith 1993) and related problems in modeling and querying graphs in databases (Hoel et al. 2005; Güting 1994), are not suitable for modeling the mentioned travel-time big data. They are most convenient when the property being represented can be completely decomposed into properties of binary relations. In other words, they are not suitable for representing the previously described holistic properties. Current related work would represent our previous signal coordination scenario using the following two data structures (see Fig. 8): one containing travel-time on individual segments (binary relations) S-B, B-C, C-E, and E-D; the second containing the delays and the traffic controlled by signals SG1, SG2, and SG3 (also binary). However, this is not convenient as non-local interactions affecting travel-times on some journeys (e.g. S-B-C-E-D) are not expressed explicitly. Note that this representation would have been good enough if SG1, SG2 and SG3 had not been coordinated.

Fig. 8 Related work representation

Road	Typical Travel Time
S-B	3mins
B-C	8mins
C-E	5mins
E-D	5mins

Signal	Incoming Traffic	Outgoing Traffic	Max Delay
SG1	S-B	B-C	90secs
SG2	B-C	C-E	90secs
SG3	C-E	E-D	90secs

Fig. 9 Our proposed representation

Journeys with non-local interactions	Typical travel-time Experienced
S-B +delay at SG1	3 mins --- 4 mins 30 sec
S-B +SG1+ B-C +SG2	11 mins -- 12 mins 30 sec
S-B +SG1+ B-C +SG2+ C-E +SG3	16 mins -- 17 mins 30 sec

Ideally, the representation model should express the non-local interactions more explicitly. Figure 9 illustrates our proposal (Gunturi and Shekhar 2014) for the previous signal coordination scenario. Here, we propose to represent the journey as a series of overlapping sub-journeys, each accounting for a non-local interaction. The first entry in the figure corresponds to travel-time experienced on the sub-journey containing road segment S-B (3 min) and delay at SG1 (max delay 90 s). This would be between 3 min and 4 min 30 s (no non-local interactions in this case). Next we would store travel-time experienced on the sub-journey containing road segment S-B (3 min), delay at SG1 (max delay 90 s), segment B-C (8 min) and delay at SG2 (max 90 s) as between 11 min and 12 min 30 s. Note that we did not consider the delay caused by SG2 due to non-local interaction from SG1. This process continues until all the possible non-local interactions are covered.

3.1.1 Potential Next Steps for Research

A key task while developing a logical model for travel-time big data would be to balance the tradeoff between computational scalability and accuracy of representation of important travel related concepts such as non-local interactions. Further, the proposed model should provide a seamless integration of TD roadmaps, traffic signal data and experiences of commuters through GPS traces. This would allow the routing framework to explore candidate travel itineraries across multiple sources to get richer results. For example, routes from TD roadmaps, which recommend based on historic congestion patterns, can be compared (during the candidate exploration phase itself rather than offline) against commuter preferences for factors like fuel efficiency and "convenience".

In addition to seamless integration, one also has to develop a set of data-types and operations for queries on this representational model. This would involve addressing questions such as: What would be minimal set of operators expressive enough to represent spatio-temporal routing queries about a traveler's experience? Do they facilitate the design of dynamic programming based methods? How should the set be refined to address the conflicting needs of expressive power and support for efficient algorithms? And lastly, we would also have to explore the computational structure of the various use-case queries. A sample research question could be: Does the classical dynamic programming for the fastest query hold in the face of new features like fuel efficiency, signal delays and travel-time information at multiple departure-times?

4 Conclusion

BSTN data analytics has value addition potential for societal applications such as urban navigational systems and transportation management and planning. It does however present challenges for the following reasons: First, the linear nature of the patterns in BTSN data (e.g., vehicle measurement big data) raises significant semantic challenges to the current state-of-the-art in the area of mining statistically significant spatial patterns which mostly focus on geometric shapes, e.g., circles and rectangles. And second, holistic properties increasingly being captured in the BSTN data (travel-time big data) raise representational challenges to the current state-of-the-art data-models for spatio-temporal networks. In future, we plan to explore the research questions emerging from this paper towards formalizing the notion of BSTN data analytics.

Acknowledgment This work was supported by: NSF IIS-1320580 and 0940818; USDOD HM1582-08-1-0017 and HM0210-13-1-0005; IDF from UMN. We would also like to thank Prof William Northrop and Andrew Kotz of University of Minnesota for providing visualizations of the vehicle measurement big data and an initial insight into interpreting it. The content does not necessarily reflect the position or the policy of the government and no official endorsement should be inferred.

References

Agnarsson G, Greenlaw R (2006) Graph theory: modeling, applications, and algorithms. Prentice Hall, Upper Saddle River, NJ. ISBN 0131423843
Ahuja RK, Magnanti TL, Orlin JB (1993) Network flows: theory, algorithms, and applications. Prentice Hall, Upper Saddle River, NJ
Ali RY et al (2015) Discovering non-compliant window co-occurrence patterns: a summary of results. In: International symposium on spatial and temporal databases. Springer, New York
Anselin L, Getis A (2010) Spatial statistical analysis and geographic information systems. In: Perspectives on spatial data analysis. Springer, New York, pp 35–47

Boyle DE, Yates DC, Yeatman EM (2013) Urban sensor data streams: London 2013. IEEE Internet Comput 17(6):12–20

Ernst M, Lang M, Davis S (2011) Dangerous by design: solving the epidemic of preventable pedestrian deaths. Transportation for America: Surface transportation policy partnership. The National Academies of Sciences, Engineering, and Medicine, Washington, DC

George B, Shekhar S (2006) Time-aggregated graphs for modeling spatio-temporal networks. In: Advances in conceptual modeling-theory and practice. Springer, New York, pp 85–99

George B, Shekhar S (2007) Time-aggregated graphs for modelling spatio-temporal networks. J Data Semantics XI:191

George B, Shekhar S (2008) Time-aggregated graphs for modeling spatio-temporal networks. J Data Semantics XI:191–212

George B, Kim S, Shekhar S (2007) Spatio-temporal network databases and routing algorithms: a summary of results. In: Advances in spatial and temporal databases. Springer, New York, pp 460–477

Gunturi VMV, Shekhar S (2014) Lagrangian Xgraphs: a logical data-model for spatio-temporal network data: a summary. In: Advances in conceptual modeling. Springer, New York, pp 201–211

Gunturi VMV et al (2011) A critical-time-point approach to all-start-time Lagrangian shortest paths: a summary of results. In: Advances in spatial and temporal databases. Springer, New York, pp 74–91

Gunturi VM, Shekhar S, Yang K (2015) A critical-time-point approach to all-departure-time Lagrangian shortest paths. IEEE Trans Knowl Data Eng 27(10):2591–2603

Güting RH (1994) GraphDB: modeling and querying graphs in databases. In Proceedings of the 20th international conference on very large data bases, pp 297–308

Hoel EG, Heng W-L, Honeycutt D (2005) High performance multimodal networks. In: Advances in spatial and temporal databases. Springer, New York, pp 308–327

Kargupta H et al (2006) On-board vehicle data stream monitoring using minefleet and fast resource constrained monitoring of correlation matrices. N Gener Comput 25(1):5–32

Kargupta H, Gama J, Fan W (2010) The next generation of transportation systems, greenhouse emissions, and data mining. In Proceedings of the 16th ACM SIGKDD international conference on knowledge discovery and data mining, pp 1209–1212

Kaufman DE, Smith RL (1993) Fastest paths in time-dependent networks for intelligent vehicle-highway systems application. I V H S J 1(1):1–11

Kelmelis J, Loomer S (2003) The geographical dimensions of terrorism, Chapter 5.1, Table 5.1.1. Routledge, New York

Köhler E, Langkau K, Skutella M (2002) Time-expanded graphs for flow-dependent transit times. In: Proceedings of the 10th annual European symposium on algorithms. ESA'02. Springer, London, pp 599–611

Kulldorff M (1997) A spatial scan statistic. Commun Stat Theory Methods 26(6):1481–1496

Kulldorff M (1999) Spatial scan statistics: models, calculations, and applications. In: Scan statistics and applications. Springer, New York, pp 303–322

Kulldorff M et al (1998) Evaluating cluster alarms: a space-time scan statistic and brain cancer in Los Alamos, New Mexico. Am J Public Health 88(9):1377–1380

Lee U et al (2009) Dissemination and harvesting of urban data using vehicular sensing platforms. IEEE Trans Veh Technol 58(2):882–901

Liu H, Hu H (2013) SMART-signal phase II: arterial offset optimization using archived high-resolution traffic signal data. Technical report # CTS 13-19. Center for Transportation Studies, University of Minnesota

Neill DB, Moore AW (2004) Rapid detection of significant spatial clusters. In: KDD, pp 256–265

Oliver D et al (2014) Significant route discovery: a summary of results. In: Geographic information science - 8th International conference, GIScience 2014 proceedings, vol 8728, LNCS. Springer, New York, pp 284–300

Shekhar S et al (2012) Spatial big-data challenges intersecting mobility and cloud computing. Proceedings of the 11th ACM international workshop on data engineering for wireless and mobile access, pp 1–6. http://dl.acm.org/citation.cfm?id=2258058. Accessed 5 Jun 2014

Trudeau RJ (2013) Introduction to graph theory. Courier Corporation, North Chelmsford. ISBN 0486678709

US Environmental Protection Agency (2015) Heavy-duty highway compression-ignition engines and urban buses -- exhaust emission standards. US Environmental Protection Agency, Washington, DC

Wolfson O, Xu B (2014) A new paradigm for querying blobs in vehicular networks. IEEE MultiMedia 21(1):48–58

Yang K et al (2011) Smarter water management: a challenge for spatio-temporal network databases. In: Advances in spatial and temporal databases, Lecture notes in computer science. Springer, Berlin, pp 471–474

Yuan J et al (2010) T-drive: driving directions based on taxi trajectories. In Proceedings of the SIGSPATIAL international conference on advances in geographic information systems. GIS'10, pp 99–108

Zheng Y, Zhou X (2011) Computing with spatial trajectories. Springer, New York

A Review of Heteroscedasticity Treatment with Gaussian Processes and Quantile Regression Meta-models

Francisco Antunes, Aidan O'Sullivan, Filipe Rodrigues, and Francisco Pereira

Abstract For regression problems, the general practice is to consider a constant variance of the error term across all data. This aims to simplify an often complicated model and relies on the assumption that this error is independent of the input variables. This property is known as homoscedasticity. On the other hand, in the real world, this is often a naive assumption, as we are rarely able to exhaustively include all true explanatory variables for a regression. While Big Data is bringing new opportunities for regression applications, ignoring this limitation may lead to biased estimators and inaccurate confidence and prediction intervals.

This paper aims to study the treatment of non-constant variance in regression models, also known as heteroscedasticity. We apply two methodologies: integration of conditional variance within the regression model itself; treat the regression model as a black box and use a meta-model that analyzes the error separately. We compare the performance of both approaches using two heteroscedastic data sets.

Although accounting for heteroscedasticity in data increases the complexity of the models used, we show that it can greatly improve the quality of the predictions, and more importantly, it can provide a proper notion of uncertainty or "confidence" associated with those predictions. We also discuss the feasibility of the solutions in a Big Data context.

Keywords Heteroscedasticity • Gaussian processes • Quantile regression • Confidence intervals • Prediction intervals

---

F. Antunes (✉)
Center of Informatics and Systems of the University of Coimbra, Coimbra, Portugal
e-mail: fnibau@dei.uc.pt

A. O'Sullivan
Singapore-MIT Alliance for Research and Technology, Singapore, Singapore

F. Rodrigues • F. Pereira
Technical University of Denmark, Lyngby, Denmark

1 Introduction

Since the beginning of the twenty-first century, alongside the IT revolution, we have been witnessing a deluge of data, at some point referred to as "Big Data": a very large, complex and heterogeneous collection of information, either structured or unstructured, which increases at a high rate over time. One of the main reasons for this data explosion is the proliferation of ubiquitous sensing-systems. For example, nowadays, even the simplest smartphone is usually equipped with, at least, GPS and accelerometer sensors. Similarly, many cities already have sensors gathering information from traffic, weather, security cameras, emergency systems, and so forth, thus making dataset sizes reach unprecedented levels.

For practical reasons, data is often treated as having more signal than noise, and a well-behaved noise structure. This structure usually relies on assuming non-biased models and constant variance, typically formulated as a "white noise" Gaussian distribution, $\mathcal{N}(0, \sigma^2)$. The problem is that, more often than not, reality is not as "well behaved" and such assumptions may become unrealistic and inappropriate. Consider the problem of travel time prediction using a few typical variables (e.g. time of day, day of week, latest observed travel times from sensors, flow data from sensors, weather status). For many urban areas, assuming a homoscedastic noise model amounts to saying that the travel time only depends on those special c variables. This assumption is seldom realistic, since in practice fluctuations exist due to a myriad of phenomena (excess/lack of demand, traffic incidents, special events, errors in sensors). Indeed, under certain circumstances (e.g. heavy rain, working day), the travel times inherently vary much more than others (e.g. clear skies, weekend night) and the system is too complex to efficiently capture all components.

In this paper we demonstrate the critical importance of taking into account the heteroscedasticity in the data, by comparing the performance of homoscedastic and heteroscedastic approaches on different problems. We show the problems that arise when the assumption of constant variance is violated, and we propose some guidelines for the treatment of heteroscedasticity.

Mathematical approaches for this problem range from the classic statistical models to the more recent machine learning techniques. In a very general way, the majority of the regression models are of the form

$$y = f(x) + \varepsilon,$$

which establishes a relationship between the output variable we want to study y, with other (input) variables $x = [x^1, \ldots, x^D]^T$, where D denotes the number of explanatory variables, by means of a certain function f. We usually rewrite f as $f(x, w)$, where $w = [w_1, \ldots, w_D]^T$ are the model parameters. In this kind of model, ε is the error or residual, which is usually assumed to be a random variable, with mean zero and some variance. Whenever this variance is constant, the model is considered homoscedastic, if it is assumed to vary with x, it is considered heteroscedastic. The most common option is to assume a homoscedastic variance, with $\varepsilon \sim \mathcal{N}(0, \sigma^2)$. Most of these models appear in the form of regression/classification or time series analysis.

Although it may be acceptable to expect a roughly unbiased model (i.e. error mean $= 0$), it is often unrealistic to assume constant variance. A possible way to get around this issue is to consider heteroscedastic models. Assuming variable error along our input variables x should result in better predictions and more accurate confidence intervals.

While the improvements in accuracy may be negligible with respect to the mean (if the homoscedastic model has little bias), we have much to gain in terms of confidence intervals. For example, from the user's point of view, it is completely different to know that the bus arrival time will be 10 min with a confidence interval of 10 min, or with a confidence interval of just 1 min.

This paper is organized as follows: Section "Linear models for regression" discusses general regression methods and the treatment of heteroscedasticity with emphasis on linear models. In section "Gaussian Process Regression", we move the discussion to more powerful non-linear regression models: Gaussian Processes. Section "Quantile Regression" presents quantile regression approaches for the problem of non-constant variances. We compare the different approaches in section "Experiments" using different datasets, and we present the conclusions in section "Conclusions and Future Work".

2 Linear Models for Regression

We now review the traditional Multiple Linear Regression (MLR) model and its problems with heteroscedasticity. This method has been widely studied and it continues to be a statistical analysis standard in the majority of fields, mainly because of its interpretation simplicity and computational speed. The MLR model is defined as:

$$y_i = w_0 + w_1 x_{i1} + w_2 x_{i2} + \ldots + \varepsilon_i, \; i \in \{1, \ldots, N\},$$

or in matrix notation,

$$\begin{aligned} y &= f(X, w) + \varepsilon \\ &= X^T w + \varepsilon \end{aligned}$$

where $y = [y_1, y_2, \ldots, y_N]^T$, X is called the design matrix, a $N \times (D+1)$ matrix such that each column $j \in \{1, \ldots, D\}$ is $x_j = [1, x_{1j}, x_{2j}, \ldots, x_{Nj}]^T$, and $\varepsilon = [\varepsilon_1, \varepsilon_2, \ldots, \varepsilon_N]^T$. So, the MLR states that y is a weighted linear combination of the input x. Alongside a few more assumptions stated by the Gauss-Markov Theorem (Chipman 2011), ε is assumed to follow the standard Gaussian distribution with fixed variance σ^2, that is, $\varepsilon \sim \mathcal{N}(0, \sigma^2)$. Under this assumption, maximizing the (log) likelihood function w.r.t to the parameters w is equivalent to minimizing the least-squares function (Bishop 2006), that is

$$\hat{w} = arg\ max_w = \sum_{i=1}^{N} \log \mathcal{N}(y_i | x_i^T w, \sigma^2 I) = arg\ min_w \sum_{i=1}^{N} (y_i - w^T x_i)^2. \quad (1)$$

The solution of (1) is called the Ordinary Least-Squared (OLS) estimator and it is given by

$$\hat{w} = (XX^T)^{-1} Xy. \quad (2)$$

This estimator has the following properties for the mean and variance:

$$\mathbb{E}[\hat{w} | X] = \mathbb{E}\left[(XX^T)^{-1} Xy\right] = \mathbb{E}\left[(XX^T)^{-1} X(X^T w + \varepsilon)\right] \quad (3)$$

$$= \mathbb{E}\left[(X^T)^{-1} X^{-1} X(X^T w + \varepsilon)\right] = \mathbb{E}[w] = w$$

$$\mathbb{V}[\hat{w} | X] = (XX^T)^{-1} X \Phi X^T (XX^T)^{-1}, \quad (4)$$

where Φ is a diagonal matrix with $\Phi_{ii} = \mathbb{V}[\varepsilon_i] = \sigma^2$. Since the error term is homoscedastic, i.e., $\mathbb{V}[\varepsilon_i] = \sigma^2$ for all i, then $\Phi = \sigma^2 I$, where I is the identity matrix. So (4) can be simplified to

$$\mathbb{V}[\hat{w} | X] = \sigma^2 (XX^T)^{-1}.$$

Under the assumptions of the Gauss-Markov theorem, (2) is the best linear unbiased estimator of the covariance matrix of w. From (3) and (4) we can see that $\hat{w}$ is an unbiased estimator and its variance is influenced by the error variance. Notice that, under the presence of heteroscedastic residuals, $\hat{w}$ will stay untouched, in contrast to its variance, which will be a function of σ_i^2. Hence, the associated tests of significance and confidence intervals for the predictions will no longer be effective. In Osborne and Waters (2002) the reader can find a brief but practical guide to some of the most common linear model assumption violations, including the usual assumption of homoscedastic residuals.

To overcome this problem, White (1980) suggested a heteroscedasticity-consistent covariance matrix. In its simplest form, this amounts to setting Φ to be a $N \times N$ diagonal matrix whose elements are $\Phi_{ii} = \mathbb{E}(\varepsilon_i^2) = \mathbb{V}[\varepsilon_i]$. Here the problem is that we do not know the form of $\mathbb{V}[\varepsilon_i]$, so we need to estimate it. Under certain conditions, this can be achieved by constructing the consistent estimator

$$\hat{\mathbb{V}}[\varepsilon_i] = \frac{1}{N}\sum_{i=1}^{N}\varepsilon_i^2 X_i X_i^T.$$

However, ε_i^2 it is not observable. Fortunately, it can be estimated by

$$\hat{\varepsilon}_i^2 = y_i - x_i^T\hat{w},$$

thus leading us to consider the following estimator for $(X\Phi X^T/N)$,

$$\frac{1}{N}\sum_{i=1}^{N}\hat{\varepsilon}_i^2 x_i x_i^T \qquad (5)$$

White proved that (5) is a heteroscedastic-consistent covariance matrix estimator of the "asymptotic covariance matrix" given by (4). For an extensive detailed reading on the subject besides White (1980), we suggest MacKinnon and White (1983); MacKinnon (2012); Robinson (1987) and the more recent Long and Ervin (1998); Leslie et al. (2007); Zeileis and Wien (2004).

The MLR can be extended to incorporate linear combinations of non-linear functions of the input, thus allowing y to be a non-linear function of x, but still a linear function of the weights w. Despite this generalization, the demand for pure non-linear models continued.

Specially concerning the treatment of non-constant variance, many non-linear heteroscedastic models have been presented since the beginning of the 1980s, mainly applied to finance data (Engle 1982), where time-dependent volatility takes its upmost form. There were suggested models like the ARCH, ARMA-CH and its variations (Chen et al. 2011; Tsekeris and Stathopoulos 2006; Zhou et al. 2006) and, more recently, regarding regression by Gaussian Processes (Rasmussen and Williams 2006), there have been developments, with the introduction of the Heteroscedastic Gaussian Process concept (Goldberg et al. 1998; Gredilla and Titsias 2012; Kersting et al. 2007), which models the error term with another Gaussian Process dependent on the input (see section "Gaussian Process Regression").

Causes for heteroscedasticity vary from case to case, but most of them are related to the model misspecification, measurement errors, sub-population heterogeneity, noise level or it is just a natural intrinsic property of the dataset. To determine the application feasibility of heteroscedastic models, we should first test its presence. There are a number of statistical tests that allow, to some extent, checking the presence of heteroscedasticity in data. This includes Breuch-Pagan (1979), Cook-Weisberg (1983), White (1980) and Goldfeld-Quandt (1965) tests, among others. Another less statistically significant test, although very practical, is the visual inspection of the model residuals. The scatter-plot of the standardized residuals against the predicted values of y should lie uniformly around the zero horizontal line. Otherwise, heteroscedasticity should be suspected.

3 Gaussian Process Regression

Gaussian Processes (GP) propose a very different angle to regression than MLR, allowing for modeling non-linear relationships and online learning. Their non-parametric Bayesian nature gives them a sufficiently good flexibility to be used in a vast variety of regression and classification problems. In the following sections we will present the standard GP model and a heteroscedastic variant.

3.1 Standard GP

As seen in Rasmussen and Williams (2006), a Gaussian Process is a (possibly infinite) collection of random variables, $f = (f_t, t \in \mathcal{T})$, any finite number of which have a joint Gaussian distribution. $\mathcal{T}$ is a set of indexes and if it is finite, the GP reduces to joint Gaussian distribution. Each GP is completely determined by its mean and covariance (or kernel) functions. These functions, respectively $m_f(x)$ and $k_f(x, x')$, can be de defined as:

$$m_f(x) = \mathbb{E}[f(x)]$$
$$k_f(x, x') = \mathrm{Cov}(f(x), f(x')) = \mathbb{E}\big[(f(x) - m_f(x))(f(x') - m_f(x'))\big]. \quad (6)$$

Thus we can simply denote a Gaussian Process as $\mathcal{GP}(m_f(x), k_f(x, x'))$. In order to keep things simple, $m_f(x)$ is often considered zero (or constant). This is a common practice and it is not a very restrictive one, since the mean of the posterior process is not confined to the mean function of the GP.

The standard Gaussian Process regression is a Bayesian framework, which assumes a GP prior over functions, i.e.

$$y = f(x) + \varepsilon, \quad (7)$$

where $\varepsilon \sim \mathcal{N}(0, \sigma^2)$ and $f(x) \sim \mathcal{GP}(m_f(x), k_f(x, x'))$.

Assuming $m_f(x) = 0$, the prior over the latent function values is then given by:

$$p(f | x_1, x_2, \ldots, x_n) = \mathcal{N}(0, K_f),$$

where $f = (f_1, f_2, \ldots, f_n)^T$, $f_i = f(x_i)$ and K_f is the covariance matrix, with its elements given by $[K_f]_{ij} = k_f(x_i, x_j)$. Many forms for the covariance functions can considered, each one of them typically having a number of free hyper-parameters, which we refer to as θ. For a particular application, we need to fix, a priori, a family of covariance functions and optimize the kernel w.r.t. the hyper-parameters. One way of setting the hyper-parameters is to maximize the marginal likelihood given by:

$$\log p(y|X, \theta) = -\frac{1}{2} y^T K_y^{-1} y - \frac{1}{2} \log|K_y| - \frac{N}{2} \log(2\pi),$$

where $K_y = K_f + \sigma^2 I$ and K_f are, respectively, the covariance matrix for the noisy targets y and noise-free latent variable f.

Having set the covariance function and its corresponding hyper-parameters, the conditional distribution of a new test point x_* is given by:

$$f_*|X, y, x_* \sim \mathcal{N}(\mathbb{E}[f_*], \mathbb{V}[f_*]),$$

with

$$\mathbb{E}[f_*] := \mathbb{E}[f_*|X, y, x_*] = k_{f_*}^T [K_f + \sigma^2 I]^{-1} y,$$

$$\mathbb{V}[f_*] := k_{f_{**}}^T - k_{f_*}^T [K_f + \sigma^2 I]^{-1} k_{f*},$$

where we introduced the notation $k_{f_*} = k_f(X, x_*)$ and $k_{f_{**}} = k_f(x_*, x_*)$.

Thus we can construct a $(1 - \alpha)\%$ Confidence Interval (CI) for y_* as follows:

$$k_{f_*}^T [K_f + \sigma^2 I]^{-1} y \pm z_{\frac{\alpha}{2}} \sqrt{k_{f_{**}}^T - k_{f_*}^T [K_f + \sigma^2 I]^{-1} k_{f*}},$$

where $z_{\frac{\alpha}{2}}$ is the $\frac{\alpha}{2}$th order quantile of $\mathcal{N}(0, 1)$ and $\alpha \in \,]0, 1[$.

This allows us to construct confidence intervals under the constant variance assumption. In the next section we extend this model in order to relax this assumption.

3.2 Heteroscedastic GP

In this section we will closely follow the work of Gredilla and Titsias (2012). As seen in the previous section, the standard GPs assume a constant variance, σ^2, throughout the input, x. We will now relax this assumption.

To define a Heteroscedastic GP (HGP), besides placing a GP prior on $f(x)$ as in (7), we also place a GP prior on the error term, so that we have:

$$y = f(x) + \varepsilon,$$

with $f(x) \sim \mathcal{GP}(0, k_f(x, x'))$ and $\varepsilon \sim \mathcal{N}(0, r(x))$, where $r(x)$ is an unknown function. To ensure positivity and without losing generality, we can define $r(x) = e^{g(x)}$, where

$$g(x) \sim \mathcal{GP}(\mu_0, k_g(x, x')).$$

After fixing both covariance functions, k_f and k_g, the HGP is fully specified and depends only on its hyper-parameters. Unfortunately, exact inference in the HGP is no longer tractable. To overcome this issue, the authors propose a variational inference algorithm which establishes a Marginalized Variational (MV) bound, for the likelihood function, given by:

$$F(\mu, \Sigma) = log \mathcal{N}(y|0, K_f + R) - \frac{1}{4} tr(\Sigma) - KL(\mathcal{N}(g|\mu, \Sigma) || \mathcal{N}(g|\mu_0 1, K_g)),$$

where R is a diagonal matrix such that $R_{ii} = e^{\mu_i - \frac{[\Sigma]_{ii}}{2}}$, K_g is the covariance matrix resulting from the evaluation of $k_g(x, x')$, μ and Σ are the parameters from the variational distribution $q(g) = \mathcal{N}(g|\mu, \Sigma)$, and KL is the Kullback-Leibler divergence. The hyper-parameters of this Variational Heteroscedastic GP (VHGP) are then learned by setting the following stationary equations

$$\frac{\partial F(\mu, \Sigma)}{\partial \mu} = 0, \quad \frac{\partial F(\mu, \Sigma)}{\partial \Sigma} = 0.$$

The solution for this system is a local or global maximum and it is given by:

$$\mu = K_g \left(\Lambda - \frac{1}{2} I \right) + \mu_0 1, \quad \Sigma^{-1} = K_g^{-1} + \Lambda,$$

for some positive semidefinite diagonal matrix. We can see that both μ and Σ^{-1} depend on Λ, letting us rewrite the MV as a function of Λ,

$$F(\mu(\Lambda), \Sigma(\Lambda)) = F(\Lambda),$$

which needs to be maximized w.r.t. to the N variational parameters in Σ. At the same time, it is possible to maximize F w.r.t to the model hyper-parameters θ. After learning the variational parameters and model hyper-parameters, (Λ^*, θ^*), by maximizing F, we need to approximate the predictive distribution $p(y_*|x_*, X, y)$ for a new test input-output point (x_*, y_*), which again is intractable. So, within the variational approach, the predictive distribution of y_* can be approximated by:

$$q(y_*) = \int \mathcal{N}(y_*|a_*, c_*^2 + e^{g^*}) \mathcal{N}(g_*|\mu_*, \sigma_*^2) dg_* \qquad (9)$$

with $a_* = k_{f*}^T (K_f + R)^{-1} y$, $c_*^2 = k_{f**} - k_{f*}^T (K_f + R)^{-1} k_{f*}$, $\mu_* = k_{g*}^T (\Lambda - \frac{1}{2}I)1 + \mu_0$ and $\sigma_*^2 = k_{g**} - k_{g*}^T (K_g + \Lambda)^{-1} k_{g*}$. For a more detailed derivation, please refer to Gredilla and Titsias (2012). For an introduction on the dense theory of the variational approximation framework, we recommend Tzikas et al. (2008) and Fox and Roberts (2012).

Although (9) is not analytically tractable, its mean and variance can be computed exactly as:

$$\mathbb{E}_q[y_*|x_*,X] = a_*$$

$$\mathbb{V}_q[y_*|x_*,X] = c_*^2 + e^{\mu_* + \frac{\sigma_*^2}{2}}.$$

We are then able to construct the following $(1-\alpha)\%$ heteroscedasticity-consistent CI for y_*,

$$a_* \pm z_{\frac{\alpha}{2}} \sqrt{c_*^2 + e^{\mu_* + \frac{\sigma_*^2}{2}}},$$

with $\alpha \in \,]0,1[$.

3.3 GPs for Big Data

The biggest drawback of Gaussian Processes is that they suffer from computational intractability issues for very large datasets. As seen in Rasmussen and Williams (2006), typically the complexity scales around $\mathcal{O}(N^3)$ so, for $N > 10000$, both storing and inverting the covariance matrix proves to be a prohibitive task on most modern machines. More recently, and specifically concerning the treatment of Big Data, Hensman et al. (2013) proposed a stochastic variational inference algorithm that is capable of handling millions of data points. By variationally decomposing the GP and making it only dependent on a specific set of relevant M inducing points, the authors lowered the complexity to $\mathcal{O}(M^3)$. Also concerning the application of GPs to large datasets Snelson and Ghahramani (2007) suggested a new sparse approximation resulting from the combination of both global and local approaches. There is a vast literature regarding GP approximations. For an overview on some of these approximation methods we suggest Quinonero-Candela et al. (2007).

In this work we address the challenge posed by Big Data from the variance treatment point-of-view. The non-constant variance which is usually associated to the presence of a great amount of unstructured noise is also an intrinsic characteristic of Big Data that should be taken into account and treated accordingly with the right heteroscedastic models

4 Quantile Regression

Since, in some cases, integrating a heteroscedastic component in the model may not be practical due to overwhelming complexity or computational intractability, an alternative is to do post-model analysis. In this case, we analyze the output performance of a certain prediction process, whose internal details are not necessarily known or understood, and model its observed error. This is the approach proposed by Taylor and Bunn (1999) and Lee and Scholtes (2014), where the authors focus on time series prediction. In Pereira et al. (2014), we extended this approach with more complex Quantile Regression (QR) models applied to general predictors. Our algorithm treats the original prediction model as a black box and generates functions for the lower and upper quantiles, respectively providing the lower and upper bounds of the prediction interval. Despite this flexibility, being downstream to the model itself comes with a cost: this approach will hardly correct earlier modeling mistakes and will only uncover such limitations (by presenting very wide prediction bounds).

In contrast to least-squared based regression approaches, QR fits the regression parameters to the quantiles, instead of fitting them to the conditional mean. By doing so, this type of regression tends to be more robust against outliers and does not depend on the common assumptions of normality or symmetry of the error. It is however worth mentioning that this approach can suffer from inconsistency as the resulting quantiles may cross. In practical terms this is not usually a problem, although it is important to be aware of it. Following Koenker and Hallock (2001) and Fargas et al. (2014), we can denote the conditional quantile function as:

$$Q_y(\tau|X),$$

where τ is the quantile order we want to estimate.

In its simplest approach, if Q_y has a parametric structure, i.e., if $Q_y = \xi(X, w)$, to estimate the desired quantile τ, we proceed to the following optimization:

$$min_w \sum_{i=1}^{N} \rho_\tau(y_i - \xi(X, w)),$$

where $\rho(.)$ is the tilted loss function represented in Fig. 1.

As we can see, Q_y can have a variety of functional structures, either parametric or non-parametric. In particular, we can define the Gaussian Processes for Quantile Regression (GPQR), by placing a GP prior over Q_y, that is,

$$Q_y(\tau|X) \sim \mathcal{GP}(0, k(X, X')),$$

where k is the covariance function. Contrary to the standard QR, the GPQR is, of course, a Bayesian approach to the quantile estimation.

Fig. 1 Tilted loss function.
Source: Koenker and Hallock (2001)

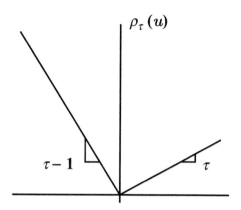

In this paper, we will use the meta-model perspective suggested by Pereira et al. (2014) and Fargas et al. (2014). For that, consider a random black box model that generates predictions, $\hat{y}_i$. We will focus on the error of such predictions in terms of the deviations $d_i = y_i - \hat{y}_i$, where y_i is the series of the observed values. The aim of this model is to associate lower and upper bounds for the predicted target value, i.e., to determine the functions $d^{\tau_-}(X)$ and $d^{\tau_+}(X)$ which provide, respectively, the lower and upper quantiles, τ_- and τ_+, defined a priori. This explains why the QR is performed on the residuals instead of directly on the data itself, leading us to the construction of another type of intervals, as it will be forwardly described.

When seeking for the quantile τ, the loss of choosing d^τ, instead of d, is quantified by:

$$\rho_\tau(u) = u(\tau - 1_{<0}),$$

with $u = d - d^\tau$, which is equivalent to

$$\rho(d - d^\tau) = \begin{cases} \tau(d^\tau - d), & d \geq d^\tau \\ (1-\tau)(d - d^\tau), & d < d^\tau. \end{cases}$$

Within the GPQR approach, the posterior distribution of the quantile function Q_y, is:

$$p(Q_y|\mathcal{D}, \theta) = \frac{1}{Z} p(\mathcal{D}|Q_y, \theta) p(Q_y|\theta),$$

where θ is the set of the covariance function hyper-parameters, Z a normalization factor and $\mathcal{D} = (X, y)$ the training set.

The hyper-parameters are estimated by maximizing the likelihood function,

$$arg\ max_\theta p(Q_y|\mathcal{D},\theta),$$

which is proved to be equivalent to minimizing the tilted loss function. Hence, the likelihood becomes:

$$p(\mathcal{D}|Q_y) = \left(\frac{\tau(1-\tau)}{\sigma}\right)^N e^{\left[-\sum_{i=1}^{N}\frac{\mu_i}{\sigma}(\tau-1_{\mu_i<0})\right]},$$

where σ is the standard deviation of the Asymmetric Laplace distribution and $u_i = d_i - d_i^\tau$.

With this meta-model approach, we introduced a new kind of interval: the Prediction Interval (PI). Although the differences between both CI and PI are subtle, it is important to distinguish the two concepts. If on the one hand, a $(1-\alpha)$% CI is expected to contain a certain population parameter (e.g. mean, variance) at least $(1-\alpha)$% of the times, a prediction interval should cover the next predicted value, at least, $(1-\alpha)$% of the times.

5 Experiments

In this section we will compare the performances of the GP, VHGP and GPQR approaches to the problem of the heteroscedasticity, using two one-dimensional datasets. From now on we will refer to the GPQR only by QR. More than assessing the pointwise quality of the predictions using the standard performance measures, we are mostly interested in evaluating how accurately the constructed CIs and QR prediction intervals handled the volatility in the data.

5.1 Performance Measures

For the single-point predictions we will use the following standard and well accepted measures: RMSE (Root Mean Squared Error), MAE (Mean Absolute Error), RAE (Root Absolute Error), RRSE (Root Relative Squared Error) and COR (Pearson's linear correlation coefficient)

For the interval quality evaluation we will consider some of the performance measures suggested in Pereira et al. (2014), although with slight modifications in order to extend their applicability to both kind of intervals, CI and PI. Let l_i and u_i be the series of the lower and upper bounds of a confidence/prediction interval for $\hat{y}_i$, respectively. Consider also the series of the real target values y_i. Then we can de ne the following measures:

- Interval Coverage Probability (ICP),

$$\frac{1}{N}\sum_{i=1}^{N} 1_{y_i \notin [l_i, u_i]},$$

- Relative Mean Interval Length (RMIL),

$$\frac{1}{N}\sum_{i=1}^{N} \frac{(u_i - l_i)}{|y_i - \widehat{y}_i|},$$

- Coverage-length-based Criterion (CLC2)

$$e^{(-RMIL(ICP-\mu))},$$

with μ a controlling parameter.

In our case, for both CI and QR bounds, ICP should be as close to $(1-\alpha)$ = $(\tau^+ - \tau^-) = 0.95$ as possible. RMIL intrinsically expresses the idea that, for a large observed error, we need to allow large intervals, so that the predicted value can be covered. It can be seen as a weighted average relative to the actual observed error. For CLC2, we can see that μ represents the ideal value for ICP. When $\mu = $ ICP, CLC2 reaches its optimal value, 1. In the same exponent, RMIL can be regarded as an amplification factor of the "importance" of being close to the desired ICP. CLC2 is a simplification of CLC Khosravi et al. (2011) that replaced an arbitrary weight, η, for RMIL as just described.

5.2 Datasets

We used the motorcycle dataset from Silverman (1985) composed of 133 data points and a synthetically generated toyset consisting of 1000 data points exhibiting a linear relationship with local instant volatility. For each dataset we ran three models: GP, VHGP and QR. The QR model used as input the predictions from the two others. For both GP and VHGP we generated predictions in a tenfold cross-validation fashion, using the code freely available from Rasmussen and Williams (2006) and Gredilla and Titsias (2012). For the QR, we used the code available from Boukouvalas et al. (2012). We fixed $\alpha = 0.05$, $\tau^- = 0.025$ and $\tau^- = 0.975$ so that we have equivalent interval coverages.

5.3 Results

Table 1 summarizes the overall obtained results. The variance evolution of the GP and VHGP are plotted in Figs. 2 and 3. Figures 4, 5, 6, and 7 compare the three different approaches within each dataset.

Table 1 Performance measures results table

Model	Dataset	ICP	RMIL	CLC2	RMSE	MAE	RAE	RRSE	COR
GP	Motorcycle	0.933	29.508	1.684	23.343	17.525	0.454	0.485	0.875
VHGP	Motorcycle	0.955	14.562	0.931	23.666	17.725	0.459	0.491	0.871
QR over GP	Motorcycle	0.955	12.287	0.941	n.a	n.a	n.a	n.a	n.a
QR over VHPG	Motorcycle	0.947	8.526	1.023	n.a	n.a	n.a	n.a	n.a
GP	Toyset	0.931	329.750	525.978	253.431	152.363	0.202	0.285	0.958
VHGP	Toyset	0.940	88.707	2.428	253.656	149.791	0.198	0.286	0.959
QR over GP	Toyset	0.910	85.745	30.870	n.a	n.a	n.a	n.a	n.a
QR over VHGP	Toyset	0.956	19.762	0.888	n.a	n.a	n.a	n.a	n.a

A Review of Heteroscedasticity Treatment with Gaussian Processes and... 155

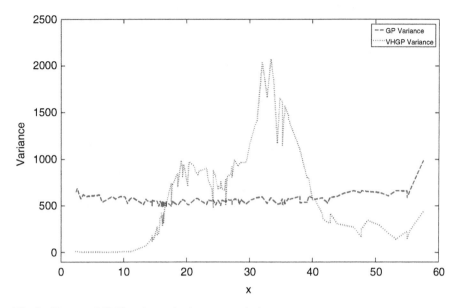

Fig. 2 GP versus VHGP variances in the motorcycle dataset

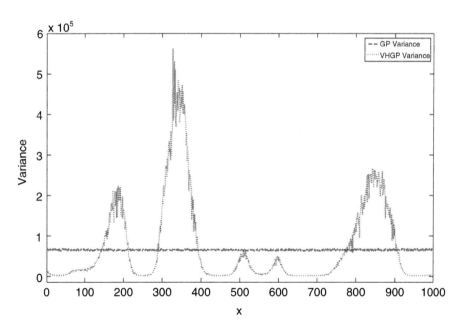

Fig. 3 GP versus VHGP variances in the toyset

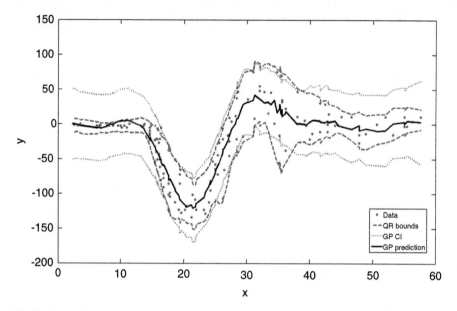

Fig. 4 Comparison between GP and QR over GP predictions for the motorcycle dataset

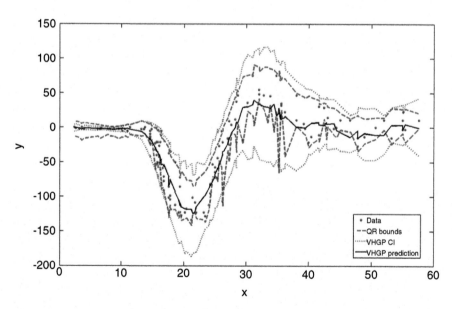

Fig. 5 Comparison between VHGP and QR over VHGP predictions for the motorcycle dataset

In terms of interval coverage, we can see that all models performed well, although the VHGP and QR had an ICP slightly closer to the ideal value, except for the toyset. It is clear that the homoscedastic model (GP) was able to reach such

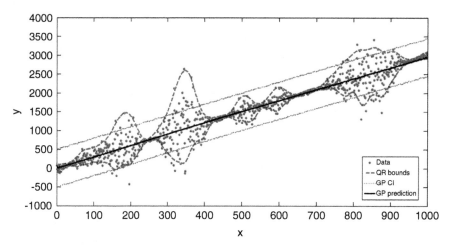

Fig. 6 Comparison between GP and QR over GP predictions for the toyset

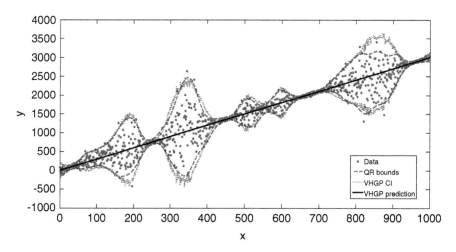

Fig. 7 Comparison between VHGP and QR over VHGP predictions for the toyset

ICP values because of its large, but constant, variance (see Figs. 2 and 3). However, ICP cannot be regarded independently from the others. High values of ICP do not necessarily imply good intervals, they merely mean that the constant variance assumed for the model is sufficiently large for the CIs to cover almost every data point. So, the GP did achieve, in fact, reasonable good marks for ICP, but it did it so at the expense of larger (and constant) interval ranges. As a result these values are best considered in combination with RMIL, which relates the interval bounds and the actual observed error. Unsurprisingly, the RMIL values for the GP are the larger ones, which means that, on average, the considered intervals were larger than required to cover the data or too narrow to handle its volatility. On the other hand, the heteroscedasticity-consistent approaches, VHGP and QR, were able to dynamically adapt their error variances to the data volatility, leading to tighter intervals that more accurately reflect the uncertainty in the data, as we can see from the values of RMIL and CLC2.

The results obtained for the toyset are consistent with and show a similar pattern to those of the motorcycle dataset. The values of RMIL and CLC2 obtained for the VHGP show a great improvement over the standard homoscedastic GP as observed in the toyset and as would be expected given the heteroscedasticity present in the data. Again, as observed in the toyset, the addition of a quantile regression metamodel approach to the standard homoscedastic GP results in much better performance in terms of RMIL and CLC2. The RMIL in particular is now comparable to the value obtained using VHGP. There is some reduction in ICP which is undesirable but we still obtain a high value, albeit slightly less than desired. The QR-VHGP combination again exhibits the best overall performance in comparison to the three previous methods.

6 Conclusions and Future Work

In this paper we have explored the problem of heteroscedasticity for regression problems. Using data with this characteristic property, both real and synthetic, we have compared the performance of a GP model which has been augmented to handle heteroscedasticity with that of a standard GP. The results obtained highlight the improved performance of this more complex model and provide motivation for its use in such settings. Given the current popularity of GPs as an advanced machine learning tool for Big Data problems, this is an important result that demonstrates that while the GP framework is extremely flexible and powerful, the problem of heteroscedasticity is an issue that must be considered and handled using appropriate tools. With this in mind we also studied the use of quantile regression as a post processing means of taking heteroscedasticity into account even when we cannot or do not want to make structural changes in our model.

The results show that the post-processing approach can improve the results of a homoscedastic model to the point of being comparable with a fully heteroscedastic model. This means that in some cases we may have proper alternatives that do not involve having a more complex model to handle heteroscedastic error terms. Since post-processing approaches can be faster, easier to implement and depend on very few weak assumptions, they constitute a promising area for Big Data applications that warrants further exploration and will be a direction of future work. Furthermore, with QR we can now have prediction intervals for any kind of machine learning algorithms, even non-probabilistic ones.

We showed that accounting for heteroscedasticity can greatly improve our confidence over traditional pointwise predictions by generating heteroscedasticity-consistent confidence and prediction intervals using different approaches. As mentioned in this paper, within the transportation system context and from the user's perspective, it is often of extreme importance to have precise estimates which can incorporate the instant volatility that can occur at any moment. Hence, as a future line of work, we will apply these procedures to real data such as public transport travel time data.

References

Bishop CM (2006) Pattern recognition and machine learning. Springer, New York
Boukouvalas A, Barillec R, Cornford D (2012) Gaussian process quantile regression using expectation propagation. In: Proceedings of the 29th international conference on machine learning (ICML-12), pp 1695–1702
Breusch TS, Pagan AR (1979) A simple test for heteroscedasticity and random coefficient variation. Econometrica 47(5):1287–1294
Chen C, Hu J, Meng T, Zhang Y (2011) Short-time traffic flow prediction with ARIMA-GARCH model. In: Intelligent vehicles symposium (IV), IEEE, pp 607–612
Chipman JS (2011) International encyclopedia of statistical science. Springer, Berlin, pp 577–582
Cook RD, Weisberg S (1983) Diagnostics for heteroscedasticity in regression. Biometrika 70(1):1–10
Engle RF (1982) Autoregressive conditional heteroscedasticity with estimates of the variance of United Kingdom inflation. Econometrica 50(4):987–1007
Fargas JA, Ben-Akiva ME, Pereira FC (2014) Prediction interval modeling using gaussian process quantile regression. Master's Thesis, MIT, pp 1–65
Fox CW, Roberts SJ (2012) A tutorial on variational Bayesian inference. Artif Intell Rev 38(2):85–95
Goldberg P, Williams C, Bishop C (1998) Regression with input-dependent noise: a Gaussian process treatment. Adv Neural Inf Process Syst 10:493–499
Goldfeld SM, Quandt RE (1965) Some tests for homoscedasticity. J Am Stat Assoc 60:539–547
Gredilla LG, Titsias MK (2012) Variational heteroscedastic Gaussian process regression. In: 28th international conference on machine learning
Hensman J, Fusi N, Lawrence ND (2013) Gaussian processes for big data. In: Proceedings of the 29th conference annual conference on uncertainty in artificial intelligence (UAI-13), pp 282–290
Kersting K, Plagemann C, Pfaff P, Burgard W (2007) Most likely heteroscedas-tic Gaussian process regression. In: Proceedings of the International Machine Learning Society, pp 393–400
Khosravi A, Mazloumi E, Nahavandi S, Creighton D, Van Lint JWC (2011) Prediction intervals to account for uncertainties in travel time prediction. IEEE Trans Intell Transp Syst 12(2):537–547
Koenker R, Hallock KF (2001) Quantile regression. J Econ Perspect 15(4):143–156
Lee YS, Scholtes S (2014) Empirical prediction intervals revisited. Int J Forecast 30(2):217–234
Leslie DS, Kohn R, Nott DJ (2007) A general approach to heteroscedastic linear regression. Stat Comput 17(2):131–146
Long JS, Ervin LH (1998) Correcting for heteroscedasticity with heteroscedasticity-consistent standard errors in the linear regression model: small sample considerations, Working Paper, Department of Statistics, Indiana University
MacKinnon JG (2012) Thirty years of heteroskedasticity-robust inference, Working Papers, Queen's University, Department of Economics
MacKinnon JG, White H (1983) Some heteroskedasticity consistent covariance matrix estimators with improved finite sample properties, Working Papers, Queen's University, Department of Economics

Osborne J, Waters E (2002) Four assumptions of multiple regression that researchers should always test. Pract Assess Res Eval 8(2):1–9

Pereira FC, Antoniou C, Fargas C, Ben-Akiva M (2014) A meta-model for estimating error bounds in real-traffic prediction systems. IEEE Trans Intell Trans Syst 15:1–13

Quinonero-Candela J, Rasmussen CE, Williams CKI (2007) Approximation methods for Gaussian process regression, Large-scale kernel machines, pp 203–223

Rasmussen CE, Williams C (2006) Gaussian processes for machine learning. MIT Press, Cambridge, MA

Robinson PM (1987) Asymptotically efficient estimation in the presence of heteroskedasticity of unknown form. Econometrica 55(4):875–891

Silverman BW (1985) Some aspect of the spline smoothing approach to non-parametric regression curve fitting. J R Stat Soc 47(1):1–52

Snelson E, Ghahramani Z (2007) Local and global sparse Gaussian process approximations. In: International conference on artificial intelligence and statistics, pp 524–531

Taylor JW, Bunn DW (1999) A quantile regression approach to generating prediction intervals. Manag Sci 45(2):225–237

Tsekeris T, Stathopoulos A (2006) Real-time traffic volatility forecasting in urban arterial networks. Transp Res Rec 1964:146–156

Tzikas DG, Likas AC, Galatsanos NP (2008) The variational approximation for Bayesian inference. IEEE Signal Process Mag 25(6):131–146

White H (1980) A heteroskedasticity-consistent covariance matrix estimator and a direct test for heteroskedasticity. Econometrica 48(4):817–838

Zeileis A, Wien W (2004) Econometric computing with HC and HAC covariance matrix estimators. J Stat Softw 11(10):1–17

Zhou B, He D, Sun Z (2006) Traffic predictability based on ARIMA/GARCH model. In: 2nd conference on next generation internet design and engineering, pp 207–214

Part III
Changing Organizational and Educational Perspectives with Urban Big Data

Urban Informatics: Critical Data and Technology Considerations

Rashmi Krishnamurthy, Kendra L. Smith, and Kevin C. Desouza

Abstract Cities around the world are investing significant resources toward making themselves smarter. In most cases, investments focus on leveraging data through emerging technologies that enable more real-time, automated, predictive, and intelligent decision-making by agents (humans) and objects (devices) within the city. Increasing the connectivity between the various systems and sub-systems of the city through integrative data and information management is also a critical undertaking towards making cities more intelligent. In this chapter, we frame cities as platforms. Specifically, we focus on how data and technology management is critical to the functioning of a city as an agile, adaptable, and scalable platform. The objective of this chapter is to raise your awareness of critical data and technology considerations that still need to be addressed if we are to realize the full potential of urban informatics.

Keywords Smart cities • Data • Technology • Urban informatics • Open data • Big data • Mobile data • Intelligent cities • Platform

1 Introduction

Urban planners around the world are investing significant resources towards making their city's infrastructure 'smarter' (or 'more intelligent') through leveraging data. By 2020, $400 billion a year will be invested in building smart cities (Marr 2015); and it is estimated that by 2016, smart cities will become a $39.5 billion market (Stewart 2014) and a $1 trillion market by 2020 (Perlroth 2015). Leveraging data and information—the informatics element—is critical towards creating smarter cities. Urban informatics is "the study, design, and practice of urban experiences across different urban contexts that are created by new opportunities for real-time, ubiquitous technology and the

---

R. Krishnamurthy (✉) • K.C. Desouza
School of Public Affairs, Arizona State University, Phoenix, AZ, USA
e-mail: rkrish19@asu.edu; kev.desouza@gmail.com

K.L. Smith
Morrison Institute for Public Policy, Arizona State University, Phoenix, AZ, USA
e-mail: klsmit40@asu.edu

augmentation that mediates the physical and digital layers of people, networks, and urban infrastructures" (Foth et al. 2011, p. 4).

A focus on urban informatics requires us to consider: how can the planning, design, governance, and management of cities be advanced through the innovative use of information technologies (ITs). Towards this end, it is important to focus on both the information and the technical elements of cities. Cities generate and disseminate large quantities of data and information. While most of the data and information flows through technical systems, the social (human) component is also critical. Hollands (2008) argues that smart cities are 'wired cities' that include both human and technical infrastructures. In this regard, employing ITs alone is not sufficient to transform cities; rather, how humans (social) leverage technologies for managing and governing in the urban sphere makes cities smarter (Kitchin 2014).

The concept of smart cities promotes the idea of creating a city where diverse stakeholders leverage interconnected technologies to create innovate solutions and address complex urban problems. Today, anyone can participate in the design and building of solutions for cities. For example, Deck5, a software company, created the 312 app that helps people new to Chicago navigate the city. This app uses GPS and open source data from Chicago's city portal to allow users to obtain information about city neighborhoods; additionally, and more importantly, the app was created to promote culture and engagement by letting people know interesting things that are happening in other neighborhoods (Pollock 2013).

Technology enthusiasm has opened up new avenues for citizens and organizations to actively participate in the transformation of their urban spaces. However, as we know, technology is no panacea. Several challenges remain in our quest towards having smarter cities. First, from a fabric and structural perspective, cities need to be open and capable to integrating multiple solutions. This is no easy feat to accomplish as issues of standardization, alignment of interests, and even compatibility of solutions play major roles here. Second, cities are comprised of multiple systems (and sub-systems) that are often not optimally integrated. Given that there are significant legacy issues that need to be managed here (i.e. historic investments made into the design and operation of these systems) and the pace of technological innovations, cities are often faced with a conundrum on whether to completely abandon existing solutions and designing with a clean slate or trying to retrofit and tweak old systems. Third, new technology-enabled economic models are being introduced into cities that fundamentally impact how assets and infrastructure within cities are utilized. Consider the impact of new technologies such as sharing economy mobile apps like Airbnb and Uber that modify longstanding, traditional government roles such as tax collecting, regulating, and protecting citizens (Desouza et al. 2015). Cities have to continuously adapt their governance and business models. However, most cities have age-old bureaucracies that need to be navigated which stifle rapid innovation. Fourth, we must always remember that significant access issues and variances in how technology is used by citizens need to be accounted for when designing solutions for urban management.

In this chapter, we bring to the forefront several issues that lie at the intersection of data and technology as they pertain to urban informatics. To do so, we frame cities as platforms that facilitate social and technical interactions as well as constrain people, technology, and processes. City as a platform is comprised of technical, social, and socio-technical spheres. The technical sphere includes both physical and virtual infrastructures that enable smooth functioning of city's operations. The social sphere consists of individuals, organizations, and communities that leverage technologies to communicate and connect with each other and the city. The interaction between technical and social spheres produces a socio-technical sphere, which includes policies, procedure, and norms that govern the city and its habitants. The glue that holds these components together and enables integration and coordination is data and information management. The effective and efficient management of data and information is not only critical to ensure that each of the components operate optimally but also ensures that the overall system—the city—achieves its objectives of being livable, sustainable, and resilient (Desouza 2014b). A platform view allows us to study the interdependencies that exist between the technical and social dimensions of a city. Additionally, through this perspective, we can look at how data and technology, two critical ingredients, impact components of a city both individually and collectively.

The objective of this chapter is to *raise awareness of critical data and technology considerations that still need to be addressed if we are to realize the full potential of urban informatics*. We will draw on over 3 years of research in smart cities and urban informatics.[1] During these projects, we collected and analyzed both primary (over 45 case studies and 60 interviews) and secondary (content analysis of policy documents and examination of over 70 technology solutions that span mobile apps, online crowdsourcing platforms, sensors, analytical and visualization technologies) data.

The rest of this chapter is organized as follows. We begin by outlining the growing potential of emerging technologies as it pertains to transforming how we design, experience, and live in urban environments. Then, we discuss the three data elements—open, big, and mobile data—that are powering smart cities and their challenges. Next, the concept of the city as a platform is explained as a constant interaction between interconnected data, technical, and social systems. Next, a discussion on the critical issues facing urban informatics will focus on challenges of data, inequality, governance, future technology, project management, privacy, and innovation. Finally, in light of the critical issues raised, recommendations for urban planners will be made to help planners unlock the value of smart cities in ways that suit their city.

---

[1] For more information, see—Desouza (2012a, b, 2014a, b, c); Desouza and Bhagwatwar (2012a, b, 2014); Desouza and Flanery (2013); Desouza and Schilling (2012); Desouza and Simons (2014); Desouza and Smith (2014a); Desouza et al. (2014); Desouza et al. (2015)

2 Emerging Technologies

Emerging technologies will increasingly connect cities and citizens through more access and more data. For instance, Facebook recently announced that they are piloting a drone that will provide Internet access around the world, even in the most remote pockets of the world (Goel and Hardy 2015). Through greater access, more data will be generated and collected on citizens' actions, by both private and public organizations. The joining of access and data will enable cities to become 'smarter' and even more interconnected. In addition, the rise of automation, greater real-time data collection, and predictive analytic capabilities, are significantly modifying who are making decisions and how decisions are being made.

Let us consider one emerging technology, automated vehicles (AVs). It has been predicted by car and industry leaders that AVs will enter the market in the next 10 years (Mack 2014; Cheng 2014). These vehicles are totally driven by monitors, sensors, and an intelligent system that makes decisions without human input on the speed to drive, turns to make, and when to stop; all actions that, in a vehicle, can be a matter of life and death. For instance, in an AV that Google has been testing, a scenario showed the power of removing human decision-making from driving. At a light, three cars—two human driven cars and one AV—were preparing to proceed through a green light when a cyclist coming from another direction ignored and tried to speed through their red light (Gardner 2015). The AV's sensors picked up the cyclists speed and did not go through the green light while the two human drivers proceeded through the intersection forcing the cyclist to swerve. AVs are designed to override human decisions, and more importantly, human mistakes. AVs as a technology will not only impact how we traverse our physical spaces within cities, but will also impact the social and economic dimensions as well. With the adoption of AVs, the number of cars on the road is likely to be reduced significantly because, due to the sharing economy, fewer people will own cars and opt for more economical options such as ridesharing. Morgan Stanley estimated that AVs will help the U.S. experience $1.3 trillion in annual savings by decreasing accident and fuel consumption cost while increasing productivity (Morgan Stanley 2013). PricewaterhouseCoopers predicts that when AVs are commonplace in the U.S., car ownership will reduce from 245 million to 2.4 million (PWC 2013).

Further, vehicle-to-vehicle (V2V) and vehicle-to-infrastructure (V2I) technologies offer the capacity for vehicles to communicate with other vehicles and infrastructure to increase safety, improve mobility, and lessen environmental degradation (U.S. Department of Transportation n.d.). For instance, V2I capabilities could alert drivers to crashes or closed roads while redirecting them to other paths, reducing traffic congestion, lowering carbon emissions and increasing fuel-efficiency. It could in turn, also lead to a reduction in public services such as police and fire and rescue being required to manage accidents and traffic. Since excess fuel costs and road congestion cost upwards of $100 billion in the U.S. annually, this type of efficiency creates important benefits for cities (Dobbs et al. 2013). In Glasgow, Scotland intelligent streetlights, networks of sensors and CCTV cameras have been installed around the city to increase wireless monitoring and connectivity (Macdonell 2015). Data from road sensors feed into analytical engines that adjust traffic lights to reduce bottlenecks. Further, detailed maps of the city with outlined areas for walking tours and cycling are made available. Glasgow's smart city technology is even used by citizens for social services. As a city whose citizens have low life expectancy, a 2011 study found that Glasgow citizens are 30 % more likely to die young than their counterparts in Manchester and Liverpool (Glasgow Centre for Population Health 2011). 60 % of the deaths of those that will die young in Glasgow are due to excesses in alcohol, violence, and suicide. An example of Smart city connectivity helping with this is by allowing recovering alcoholics access to city smart maps, navigating them away from harmful places like bars and clubs.

Emerging technologies are fundamentally altering the economic models within cities (Desouza et al. 2015). For instance, peer-2-peer apps and on-demand services are designed to exchange goods directly without an intermediary. The on-demand economy allows consumers to place demands for goods and/or services and a freelancer immediately fulfills that demand. For instance, TaskRabbit is an on-demand app that allows consumers to input small tasks (i.e. house cleaning, handyman, delivery, parties and events) they need completed, the price they're willing to pay for the service, and freelancers bid to complete the job. Peer-2-peer and on-demand technologies have already spurred a new direction in employment. A recent report revealed that 53 million Americans are working as freelancers (Horowitz and Rosati 2014). Many of these freelancers are people that use online platforms such as Uber, TaskRabbit, and Amazon Mechanical Turk to find new contracts and tasks. However, despite their benefits, these jobs lack employee benefits such as insurance and retirement plans, which leaves urban planner's to figure out how to operate amidst these new realities (Howard 2015).

Concerns are plentiful as cities are becoming more connected and guided by emerging technologies. Issues of dependency, security, safety, and privacy are non-trivial. In 2015, it was discovered that several of Chrysler's models of car were vulnerable to hacking through the Internet. That is, hackers could cut brakes, shut down a car's engine, and even drive a car off the road remotely (Pagliery 2015). Hackers had this access to the vehicles through a wireless service called Uconnect that connects the cars with the Sprint network. This happened because Chrysler unwittingly left a communication channel open that granted outside access to car controls. As a result, Chrysler has recalled 1.4 million cars and trucks to update software protection to prevent hacking.

Additionally, emerging technologies are released and consumed so quickly by citizens that governance and regulatory procedures have not kept pace with them. One such technology is the use of drones or unmanned aerial vehicles (UAVs). Traditionally, drones have been used for military and special operations for such purposes as reconnaissance, armed attacks, and research. Now, drones are available to businesses, media, and citizens for varied purposes. Alistar 2014, $14 million worth of drones were traded on E-bay alone (Desouza et al. 2015). Drones are now being used in urban spaces and causing problems for citizens and cities. For instance, the U.S. Federal Aviation Administration revealed that there were 25 near collisions between drones and airlines in 2014. However, the agency has fined only five people due to illegal use and undefined regulations (Whitlock 2015). There was another incident where a Seattle woman claimed that a drone was looking into her window but law enforcement were not legally able to do anything as she was unable to prove it (Bever 2014). Because there are inadequate rules and regulations developed for drones, city administrators and urban planners' hands are tied when attempting to protect citizens' privacy (Queally 2014); an issue sure to continue as more technology becomes more accessible to the masses.

3 Data

Data are invaluable aspects of smart city functioning. Data can originate from varied sources in both predictable and emergent manners with both intended and unintended consequences. Three data movements have become essential for smart cities: *open*, *big*, and *mobile* data.

The focus of open data efforts is to promote transparency and increase civic participation (Noveck 2012). Open data movements are becoming more prevalent all across the globe (Davies 2013; Manyika et al. 2013). Public agencies are increasingly moving towards 'open by default' mandates for their public data. During the 2013 G8 summit, the US, the UK, France, Canada, Germany, Russia, Italy, and Japan signed the Open Data Charter to increase data sharing efforts, improve the quality of data, and consolidate efforts to build a data repository (Sinai and Martin 2013). Agencies are making data available to the public about all facets of a city from transit to crime and in addition, are liberating data that were traditionally locked up within administrative systems. Open data efforts have spurred the development of (a) mobile applications to navigate urban spaces and public institutions (Desouza and Bhagwatwar 2012a), (b) incentivized innovation competitions/contests (Mergel and Desouza 2013; Desouza 2012a), and (c) information system innovations to realize administrative and process efficiencies.

However, open data has its drawbacks. Even though open data policies have largely been developed to enhance transparency, open data can misguide the public, either purposely or inadvertently. Open data that misguides can be attributed to incomplete data or data that is poorly curated for the reader. Desouza and Smith (2014c) discuss this as the 'transparency tragedy' where data is so low quality that being able to draw any conclusions from it extremely difficult. Further, the release and use of open data can create or sustain inequalities. For instance, predictive policing is the use of data and analytics to find patterns in criminal activity to proactively fight crime. While decent in its intention, the ethicality of data used to profile individuals (especially without their knowledge) that have previously committed crimes is questionable at the least and ethically wrong at worst (Desouza and Smith 2014a).

The primary focus of big data efforts is to build a capacity for predictive analytics that promotes real-time sensing of environments to increase situational awareness, automated and intelligent decision-making, and securing administrative and process efficiencies through information management. Public agencies around the world are investing in big data analytics (Manyika et al. 2011). In 2012, the U.S. announced plans to invest $200 million in developing big data analytics in the public sector (U.S. Office of Science and Technology Policy 2012). The U.K. Government has entered into a partnership with IBM to invest £113 million (approximately $175.8 million) in to the Hartree Centre in big data analytics over next 5 years (Consultancy.uk 2015). Further, widespread use of social networking sites such as Facebook, Twitter, and YouTube are also producing unstructured data streams. Individuals and organizations are increasingly using these sites to connect, exchange, share, and learn about information. Cities are increasingly using social media sites to disseminate information and engage people (Oliveira and Welch 2013). Data produced during this process provides rich situational and behavioral information in real-time (Harrison et al. 2010).

Big data efforts have led to innovations in (a) urban modeling of systems, (b) algorithmic regulations, and (c) management of public agencies. Big data analytics helped the New York City Department of Environmental Protection crack down on a long-term problem of restaurants illegally dumping cooking oils and grease into neighborhood sewers. To find the culprits, data was utilized from a city agency that previously certified local restaurants' employ of services that hauled away their grease. Through a few calculations, the research team was able to compare restaurants that did not employ a grease hauler with geo-spatial sewer data to come up with a list of possible culprits. The analysis resulted in a 95 % success rate of collaring dumpers (Feuer 2013).

The focus of mobile data efforts is to improve connectivity and reach remote populations. Public agencies are leveraging mobile technologies to provide e-services, engage citizens, and track urban movement (OECD Report on M-Government, 2011). In the first quarter of 2015, worldwide mobile subscriptions reached 7.2 billion; Africa and Asia account for three quarters of added mobile subscriptions (Pornwasin 2015). Mobile phones are increasingly becoming an integral part of individuals' lives, especially in developing countries. By tapping into mobile networks, cities can reach people at a lower cost and even in areas that lack infrastructures (Rotberg and Aker 2013). Mobile data efforts provide opportunities to (a) improve connectivity between agents and organizations, (b) empower greater public participation, (c) effective navigation of the urban space both from a physical and resources viewpoint, and (d) for greater personalization of services based on time, location, and behavioral contexts (Desouza and Smith 2014b; Desouza 2014c). The State of California is using mobile data to communicate with citizens to be safe drivers and avoid driving while impaired. Statewide, they released the Be My Designated Driver (BeMyDD) App that gives discounts and exclusives offers to designated drivers as they wait to drive inebriated persons home (Tatro 2014). The app developer's plan to expand their services by offering cheap driver deals not just for late-night partying but also for other needs such as airport transportation and medical visits.

4 Conceptualizing City as a Platform

Previous literature on platform thinking views a platform as a complex system comprising of heterogeneous interacting sub-systems (Baldwin and Woodard 2009). Platform architecture can be defined as a "conceptual blueprint that describes how the ecosystem is partitioned into a relatively stable platform and a complementary set of modules that are encouraged to vary, and the design rules binding on both" (Tiwana et al. 2010, p. 677). Platforms consist of heterogeneous sub-systems that interact, adapt, and evolve in response to changing eco-systems. The process of evolution and adaptation is controlled by platforms. In other words, platform architecture possesses the power to determine which parts of the eco-system will change and which parts will remain (Baldwin and Woodard 2009). Consider the case of Apple's iPhone. Apple's iPhone as a platform enables developers to build applications for their device that increases the value of the device. As more devices were sold, developers gained more revenue for their apps, which attracted more developers to build apps for iPhone. This created a network effect where more users attracted more applications and vice-versa (Sarver 2013). Thus, a platform facilitates controlled interaction between sub-systems and adds value that benefits each sub-system. In a way, platforms nurture its sub-systems to create value and positive benefits (Evans et al. 2006).

Based on this perspective, viewing city as a platform is advantageous for several reasons including its capacity to develop, manage, and respond to changing environments. First, cities are complex systems that are comprised of social and technical sub-systems that interact and produce outcomes (Batty 2007; Foth et al. 2011). Second, unprecedented growth in sub-systems (e.g. technological advancement) often produces unforeseeable outcomes. Third, city as platform architecture controls the interaction between social and technical sub-systems by designing formal (e.g. policies) and informal (e.g. norms) rules. Thus, the platform view allows us to break down its core components that can be modified or built to generate products, accommodate diverse demands, and respond to changing environments. Furthermore, this view allows us to consider an architectural interface that evolves to accommodate unpredictable changes that were not accounted for during the initial design stages (Baldwin and Woodard 2009).

A city's *technical* sphere includes both physical and virtual infrastructures that provide real-time information about the city (Yigitcanlar and Velibeyoglu 2008). Technologies are installed into the fabric of the city, including hardware, software, processes, and networks (Harrison et al. 2010; Yigitcanlar and Velibeyoglu 2008). To this end, a city becomes increasingly instrumented through sensors, meters, mobile devices, GPS, and cameras etc. (Harrison et al. 2010). Thus, a technical system combines physical and virtual infrastructures for meeting the demands of individuals, organizations, and communities (Nam and Pardo 2011). The integration of physical and virtual infrastructures allows a city to collect, merge, analyze, and manage data from different sources to optimize decisions in accordance with

Fig. 1 Conceptualizing city as a platform

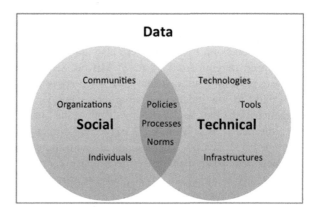

the environment (Harrison et al. 2010). Figure 1 provides a visual representation of city as a platform and its components.

A city's *social* sphere consists of individuals, organizations, and communities. The city facilitates interaction between these groups to drive innovation and growth. Individuals or communities who are innovative and creative can be instrumental in developing solutions that transform a city. Smart cities pay particular attention to how people relate to each other and their cities. Based on this perspective, individuals within a city leverage technologies to connect, communicate, and exchange with others (Williams et al. 2009). Individuals also use technology as a medium to navigate the city (Kitchin 2014) and all of these interactions are critical in developing knowledge and understanding social behavior towards smooth functioning smart cities (Nam and Pardo 2011). The use of emerging technologies will succeed when it is embedded into the everyday lives of people (Roche et al. 2012).

In addition to the technical and social sub-systems, the city also creates and promotes an open, transparent, and agile ecosystem—the *socio-technical* sphere. The environment is not only conducive to helping individuals, organizations, and communities locate resources that promotes economic growth and development, but also adopts newcomers and merges technologies into the social and economic fabric of the city (Florida 2004). Further, the city creates urban spaces for people to connect, exchange, share, and collaborate on ideas for developing novel solutions to complex urban challenges. Policies and procedures encourage public agencies, businesses, and NGOs to collectively work together. The city as platform supports this by (a) creating rules, regulations, and policies that encourage collaboration between public agencies, businesses, NGOs, and citizens and (b) promoting norms and social behavior that foster open and transparent knowledge sharing.

Cities should be designed as open platforms that promote and facilitate seamless integration across all three spheres—technical, social, and socio-technical. One of the key features of city as a platform architecture is its ability to provide an interface that allows people (e.g. businesses) to modify and build upon existing technical systems to generate and produce new products and services. For instance, the City of New York partnered with Bigbelly Waste Management Company to convert trashcans and recycling bins into Wi-Fi hotspots. The waste management company upgraded two of its recycling stations in Manhattan into Wi-Fi hotspots. City recycling bins and trashcans were also converted into free public Wi-Fi hotspots with speed ranging from 50 to 75 MB per second. The company is planning to convert recycling stations into Internet hotpots in all five New York boroughs, especially in low-income neighborhoods. Further, it is anticipated that by using these recycling bins as advertising displays, the bins will generate revenue to cover the Internet access. Moreover, Bigbelly Company's recycling bins are already equipped with sensors to communicate with the city's department of waste management and solar planners to decompose certain materials (McDonald 2015). Thus, a city as an architectural platform provides avenues to integrate and build on existing technical components.

Further, a city as a platform has the capacity to exploit positive interaction between social and technical systems to create network effects (Evans et al. 2006). Cities can open up its data reservoirs to encourage people to create apps that help solve urban challenges. By providing APIs, cities can generate economic and social value. For instance, BrightScope, a financial information company, helps individuals and corporations make better decisions about wealth management and retirement plans. Using 401(k) plans and data made available by the Department of Labor and other government agencies, BrightScope advises the public on retirement plans. Using this business model, BrightScope has raised venture capital and expanded its workforce (Noveck 2011). Thus, cities can provide avenues for people to leverage technologies for creating economic value. Further, cities can nurture and develop linkages between all three spheres for generating new products and accommodating diverse user needs.

5 Critical Issues in Urban Informatics

There is no doubt that emerging technologies and the deluge of data have the potential to transform cities into smart, livable, sustainable, and resilient cities. Policymakers, big technology firms, and NGOs are excited about the potential of emerging technologies in addressing complex urban challenges. While significant resources are invested in deploying technologies for building and transforming cities, it is important to consider how interaction between social and technical systems will present emergent challenges and issues for cities. Further, cities have legacies of governance, management, and culture that create path dependencies that dictate how emerging technologies will be embedded into the urban sphere. To truly realize the objective of developing smart cities, we need to critically evaluate, understand, and address challenges associated with data and technology. These challenges can be broadly classified into technical, social, and socio-technical. Table 1 highlights these key challenges faced by cities as they transform into becoming smart.

Table 1 Critical issues in urban informatics

	Issue	Data and technology challenges
Technical	Designing predictive failures	Highly interconnected systems increase interdependencies between critical infrastructures increasing chances of whole system failure
	Securing systems and systems of systems	Interconnected cities are prone to threats and attacks from intruders, where disruption could lead to cascading negative effects.
	Developing agile governance for regulating on-demand economy:	Technological advancement are creating new types of marketplace that requires development of agile regulations that can monitor and regulate without hindering growth
Social	Managing citizens' right and unintended consequences	Social context within which technologies are embedded determines its use. Cities are realizing that providing access to data alone cannot address social problems. Citizens need to be educated and trained to effectively use data without harming sub-sections of society.
	Innovation in knowledge discovery and experimentation	Detecting patterns will depend upon the ability of researchers' to separate noise from data. Analysis of large volumes of data requires new kinds of significance tests and other validation techniques to discover patterns and relationships.
Sociotechnical	Managing urban infrastructure projects	Public agencies have a poor track record when it comes to investing and maintaining large scale IT projects. Cities often grapple with managing stakeholder expectations, planning timelines, and budgeting accurate estimates.
	Balancing personalization and privacy	Collecting data about individuals are central to developing urban space and personalized solutions. As cities collect more sensitive information about citizens, issues of privacy and security concerns have increased.
	Navigating data sharing contracts	Cities are experiencing a change in their role as owners of data. Increasingly cities are relying on private companies for accurate data. Developing adequate information, sharing contracts and accessing accuracy of data are presenting new challenges to cities. Cities are yet to learn how to develop long-lasting partnerships.
	Mindful implementation of technology for equal and inclusive growth	Engaging citizens in the process of policy making is the fundamental ethos of democracy. Cities are leveraging technology to empower and engage citizens. Issues such as poverty, inequality, and access can result in an unequal society, where some sections of the population may have more access than others.

(continued)

Table 1 (continued)

Issue	Data and technology challenges
Do not forget small innovation and developing countries	Developing solutions that match the needs of the local marketplace is critical. Cities in developing countries are experiencing challenges when it comes to adopting solutions for smart cities in developed economies. While frugal innovation offers solutions based on environmental constraints, cities are yet to understand the value of frugal innovation. Gaining social support for frugal innovation projects is critical.

5.1 Designing Predictive Capacity Failure

One of the core concepts of smart cities is developing interconnected and instrumented infrastructures (Harrison et al. 2010). While interconnected infrastructures are critical for effectively managing, regulating, and monitoring urban areas, failure in one system can potentially disrupt a city's network affecting city livability. For example, a blackout in Bukit Panjang in Singapore affected not only affected 19 blocks of residential houses, but also impacted elevators, streetlights, and traffic signals (Heinimann 2015). While the power supply was restored in an hour, during the disruption period, the traffic police had to hand direct traffic. As urban planners invest resources and deploy technologies to connect every aspect of cities for smart operations, planning, and functioning, they need to think critically about developing predictive capacity for anticipating and responding to failures in one system before it disrupts other interconnected parts of the system. Doing so will require urban planners to get a better handle on how information and data flows between systems, the cascading dynamics of failures across systems, and innovative computational models to test the resilience of systems.

5.2 Securing Systems and Systems of Systems

In addition to managing high interdependencies among critical infrastructures, urban areas increasingly face threats and attacks from hackers. Interconnected systems are excellent targets for cyber attackers and terrorists, as disruption in one system could potentially damage other systems. For instance, in 2013, unknown assailants entered an underground vault to cut telephone cables in California. Then, snipers opened fire and knocked out 17 transformers that funnel power to Silicon Valley in about 19 min. It took 27 days for utility workers to repair the station and energy from other substations to power the area was used in the interim. After the attack, the U.S. Federal Energy Regulating Commission's chairman noted that if this type of attack happened across the country, it could potentially take down the electric grid and lead to a blackout across the country (Smith 2014). Moreover, pervasive and interconnected technologies make cities more vulnerable to cyber attacks. In 2014, Cesar Cerrudo, Chief Security Officer at IOActive Labs, showed that 200,000 traffic sensors installed in major cities in the U.S., Australia, and France where unprotected and vulnerable to cyber attacks. He demonstrated that these sensors could be intercepted from 1500 ft away. A hacker could potentially use drones to attack these systems because they were unencrypted. When he conducted a test in the city of San Francisco in 2015, he found that the city has yet to encrypt their traffic signals (Perlroth 2015).

5.3 *Developing Agile Governance for Regulating On-Demand Economy*

The growth of the on-demand economy is changing the landscape of economic marketplace. The on-demand economy builds upon existing infrastructures and adds a digital layer to provide services (Jaconi 2014). Online platforms such as Uber, Lending Club, and TaskRabbit.com are matching people with jobs, loans, and opportunities that meet their demands without the government as an intermediary. These online platforms could add $2.7 trillion to the world economy by 2025 by modifying how people find jobs, the hours they work, where they attain credit, how they travel, and how often they choose to work. In essence, they are creating a new kind of market where people are working as contractors (Manyika et al. 2015) and are in more control of their time and resources. The on-demand economy is evidence that cities will need to adopt an agile governing system that can adapt to changing urban landscapes. Developing effective strategies to manage, regulate, and monitor these emerging economies will not be easy. The ride-sharing app Uber has clashed with local and state government about the legality of its operations. Uber clashed with New York City leaders regarding regulation attempts to cap the number of drivers they can have. More than 2000 drivers are added every month in New York. In a year, more than 25,000 new-hire vehicles will be added in New York, which is two times more than the number of Yellow Taxi cabs in the city. City administrators expressed concerns about increasing congestion in New York City (De Blasio 2015). In response, Uber mobilized its users to speak out against New York City leaders' proposed regulation. They mailed out pamphlets and organized rallies. The company even added visuals to the apps so that every time a rider accessed Uber app to book a ride, they were shown how this legislation would increase wait times for riders (Badger 2015). Urban planners cannot ignore the growth of the on-demand economy and need to develop regulations and rules to manage and monitor this growing market. However, achieving this will require developing agile governing systems that accommodate growth in the technology-driven economy such as with on-demand services, and stabilize and manage the unpredictable outcomes.

5.4 Managing Citizens' Rights and Unintended Consequences

In addition to cities acquiring new technologies, citizens are also acquiring technologies of their own that are creating new challenges for cities. The unprecedented growth of information and computation technologies has reduced the cost and barriers to accessing emerging technologies.

Through the provision of data on activities and behaviors, we might increase the level of intelligent decision-making by individuals and organizations in a city. For example, the electrical company OPOWER sends energy consumption bills to households that include information about household's energy consumption in comparison to other similar households. The company found that a simple intervention such as providing information about neighbors' energy consumption, on average resulted in 2 % reduction in energy consumption. If scaled across the U.S., this program could reduce energy consumption and provide net benefits of $2.2 billion per year (Allcott and Mullainathan 2010). Yet, we have to watch against the unintended consequences of publicizing and sharing data. For example, in the U.S., homeowners are increasingly seeking the advice of private firms that offer one-on-one advice when it comes buying homes. Individuals have reported that these firms provide accurate information about localities based on their customer preferences. While this is good news for people searching for homes, this raises challenges for cities. Residents have a fundamental right to information; however, some information may result in creating situations where residents are likely to choose "people like us" (Pervost 2014). This was illustrated in London when a map was created that overlaid the last name of residents on top of the city's geography. The map revealed a clear picture of the racial landscape of London (Alistar 2014).

While planners can't manage all of the uses of the data it releases, they must think deeply about how to mitigate the challenges that might weaken the social system through data. Open data was created for transparency purposes but that transparency can also create brand new challenges for cities. For instance, the public transportation system BART in the Bay Area has a mobile security app that allows riders to send text message and photo alerts to BART police about crimes and non-crimes happening on their ride. It was found that riders were disproportionately sending messages to report African Americans and the homeless. In further analysis of the complaints, out of 763 alerts, 198 included some mention of the alleged offender's African American race while only 37 alerts mentioned the race of white alleged offenders (BondGraham 2015). The criminalization of the homeless and African Americans could result in more arrests based on these text alerts.

5.5 Innovation in Knowledge Discovery and Experimentation

A key challenge facing societies for generations is communicating knowledge between policy makers, scientists, businesses, and citizens. Advancements in computational technologies make it easier to analyze large volumes of data about urban activities for helping people make smart choices (Harrison et al. 2010). For instance, the advancement in predictive analytics and the deluge of data triggered debates about the value of traditional research design. Proponents of data analytics strongly favor emerging technologies and point out the superiority of these technologies in revealing hidden insights and patterns. In 2010, when data scientists at Google accurately estimated the actual number of people who contracted swine flu, weeks before the official announcement from the US Centers for Disease Control and Prevention (CDC), many questioned the CDC's method of disease prediction (Loukides 2010). Not surprisingly, CDC's traditional ways of data collection were deemed outdated. However, a recent study has revealed that the Google team has been overestimating flu outbreaks since 2011—their prediction is two points off compared to the CDC estimates. Additionally, by employing its traditional methods of estimation, the CDC has been accurately predicting flu outbreaks (Lazer et al. 2014). Does this mean that traditional scientific analysis is superior to predictive analytics?

Without a doubt, debates about the value of traditional scientific design versus predictive analytics are important. However, from the perspective of addressing society's intricate challenges, the most appropriate question is to understand what types of data and analysis could aid urban planners in designing policies that enhance livability, improve sustainability, and increase the resilience of cities. That is not to say that debates about scientific enquiry are unnecessary, rather we need to focus more on utilizing and experimenting with different approaches for improving our understanding about urban space and its needs. For example, the widespread use of social network sites offers new streams of contextual and behavioral data in real-time. Networks of people use these platforms to discuss, express, share, and view information about wide ranging issues that they are interested in. Collecting data about user experience and perception can reveal critical insights about human behavior. However, researchers must pay critical attention to issues such as who participates in these online platforms? Is this representative of the population? How are unstructured data different from traditional structured data? Researchers should apply methodological rigor grounded in sound theories and be transparent about the process of data analysis (Boyd and Crawford 2011; Keil 2013). Analysis of large volumes of data requires new kinds of significance tests and other validation techniques that gauge the temporal variability in order to discover patterns and relationships.

Further, we need to develop urban spaces into living laboratories, where cities and urban areas are designed to help people think about choices and change their behavior for sustainable smart outcomes. For example, the City as Living

Laboratory for Sustainability in Urban Design (CaLL) piloted a project in the city of New York to access how people connect with their urban environment. The project explored how arts can facilitate conversations between citizens and scientists about sustainable choices. The CaLL project team installed 'tactic art' in Montefiore Park, New York for people to think about connection between city and its critical infrastructures such as streetlights, hydrants, and manhole covers. People were encouraged to ponder about connections between critical infrastructures and cities. The project aimed to access how people who shape and use urban environment think about their role in building sustainable cities (Fraser and Miss 2012).

5.6 Managing Urban Infrastructure Projects

In many cases, cities lack the capacity to effectively plan, manage, and implement large scale IT projects. Public agencies have poor track records when it comes to planning and implementing large scale IT projects. Many projects have overrun costs and time, and worse, some are abandoned after spending significant public resources. Consider these examples: the Boston Big Dig is one of the most expensive highway projects, costing $14 billion and taking 32 years to complete. Initially the project was estimated to cost $2.4 billion. In addition to cost and time overrun, the project experienced several construction flaws (Hofherr 2015). The Seattle Monorail Project (SMP) was shut down after 3 years of planning and research in 2005 (Yuttapongsontorn et al. 2008). In 2005, Victoria State in Australia began a project to develop smartcard-ticketing systems. The government awarded $500 million to the Keane Australia Micropayment Consortium (Kamco) and expected to launch this new system on March 1, 2007. The project experienced delays and ran $500 million over-budget (Charette 2010). We need to increase our track record with urban projects if we are to stand any chance of realizing the true potential of technological and social innovations at scale.

5.7 Balancing Personalization and Privacy

Another critical concern plaguing investment in technologies is about privacy and security. Media reports are filled with headlines about security breaches at supermarkets, health insurance companies, financial institutions, and mobile operators (Hardekopf 2014). It is not surprising to hear debates about increases in negative sentiments about data and security breaches. For example, in 2015, Pew survey results reported that 93 % of Americans want to control who gets information about them; and 90 % wants to control what information is collected about them (Madden and Rainie 2015). While we are witnessing an increase in concerns about privacy and security issues, at the same time people are willing to share personal information for receiving personalized services such as through apps and Internet usage.

The critical question for urban planners is to manage the trade-off between offering personalized services and privacy concerns. Are people willing to share information in return for personalized services? How much personal information can be collected? For instance, smart infrastructures such as smart metering, electronic tolling, and smart parking offer several benefits to citizens. Citizens can use these infrastructures to manage travel, monitor consumption, save time, etc. At the same time, this information can be used to predict patterns about user behavior. Electronic toll collection programs such as EZPass allows user to purchase toll passes online to avoid hassles and improves travel experiences. However, the information collected about user travel can reveal where people live, how often they travel and so on (Humphries 2013). While concerns about security and privacy are critical, collecting more information and data about people and their activities is also critical for developing personalized solutions.

5.8 Navigating Data Sharing Contracts

We are witnessing a shift in data ownership where private companies are collecting, managing, and analyzing urban data. Inrix, a global transportation provider has launched a program to track movement of connected cars. The company uses GPS data from 250 million cars and devices to collect data about people's movement around a city. Inrix data can provide insight into how many vehicles pass through a location and at what time. Cities can use this data to (a) understand and predict population movement across urban space and (b) plan and prioritize transit and their city's transportation infrastructure (Traffic Technology Today.com 2015). Waze is a mobile application that allows users to create and use live maps and real-time traffic updates for navigation. Waze collects a wide range of data about its users including date, origin and destination, route, and speed. Additionally, users can also report about accidents, road conditions, and speed traps. The City of Rio de Janeiro, Brazil and the City of Jakarta are combining their own traffic data with Waze data to gain better situational awareness of their roads and citizen safety. Cities use these incident reports and real-time updates to repair roads, divert traffic flows, and prepare emergency responses (Dembo 2014).

For smart cities, access to reliable and accurate data about people's movement offers an unprecedented advantage to design urban spaces and address complex urban challenges such as transportation, energy, and water consumption. At the same time, these partnerships raise critical questions about data ownership and access (Smith and Desouza 2015). Traditionally, public agencies were collecting information and data about citizens for urban planning and governing purposes. However, we are now witnessing a trend where private companies such as mobile operators, social networking sites, and smartphone apps are collecting more data and information about every day activities of people (Leber 2013). For instance, whereas cities could tap into taxi cab operations to understand travel patterns in their city through reports taxi drivers must file, they miss a whole segment of the population because many citizens use Uber or Lyft to get around town; private companies that are not required to share travel information with government. Cities need to collaborate with private companies to obtain data about urban use. These partnerships in some cases have led to improvement in service delivery, while in others increased concerns about data insecurity. The success of these efforts will depend upon developing rules, norms, practices, and new partnerships.

Urban planners are increasingly developing partnerships with businesses to provide urban services that meet the demands of the citizens. Clearly, cities developing data partnerships and designing data sharing contracts is new territory for planners. These partnerships present several opportunities and challenges to urban planners. We do not know how cities can develop lasting-partnerships and avoid losses (Smith and Desouza 2015). Urban planners will need to share their experiences about developing and managing partnerships with private companies.

5.9 Mindful Implementation of Technology for Equal and Inclusive Growth

While cities are making significant efforts towards developing partnerships with private companies, issues of inequality and access becomes central concerns for cities investing in technology driven solutions to urban challenges. Access to technologies often dictates whose voices are heard. Many of the smart cities initiatives are built on the assumption that citizens have access to smart phone or Internet. Yet, one has to mindful when infusing technology without care to issues such as access, knowledge to use, adoption rates across segments of the population, etc. Incidents such as Boston's Street Bump apps clearly indicate that access to technologies determine whose complaints are heard. When the city of Boston introduced the Street Bump app that automatically detects potholes and sends reports to city administrators, they found that the program directed crews to mostly wealthy neighborhoods because those residents were more likely to have access to smartphones (Rampton 2014).

5.10 Do Not Forget Small Innovation and Developing Countries

Over-emphasis on technologies and its ability to solve urban challenges becomes more problematic in emerging economies. Developing countries lack infrastructure and significant resources for investing in technologies to transform cities. However, cities in developing countries often face tremendous strain on their critical infrastructures due to massive growth in populations. Frugal innovation, or frugal engineering, has been developed as a response to infrastructure gaps in both developed and developing economies. Frugal engineering is based on the principal of developing products and services that meet functional needs of the environment (Desouza 2014b). Oftentimes, expensive technological fixes to urban challenges such as retrofitting builds or building brand new structures are not feasible in developing countries. Solutions to urban challenges need to take into considerations environmental constraints, needs, and market capabilities (Desouza 2014b). For instance, in Kenya, consistent and safe banking services are not guaranteed to all residents. M-Pesa, a mobile-based money transfer service was introduced in 2007 to allow those who do not have regular access to financial services, mobile banking. Small retailers and vendors can use this service to transfer, withdraw, and deposit cash electronically. During peak hours, more than 100 transactions per second are conducted. The M-Pesa network accounts for 20 % of Kenya's gross domestic product (Townsend 2013). Cities around the world, especially in developing countries need to develop better business models and incentives for encouraging frugal innovation. Much of the work on frugal innovation is in early stages. If leveraged effectively, frugal innovation can promote growth and development in urban areas that lack infrastructures.

6 Discussion: A Way Forward

Undoubtedly, cities will continue to invest significant resources in technological solutions to transform urban areas. However, as noted in this chapter, urban planners need to pay special attention to the social context within which technologies are embedded. Technologies do not exist in a vacuum and how they are infused into the modalities of the everyday lives of citizens determine their success (Roche et al. 2012). That does not mean that technologies have no value, but rather the interaction between social and technical systems create complex solutions and outcomes (Nam and Pardo 2011).

The concept of smart cities is of recent origin. We must caution against prematurely judging the success or failure of emerging technologies in creating smart cities. We still do not know much about how the deployment of new technologies will help urban landscapes above and beyond the use of traditional technologies. Further research is needed to outline how these technologies interact with human systems to produce outcomes that make cities more livable, sustainable, and resilient. Will these technologies make cities more livable but adversely affect cities' sustainability? What are some of the metrics to measure the impact of smart cities? Do cities need to invest in developing and upgrading all of its infrastructures to become smarter? How will smart cities in developing countries vary from developed countries? How does cultural difference influence the adoption and use of emerging technologies? How can cities develop capacity of its human capital to harness collective intelligence? These are some of the unexplored questions for future research. Exploring these questions will provide further insights into how cities can leverage emerging technologies to be transformed into smart cities.

Urban planners need to pay attention and experiment with different approaches rather than focusing on technologies as solutions for addressing urban challenges. For instance, the concept of developing smart grids has gained traction around the world for reducing energy consumption. However, recent research has revealed that installation of cogeneration technologies, solar heater pumps, and building insulation practices can potentially reduce New York's carbon emissions by 50 %. Developing old-school solutions may provide similar outcomes rather than investing in expensive smart technologies (Hammer 2010).

Further, evidence from behavioral economics experiments such as the OPOWER example suggests that simple behavioral tweaks can produce significant return on investments similar to R&D subsidies (Allcott and Mullainathan 2010). As the evidence suggests, blindly investing in technologies will not result in transforming cities that are livable, sustainable, and resilient. Further research is needed to explore and examine how these behavioral insights can be scaled up to national levels to promote alternative measures to make cities livable and sustainable.

Given the magnitude of urban challenges, cities need to develop partnerships with businesses, universities, and NGOs for effectively addressing pressing challenges of urban areas. Private companies are increasingly developing novel solutions. For example, environmental health startup Aclima partnered with Google Earth Outreach and the Environmental Protection Agency (EPA) to equip Google street cars with sensors for collecting air quality data. In 2014, they piloted the program in Denver, where the three cars collected over 150 million air quality data in a month. The EPA's research directed by Dan Costa who noted that this imitative provides a perfect opportunity to update and move forward the science of monitoring air quality (CNNMoney 2015). Further research is needed to understand how partnerships with private companies can help cities upgrade their traditional data collection strategies. Additionally, cities also need to experiment with analytics for understanding and exploring how this improved information will aid in urban planning and developing sustainable cities.

Urban planners also need to think critically about the challenges associated with developing technology-based solutions to urban challenges. They should also carefully consider the issues of inequality and access that can arise when deploying urban solutions. How can cities promote inclusive urban growth? One viable solution could be investments in frugal innovation. Frugal innovations pay particular attention to environmental constraints and are best suited for addressing challenges faced by low-income families. As cities increasingly experience resource constraints (e.g. financial), they need to develop their capacity for finding inexpensive solutions for sustainable outcomes. They also need to be mindful of investing massive resources in technologies that could potentially disrupt cities' well-being (Roche et al. 2012).

7 Conclusion

In this chapter, we outlined the city as a platform consisting of social and technical components and the interactions between them. Data and information acts as glue that enables interaction, communication, and exchange between these components. We have enumerated several critical considerations that merit further discussion, debate, and scientific investigations to advance the field of urban informatics. Urban planners must develop sophisticated technologies and enhance human capacity for proactively addressing challenges associated with the application of urban informatics.

References

Alistar (2014) Big data is our generation's civil rights issue, and we don't know it. [WWW Document]. http://solveforinteresting.com/big-data-is-our-generations-civil-rights-issue-and-we-dont-know-it/. Accessed 27 Jul 2015

Allcott H, Mullainathan S (2010) Behavioral science and energy policy. Science 327:1204–1205

Badger E (2015) Uber's war with New York is so serious it's giving out free hummus. [WWW Document]. The Washington Post. http://www.washingtonpost.com/news/wonkblog/wp/2015/07/21/ubers-war-with-new-york-is-so-serious-its-giving-out-free-hummus/. Accessed 27 Jul 2015

Baldwin CY, Woodard CJ (2009) The architecture of platforms: a unified view. Harvard Business School Finance Working Paper

Batty M (2007) Cities and complexity: understanding cities with cellular automata, agent-based models, and fractals. The MIT press, Cambridge

Bever L (2014) Seattle woman spots drone outside her 26th-floor apartment window, feels "violated." The Washington Post. [WWW Document]. http://www.washingtonpost.com/news/morning-mix/wp/2014/06/25/seattle-woman-spots-drone-outside-her-26th-floor-apartment-window-feels-violated/. Accessed 27 Jul 2015

BondGraham D (2015) BART riders racially profile via smartphone app [WWW Document]. East Bay Express. URL http://www.eastbayexpress.com/oakland/bart-riders-racially-profile-via-smartphone-app/Content?oid=4443628. Accessed 5 Aug 2015

Boyd D, Crawford K (2011) Six provocations for big data, a decade in internet time: symposium on the dynamics of the internet and society. [WWW Document]. http://ssrn.com/abstract=1926431. Accessed 27 Jul 2015

Charette R (2010) Australia's AU$1.3 Billion Myki ticketing system introduction marred by multiple missteps [WWW Document]. http://spectrum.ieee.org/riskfactor/computing/it/australias-au13-billion-myki-ticketing-system-introduction-marred-by-multiple-missteps. Accessed 27 Jul 2015

Cheng R (2014) General Motors President sees self-driving cars by 2020 [WWW Document]. CNET. http://www.cnet.com/news/general-motors-president-sees-self-driving-cars-by-2020/. Accessed 5 Aug 2015

CNNMoney (2015) Google Street View cars will soon measure pollution [WWW Document]. CNNMoney. http://money.cnn.com/2015/07/30/technology/google-aclima-air-pollution/index.html. Accessed 5 Aug 2015

Consultancy.uk 2015 UK GOV partners with IBM to boost Big Data research [WWW Document]. Consultancy.uk. http://www.consultancy.uk/news/2128/uk-gov-partners-with-ibm-to-boost-big-data-research. Accessed 5 Aug 2015

Davies T (2013) Open Data Barometer 2013 global report. Open Data Barometer

De Blasio B (2015) A fair ride for New Yorkers: how the city should respond to the rapid rise of Uber [WWW Document]. NY Daily News. http://www.nydailynews.com/opinion/bill-de-blasio-fair-ride-new-yorkers-article-1.2296041. Accessed 25 Jul 2015

Dembo M (2014) The power of public-private partnerships: mobile phone apps and municipalities [WWW Document]. Planetizen: The Urban Planning, Design, and Development Network. http://www.planetizen.com/node/70934. Accessed 25 Jul 2015

Desouza KC (2012a) Leveraging the wisdom of crowds through participatory platforms: designing and planning smart cities. Planetizen: planning, design & development

Desouza KC (2012b) Designing and planning for smart(er) cities. Pract Plann 10:12

Desouza KC (2014a) Our fragile emerging megacities: a focus on resilience [WWW Document]. Planetizen: the urban planning, design, and development network. http://www.planetizen.com/node/67338. Accessed 27 Jul 2015

Desouza KC (2014b) Realizing the promise of big data | IBM Center for the Business of Government. IBM Center for the Business of Government, Washington, DC

Desouza KC (2014c) Intelligent cities. In: Atlas of cities. Princeton University Press, Princeton, NJ

Desouza KC, Bhagwatwar A (2012a) Citizen apps to solve complex urban problems. J Urban Technol 19:107–136

Desouza KC, Bhagwatwar A (2012b) Leveraging technologies in public agencies: the case of the US Census Bureau and the 2010 Census. Public Adm Rev 72:605–614

Desouza KC, Bhagwatwar A (2014) Technology-enabled participatory platforms for civic engagement: the case of US cities. J Urban Technol 21:25–50

Desouza KC, Flanery TH (2013) Designing, planning, and managing resilient cities: a conceptual framework. Cities 35:89–99

Desouza KC, Schilling J (2012) Local sustainability planning: harnessing the power of information technologies. PM Magazine 94

Desouza KC, Simons P (2014) Society for Information Management. Society for Information Management - Advanced Practices Council, Mount Laurel, NJ

Desouza KC, Smith K (2014a) Big data for social innovation (SSIR). Stanford Soc Sci Rev 12:38–43

Desouza KC, Smith K (2014b) Finding a fair and equitable use of citizen data: the case of predictive policing [WWW Document]. The Brookings Institution. http://www.brookings.edu/blogs/techtank/posts/2014/10/15-police-citizens-data. Accessed 5 Aug 2015

Desouza KC, Smith K (2014c) The transparency tragedy of open data [WWW Document]. The Brookings Institution. http://www.brookings.edu/blogs/techtank/posts/2014/11/5-transparency-tragedy. Accessed 5 Aug 2015

Desouza KC, Swindell D, Koppell J, Smith K (2014) Funding smart technologies: tools for analyzing strategic options. Smart Cities Council, Washington, DC

Desouza KC, Swindell D, Smith KL, Sutherl A, Fedorschak K, Coronel C (2015) Local government 2035: strategic trends and implications of new technologies (No. 27). Issues in Technology Innovation. Brookings, Washington, DC

Dobbs R, Phol H, Lin D-Y, Mischke J, Garemo N, Hexter J, Matzinger S, Palter R, Nanavatty R (2013) Infrastructure productivity: how to save $1 trillion a year. McKinsey Global Institute, London

Evans DS, Hagiu A, Schmalensee R (2006) Invisible engines: how software platforms drive innovation and transform industries. MIT Press, Cambridge, MA

Feuer A (2013) Mayor Bloomberg's Geek Squad. The New York Times. [WWW Document]. http://www.nytimes.com/2013/03/24/nyregion/mayor-bloombergs-geek-squad.html?pagewanted=all. Accessed 27 Jul 2015

Florida R (2004) The rise of the creative class and how it's transforming work, leisure, community and everyday life (Paperback Ed.). Basic Books, New York

Foth M, Choi JH, Satchell C (2011) Urban informatics. In: Proceedings of the ACM 2011 conference on computer supported cooperative work. ACM, pp 1–8

Fraser J, MIss M (2012) City as living laboratory for sustainability in urban design. New Knowledge Organization, New York

Gardner G (2015) Google tests self-driving cars in tricky situations [WWW Document]. Detroit Free Press. http://www.freep.com/story/money/2015/07/22/google-car-self-driving/30514747/. Accessed 27 Jul 15

Glasgow Centre for Population Health (2011) Scottish "excess" mortality: comparing Glasgow with Liverpool and Manchester [WWW Document]. http://www.gcph.co.uk/work_themes/theme_1_understanding_glasgows_health/excess_mortality_comparing_glasgow. Accessed 27 Jul 2015

Goel V, Hardy Q (2015) A Facebook project to beam data from drones is a step closer to flight. The New York Times. [WWW Document]. http://www.nytimes.com/2015/07/31/technology/facebook-drone-project-is-a-step-closer-to-flight.html?_r=0. Accessed 5 Aug 2015

Hammer S (2010) The smart grid may not be the smartest way to make cities sustainable [WWW Document]. Harvard Business Review. https://hbr.org/2010/09/smart-energy-for-smart-cities. Accessed 5 Aug 2015

Hardekopf B (2014) This week in credit card news: massive data breach at chase, the value of stolen medical data [WWW Document]. Forbes. URL http://www.forbes.com/sites/moneybuilder/2014/10/03/this-week-in-credit-card-news-massive-data-breach-at-chase-the-value-of-stolen-medical-data/. Accessed 5 Aug 2015

Harrison C, Eckman B, Hamilton R, Hartswick P, Kalagnanam J, Paraszczak J, Williams P (2010) Foundations for smarter cities. IBM J Res Dev 54:1–16

Heinimann HR (2015) Strengthening a city's "backbone" [WWW Document]. The Straits Times. http://www.straitstimes.com/opinion/strengthening-a-citys-backbone. Accessed 27 Jul 2015

Hofherr J (2015) Can we talk rationally about the big dig yet? [WWW Document]. Boston.com. http://www.boston.com/cars/news-and-reviews/2015/01/05/can-talk-rationally-about-the-big-dig-yet/0BPodDnlbNtsTEPFFc4i1O/story.html. Accessed 27 Aug 2015

Hollands RG (2008) Will the real smart city please stand up? Intelligent, progressive or entrepreneurial? City 12:303–320

Horowitz S, Rosati F (2014) 53 million Americans are freelancing, new survey finds [WWW Document]. Freelancers Union. https://www.freelancersunion.org/blog/dispatches/2014/09/04/53million/. Accessed 27 Jul 2015

Howard A (2015) How digital platforms like LinkedIn, Uber and TaskRabbit are changing the on-demand economy [WWW Document]. The Huffington Post. http://www.huffingtonpost.com/entry/online-talent-platforms_55a03545e4b0b8145f72ccf6. Accessed 25 Jul 2015

Humphries C (2013) The too-smart city [WWW Document]. The Boston Globe. https://www.bostonglobe.com/ideas/2013/05/18/the-too-smart-city/q87J17qCLwrN90amZ5CoLI/story.html. Accessed 25 Jul 2015

Jaconi M (2014) The "On-Demand Economy" is revolutionizing consumer behavior — here's how [WWW Document]. Business Insider. http://www.businessinsider.com/the-on-demand-economy-2014-7. Accessed 25 Jul 2015

Keil P (2013) Data-driven science is a failure of imagination. [WWW Document]. http://www.petrkeil.com/?p=302. Accessed 25 Jul 2015

Kitchin R (2014) The real-time city? Big data and smart urbanism. GeoJournal 79:1–14

Lazer D, Kennedy R, King G, Vespignani A (2014) The parable of Google flu: traps in big data analysis. Science 343:1203–1205. doi:10.1126/science.1248506

Leber J (2013) How Verizon and other wireless carriers are mining customer data [WWW Document]. MIT Technology Review. http://www.technologyreview.com/news/513016/how-wireless-carriers-are-monetizing-your-movements/. Accessed 25 Jul 2015

Loukides M (2010) What is data science? [WWW Document]. O'Reilly Media. https://beta.oreilly.com/ideas/what-is-data-science. Accessed 5 Aug 2015

Macdonell H (2015) Glasgow: the making of a smart city [WWW Document]. The Guardian. http://www.theguardian.com/public-leaders-network/2015/apr/21/glasgow-the-making-of-a-smart-city. Accessed 27 Jul 2015

Mack E (2014) Elon Musk: don't fall asleep at the wheel for another 5 years [WWW Document]. CNET. http://www.cnet.com/news/elon-musk-sees-autonomous-cars-ready-sooner-than-previously-thought/. Accessed 5 Aug 2015

Madden M, Rainie L (2015) Americans' attitudes about privacy, security and surveillance. Pew Research Center: Internet, Science & Tech, Washington, DC

Manyika J, Michael C, Brown B, Bughin J, Dobbs R, Roxburgh C, Byers AH (2011) Big data: the next frontier for innovation, competition, and productivity. McKinsey Global Institute, London, [WWW Document] http://www.mckinsey.com/insights/mgi/research/technology_and_innovation/big_data_the_next_frontier_for_innovation. Accessed 5 Aug 2015

Manyika J, Chui M, Farrell D, Kuiken SV, Groves P, Doshi EA (2013) Open data: unlocking innovation and performance with liquid information. McKinsey Global Institute, London, [WWW Document]. http://www.mckinsey.com/insights/business_technology/open_data_unlocking_innovation_and_performance_with_liquid_information. Accessed 5 Aug 2015

Manyika J, Lund S, Robinson K, Valentino J, Dobbs R (2015) Connecting talent with opportunity in the digital age. McKinsey & Company, London

Marr B (2015) How big data and the internet of things create smarter cities [WWW Document]. Forbes. http://www.forbes.com/sites/bernardmarr/2015/05/19/how-big-data-and-the-internet-of-things-create-smarter-cities/. Accessed 23 Jul 2015

McDonald G (2015) NYC is turning trash cans into Wi-Fi hotspots [WWW Document]. DNews. http://news.discovery.com/tech/gear-and-gadgets/nyc-is-turning-trash-cans-into-wi-fi-hotspots-150717.htm. Accessed 5 Aug 2015

Mergel I, Desouza KC (2013) Implementing open innovation in the public sector: the case of Challenge. gov. Public Adm Rev 73:882–890

Morgan Stanley (2013) Autonomous cars: self-driving the new auto industry paradigm. Morgan Stanley blue paper

Nam T, Pardo TA (2011) Conceptualizing smart city with dimensions of technology, people, and institutions. In: Proceedings of the 12th annual international digital government research conference: digital government innovation in challenging times. ACM, pp 282–291

Noveck B (2011) Why cutting E-Gov funding threatens American jobs [WWW Document]. Huffington Post. http://www.huffingtonpost.com/beth-simone-noveck/why-cutting-egov-funding-_b_840430.html. Accessed 5 Aug 2015

Noveck B (2012) Open data - the democratic imperative [WWW Document]. Crooked Timber. http://crookedtimber.org/2012/07/05/open-data-the-democratic-imperative/. Accessed 28 Jul 2015

OECD, International Telecommunication Union (2011) OECD Report on M-Government. 2011. M-Government: mobile technologies for responsive governments and connected societies. OECD Publishing, Paris, [WWW Document]. http://dx.doi.org/10.1787/9789264118706-en. Accessed 27 Jul 2015

Oliveira GHM, Welch EW (2013) Social media use in local government: linkage of technology, task, and organizational context. Gov Inf Q 30:397–405

Pagliery J (2015) Chryslers can be hacked over the Internet [WWW Document]. CNN. http://money.cnn.com/2015/07/21/technology/chrysler-hack/index.html?iid=ob_homepage_desk recommended_pool&iid=obnetwork. Accessed 27 Jul 2015

Perlroth N (2015) Smart city technology may be vulnerable to hackers [WWW Document]. Bits Blog. http://bits.blogs.nytimes.com/2015/04/21/smart-city-technology-may-be-vulnerable-to-hackers/. Accessed 25 Jul 2015

Pervost L (2014) The data-driven home search. [WWW Document]. The New York Times. http://www.nytimes.com/2014/07/20/realestate/using-data-to-find-a-new-york-suburb-that-fits.html. Accessed 25 Jul 2015

Pollock M (2013) Five Chicago Apps that make city life a little less annoying [WWW Document]. Chicago magazine. http://www.chicagomag.com/city-life/October-2013/Chicago-App-Roundup/. Accessed 5 Aug 2015

Pornwasin A (2015) The world is becoming more mobile and networked [WWW Document]. The Nation. http://www.nationmultimedia.com/politics/The-world-is-becoming-more-mobile-and-networked-30265200.html. Accessed 27 Jul 2015

PWC (2013) Autofacts. [WWW Document]. http://www.detroitchamber.com/wp-content/uploads/2012/09/AutofactsAnalystNoteUSFeb2013FINAL.pdf. Accessed 5 Aug 2015

Queally J (2014) Seattle woman says drone wasn't spying on her after all. Los Angeles Times. WWW Document]. http://www.latimes.com/nation/nationnow/la-na-nn-seattle-drone-update-20140625-story.html. Accessed 27 Jul 2015

Rampton R (2014) White House looks at how "Big Data" can discriminate. [WWW Document] http://uk.reuters.com/article/2014/04/27/uk-usa-obama-privacy-idUKBREA3Q00S20140427. Accessed 5 Aug 2015

Rotberg RI, Aker JC (2013) Mobile phones: uplifting weak and failed states. Wash Q 36:111–125

Roche S, Nabian N, Kloeckl K, Ratti C (2012) Are "smart cities" smart enough. Presented at the Global Geospatial Conference 2012, Spatially Enabling Government, Industry and Citizens, Québec City, Canada, pp. 215–235

Sarver R (2013) What is a Platform? Ryan Sarver

Sinai N, Martin M (2013) Open data going global [WWW Document]. The White House. http://www.whitehouse.gov/blog/2013/06/19/open-data-going-global Accessed 28 Jul 2015

Smith R (2014) Assault on California Power Station raises alarm on potential for terrorism. [WWW Document]. Wall Street J. http://www.wsj.com/articles/SB10001424052702304851104579359141941621778. Accessed 23 Jul 2015

Smith K, Desouza KC (2015) How data privatization will change planning practice [WWW Document]. Planetizen: the urban planning, design, and development network. http://www.planetizen.com/node/79680/how-data-privatization-will-change-planning-practice. Accessed 25 Jul 2015

Stewart E (2014) A truly smart city is more than sensors big an all-seeing internet [WWW Document]. The Guardian. http://www.theguardian.com/sustainable-business/2014/nov/21/smart-city-sensors-big-data-internet. Accessed 23 Jul 2015

Tatro S (2014) New App rewards designated sober drivers [WWW Document]. NBC 7 San Diego. http://www.nbcsandiego.com/news/local/New-App-Rewards-Designated-Drivers-286583491.html. Accessed 5 Aug 2015

Tiwana A, Konsynski B, Bush AA (2010) Research commentary-platform evolution: coevolution of platform architecture, governance, and environmental dynamics. Inf Syst Res 21:675–687

Townsend AM (2013) SMART CITIES: big data, civic hackers, and the quest for a new utopia [WWW Document]. Stanford Social Science Review. http://www.ssireview.org/articles/entry/smart_cities_big_data_civic_hackers_and_the_quest_for_a_new_utopia. Accessed 27 Jul 2015

Traffic Technology Today.com (2015) New "Big Data" analytics platform can aid urban planning [WWW Document]. Traffic Technology Today.com. http://www.traffictechnologytoday.com/news.php?NewsID=68829. Accessed 25 Jul 2015

U.S. Department of Transportation (n.d.) The vehicle-to-vehicle and vehicle-to-infrastructure technology Test Bed – Test Bed 2.0: Available for Device and Application Development

U.S. Office of Science and Technology Policy (2012) Obama administration unveils "Big Data" initiative: announces $200 million in new R&D investments [WWW Document]. https://www.whitehouse.gov/sites/default/files/microsites/ostp/big_data_press_release.pdf. Accessed 27 Jul 2015

Whitlock C (2015) How crashing drones are exposing secrets about U.S. war operations. The - Washington Post. [WWW Document]. https://www.washingtonpost.com/world/national-security/how-crashing-drones-are-exposing-secrets-about-us-war-operations/2015/03/24/e89ed940-d197-11e4-8fce-3941fc548f1c_story.html. Accessed 27 Jul 2015

Williams A, Robles E, Dourish P (2009) Urbane-ing the city: examining and refining the assumptions behind urban informatics. In: Handbook of research on urban informatics: the practice and promise of the real-time city. IGI, Hershey, PA, pp 1–20

Yigitcanlar T, Velibeyoglu K (2008) Knowledge-based urban development: the local economic development path of Brisbane, Australia. Local Econ 23:195–207

Yuttapongsontorn N, Desouza KC, Braganza A (2008) Complexities of large-scale technology project failure: a forensic analysis of the Seattle popular monorail authority. Public Perform Manage Rev 31:443–478

Digital Infomediaries and Civic Hacking in Emerging Urban Data Initiatives

Piyushimita (Vonu) Thakuriah, Lise Dirks, and Yaye Mallon Keita

Abstract This paper assesses non-traditional urban digital infomediaries who are pushing the agenda of urban Big Data and Open Data. Our analysis identified a mix of private, public, non-profit and informal infomediaries, ranging from very large organizations to independent developers. Using a mixed-methods approach, we identified four major groups of organizations within this dynamic and diverse sector: general-purpose ICT providers, urban information service providers, open and civic data infomediaries, and independent and open source developers. A total of nine types of organizations are identified within these four groups.

We align these nine organizational types along five dimensions that account for their mission and major interests, products and services, as well activities they undertake: techno-managerial, scientific, business and commercial, urban engagement, and openness and transparency. We discuss urban ICT entrepreneurs, and the role of informal networks involving independent developers, data scientists and civic hackers in a domain that historically involved professionals in the urban planning and public management domains.

Additionally, we examine convergence in the sector by analyzing overlaps in their activities, as determined by a text mining exercise of organizational webpages. We also consider increasing similarities in products and services offered by the infomediaries, while highlighting ideological tensions that might arise given the overall complexity of the sector, and differences in the backgrounds and end-goals of the participants involved. There is much room for creation of knowledge and value networks in the urban data sector and for improved cross-fertilization among bodies of knowledge.

P. Thakuriah (✉)
Urban Studies and Urban Big Data Centre, University of Glasgow, Glasgow, UK
e-mail: Piyushimita.Thakuriah@glasgow.ac.uk

L. Dirks
Urban Transportation Center, University of Illinois at Chicago, Chicago, IL, USA
e-mail: ldirks1@uic.edu

Y.M. Keita
Department of Urban Planning and Policy, University of Illinois at Chicago, Chicago, IL, USA
e-mail: ykeita2@uic.edu

Keywords Digital infomediaries • Civic hacking • Urban Big Data • Open data • Text mining

1 Introduction

There has been a surge of interest recently in urban data, both "Big Data" in the sense of large volumes of data from highly diverse sources, as well as "Open Data", or data that are being released by government agencies as a part of Open Government initiatives. These sources of data, together with analysis methods that aim to extract knowledge from the data, have attracted significant interest in policy and business communities. While public and private organizations involved in planning and service delivery in the urban sectors have historically been users of urban data, recent developments outlined below have opened up opportunities for policy and planning reform in public agencies and for business innovation by private entities. Such opportunities have also attracted innovative new organizations and facilitated new modes of ICT entrepreneurship. The objective of this paper is to examine the diverse organizations and networks around such emerging sources of urban "Big Data" and "Open Data".

While there are many explanations of Big Data, it is the term being applied to very large volumes of data which are difficult to handle using traditional data management and analysis methods (Thakuriah and Geers 2013; Batty 2013), and which can be differentiated from other data in terms of its "volume, velocity and variety" (Beyer and Laney 2012). It has also stimulated an emphasis on data-driven decision making based on analytics and data science (Provost and Fawcett 2013) that has the potential to add to the value brought about by traditional urban and regional modeling approaches.

Urban Big Data can be generated from several sources such as sensors in the transportation, utility, health, energy, water, waste and environmental management infrastructure, and the Machine-to-Machine (M2M) communications thereby generated. The increasing use of social media, using Web 2.0 technologies, personal mobile devices and other ways to connect and share information has added to the vast amounts of socially-generated user-generated content on cities. Open Data initiatives adopted by city governments are leading to an increasing availability of administrative and other governmental "open data" from urban management and monitoring processes in a wide variety of urban sectors. These initiatives have the potential to lead to innovations and value-generation (Thorhildur et al. 2013). Privately-held business transactions and opinion-monitoring systems (for example, real estate, food, or durable goods transactions data, data on household energy or water consumption, or customer reviews and opinions) can yield significant insights on urban patterns and dynamics.

The increasing availability of such data has generated new modes of enquiry on cities, and has raised awareness regarding a data-driven approach for planning and decision-making in the public, private and non-profit sectors. Urban Big Data has stimulated new data entrepreneurship strategies by businesses, independent developers, civic technologists, civic hackers, urban data scientists and the like, who are using data for civic activism, citizen science and smart urban management in novel new ways. However, very little has been written about ways in which such data-centric developments are being organized and delivered, particularly the organizations and networks involved.

In this paper, we qualitatively review organizations in the emerging urban data sector, with the purpose of understanding their involvement in production and service delivery using the data. We use the term "urban digital infomediaries" to describe enabling organizations and networks which are fostering ICT-based data-centric approaches to studying, managing and engaging in cities. Many organizational modes are prevalent in this landscape, including well-established multinational ICT-focused businesses as well as informal networks of individuals or freelance ICT developers with interests in data and cities. Hence the traditional definition of an infomediary as an "organization" may need to be expanded to include these unstructured and informal activities.

Given the dynamic growth that is being experienced in the urban data sector, several new cross-cutting technology solutions and service delivery processes are emerging. These solutions are being adopted by a wide spectrum of infomediaries with vastly different mission and customer/client focus, leading to similarity in products and services, skills of the workforce involved, and to transcending and blurring of traditional functional lines and organizational/market boundaries. More and more ICT products use similar intermediate inputs over time and meet similar demands, greatly facilitating such convergence (Xing and Ye 2011; Stieglitz 2003). Convergence in ICT companies is a well-established area of research, where authors have used conceptual frameworks and case studies to analyze the mechanism of the convergence and its implications. The analysis undertaken here is partly driven by this strand of literature, and we attempt to understand where traditional organizational boundaries are blurring as new opportunities for collaboration arise in the urban data sector.

The paper is motivated by a need to understand city-centric ICT services in order to discern trends relevant for governance, business development and service provision. We are also motivated by convergence in processes for potential partnership building and alliances in providing urban services. The analysis is composed of two major components. First, we undertake a broad, qualitative assessment of organizations and their formal and informal activities regarding urban Big Data. Second, we use a mixed-methods approach to take an in-depth look at a sample of organizations to understand organizational mission and stated objectives, major interests and functional activities, skills and interests of the workforce involved, and services and products. The end result is a description of organizations in this emerging sector along multiple dimensions, as opposed to the traditional categorization such as industrial classification systems.

This paper is organized as follows. In Section 2, we describe our research approach, followed by the analysis of urban infomediaries in Section 3. Further results and discussion are presented in Section 4, 5, 6, and 7. Conclusions are given in Section 8.

2 Analysis of Organizations: Research Approach

Our objective is to examine organizations that are involved in urban data and the myriad activities that relate to the urban data infrastructure. A mixed-methods approach is utilized, consisting of qualitatively examining the literature and using personal experience and by communicating with experts, as well as quantitatively analyzing material from websites of selected organizations using text mining. The qualitative assessment helped to identify the major groups of stakeholders in the urban data landscape, types of products or services generated, skillsets of professionals involved and evolving professional networks.

In order to understand the work of the urban digital infomediaries in greater detail, a database of webpages of 139 public, private and non-profit ICT-focused organizations and informal ICT entities was constructed. The webpages were collected using snowballing techniques, starting with a list of organizations involved in city-related activities known to the authors. Additional organizations were identified through Internet search using keywords such as "open data", "smart cities", "big data", "civic", "open source", "advocacy" and so on, as well as with keywords relating to modes such as "participatory sensing", "crowdsourcing", "civic engagement", "public engagement" and related terms. Pages within websites retrieved for this purpose include, among others: (1) about us, mission statement, products or services, or similar page(s) of organizations which describes the organization; and (2) terms of service, privacy policy or related pages which describe the organization's terms or policies regarding information use and data sharing.

The first step was to manually label and categorize the type of organizations, sector (public, private, non-profit), major functional interest or domain area and types of services offered, as well as policies and markets. It also allowed us to make an assessment of the skills sets of the urban-centric workforce involved in the organizations, although this aspect was informed by additional reviews and judgment. This led to the identification of four major groups of infomediaries, which were then organized into nine subgroups based on the stated missions and interests of the organizations.

One use of the database was to understand, using text mining, the emphasis of the organizations with regard to their activities and processes that may not be apparent from their stated mission and objectives. For example, an ICT-focused organization may indicate that it is in the business of "smart cities"; it is possible that it is involved in the smart cities agenda by helping to empower residents to connect to city governments or it may be focused to a greater extent on building the technologies to make such empowerment technically possible. The assumption is that specific focus in the work of "serving communities through smart city technology" can be discerned from the text contained in the websites of organizations.

A statistical clustering approach was applied to the data retrieved from the webpages. This process involved the following steps: (1) creation of a database of words extracted from the webpages of organizations after standard preprocessing to convert text to data (i.e., where each word in every document becomes a column of a rectangular database, with the rows being the organization ID, and with the cells of the final database giving a 1 indicating the occurrence of a word in the documents of a specific organization, and 0 otherwise); (2) the use of a lexicon (WordNet) to convert words into hypernyms or supersets of words to capture the concepts expressed within the organization's webpages; (3) development of decision-rules for the retention of hypernyms that are within the topic of interest to us, by the use of a-priori specification of hypernyms of interest, e.g., "computation", "platform", "hacking" and so on; a total of 40 hypernyms were retained for further analysis; (4) determining the weight given within an organization's documents to a concept by calculating the percent occurrences of that hypernym out of the total of all hypernyms—this is denoted by the variable name *hypernym_weight*; (5) clustering organizations based on the hypernym occurrence percent, using a k-means clustering method, where an Iteratively Reweighted Least Squares minimizes the root mean square difference between the data and the corresponding cluster means.

Various metrics were used to determine the final choice of number of clusters and spatial disjointness of the clusters, including the overall R^2 of 0.49 and Sarle's Cubic Clustering Criterion (CCC) of 1.2, lending further evidence that the clusters are spatially separated enough to provide meaningful groupings; and (6) labeling clusters according to the relative values of the hypernym_weight within the cluster.

3 Analysis of Urban Digital Infomediaries

In our approach, there are three aspects to understanding the emerging urban data sector: identifying the types of entities that are active in this space; understanding the scope of what the organizations do; and, assessing the extent to which these boundaries are blurring over time towards the goal of inferring trends towards convergence. We identify four major groups of urban digital infomediaries consisting of nine specific organizational types, based on their stated mission and objectives, and the products and services delivered. Table 1 shows these four groups, with a description of the specific types of organizations within each group, and the sectors (public, private, non-profit, informal) to which they belong. The table also displays the percentage of the total sample which a specific type of organization comprises of.

Table 1 Urban digital infomediaries and dominant sector

Organization type		Description	Dominant sector	% of total sample (N = 139)
General-purpose ICT infomediaries				
SCC	Smart City Companies (includes units of comprehensive ICT businesses)	Companies or business units focused on improved performance and efficiency of city ICT systems	Private	18
MSICTC	Multiple-service ICT Companies	Organizations providing multiple hardware, software, and communications services targeted to location-based information, information-sharing, collaborative tools and related products	Private	6
Urban information service provider infomediaries				
CIS	City Information Services	Organizations providing directory services or other information for residents to connect to social, entertainment and other aspects of cities	Private, informal	4
LBS	Location-Based Services	Organizations providing generic location-focused services including navigation, retail, health and wellbeing based on location as well as social interaction and urban engagement opportunities based on location-based social networks	Private, informal	21
Urban open and civic data infomediaries				
ODO	Open Data Organizations	Organizations publishing open data for further analytics and use	Public, non-profit, informal	30
CHO	Civic Hacking Organizations	Organizations analyzing and distributing civic statistics, maps and other information of interest for civic and public discourse	Non-profit, private, informal	3
CBISO	Community-Based Information Service Organizations	Organizations and individuals connecting information services to specific cities, communities and neighborhoods through analytics, content creation, visualization, mapping and other methods	Public, non-profit, informal	11

(continued)

Table 1 (continued)

Organization type		Description	Dominant sector	% of total sample (N = 139)
Independent and open source applications, software and content developer infomediaries				
IAD	Independent App Developers	Individuals primarily focused on developing software and apps to link citizens to information	Informal	3
OSD	Open Source Developers	Organizations and entities creating open source software, social coding accounts, developer networks and other open source ways to allow access to Big Data and Open Data	Private, informal	4

To a certain extent, the distinctions we have drawn among organizations are already product or service-based. Nevertheless, it is useful to examine the range of products and services with which urban data organizations are involved and the extent to which organizations rank high or low in being a producer versus a user of specific products and services. Four types of product-service mix can be identified for the organizations examined: (1) data generation, communication and management technologies such as sensor systems, wired and wireless communication systems, information processing and database management systems, positioning systems, web services, and associated hardware and software to manage data, (2) software solutions including data platforms and tools to enable further use of urban data; (3) analytics and knowledge-discovery services including data mining, urban and regional modelling, mapping and visualization and human interpretation of the results, towards understanding and exploring cities and communities, and (4) end-user services including collaborative community decision support tools, user apps for numerous functions and various web-based information services. These are described in Table 2 with our qualitative ranking of the infomediaries on the four product-service mix considered, with "H" indicating our assessment of an organizational type ranking "High", "M" for "Medium" and "L" for "Low".

Based on the information above, the urban ICT infomediaries are as follows:

General Purpose ICT Infomediaries: These organizations are most likely to provide services towards building and managing the intelligent infrastructure in networked cities, and is composed almost entirely of private firms. The primary business models for this group are Business-to-Government (B2G) and Business-to-Business (B2B), although the value of their foundational information infrastructure ultimately benefits myriad of end users. The group is composed of two types of business organizations which are distinguished by the extent of urban emphasis in

Table 2 Organizational products and services

Infomediary type	Data management and communication tools — Technologies to generate, network, manage, integrate and communicate information including associated hardware and software to manage data	Platforms and tools to connect data to communities — Technologies to disseminate information for further applications development, and end users, e.g., open data portals, city dashboards, social coding services	In-house analytics and knowledge-discovery — Use of tools for data mining, urban and regional modelling, mapping and visualization and human interpretation of the results on cities and communities	End-user services — ICT services to be used by communities including collaborative decision support tools, apps and various web-based information services
SCC	H	H	H	L
MSICTC	H	L	L	L
CIS	L	L	L	H
LBS	H	L	M	H
ODO	H	H	M	L
CHO	L	M	H	L
CBISO	L	L	H	M
IAD	M	L	M	H
OSD	H	M	L	L

their overall product–service mix (for example, the percentage of total business geared to urban ICT products and services).

(1) Smart City Companies (or Units): These entities are focused on improved performance and efficiency of city ICT systems. While there are many definitions of smart cities, the overall vision is one of having ICT-focused solutions be an integral part of urban development, driving economic competitiveness, environmental sustainability, and general livability. As noted by Thakuriah and Geers (2013), the term "smart city" is championed by commercial entities and the expectation is that the networking and integration of multiple urban sectors will enable cross-agency efficiencies for a range of services (such as traffic management, utilities, law enforcement, garbage disposal, emergency services, aged care, etc). In some cases, the entire SCC business is focused on city-centric applications of intelligent infrastructure, data management and analytics, and in other cases, the smart city business is handled by specific units within comprehensive ICT businesses which are involved in many other ICT sectors such as health, finance, energy and so on. This group also includes consulting firms which have historically offered services in Intelligent Transportation

Systems, smart energy and water management and other sectors. As shown in Table 2, SCCs are high on data management and communications tools and in developing platforms and tools for further processing of data. They are likely to be involved in analytics and knowledge discovery processes for the purpose of business solutions, but are likely to be involved in end-user services to a lesser extent.

(2) Multiple-Service ICT Companies: These are business organizations providing foundational, general-purpose hardware, software, and communications services targeted to location-based information, information-sharing, search engines, map databases, web services, sensing technologies, collaborative tools, social media, Web 2.0 and related products. They provide general ICT services in the telecommunications and information technology sector, without being focused on urban applications such as smart cities, or are focused on them only in incidental ways, without making the urban focus a core aspect of business; yet, the information infrastructure they provide are vital to urban data initiatives. For example, they are owners of cell phone data which has a wide variety of urban mobility analysis applications. As in the case of SCC, MSICTC can range from small private firms to large multinationals.

Urban Information Service Infomediaries: The second group consists of organizations which deliver end-user ICT services to urban residents to explore cities and communities, and to connect citizens to social, entertainment, economic and commercial opportunities. The services provided are connected to specific business models such as mobile commerce and location-based advertising or banner ads. While the majority of the organizations we examined are established private businesses, some of these services are also being offered by informal, independent developers. Two specific digital infomediaries can be differentiated by the types of information they provide, the degree to which location and real-time information streams are explicit and front-and-center in their products, and the extent to which social networking processes are utilized.

(3) City Information Services (CIS): CIS organizations provide directory services, question and answer databases or recommender systems for residents to engage in social, commercial, entertainment and other aspects of cities. As shown in Table 2, CIS are generally focused on end-user services. These organizations often utilize crowdsourced information by means of user reviews and ratings of businesses, entertainment services, restaurants and other retail and commercial entities, thereby creating user communities who may form a social network. CIS infomediaries tend to be private companies or informal organizations and independent developers. They are distinguished from Location-Based Services by not requiring explicit positioning and navigation capability, which call for additional (positioning and sensor) technologies.

(4) Location-Based Services (LBS): The LBS industry is either private and to a limited degree, informal, and has been studied extensively. LBS are information services that capitalize on the knowledge of, and are relevant to, the mobile user's current or projected location. Examples of LBS include resource

discovery services in response to spatial queries such as "Where am I?" or "What's around me?" Other examples are directory assistance and service location (for example, find the nearest gas station with cheap gas), Points of Interest locations (for example, find the social services building), routing and navigation and many others. The LBS industry has been noted to be highly heterogeneous, with many different types of players. These include mobile operators, content providers, content aggregators, wireless application service providers. One specific group of LBS are Location-Based Social Networks (LBSN) which explicitly connect location to social networks not only by adding location information to an existing social network so that people in the social structure can share location-based information, but also new, perhaps ad-hoc social networks made up of individuals connected by locational proximity (Zheng 2011). Due to their utilization of location-aware technologies, real-time and heterogeneous information sources, LBS companies are likely to be high on data management aspects, as well as end-user solutions.

Urban Open and Civic Data Infomediaries: This group consists primarily of government, non-profit and informal entities, but with a sprinkling of private organizations (in this highly dynamic environment, it is also possible that the informal entities are looking for business models and private equity funds). The major goal of this group, which consists of three types of infomediaries, is to make government data available for further use or to work with administrative and other data to enable public discourse regarding government transparency, community involvement and civic engagement.

(5) Open Data Organizations (ODO): These are primarily ICT-focused units within government agencies publishing Open Data for further analytics and use. Many cities now have data portals where government agencies upload digital information that is license-free and in non-proprietary formats. Data that are being released in portals are on transportation, public, health, crime, public works and public services, education, and economic development investments and business affairs. Recent trends in government-supported APIs and mashups have enabled the ability to tap into such data in a growing number of ways. ICT entrepreneurs in government agencies and civic leaders pushing the agenda of open standards and open source software has been catalysts in Open Data initiatives. Table 2 shows that ODSs are also likely to be involved in data management and communications aspect, and due to their objective of making data accessible for end-uses, also on platforms and tools for further use.

(6) Civic Hacking Organizations (CHO): CHOs are involved in data-centric activism relating to civic and community issues of interest to them, through analytics, visualization and knowledge-discovery of administrative data. This information is shared as civic statistics, maps and other media with government and community leaders, other decision-makers and the general public, thereby generating informal urban analytics on civic or governance matters of value to specific user communities. Within this sphere of work, informal analysis relating to government transparency, governmental funding priorities, and

related topics have gained much attention. As noted previously, in addition to being organized into CHOs, civic hackers may act independently, without being a part of a formal organization. CHOs may also be involved to a lesser in the creation of collaborative, interactive solutions to present data for additional processing by others.

(7) Community-Based Information Service Organizations (CBISCO): These are typically established community organizations providing a range of services in addition to connecting data and information services to citizens of specific communities and neighbourhood. The information that is disseminated include the output of analytics, content creation, visualization, mapping and other methods. CBISOs are, like CHOs, highly involved in analytics about their area, but the analytics work may not be done by the organization itself; their value in the urban data chain is to filter and package appropriate information for use by their stakeholders. They are also likely to host community decision-support and collaborative tools for participatory neighbourhood problem-solving.

Independent and Open Source Applications, Data and Content Developer Infomediaries: This group of ICT-focused infomediaries develop cross-cutting urban applications and user apps but can be distinguished from the previous groups in that their work takes place in primarily independent and informal ways, in contrast to established organizations.

(8) Independent App Developers (IAD): This group is composed of entrepreneurial ICT developers, who are primarily focused on developing apps to link citizens to information primarily in the Business-to-Customer (B2C) sector. Ferraro and Aktihanoglu (2011) have identified several approaches to building user communities and revenue models for B2C commerce, which are particularly relevant to IADs, including "freebie", advertising-supported, and "premium" versions. IADs are involved in end-user solutions, and in the process of doing so, they may be a part of data management and analytics solutions.

(9) Open Source Developers (OSD): Of great value to the open data community are open source software and social coding accounts, which makes data, coding and analytics tools freely available. There has been a general shift in the urban data communities to open standards, software and resources. For example, major Big Data management tools such as Hadoop and Pig are open source. Urban-focused OSD develop or modify open source software for the creation of civic statistics, contribute lines of code to social coding accounts, and develop open-source developer API. OSDs also contribute to urban data through developer networks and other networked approaches to allow access to urban data. OSDs are likely to be a part of data management and communications activities, as well as activities relating to the development of solutions for further processing and analytics of data by others.

4 Activity Clusters: Approaches and Processes for Convergence

One of the questions raised earlier is the potential blurring of activities among different types of infomediaries with different mission/objectives. The text mining exercise described in Sect. 2 was used to extract information on the types of activities undertaken by the 139 organizations.

This led to the identification of seven Activity Clusters given in Table 3 which describe processes and approaches undertaken to operationalize organizational objectives: (1) data, computation and tool-building; (2) accessible, advisory, citizen-oriented; (3) economically efficient and resilient urban communities;

Table 3 Activity clusters of urban digital infomediaries

Activity cluster no.	Activity focus cluster label	Description of activities	Exemplar hypernames with high ranking
1	Data, computation and tool-building	Involved in intelligent infrastructure and computationally-intensive application development	Computational and data management constructs such as "algorithm", "code", "intelligent", "computation", "platform"
2	Accessible, advisory, citizen-oriented	Work include a focus on citizen-oriented accessibility and equity concerns, possibly on the social justice aspects of the use of ICT	"Accessibility", "advisory", "citizen", "equity" and related terms
3	Economically efficient and resilient urban communities	Focus on ICTs from the perspective of economic development, resiliency and sustainability	"Urban", "communities", "neighborhood", "economic efficiency", "effectiveness", "collaborative"
4	ICT-focused urban systems management	Focus on urban systems management, but with some urban engagement, e.g., technology for resource discovery, community monitoring and sharing	"System", "efficiency", "management", "urban", "intelligent", "engagement"
5	Smart and sustainable communities	Focus not unlike Clusters 2 and 3, but the with an additional focus on smart and sustainable urban communities	"Accessibility", "advisory", "citizen", "communities", "economic efficiency", "smart", "effectiveness", "collaborative"
6	Community information and location services	Involved in community-based information and ICT support	"Community", "apps", "location", "collaborative", "participation" and "platform"
7	Accountability, advocacy and data activism	Focus on data-centric activities for civic activism and advocacy	"Accountability", "transparency", "activism", "participatory" and "collaborative"

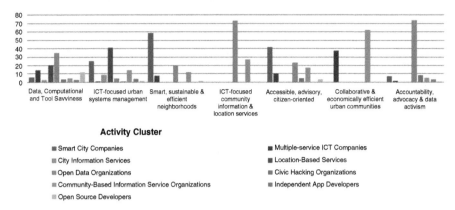

Fig. 1 Distribution of organization type across activity clusters

(4) ICT-focused urban systems management; (5) smart and sustainable communities; (6) community information and location services; and (7) accountability, advocacy and data activism.

The spread of the nine different organizational types across these seven Activity Clusters is given in Fig. 1. Given the overall focus on technology, it is not surprising that all nine organizational types have at least some organizations that are dominant in Activity Cluster 1, "Data, Computational and Tool Savviness". However, for close to 35 % of ODO organizations, this is the dominant activity, as determined by hypernyms, followed by LBS and MSICTs, indicating that such organizational types with their interest in data sharing and publication are pushing the envelope in technology development around urban data.

LBS and SCCs dominate in the "ICT-focused urban systems management cluster". In the case of this activity cluster, as in the case of first cluster, it is somewhat surprising that Community-Based Information Service Organizations dominate quite strongly in terms of overall focus, indicating that very different types of organizations are carrying out functionally similar activities. The dominance of SCC, ODO and CBISO organizations within the "Accessible, advisory and citizen-oriented" activity cluster similarly suggests increased convergence in the use of data and technology to serve cities.

Similar trends regarding the dominance of very different types of organizations within other activity clusters seem to indicate increasing convergence of focus. Studies of convergence go back to the analysis of the machine tool industry by Rosenberg (1963) and recently, the ICT sector has received considerable interest— convergence not only in technologies from the supply side used but also in products from the demand side (Stieglitz 2003). It is likely that as urban digital infomediaries in historically different industries increasingly use urban data to produce similar products and services, convergence may be stimulated, leading to potentially useful networks, alliance strategies and partnerships.

5 Skills and Interests of the Workforce Involved

Within these organizations are a wide spectrum of professionals with interests in urban data and related analysis and operations, and backgrounds and experiences partly shape the activities of the infomediaries. One group of professionals is "data scientists". One of many definitions of data scientists is that they are individuals "with the training and curiosity to make discoveries in the world of Big Data" (Patil and Hammerbacker n.d.). Data scientists are likely to be drawn into urban analytics due to the data-rich characteristics of cities, and the potential for scientific discoveries and commercial innovations.

A second group which is involved in urban data are ICT developers. Developers perform myriad functions within this environment, including the integration of ICT systems of different urban sectors, development of city or community dashboards or "apps". This subgroup may also include ICT-oriented individuals social entrepreneurs (civic hackers) who produce curated information including mashups that convey relevant events and apps for others to use ("apps for democracy") and ordinary citizens who access the data to understand more about transportation and other conditions in their communities.

A third group consists of analysts trained in urban and regional planning or related disciplines such as geography, economics, or civil engineering. Some within this subgroup are professionals with an interest in urban models and simulations for quantitative analysis of urban areas, but in contrast to the models of data scientists, who tend to be "data-driven", analytical tools of urban and regional planning professionals are driven by a long history of various aspects of urban theory (for example, regarding urban economies, mobility and so on) and in the quantitative analysis of various aspects of cities, such as transportation, regional economic analysis, public health, environmental planning and so on, and in the theoretical and methodological aspects of their subjects.

A fourth group consists of managers who lead data-centric projects in cities or communities. Examples include private sector managers responsible for delivering a Big Data sensing project to a government agency or public managers responsible for bringing Open Data Portals online. Other examples include community planners who are establishing neighbourhood dashboards with community information and civic statistics, or transit managers who are responsible for delivering a real-time transit arrival system. Urban data project managers may be required to take on multiple leadership roles in order to see a project from start to finish, including project championing (Lam 2005), having the social skills to communicate sometimes highly technical information to non-technical audience, as well as technical skills, team building and project management skills (Gil-Garca and Pardo 2005).

General Purpose ICT Infomediaries are most likely to have developers, data scientists and ICT managers and administrators in their mix of employees. The Urban Information Service Provider Infomediaries group too is likely to be staffed with developers and data scientists. They are also likely to utilize the work of citizens who generate information via sensing and crowdsourcing systems, which are then shared with other service users. Urban Open and Civic Data Infomediaries are likely to employ personnel formally trained in disciplines in the urban and regional planning domains, whether as methodologists or generalists interested in civic and community issues. Within this group, informal civic hackers with no formal ties to any of the organizations discussed here may be actively involved. This group is also likely to utilize the work of ordinary citizens in the monitoring and reporting of events within their communities.

Independent App Developers or Open Source Developers are very likely to be developers or data scientists. It is also possible that within the IAD group, citizens with no formal training in informatics or the urban disciplines, but who have self-taught the use of social software and open source tools, are playing a part. Data-centric managers, as described above, are likely to be employed in all infomediary groups considered, since broad-based ICT technical, managerial and entrepreneurial skills are needed in virtually all urban data sectors.

6 Role of Networks

As noted previously, many of the activities driving urban data are not occurring within traditional organizational boundaries, but through formal and informal networks involving the actors discussed above. We discuss this aspect here very briefly, with the note that this topic is a significant research area in its own right.

Informal networks of informed citizens, civic technologists and civic hackers have become an important aspect of urban data and it is possible that they are attracting individuals from a wide variety of ICT organizations, although to the best of our knowledge, we have not seen the results of any study on this particular topic. In the area of civic hacking, ongoing networks are emerging due to the use of social networking (both online and face-to-face meetings via Meetup groups) to exchange knowledge among civic hackers who are more tech-savvy developers and the less technically savvy, and to discuss developments in software, data and policy and civic issues, is becoming increasingly important. "Hackathons" that are being sponsored by government agencies and non-profits, as well as design and crowdsourced competitions for citizen apps, are giving increasing identity, visibility and legitimacy to these activities. A slightly different type of network are those spearheaded by primarily established ICT companies, and focused to a greater degree on urban data and urban management applications, in contrast to civic and government transparency issues.

As indicated earlier, members of the organizations we discussed tend to populate ongoing networks relating to data, communication and other standards, and other technical aspects related to urban data. In contrast to ongoing networks, there are also project-based networks in the urban data sector. These are formed primarily around technology-focused city projects (smart city projects, field operational tests of intelligent transportation, smart energy systems and so on) primarily by members of government agencies who are sponsoring the project, businesses that are providing the service, and affiliate members consisting of other planning and administrative agencies, non-profits, higher education and research institutions. These types of networks are often formed as result of project requirements for partnerships or public participation, and their work typically end when the project is complete, although they may continue to come together well after that to follow up on evaluation results and to develop strategies regarding lessons learned.

7 Urban Data Infomediaries and Functional Dimensions

In the previous discussion, several different measures were considered in understanding urban data infomediaries: i.e., their mission, products and services, major activity clusters, and type of workforce involved. Based on the information contained in the measures, urban data infomediaries can be considered as having five dimensions of functional interest: techno-managerial, scientific, business and commercial, urban engagement, and openness and transparency. The alignment between the organizational type and the dimension of functional interest is given in Table 4.

One dimension is *techno-managerial* and focuses on how urban data can foster effective management of cities through interconnectivity among different urban sectors, collaboration with citizens and communities, and dynamic resource management. A second dimension of interest is *scientific* focusing on quantified urbanism with the view that urban data can help learn about cities in new ways thereby creating new scientific understanding and knowledge discovery; however, scientific does not mean a purely research-level endeavor, as scientific discoveries can be put

Table 4 Functional interest dimensions of urban data infomediaries

Infomediary type	Functional interest dimension
Smart City Companies	Techno-managerial; business and commercial; scientific
Multiple-service ICT Companies	Business and commercial; scientific
City Information Services	Business and commercial; urban engagement
Location-Based Services	Business and commercial; techno-managerial; scientific
Open Data Organizations	Openness and transparency; urban engagement; techno-managerial; scientific
Civic Hacking Organizations	Openness and transparency; urban engagement; scientific
Community-Based Information Service Organizations	Urban engagement; techno-managerial; urban engagement
Independent App Developers	Business and commercial; urban engagement; scientific
Open Source Developers	Openness and transparency; scientific

to operational and policy use. This dimension also views data on city dynamics as offering interesting opportunities and test-bed to address communications, information processing, data management, computational, and analytics challenges posed by urban Big Data. Another aspect of the scientific interest is to make advances in open source and social software, technologies for privacy preservation and information security, and other challenges associated with urban data.

A third dimension is *business and commercial* where previous modes of e-commerce are being augmented with location-based social networks for mobile commerce, user-generated content for reviews and recommender systems, crowdsourcing of input for idea generation and other business product development, and other commercial purposes in cities, ultimately leading to participant-generated information on the social, recreational and entertainment aspects of cities.

A fourth dimension of interest is *urban engagement and community well-being*, with a focus on civic participation and citizen involvement. One aspect of this strand is community-based information and monitoring, which may involve technologies similar to techno-managerial strand.

A fifth dimension is *openness and transparency* of government information towards more interactive, participatory urban governance and bearing close similarities to other "open" movements including open access, open source, open knowledge and others. One aspect of this strand of interest is likely to overlap with the second aspect of the scientific dimension, i.e., on computational and data management aspects.

8 Conclusions

Our objective was to make a qualitative assessment of public, private, non-profit and informal infomediaries who are pushing the agenda of the urban data sector. Using a mixed-methods approach, we identified four major groups of organizations: general-purpose ICT infomediaries, urban information service provider infomediaries, urban open and civic data infomediaries, and independent and open source developer infomediaries.

A total of nine organizational types were highlighted within these four groups. Organizations are found to have the following seven areas of focus regarding their activities and process: data, computation and tool development; accessible, advisory, citizen-oriented; economically efficient and resilient urban communities; urban systems management; smart and sustainable communities; community information and location services; and accountability, advocacy and data activism focus. A variety of professionals are involved within these organizations in urban data including urban and regional planning professionals, data scientists, developers, and ICT-trained or urban-trained project managers. The work of the organizations also involves others such as civic hackers and citizen scientists, in a mixed of paid work and volunteer efforts.

The urban data sector is highly dynamic and involves a significant informal entrepreneurial sector which is entering the domain given the opportunities and the overall challenges involved. However, for these opportunities to be realized, a broad-based strategy is needed that reflects research and policy deliberations regarding the social, policy, behavioral and organizational implications of the data management and dissemination processes. This entrepreneurial sector is itself highly diverse and includes developers passionate about computational and technological challenges, and data scientists who are interested in analyzing complex data, as well as in open source software.

Urban data are also increasingly being seen by professionals in the urban planning and public management domains as being important towards urban management. They also include the work of civic hackers and civic technologists who value openness and transparency and open source technologies, and are interested in data and analytics to address civic and urban challenges. While some entrepreneurs work in the private firms, civic and public agencies, or in non-profits, others freelance in ICT development work.

Using all the different measures, the dimensions of interest addressed by urban data infomediaries are fivefold: techno-managerial, scientific, business and commercial, urban engagement, and openness and transparency. Using these dimensions, it may be possible to predict aspects of city operations and management where innovations may result from the work of urban data infomediaries.

Informal networks are playing an important role in creating and sharing knowledge regarding technical skills and urban and governance issues. Social networks that have formed among developers, data scientists and others involved in civic hacking are important in this regard, but require greater involvement of professionals from the urban research community. The participation of the latter group is particularly important, since what is sometimes presented as novel digital modes of urban planning, are effectively practices and strategies that are well-established in the urban domain. In much the same way, much can be learned about emerging technology and analytics solutions by the urban community. This indicates that there should overall be greater cross-fertilization of knowledge among the various professional domains involved.

One issue we were interested in is the idea of convergence, which is a general trend in the ICT sector, as noted by many authors. We found evidence of *technical convergence* because many different types of organizations use similar technologies and offer similar urban data products and services. We also found through our clustering analysis that several different types of organizations across the four infomediary groups are focused on similar ICT-focused activities relating to urban management, accountability and data activism.

However, less evident is *ideological convergence*. Ideological tensions occur with differences in viewpoints regarding the way things should be, or should be done. There are several examples in the urban data landscape. A far from complete list is on ideological tensions regarding what open government should mean: transparent government, innovative or collaborative government. These questions are important to consider because they have implications for policy and stakeholder generation, as well as for practice and bodies of knowledge in these areas.

The study has several limitations. The emerging nature of the sector necessitated an exploratory study. Aside from the informal nature of the sample of organizations examined, another limitation is that it consists of only those entities that have a formal presence (websites) in the Internet. Informal digital infomediaries who do not have websites but are active through blogs, social networking sites such as Facebook, or have a social web presence via social coding services such as Github, SourceForge and so on, or who have no presence in the Internet at all, are not included in this study. This potentially excludes a significant share of informal digital infomediaries including civic hackers, many of whom are one-person entities contributing without an established organizational presence. However, we were able to identify some organizations which are undertaking civic hacking activities and these are included in the sample. At the time of writing this paper, we are administering a survey instrument to gather data on such the independent data activists and civic hackers. Another limitation of the sample is that it excludes higher education and research institutions, some of which are focusing heavily on Big Data and Open Data research.

References

Batty M (2013) Big data: big issues. Geographical magazine of the royal geographical society. p 75
Beyer MA, Laney D (2012) The importance of 'Big Data': a definition. Gartner
Ferraro R, Aktihanoglu M (2011) Location-aware applications. Manning, Greenwich
Gil-Garca JR, Pardo TA (2005) E-government success factors: mapping practical tools to theoretical foundations. Gov Inf Q 22:187–216
Lam W (2005) Barriers to e-government integration. J Enterp Inf Manag 18:511–530
Patil DJ, Hammerbacker J (n.d.) Building data science teams. http://radar.oreilly.com/2011/09/building-data-science-teams.html#what-makes-data-scientist. Accessed 1 Aug 2014
Provost F, Fawcett T (2013) Data science for business: what you need to know about data mining and data. O'Reilly Media, Sebastopol
Rosenberg N (1963) Technological change in the machine tool industry, 1840–1910. J Econ Hist 23(4):414–443
Stieglitz N (2003) Digital dynamics and types of industry convergence: the evolution of the handheld computers market. In: Christensen JF, Maskell P (eds) The industrial dynamics of the new digital economy. Edward Elgar, Cheltenham, pp 179–208
Thakuriah P, Geers DG (2013) Transportation and information: trends in technology and policy. Springer, New York. ISBN 9781461471288
Thorhildur J, Avital M, BjÃ"rn-Andersen N (2013) The generative mechanisms of open Government data. In: ECIS 2013 Proceedings. Paper 179
Xing X, Ye L-K (2011) Measuring convergence of China's ICT industry: an input–output analysis. Telecommun Policy 35(4):301–313
Zheng Y (2011) Location-based social networks: users. In: Zheng Y, Zhou X (eds) Computing with spatial trajectories. Springer, New York, pp 243–276

How Should Urban Planners Be Trained to Handle Big Data?

Steven P. French, Camille Barchers, and Wenwen Zhang

Abstract Historically urban planners have been educated and trained to work in a data poor environment. Urban planning students take courses in statistics, survey research and projection and estimation that are designed to fill in the gaps in this environment. For decades they have learned how to use census data, which is comprehensive on several basic variables, but is only conducted once per decade so is almost always out of date. More detailed population characteristics are based on a sample and are only available in aggregated form for larger geographic areas.

But new data sources, including distributed sensors, infrastructure monitoring, remote sensing, social media and cell phone tracking records, can provide much more detailed, individual, real time data at disaggregated levels that can be used at a variety of scales. We have entered a data rich environment, where we can have data on systems and behaviors for more frequent time increments and with a greater number of observations on a greater number of factors (The Age of Big Data, The New York Times, 2012; Now you see it: simple visualization techniques for quantitative analysis, Berkeley, 2009). Planners are still being trained in methods that are suitable for a data poor environment (J Plan Educ Res 6:10–21, 1986; Analytics over large-scale multidimensional data: the big data revolution!, 101–104, 2011; J Plan Educ Res 15:17–33, 1995). In this paper we suggest that visualization, simulation, data mining and machine learning are the appropriate tools to use in this new environment and we discuss how planning education can adapt to this new data rich landscape. We will discuss how these methods can be integrated into the planning curriculum as well as planning practice.

Keywords Big data • Urban planning • Analytics • Education • Visualization

---

S.P. French, Ph.D., F.A.I.C.P. (✉)
College of Architecture, Georgia Institute of Technology, Atlanta, GA 30332-0155, USA
e-mail: steve.french@coa.gatech.edu

C. Barchers • W. Zhang
School of City and Regional Planning, Georgia Institute of Technology, Atlanta, GA 30332-0155, USA
e-mail: cbarchers3@gatech.edu; wunwunchang@gmail.com

Planning methods have been the source of much discussion over the past few decades. Practitioners and researchers have examined what methods planning schools teach and how these methods are used in practice. The suite of traditional methods courses taught in planning programs—inferential statistics, economic cost-benefit analysis, sampling, and research design for policy evaluation—remains largely stagnant, despite the rapidly changing reality in which planners are expected to work. Although the focus of this paper is on the impact of big data for planning methods, other variables have also contributed to the need for additional methods to tackle planning problems. The rise of ubiquitous computing and a hyper-connected communication network as well as new private investment in data collection have created an environment in which greater amounts of data exist than ever before. The ability of the planner to analyze and use this data is no longer limited by computing power or the cost of data collection, but by the knowledge that planners possess to employ data analytics and visualization techniques.

Educating planners with skills that are useful for practice has been a key tenant of many planning programs over the years. Several studies have been conducted to understand how well planning programs are succeeding at this goal or not. Surprisingly, the most recent comprehensive investigation of planning education and skills demanded by practitioners was conducted in 1986. In this survey, four important conclusions were identified as relevant to how planners were being educated and the professional skills they would be required to use (Contant and Forkenbrock 1986). They found that the methods taught in planning programs remained highly relevant to the methods needed for practicing planners, and the authors concluded based on their survey results that planning educators were adequately preparing their students to solve planning problems in practice. They cited communication skills (writing and speaking) and analysis and research design as critical components of planning education and practice, but noted that educators needed to remain vigilant on seeking relevance (Contant and Forkenbrock 1986). The article also identified several changes that were occurring throughout the 1980s that affected the planning profession—the rise of micro-computing and the expansion of methods being offered by planning schools. Contant and Forkenbrock (1986) wrote "…there is little to suggest that planning schools are overemphasizing analytic methods, nor do they appear to be failing to any real extent in meeting the demands of practitioners interviewed. While more techniques are required than these practitioners feel that all planners should understand, it certainly is arguable that this situation is not at all bad." That survey of methods is now nearly 30 years old, and new realities exist that require educators to revise and expand the scope of methods taught in planning schools (Sawicki and Craig 1996; Goodspeed 2012).

Despite wide acknowledgement of the changing data landscape, planning curricula still resemble their traditional form. Kaufman and Simons completed a follow-up to this investigation which surveyed planning programs specifically on methods and research design. The more limited focus on this 1995 study "revealed a rather surprising lack of responsiveness among planning programs over time to practitioner demand for [quantitative research methods]" and that "planning programs do not seem to teach what practitioners practice, and not even what practitioners should practice" (Kaufman and Simons 1995). In a 2002 study focused on the use of technology within planning programs, Urey claims that the haphazard approach with which planning programs have introduced the use of technology to serve larger goals (research, analysis, modeling) might be problematic as increased microcomputing power becomes more widespread. While manual techniques serve learning objectives within planning methods courses, the use of technology is now required (Urey 2002). This leaves planning educators today with two questions relevant to big data and methods: what new methods must we now include in our curriculum, and what technology must students understand to employ these methods in an ethical, accurate, and precise way? Given these questions, we reviewed current methods requirements at planning schools to assess whether or not planning programs have begun to respond to these questions and adapt to the changing data landscape.

In a non-scientific review of methods taught at the top ten planning schools (as listed by Planetizen in 2014 [http://www.planetizen.com/education/planning]), we discovered that almost all programs require that planners be trained in statistics, economic cost-benefit analysis, and research design. Of the programs reviewed, including MIT, Cornell, Rutgers, UC Berkley, University of Illinois Urbana Champaign, UNC Chapel Hill, University of Southern California, Georgia Institute of Technology, UCLA, and University of Pennsylvania, none required students to seek additional data analysis courses outside of the planning department. Although the review of these programs was not scientific and limited to information published online for prospective students, it does suggest that planning education has yet to see value in teaching planners methods widely adopted in the fields of computer science and engineering. We argue, as Contant and Forkenbrok did 30 years ago, that maintaining the relevance of planning education to planning practice is important. Contant and Forkenbrok reminded educators to be vigilant in their understanding of skills that are in demand for practitioners—yet we have failed to do this in regards to our methods curricula.

The one big exception to the static nature of planning methods offerings is geographic information systems (GIS). Almost all of the top programs include a required course on GIS or include a significant section on GIS as a portion of a required methods course. This technology, once the province of a subset of computing nerds, has spilled out of the methods sequence and permeated the curriculum. It is now common to see planning students using GIS as a part of land use, housing, transportation and economic development courses. The adoption and use of GIS has been the most sweeping change in planning methods curriculum over the past 30 years. For a discussion of this history and how this technology is evolving, see Drummond and French (2008).

Big data, although currently a popular topic, is not new—and the concept of big data dates back to 2001, when industry analyst Doug Laney articulated the definition of big data as any data set that was characterized by the three Vs: Volume, Velocity and Variety (Laney 2001). Big data sets are characterized by containing a large number of observations, streaming and fast speed and requiring real time analytics. Big data sets are also usually mixed format combining both structured and unstructured data, joined by a common field such as time or location. In sum, any data sets that are too large and complex to process using conventional data processing applications can be defined as big data.

Several pioneers in the industry have already started to process and analyze big data (Lohr 2012, Cuzzocrea et al. 2011). For instance, UPS now tracks 16.3 million packages per day for 8.8 million customers, with an average of 39.5 million tracking requests from customers per day. The company stores more than 16 petabytes of data. Through analyzing those datasets, UPS is able to identify real time on-road traffic conditions, daily package distribution patterns and together with the latest real time GIS mapping technology, the company is able to optimize the daily routes for freight. With all the information from big data, UPS has already achieved savings in 2011 of more than 8.4 million gallons of fuel by cutting 85 million miles off of daily routes (Davenport and Dyché 2013). IBM teamed up with researchers from the health care field to use big data to predict outbreaks of dengue fever and malaria (Schneider 2013). It seems that big data, together with advanced analysis and visualization tools, can help people from a wide variety of industries explore large, complex data sets and reveal patterns that were once very difficult to discover. Given the increasing use of big data across fields that share interests with the field of city planning, planners should more deliberately explore and develop methods for using big data to develop insights about cities, transportation patterns and the basic patterns of urban metabolism.

Data analytics, as a powerful tool to investigate big data, is becoming an interdisciplinary field. There are new programs at universities across the United States that aim to teach students how to grapple with big data and analyze it using various analytic tools. For this paper, we collected and reviewed some common tools and skills that are taught in data analytics courses. We gathered course information from John Hopkins, Massachusetts Institute of Technology, University of Washington, and Georgia Institute of Technology. We noted that machine learning/data mining and data visualization are the tools that are frequently taught in these programs to prepare students to handle big data and some of them are actually quite new to urban planners.

Machine learning is a core subarea of artificial intelligence. Machine learning uses computer algorithms to create explanatory models. There are different types of learning approaches, including supervised learning, unsupervised learning, and reinforcement learning. Although some of the terminologies may be completely new to planners, the actual methods turn out to be quite familiar. For example, the regression model is one of the methods that is frequently used in supervised learning process. Planners who work with remote sensing images often apply supervised classification methods to reclassify the images into land cover images based on various color bands in the image. However, planners may not be familiar with other machine learning methodologies or algorithms, such as unsupervised learning and reinforcement learning. Unsupervised learning tries to identify regularities (or clusters or groupings) in the input datasets without correct output values provided by the supervisors. Reinforcement learning is primarily used in applications where the output of the system is a sequences of actions (e.g. playing chess).

In this case, what's important is not a single action, but a sequence of actions that will achieve the ultimate goal. When machine learning methods are applied to large databases, such as big data, it is often called data mining. Data mining tries to identify and construct a simple model with high predictive accuracy, based on the large volume of data. The model is then applied to predict future values. This is the kind of projection that planners have been doing for years with less sophisticated methods.

Most of the programs we reviewed also include data visualization components to help identify patterns in the data and communicate the results of data analysis. Some data visualization techniques, such as multivariate data representations, table and graph designs are quite conventional. However, those techniques may also be applied in innovative ways to help convey information behind data in a clearer manner. One example is the information graphics or infographics, which improve human cognition by utilizing graphics to improve the visual system's ability to extract patterns and trends (Smiciklas 2012; Few 2009). The latest trend in data visualization is to take the advantage of webs to present data in an interactive way. To effectively present big data interactively, the designer needs to be equipped with knowledge regarding how human beings interact with computers, and how different interaction types (i.e. filtering, zooming, linking, and brushing) will affect human being's cognition ability. In the example below, viewers can interact with data generated from Foursquare check-ins across Manhattan (Williams 2015). These interactive visualizations can be used on both big, and small data, but allowing interaction allows for more data to be presented to viewers (Fig. 1).

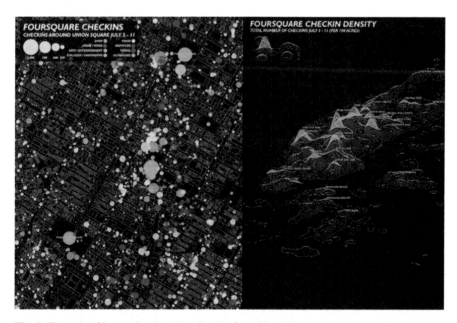

Fig. 1 Example of interactive data visualization from Here Now

In addition to the core courses, these new interdisciplinary programs require the students to master at least one programming or query language. SQL is a popular requisite and, in a survey on tools for data scientists, over 71 % of respondents used SQL (King and Magoulas 2013). Some programs also require students to understand and use open statistics software, such as R and R studio.

While these methods for analyzing data may seem somewhat out of place within a planning methods framework, they actively seek to create ways in which researchers can describe, explore, and explain data. These categories of data analysis are described in depth in Earl Babbie's *Survey Research Methods* (1990). This text serves as one of many fundamental introductions to methods for planners, and by grouping the new suite of tools available to planners and data scientists within these categories, planners can see how these tools might be useful to them. For example, data visualization is one of the key ways in which data scientists are exploring big data sets (Few 2009). Data visualization acknowledges that our typical methods of data exploration (descriptive statistics, graphing, and the like) are ill-equipped to handle larger data sets, and even less equipped to communicate information derived from those data sets to the public and to decision makers. By introducing planners to the growing field of data visualization, we can expand their ability to not only to use larger data set's but to communicate the information garnered from those data sets. As the basis for research, exploration of data sets will allow planners to ask additional questions. These additional questions will require explanatory analysis, and within this group of methods, tools such as machine learning and data mining can help planners generate predictive models from larger data sets.

Many of the data sets that planners will deal with in the future will be big data. Credit card data or web browsing histories may help planners to predict the focus of emerging public concerns. As a matter of fact in MIT's big data courses, there is a case study regarding how to utilize the Google search records to estimate the trends within the real estate industry (MIT 2014). Social media, such as Twitter and Facebook, have already become powerful information sources regarding almost every aspect of social life. Analysis of twitter feeds can help to identify the extent and intensity of hazard events. There are already studies on how to utilize information extracted from Facebook's friend list to forecast the use of airplanes. GPS or real time transportation information can help planners to calibrate and develop more accurate activity based travel demand models to forecast future travel patterns. Moreover, the real time information about energy flows such as water, sewer, and electricity flows may equip planners with critical information to design more energy efficient and sustainable cities to make built environment more resilient to natural hazards and climate change. Planning is characterized by its special affinity for place-based issues, and this focus on place will be one of the critical ways in which typical data sets can become "big data." Location is the ultimate relational field, and our ability to link data sets through location will create big data sets that are especially useful to planners. If location is the ultimate relational connector, then planning data sets will only continue to increase in size, speed, and complexity in the future. The importance of teaching planners how to effectively and accurately examine and explore this data cannot be understated, yet, our work to prepare this paper leads us to believe that planning programs have not yet taken the steps required to introduce these methods to planning students.

Big data analysis tools, such as machine learning and data visualization, can help planners to make better use of the big data sets. The Memphis Police Department has used machine learning and data mining approaches to predict potential crime based on past crime events. As a result, the serious crime rate was reduced by approximately 30 %. The city of Portland, Oregon optimized their traffic signals based on big traffic data, and was able to reduce more than 157,000 metric tons of CO_2 emissions in 6 years (Hinssen 2012). In sum, the machine learning techniques can help planners to analyze the future development of urban areas in a more accurate way to solve current problems and eliminate or at least ease some the impacts of new development. The explanatory power of machine learning will be critical for planners seeking to use big data to solve long-term challenges in cities and communities.

Data visualization has always been considered useful in the planning process, primarily as a communication method. However, it is now a critical tool for exploring large, complex data sets. Data visualization can help planners better understand how people live, work and behave within urban context. When paired with more explanatory tools such as machine learning, data visualization becomes a critical tool in the planning process. Visualization can also continue to be used as a way for planners to convey their planning concepts to corresponding stakeholders during the public participation process. In this way, visualization is used as an interpretation toolkit to help people digest the complex analysis results from big data. Planners continue to be more comfortable using traditional graphs, tables, and animation images to visualize their results. However, some planners are now using more advanced web based tools to display the information in interactive ways to encourage public participation. This trend has been on the rise for some time, and the demand for practitioners with visualization skills continues to increase (Few 2009; Sawicki and Craig 1996; Goodspeed 2012).

We argue in this paper that planners would benefit greatly from the introduction of more advanced methods of descriptive, exploratory, and explanatory data analysis in order to more effectively use an ever increasing amount of available data. When considering adding new methods to the planning curriculum, there is always the question of what will be displaced from the existing curriculum. We would urge planning educators to review their current methods carefully to see if the current offering are suitable as we move from a data poor environment to one of data abundance, At the very least, planning programs should strive to make all students aware of big data and give them some introduction to the means and methods of analyzing this data. This basic overview may be sufficient for the generalist planner, with more in depth training in big data available those who want it. This is similar to the model that was initially followed with respect to GIS—all planning students were given some basic GIS skills and vocabulary so they could communicate with spatial analysis specialists. All planning students should get some exposure to big data and its analytical techniques, but some should be able to develop more depth and the ability to collaborate with data scientists.

Two key issues for additional research emerged as we prepared this paper. The field of planning is inherently place-based, and it, therefore, has the potential to take many types of data and transform it into big data by linking mixed format information into databases based on location This suggests that planning can draw upon all types of data that is location based, including cell phone locations, license plate readers, infrastructure sensors, drone videos, and building performance data. The challenge will be how to build a theoretical framework that will allow planners to use this wealth of information. Second, the field of planning is predominantly concerned with the long-term. To date most big data applications have been used to provide insights into short term challenges. As planners, we need to be asking a larger question that relates to not just what methods can be used to analyze this data, but how this data can be employed in our search for long-term solutions. How can minute-by-minute Twitter text analysis related to planning issues allow us to reframe planning issues for years to come? How does real time transportation data help us understand how to shape transportation systems for the next generation? We did not set out to answer these questions in this paper, but we do believe that posing them will help frame the discussion of planning methods for the next generation of planning students and practitioners.

Big data represents an exciting new asset for planners who have always struggled to explore and explain patterns and trends based on limited observations of discrete data. We should make the best use of this data by giving planners the tools with which to analyze it, understand it and communicate it. Like others who have written on the topic of big data in cities, we do caution that data should not be used for data's sake. Planners are tasked with a more complex task that our data science colleagues: we must find ways in which to use the data to make existing communities better and to provide better solutions than were previously available (Sawicki and Craig 1996; Mattern 2013). In order to help planners achieve these goals, we must revamp the methods offerings in our planning programs to take full advantage of the new world of large, fast moving, ubiquitous data.

References

Babbie E (1990) Survey research methods. Wadsworth, Belmont
Contant CK, Forkenbrock DJ (1986) Planning methods: an analysis of supply and demand. J Plan Educ Res 6:10–21
Cuzzocrea A, Song I-Y, Davis KC (2011) Analytics over large-scale multidimensional data: the big data revolution! In: Proceedings of the ACM 14th international workshop on Data Warehousing and OLAP, 2011. ACM, pp 101–104
Davenport TH, Dyché J (2013) Big data in big companies. International Institute for Analytics
Drummond WJ, French SP (2008) The Future of GIS in Planning: Converging Technologies and Diverging Interests. J of Amer Plan Assoc 74:2 pp 197–209

Few S (2009) Now you see it: simple visualization techniques for quantitative analysis. Analytics Press, Berkeley

Goodspeed R (2012) The democratization of big data. http://www.planetizen.com/node/54832 2014

Hinssen P (2012) Open data, power, smart cities. How big data turns every city into a data capital. Across Technology

Kaufman S, Simons R (1995) Quantitative and research methods in planning: are schools teaching what practitioners practice? J Plan Educ Res 15:17–33

Laney D (2001) 3D data management: controlling data volume, velocity and variety. META Group Research Note, 6

Lohr S (2012) The age of big data. The New York Times, February 11, 2012

King R, Magoulas R (2013) Data Science Salary Survey: Tools, Trends What Pays (and What Doesn't) for Data Professionals. O'Reilly Media, Inc

Mattern S (2013) Methodolatry and the art of measure [Online]. Design observer. http://places.designobserver.com/feature/methodolatry-in-urban-data-science/38174/ 2014

Sawicki DS, Craig WJ (1996) The democratization of data: bridging the gap for community groups. J Am Plan Assoc 62:512–523

Schneider S (2013) 3 examples of big data making a big impact on healthcare this week. http://blog.gopivotal.com/pivotal/p-o-v/3-examples-of-big-data-making-a-big-impact-on-healthcare-this-week. Accessed 2 Oct 2013/2014

Smiciklas M (2012) The power of infographics. Using pictures to communicate and connect with your audience. Que, Indiana

Urey G (2002) A critical look at the use of computing technologies in planning education: the case of the spreadsheet in introductory methods. J Plan Educ Res 21:406–418

Williams S (2015) Here now: social media and the psychological city. http://www.spatial$32#informationdesignlab.org/projects/here-now-social-media-and-psychological-city. Accessed 14 July 2015

Energy Planning in a Big Data Era: A Theme Study of the Residential Sector

Hossein Estiri

Abstract With a focus on planning for urban energy demand, this chapter re-conceptualizes the general planning process in the big data era based on the improvements that non-linear modeling approaches provide over mainstream traditional linear approaches. First, it demonstrates challenges of conventional linear methodologies in modeling complexities of residential energy demand. Suggesting a non-linear modeling schema to analyzing household energy demand, the paper develops its discussion around repercussions of the use of non-linear modeling in energy policy and planning. Planners and policy-makers are not often equipped with the tools needed to translate complex scientific outcomes into policies. To fill this gap, this chapter proposes modifications to the traditional planning process that will enable planning to benefit from the abundance of data and advances in analytical methodologies in the big data era. The conclusion section introduces short-term implications of the proposed process for energy planning (and planning, in general) in the big data era around three topics of: tool development, data infrastructures, and planning education.

Keywords Energy policy • Residential energy demand • Non-linear modeling • Big data • Planning process

1 Introduction

According to the International Energy Outlook 2013, by 2040, world energy demand will be 56 % higher than its 2010 level, most of which is due to socioeconomic transformations in developing countries (U.S. Energy Information Administration 2013a). This increase is expected to occur despite the existence of several global agreements within the past few decades on significantly reducing greenhouse gases (GHGs) and energy demand (e.g., the Kyoto Protocol, adopted in December 1997 and entered into force in February 2005).

H. Estiri, Ph.D. (✉)
Institute of Translational Health Sciences, University of Washington, Seattle, WA 98109, USA
e-mail: hestiri@uw.edu

© Springer International Publishing Switzerland 2017
P. Thakuriah et al. (eds.), *Seeing Cities Through Big Data*, Springer Geography,
DOI 10.1007/978-3-319-40902-3_13

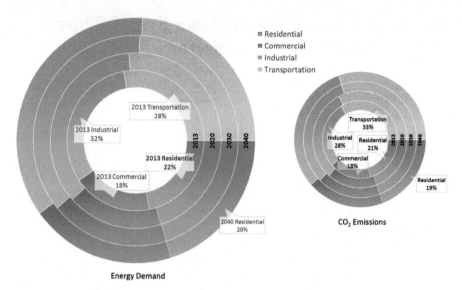

Fig. 1 U.S. energy demand and CO_2 emissions by sector, 2013, 2020, 2030, and 2040. *Data source*: U.S. Energy Information Administration (2013b)

Globally, buildings (residential and commercial) consume about 40 % of total energy (Swan and Ugursal 2009; Roaf et al. 2005; Norman et al. 2006). About 20–30 % of the total energy demand is for residential use. For example, in 2013, 22 % of the energy demand and 21 % of the CO_2 emissions production in the U.S. came from the residential sector (Fig. 1), both of which are expected to slightly diminish in their share to 20 % and 19 %, respectively, due to faster increases in industrial and commercial energy demand (U.S. Energy Information Administration 2013b). Technological improvements are expected to diminish growth rates in residential and transportation energy demand.

Most of the growth in global building energy demand is due to socioeconomic changes in developing countries. Since developed countries have greater access to up-to-date technologies, energy demand in the residential buildings is likely to increase at a slower pace in developed counties, with an average of 14 % in developed and 109 % in developing countries (Fig. 2) (U.S. Energy Information Administration 2013a).

Nevertheless, to many consumers, researchers, and policymakers, the energy consumed at homes has become an invisible resource (Brandon and Lewis 1999). A clear understanding of residential energy demand is the key constituent of effective energy policy and planning (Hirst 1980; Brounen et al. 2012). Yet, the residential sector lags behind other sectors on urban energy demand research. Two main reasons explain the uncertainties in household energy demand research and theory, obstructing the clear understanding needed for effective energy policy. First, conventional research has commonly used linear methodologies to analyze

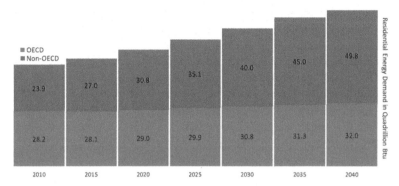

Fig. 2 World residential sector delivered energy demand, 2010–2040. *Data Source*: U.S. Energy Information Administration (2013a)

energy use in the residential sector, failing to account for its complexities. Second, lack of publicly available energy demand data for research has intensified the methodological issues in studying residential energy demand.

2 Problem: Prior Research Underestimated the Human Role

Building energy demand is an outcome of complex socio-technical processes that are driven by human activity. Due to its complexities, investigating the policy implications of behavioral determinants of residential energy demand has received little attention in prior research (Brounen et al. 2012). Traditionally, the debate on residential energy conservation has often overlooked the role of occupants' behaviors by excessively focusing on technical and physical attributes of the housing unit (Brounen et al. 2012; Kavgic et al. 2010; Lutzenhiser 1993; Kriström 2006). Since the early 1990s, energy research and policy have primarily concentrated either on the supply of energy or the efficiency of buildings, overlooking social and behavioral implications of energy demand (Lutzenhiser 1992; Aune 2007; Pérez-Lombard et al. 2008; Lutzenhiser 1994; Brounen et al. 2012). Engineering and economic approaches underestimate the significance of occupant lifestyles and behaviors (Lutzenhiser 1992).

> "Engineers and other natural scientists continue to usefully develop innovative solutions to the question of 'how we can be more efficient?' However their work does not answer the question 'why are we not more energy-efficient, when clearly it is technically possible for us to be so?'" (Crosbie 2006, p. 737)

Due to methodological or data deficiencies, even when household characteristics are incorporated in some energy demand studies, only a limited set of socio-demographic attributes are involved (O'Neill and Chen 2002). Moreover, the complexity of the human role in the energy demand process makes meaningful interpretation of modeling results rather difficult, which in turn leads to an ambiguous and limited understanding of the role of socioeconomic and behavioral determinants of residential energy demand. For example, Yu et al. (2011) suggest that because the influence of socioeconomic factors on energy demand are reflected in the effect of occupant behaviors, "there is no need to take them into consideration when identifying the effects of influencing factors" (Yu et al. 2011, p. 1409).

3 Why Has the Role of Human Been Underestimated?

3.1 Linearity vs. Non-linearity

Understanding and theorizing household energy use processes and repercussions are "a far from straightforward matter" (Lutzenhiser 1997, p. 77).

> "Household energy consumption is not a physics problem, e.g., with stable principles across time and place, conditions that can be clearly articulated, and laboratory experiments that readily apply to real world." (Moezzi and Lutzenhiser 2010, p. 209)

Linear analytical methodologies have been a research standard in understanding domestic energy demand. The assumption of linearity—where the dependent variable is a linear function of independent variables—and the difficulty of ascertaining any causal interpretations (i.e. the correlation vs. causation dilemma) are major downsides of traditional methodologies, such as ordinary multivariate regression models (Kelly 2011). As a consequence of the predominant assumption of linearity in energy demand research, "the present [conventional] energy policy still conveys a 'linear' understanding of the implementation of technology" (Aune 2007, p. 5463), while linear models cannot explain the complexities of household-level energy consumption (Kelly 2011). For better energy policies, a better understanding of the complexities of its use is needed (Aune 2007; Swan and Ugursal 2009; Hirst 1980).

3.2 Lack of Publicly Available Data

A major problem in residential energy demand research is that "the data do not stand up to close scrutiny" (Kriström 2006, p. 96). Methodological approaches lag behind theoretical advances, partly because data used for quantitative analysis often do not include the necessary socio-demographic, cultural, and economic information (Crosbie 2006). In addition, the absence of publicly available high-resolution energy demand data has hindered development of effective energy research and policy (Min et al. 2010; Kavgic et al. 2010; Pérez-Lombard et al. 2008; Lutzenhiser et al. 2010; Hirst 1980).

Even though relevant data are being regularly collected by different organizations, such data sources do not often become publicly known (Hirst 1980).

Conventional wisdom and modeling practices of energy demand are often based on "averages" derived from aggregated data (e.g. average energy demand of an appliance, a housing type, a car, etc.), which do not explicitly reflect human choice of housing and other energy consumptive goods (Lutzenhiser and Lutzenhiser 2006).

4 Non-linear Modeling

Like most urban phenomena, residential energy demand is an "outcome" of a set of complex interactions between multiple physical and behavioral factors. In complex systems, the multiplicity of causal links form a more complicated identity for the system than a single chain, encompassing an 'intricate' graph of causal networks (MacKay 2008; Phillips 2003). One main approach to modeling non-linearity is to decompose the linearity assumptions into a set of simultaneous linear sub-systems with explicit error estimates. Figure 3 illustrates one dimension of the difference between linear and non-linear approaches. A linear approach often considers the outcome as a "dependent" variable that correlates with a set of "independent" variables, which in turn, may correlate with each other as well. Clear examples of linear models are various types of multivariate regression models. In a non-linear approach, however, the outcome is the result of a set of cause-and-effect interactions between the predictor variables. This means that if one of the predictor variables changes, it will be unrealistic to assume that other variables would hold constant (a "gold standard" in reporting regression results)—with the exception of totally exogenous variables.

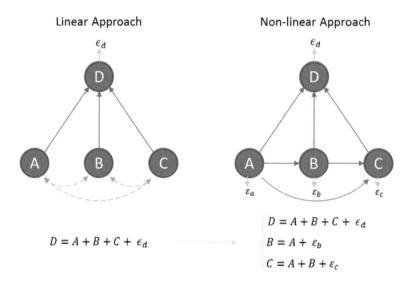

Fig. 3 Comparing linear and non-linear modeling approaches

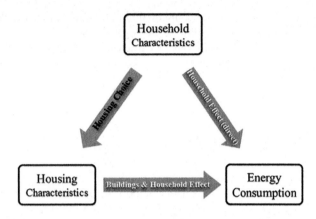

Fig. 4 A non-linear conceptual model of the impact of the household and the housing unit on energy demand. *Source*: Estiri (2014a)

This difference in the two approaches can be game changing, as the non-linear approach can reveal an often hidden facet of effects on the outcome, the "indirect" effects. Research has shown that, for example, linear approaches significantly underestimate the role of household characteristics on energy demand in residential buildings, as compared with the role of housing characteristics (Estiri 2014b; Estiri 2014a). This underestimation has formed the conventional understanding on residential energy and guided current policies that are "too" focused on improving buildings' energy efficiency.

Figure 4 illustrates a non-linear conceptualization of the energy demand at the residential sector. According to the figure, households have a direct effect on energy use through their appliance use behaviors. Housing characteristics, such as size, quality, and density also influence energy use directly. Household characteristics, however, influence the characteristics of the housing unit significantly—which is labeled as housing choice. In addition to their direct effect, through the housing choice, households have an indirect effect on energy demand, which has been dismissed with the use of linear methodologies, and so, overlooked in conventional thinking and current policies.

5 A Proposed Non-linear Modeling Schema

Energy use in the residential sector is a function of local climate, the housing unit, energy markets, and household characteristics and behaviors. A conventional linear approach to household energy use correlates all of the predictors to the dependent variable (Fig. 5). Figure 6, instead, illustrates a non-linear model that incorporates multiple interactions between individual determinants of energy demand at the residential sector. Results of the non-linear model will be of more use for energy policy.

Energy Planning in a Big Data Era: A Theme Study of the Residential Sector

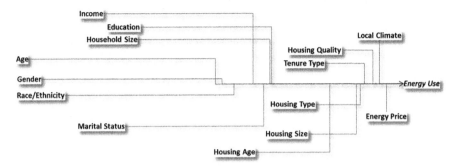

Fig. 5 Graphical model based on the linear approach. All predictors correlate with the dependent variable, while mediations and interactions among variables are neglected

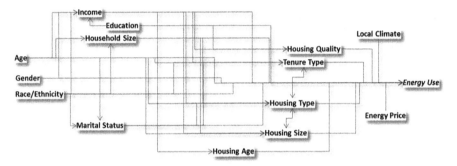

Fig. 6 Proposed graphical model based on a non-linear approach. Predictors impact both: the outcome variable and other variables

The recommended graphical model (Fig. 6) can be operationalized in form of 10 simultaneous equations with 69 parameters to be estimated, as represented in the following form:

Variable Estimates	Estimated Regression/Effect Coefficients														Variables (data)	Residuals
															Age	
															Gender	
Energy Use	β_1	β_2	β_3	β_4	β_5	β_6	β_7	β_8	β_9	β_{10}	β_{11}	β_{12}	β_{13}	β_{14}	RaceEthnicity	ϵ_1
Income	β_{15}	β_{16}	β_{17}	0	β_{18}	0	β_{19}	0	0	0	0	0	0	0	Income	ϵ_2
Education	β_{20}	β_{21}	β_{22}	0	0	0	0	0	0	0	0	0	0	0	Education	ϵ_3
Household Size	β_{23}	β_{24}	β_{25}	β_{26}	β_{27}	0	β_{28}	0	0	0	0	0	0	0	Household Size	ϵ_4
Marital Status	=	β_{29}	β_{30}	0	0	β_{31}	0	0	0	0	0	0	0	0	0	Marital Status	ϵ_5
Housing Age		β_{32}	0	β_{33}	β_{34}	β_{35}	0	0	0	0	0	0	0	0	×	Housing Age	+	ϵ_6
Housing Size		β_{36}	0	β_{37}	β_{38}	β_{39}	β_{40}	0	0	0	β_{41}	0	0	0	Housing Size	ϵ_7
Housing Type		β_{42}	0	β_{43}	β_{44}	β_{45}	β_{46}	β_{47}	0	β_{48}	0	β_{49}	0	0	Housing Type	ϵ_8
Tenure Type		β_{50}	0	β_{51}	β_{52}	β_{53}	0	β_{54}	0	β_{55}	0	0	0	0	Tenure Type	ϵ_9
Housing Quality		β_{56}	0	β_{57}	β_{58}	β_{59}	0	0	0	0	0	0	0	0	Housing Quality	ϵ_{10}
															Energy Price	
															Local Climate	

There are five exogenous variables in this model: age, gender, race/ethnicity, local climate, and energy price. All housing-related characteristics can be predicted with household characteristics (which can be improved by adding other influential variables). The parameters in these simultaneous equations can be estimated using a variety of software packages. How the estimated parameters can be used in planning and policy is yet another challenge.

6 Scientists, Planners, and Complex Modeling Outcomes

> *"For the theory-practice iteration to work, the scientist must be, as it were, mentally ambidextrous; fascinated equally on the one hand by possible meanings, theories, and tentative models to be induced from data and the practical reality of the real world, and on the other with the factual implications deducible from tentative theories, models and hypotheses."* (Box 1976, p. 792)

The better we—as individuals, planners, policy-makers—process complexities, the better decisions we'll make. Future policies need to be smarter by taking more complexities into account. With the current growing computational capacities, it is quite feasible to estimate such complex models—models can be connected to and estimated using live data, as well. Further, modern analytical algorithms can easily handle more complex models (models with increasing number or parameters). Clearly, we won't be short of tools and technologies to model more and more complexities.

However, as the models get more complicated—and ideally produce more realistic explanations for energy demand—translation of their results for policy and planning will become harder. Planners and policy-makers are not equipped with the required skillset to understand and interpret sophisticated modeling outcomes. Their strengths are in developing policies and plans that operationalize community goals. I suggest, in a big data era, planning can benefit from the abundance of data—of varying types—and the advances in computational and analytical techniques through a planning process that is accordingly modified.

7 A Modified Planning Process

The traditional planning process is not capable of directly incorporating complex scientific outcomes into policy development. The three primary steps in traditional planning process are: (1) gathering data; (2) transforming data into information; and (3) setting goals and objectives. Policies often follow explicit goals arrived at as the fourth step in the traditional planning process. There seems to be a missing link to connect complex modeling outcomes with the production of policy; perhaps an interface that can help planners and policy-makers set explicit goals for their respective communities.

The planning process needs modification to adapt to and benefit from this new big data era, with the abundance of data and growing advances in computer

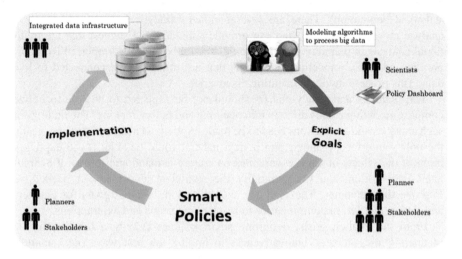

Fig. 7 The proposed modified planning process for the big data era

analytics. What is required for the outcomes of advanced complex modeling to be used in planning and policy is a paradigm shift in planning practice: a modified planning process (Fig. 7).

As I mentioned earlier, the traditional planning process often begins with data gathering. I also discussed that data unavailability is an important issue that has hindered the advancement of residential energy demand research and policy. Local utility companies are concerned about privacy issues. In addition, energy data needs to be connected to population, market, and climate data in a standardized way, to become useful for research and policy purposes.

The first step in this proposed planning process is a data collection and integration infrastructure comprised of energy, population, market, regulations, and climate data. There are various examples of federated data sharing infrastructures in health services that were developed using appropriate data governance and information architecture. Given that the bars for privacy are often set very high for health data, it should be feasible to develop similar data infrastructures for energy policy and research. Establishment of such integrated data infrastructures will require both technical and human components. Clearly, we will be needing data centers that can host the data, as well as cloud-based data sharing and querying technologies. But, technologies are only useful once the data is available—the foundation for data collection and integration are built. Here is where human role becomes important. To build a consensus among the data owners (utility companies, households, government or local agencies) multiple rounds of negotiations are conceivable. There also needs to be proper data governance in place before data can be collected, integrated, and shared with policy researchers.

New technologies (e.g., cloud computing, etc.) have made it easier to share and store data. Computer processing and analytics are also advancing rapidly, making it possible to process more data and complexities, faster and more efficiently (in its statistical denotation). There are several modern analytical approaches that can analyze more complexities, and can provide simulations. I suggest that the traditional analysis in planning process (step 2) should be enhanced/replaced by incorporating advanced modeling algorithms that are trainable and connected to live data. This process involves scientific discoveries.

Yet, planners and policy-makers should not be expected to be able to utilize complex modeling results directly into planning and policy-making. The findings of such analyses and simulations need to be made explicit via a policy interface. Using the policy interface, planners and policy-makers would be able: (1) to explicitly monitor the effects of various variables on energy demand and results of a simulated intervention, and (2) to modify the analytical algorithms, if needed, to improve the outcomes. The interface should provide explicit goals for planners and policy-makers, making it easier to reach conclusions and assumptions.

From the explicit goals, designing smart policies is only a function of the planners'/policy-makers' innovativeness in finding the best ways (i.e., smartest policies) for their respective localities to achieve their goals. Smart policies are context-dependent and need to be designed in close cooperation with local stakeholders, as all "good" policies are supposed to. For example, if reducing the impact of income on housing size by X% is the goal, then changes in property taxes might be the best option in one region, while in another region changes in design codes could be the solution. Once smart policies are implemented, the results will be captured in the data infrastructure and used for further re-iterations of the planning process.

8 Conclusion

This chapter built upon a new approach to energy policy research: accounting for more complexities of the energy demand process can improve conventional understanding and produce results that are useful for policy. I suggested that in order for planners and policy makers to benefit from the incorporation of complex modeling practices and the abundance of data, modifications are essential in the traditional planning process. More elaborations around the proposed modified planning process will require further work and collaborations within the urban planning and big data communities. Regarding the modified planning process, in the short-run, three areas of further research can be highlighted.

First is developing prototype policy interfaces. The non-linear modeling that I proposed in this work can be operationalized and estimated using a variety of software packages. More important, however, is the integration of the proposed non-linear model into the corresponding policy interface. More work needs to be done in this area using different methodologies, as well as developing more complex algorithms to understand more of the complexities in energy use in the residential sector—and perhaps, in other sectors.

Without integrated data it will be impossible to understand the complexities of energy use patterns—or any other urban phenomenon. Therefore, it is important to invest on city- and/or region-wide initiatives to securely collect and integrate data from different organizations. As the second area of future work, although establishing such initiatives and preparing the required socio-technical data infrastructure may not be a direct task for planners (for the time being), it certainly will be within the scope of work for local governments and planning / urban studies scholars.

Finally, the proposed modifications to the planning process have important implications for planning education. It will be crucial for planning practitioners or scholars in the big data era to be able to effectively play a role across one or more steps of the proposed planning process. When there is an abundance of data, planning education needs to incorporate more hands-on methodological training for planners in order to familiarize them with [at least] basic concepts of using data and data interfaces smartly. There also needs to be training around developing data architectures and infrastructures, especially for planning scholars to integrate urban data. Training options will also be helpful for planners to understand the required governance and negotiations related to obtaining and maintenance of data.

References

Aune M (2007) Energy comes home. Energy Policy 35(11):5457–5465. http://linkinghub.elsevier.com/retrieve/pii/S0301421507002066. Accessed 8 Nov 2013
Box GEP (1976) Science and statistics. J Am Stat Assoc 71(356):791–799
Brandon G, Lewis A (1999) Reducing household energy consumption: a qualitative and quantitative field study. J Environ Psychol 19(1):75–85. http://www.sciencedirect.com/science/article/pii/S0272494498901050
Brounen D, Kok N, Quigley JM (2012) Residential energy use and conservation: economics and demographics. Eur Econ Rev 56(5):931–945. http://linkinghub.elsevier.com/retrieve/pii/S0014292112000256. Accessed 26 Nov 2013
Crosbie T (2006) Household energy studies: the gap between theory and method. Energy Environ 17(5):735–753
U.S. Energy Information Administration (2013a) International energy outlook 2013. Washington, DC. http://www.eia.gov/forecasts/ieo/pdf/0484(2013).pdf
U.S. Energy Information Administration (2013b) Annual energy outlook 2014 (AEO2014) early release overview. Washington, DC. http://www.eia.gov/forecasts/aeo/er/index.cfm
Estiri H (2014a) Building and household X-factors and energy consumption at the residential sector. Energy Econ 43:178–184. http://linkinghub.elsevier.com/retrieve/pii/S0140988314000401. Accessed 23 Mar 2014
Estiri H (2014b) The impacts of household behaviors and housing choice on residential energy consumption. University of Washington, Seattle. http://search.proquest.com/docview/1529229205?accountid=14784
Hirst E (1980) Review of data related to energy use in residential and commercial buildings. Manag Sci 26(9):857–870. http://search.ebscohost.com/login.aspx?direct=true&db=bth&AN=7347856&site=ehost-live&scope=site

Kavgic M et al (2010) A review of bottom-up building stock models for energy consumption in the residential sector. Build Environ 45(7):1683–1697. http://linkinghub.elsevier.com/retrieve/pii/S0360132310000338. Accessed 7 Nov 2013

Kelly S (2011) Do homes that are more energy efficient consume less energy?: a structural equation model of the English residential sector. Energy 36(9):5610–5620. http://linkinghub.elsevier.com/retrieve/pii/S0360544211004579. Accessed 18 Nov 2013

Kriström B (2006) Residential energy demand. In: Household behaviour and the environment; reviewing the evidence. Organisation for Economic Co-Operation and Development, Paris, pp 95–115. http://www.oecd.org/environment/consumption-innovation/42183878.pdf

Lutzenhiser L (1992) A cultural model of household energy consumption. Energy 17(1):47–60. http://linkinghub.elsevier.com/retrieve/pii/036054429290032U

Lutzenhiser L (1993) Social and behavioral aspects of energy use. Annu Rev Energy Environ 18:247–289

Lutzenhiser L (1994) Sociology, energy and interdisciplinary environmental science. Am Sociol 25(1):58–79

Lutzenhiser L (1997) Social structure, culture, and technology: modeling the driving forces of household energy consumption. In: Stern PC et al (eds) Environmentally significant consumption: research directions. pp 77–91

Lutzenhiser L, Lutzenhiser S (2006) Looking at lifestyle: the impacts of American ways of life on energy/resource demands and pollution patterns. In: ACEEE summer study on energy efficiency in buildings. pp 163–176

Lutzenhiser L et al (2010) Sticky points in modeling household energy consumption. In: ACEEE summer study on energy efficiency in buildings, American Council for an Energy Efficient Economy, Washington, DC, pp 167–182

MacKay RS (2008) Nonlinearity in complexity science. Nonlinearity 21(12):T273–T281. http://stacks.iop.org/0951-7715/21/i=12/a=T03?key=crossref.126ca54ea24c1878bf924facc7197105. Accessed 31 Dec 2013

Min J, Hausfather Z, Lin QF (2010) A high-resolution statistical model of residential energy end use characteristics for the United States. J Ind Ecol 14(5):791–807. http://doi.wiley.com/10.1111/j.1530-9290.2010.00279.x. Accessed 7 Nov 2013

Moezzi M, Lutzenhiser L (2010) What's missing in theories of the residential energy user. In: ACEEE summer study on energy efficiency in buildings. pp 207–221

Norman J, MacLean HL, Kennedy CA (2006) Comparing high and low residential density: life-cycle analysis of energy use and greenhouse gas emissions. J Urban Plann Dev 132:10–21

O'Neill BC, Chen BS (2002) Demographic determinants of household energy use in the United States. Popul Dev Rev 28:53–88. http://www.jstor.org/stable/3115268

Pérez-Lombard L, Ortiz J, Pout C (2008) A review on buildings energy consumption information. Energy Build 40(3):394–398. http://linkinghub.elsevier.com/retrieve/pii/S0378778807001016. Accessed 6 Nov 2013

Phillips JD (2003) Sources of nonlinearity and complexity in geomorphic systems. Prog Phys Geogr 27:1–23

Roaf S, Crichton D, Nicol F (2005) Adapting buildings and cities for climate change: a 21st century survival guide. Architectural Press, Burlington. http://llrc.mcast.edu.mt/digitalversion/Table_of_Contents_134820.pdf

Swan LG, Ugursal VI (2009) Modeling of end-use energy consumption in the residential sector: a review of modeling techniques. Renew Sustain Energy Rev 13(8):1819–1835. http://linkinghub.elsevier.com/retrieve/pii/S1364032108001949. Accessed 11 Nov 2013

Yu Z et al (2011) A systematic procedure to study the influence of occupant behavior on building energy consumption. Energy Build 43(6):1409–1417. http://linkinghub.elsevier.com/retrieve/pii/S0378778811000466. Accessed 8 Nov 2013

Part IV
Urban Data Management

Using an Online Spatial Analytics Workbench for Understanding Housing Affordability in Sydney

Christopher Pettit, Andrew Tice, and Bill Randolph

Abstract In 2007 the world's population became more urban than rural, and, according to the United Nations, this trend is to continue for the foreseeable future. With the increasing trend of people moving to urban localities—predominantly cities—additional pressures on services, infrastructure and housing is affecting the overall quality of life of city dwellers. City planners, policy makers and researchers more generally need access to tools and diverse and distributed data sets to help tackle these challenges.

In this paper we focus on the online analytical AURIN (Australian Urban Research Infrastructure Network) workbench, which provides a data driven approach for informing such issues. The workbench provides machine to machine (programmatic) online access to large scale distributed and heterogeneous data resources from the definitive data providers across Australia. This includes a rich repository of data which can be used to understand housing affordability in Australia. For example there is more than 20 years of longitudinal housing data nationwide, with information on each housing sales transaction at the property level. For the first time researchers can now systematically access this 'big' housing data resource to run spatial-statistical analysis to understand the driving forces behind a myriad of issues facing cities, including housing affordability which is a significant issue across many of Australia's cities.

Keywords Housing affordability • Spatial statistics • Big data • Portal

1 Introduction

The world's growing and increasingly urbanized population presents significant challenges for planners and policy makers to address. In 2007 the world's population became more urban than rural, and, according to the United Nations, this trend is to continue with the percentage of the population residing in urban areas expected

---

C. Pettit (✉) • A. Tice • B. Randolph
Faculty of Built Environment, City Futures Research Centre, University of New South Wales, Kensington, NSW, Australia
e-mail: c.pettit@unsw.edu.au; andyjtice@gmail.com; b.randolph@unsw.edu.au

© Springer International Publishing Switzerland 2017
P. Thakuriah et al. (eds.), *Seeing Cities Through Big Data*, Springer Geography, DOI 10.1007/978-3-319-40902-3_14

to rise to 59.9 % and 67.2 % in 2030 and 2050 respectively (Heilig 2012). The total population is projected to be 8.3 billion people and 9.3 billion people in 2030 and 2050 respectively (Heilig 2012). This growth and the increasing trend of people moving to urban localities, predominantly cities, brings about additional pressures on services, infrastructure, housing, transport and the overall quality of life of city dwellers. Issues such as housing affordability are becoming increasingly pressing for city planners and policy makers to address.

Over the last 50 years, an increasing array of digital data has been produced relating to urban settlements resulting in the rise of the 'information city' as referred to by (Castells 1989). Some 80 % of this data can be given a location attribute and be used to create spatial databases which can then be analyzed and visualized to help understand urban growth and development and to plan for sustainable urban futures. However, as (Townsend 2013) points out, we must be conscious of the limitations of our ability to predict the future and how we use such information to engage our communities and support bottom-up participatory planning of our cities.

In this paper we introduce an online spatial analysis workbench where data representing Australian cities can be accessed, analyzed and visualized. Since 2011, the Australian Urban Research Infrastructure Network (AURIN) has made available an online workbench, which provides access to over 1800 spatial datasets from over 30 data providers from across Australia. As of August 2015 there are over eight billion data elements that cover all major cities of Australia, crossing health, housing, transport, demographics and other essential characteristics of cities. This includes historical data, current data and future data, which can be used, for example, to assess the expected population growth for major cities. In this chapter we will focus on the issue of housing affordability and how the data and analytical tools available via the AURIN online workbench (hereby referred to as 'the Workbench') can be used to understand this in the situational context of Sydney where unprecedented levels of housing unaffordability is currently being experienced (Committee for Sydney 2015).

2 Housing Affordability

Housing affordability is a persistent policy issue and has been the focus of interest to Australian housing researchers for many years (Yates et al. 2013). The recent escalation in property prices in Australia's major cities is an all too familiar feature of both popular media reports (e.g. Ting 2015) and conversations around the garden barbecue. Yet it remains a policy Cinderella, with much wringing of hands accompanied by little coherent policy intervention. The complexity of the drivers of housing markets and the range of approaches to explaining these processes militate against a coherent understanding of these drivers or proscribing policy solutions to improve affordability outcomes (Economic Review Committee 2015).

Defining what is or is not an affordable home, or how this might be best measured, is also problematic and subject to longstanding debates, both in

Australia and elsewhere (Hulchanski 1995; Stone 2006; Gabriel et al. 2005). The two basic components of any housing affordability assessment, namely housing costs and household income and wealth, defy easy measurement, especially over time. While housing affordability measures vary in detail and construction, there are two basic approaches: the housing cost to household income ratio approach in which a measure of housing costs (rent, mortgage payment, etc.) is compared to a measure of household income, and the income after housing cost or 'residual income' approach, which seeks to define the household income remaining after housing costs are deducted and then comparing this to a low income benchmark of some kind. The prevalent use of ratio measures (housing cost as a percentage of household income) stems from the more ready availability of such data and the ease of calculation and interpretation. However, the residual income method might be more appropriate for lower income households for whom the absolute gap between housing costs and household income matters the most.

The role of data in systematically analysing the issue of housing affordability is critical. Understanding how and why housing markets function and whether or not housing is affordable to various sections of the population is a fundamental prerequisite for coherent policy development. The theme of 'Urban Housing' exists as one of seven activated 'lenses' which were created by AURIN to define and then collate a range of urban data sets and develop appropriate analytical tools (Pettit et al. 2013a). AURIN's Urban Housing Lens focused on delivering the kinds of data to support both systematic monitoring of housing affordability. It also aimed to provide the base data for developing much better understanding of how the housing market works more generally and where pressures of housing costs, and therefore housing affordability, occur and for whom.

Ironically, it is not that we do not have data to help us. In fact, in Australia, details of every residential property sale and rental agreement are gathered by State and Territory governments as on-going statutory administrative requirements. These are all recorded at address level and therefore can be geo-coded to allow a precise spatial matching to the land use property cadastre. The potential for detailed spatial analysis of these data is therefore significant. In the case of property sales, these data are sold to commercial companies for on-sale, once suitably processed, to the real estate industry, media, insurance industry, banking sector and others to assist in their business activities. There are several major national private firms that gather and disseminate sales data from around Australia on a for-profit basis. However, access to property sales data by the research and policy community has been much less ubiquitous and often incurs significant expense or time consuming negotiation, often stymied by data protection concerns. This is a paradox—while the local estate agent can tell you the sales history of your house in some detail, a university researcher has significant difficulty in obtaining the same data for research purposes, unless he or she has the cash to buy it. For the first time, AURIN has developed a machine to machine data feed directly into the Australian Property Monitors housing sales databases. This is a significant step forward in supporting urban housing research across Australia in accessing this 'big' housing data asset.

For rental data, the problem is more intractable as it is held by the various jurisdictional Rental Bond Boards who have no direct interest in the market information the records contain and, to date, have shown little interest in disseminating it as useable data assets. Also, unlike the house price data, no one as yet has attempted to assemble a nationally consistent dataset on rents from the various jurisdictional departments that gather this information. Other key housing datasets, from Federal rent assistance and first home owner payments to mortgage and foreclosure data, are all collected by various government and private agencies at address level. Assembling these data together in a nationally consistent manner, geocoded to match with the property cadastre, would provide researchers and policy makers with a vastly greater capacity to study and better understand how housing markets function. This has also yet to be achieved. However, in this chapter we illustrate how such comparable data can be brought together, analysed and visualized using the AURIN workbench in the context of one jurisdiction, Sydney, for understanding various dimensions of housing affordability. It is to be hoped that a broader range of administratively and commercially held housing market data will become available for research in the not too distant future.

3 AURIN Online Spatial Analytics Workbench

AURIN is a large-scale national e-Research infrastructure project funded by the Australian Government. AURIN is tasked with providing seamless access to data sets, spatial statistical and scenario modelling tools to support urban researchers and policy and decision-makers. Importantly, AURIN also comprises the network of urban researchers from across the country, known as the Australian Urban Intelligence Network (Pettit et al. 2015a). These researchers have shaped the e-Research infrastructure requirements and are using the online spatial analytical workbench to address a number of areas of societal significance, through the concept of lenses (Pettit et al. 2013a). Lenses are essential thematic areas of interest relevant to policy and decision-making. Expert Groups have been established for seven Lenses to date including: (1) Population and demographic futures and benchmarked social indicators, (2) Economic activity and urban labour markets; (3) Urban health, well-being and quality of life; (4) Urban housing; (5) Urban transport; (6) Energy and water supply and consumption; and (7) Innovative urban design. In the context of housing related research, the expert group for Lens 4 Housing formulated a systems view of the dynamics of the housing market, which places the issue of affordability as a critical component—see Fig. 1.

The AURIN (geo)portal (hereby referred to as the "Portal") (https://portal.aurin.org.au) is the flagship application of the workbench and the focus of this particular paper. The functionality within the Portal enables end users (urban researchers, policy and decision makers) to discover, analyze and visualize a plethora of datasets which can be used to support the planning of sustainable cities (Stimson et al. 2011). It is important to emphasize that the Portal provides machine to

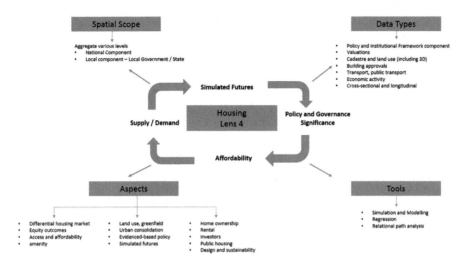

Fig. 1 Systems view of housing market in Australia with the central issue of affordability (Pettit et al. 2013a)

machine (programmatic) access to these data sets (and the data remains as close to the data custodians as technically possible). This data has been curated and the associated metadata is used to register the data feeds into the Portal. There is also a CKAN metadata search and discovery tool as part of the wider workbench. The workbench, comprising the Portal, is conceptually illustrated in Fig. 2. It has been implemented using an open source federated technical architecture (Sinnott et al. 2015). The federated data structure enables datasets from across different cities, government agencies and the private sector to feed into the workbench. A key component to the success of AURIN has been its engagement with data providers and associated stakeholders from government, industry and academia. Many of these data sets have either been siloed behind organizational firewalls and are not easily discoverable or accessible. As of August 2015 AURIN has federated over 1800 critical datasets which can be used to support evidence-based research and data-driven policy and decision-making around the challenges facing our cities. Some of this data can be considered 'big' data, such as the Australian Property Monitors (APM) data which is discussed later in this paper.

As part of the broader workbench there are also a number of spatial statistical routines and modelling tools to support urban informatics including, for example, a *Walkability Toolkit* and a *What if? Scenario Planning Tool*. The *Walkability toolkit* includes an agent based approach for calculating pedsheds (Badland et al. 2013). Pedsheds are commonly referred to as the area within walking distance from a particular destination such as a town centre or a train station. The *What if? Scenario Planning Tool* is integrating the well-known (Klosterman 1999) *What if?* GIS based planning support system (PSS) into the AURIN workbench (Pettit et al. 2013b, 2015b). Other tools within the workbench include an employment clustering tool, and a suite of spatial and statistical routines, charting and mapping visualization

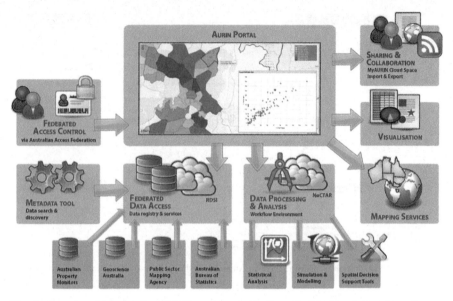

Fig. 2 System architecture for AURIN workbench Pettit et al. (2015a) adapted from Sinnott et al. (2015)

capabilities. All of these tools require access to a rich tapestry of datasets which can be shopped for via the Portal. In this paper we focus on the application of a number of spatial statistical routines which are available in the AURIN Portal.

The AURIN project has also established a number of Data Hubs and feeds across the country to programmatically access data required to support urban researchers (Delaney and Pettit 2014). These data hubs are essentially a series of distributed computer servers which reside across jurisdictions where jurisdictional data resides and where an AURIN client has been developed to be able to provide data through to the Portal. These Data Hubs close the loop between data owners and data users by ensuring each hub is established aligning to a set of core principals, defined as: (1) facilitating collaboration and interaction between end users and data custodians; (2) being held as close to the source as possible; (3) set up to serve a broad end user community, not a single project; and (4) sufficient information (including metadata) being provided for users to understand the data.

The housing affordability analysis undertaken in this research draws upon data at both aggregate census geographies from across a number of data hubs throughout Australia. However, housing data is primarily drawn from the Sydney Housing data hub where the NSW Rental Bond Board data resides and through the property data made available through the Australian Property Monitors (APM) which is a Sydney based company. At present, the NSW Rental Bond Board data available through the Portal is limited to a select number of years' worth of aggregated rental bond data, but provides median rent data for each year at the ABS's Statistical Areas 2 (SA2) level by broad dwelling type and bedroom number. Licensing agreements preclude the release of these data at the unit record level. The SA2 census tract comprises an aggregate spatial unit of approximately 15,000 properties. The APM dataset is spatial-temporal and goes back over 20 years, with monthly updates and includes descriptive information such as dwelling type (house, unit, land), number of bathrooms and bedrooms and dwelling features (including more than 20 associated dwelling characteristics such as laundry room availability, garage, BBQ facilities, harbour or beach view, etc.). This data is longitudinal and exists across 14 million land parcels across all of Australia comprising approximately 20 columns of attributes. There is in the vicinity of 5.6 billion records in total. The data has been aggregated to 12 levels of geography to align with standard geographies supported by the Australian Bureau of Statistics and Australia Post. Each level of aggregation has been created as its own unique data product to support multiple level analysis and ensure the data is in a compatible geography required to run the AURIN portal spatial-statistical tools. The APM property data has been judicially collated, cleaned internally by APM and made available spatially-temporally through a hosted Postgres/PostGIS database for distribution via GeoServer. Researchers whose credentials have been authenticated through the Australian Access Federation (Sinnott et al. 2015) can then access the APM data services via the AURIN portal.

The data also includes comprehensive information on sales, auction and rent economic cycles, including total transaction prices with corresponding statistical analysis, for example, median price, detailed price breakdown at each fifth percentile, standard deviation price, geometric mean price, as well as first and final advertisement prices and dates, settlement dates, and auction clearance rate among others. In this instance, the point to stress is that the APM data, in this form, is a highly detailed disaggregated resource drawing on a wealth of statistical assessments from a pool of much larger data. It is this form of 'big data' which is used for analysing housing affordability and trends in Australia using the suite of spatial-statistical tools available via the Portal and will be discussed further in the Sydney housing affordability uses case.

4 Understanding Housing Affordability in Sydney

This section utilises the range of data currently available within the Portal to show how a much more spatially disaggregated analysis of local housing affordability and related housing market statistics can be derived, using Sydney as a case study. As of 2014, the Australian Bureau of Statistics (ABS) reports that Sydney's population stands at some 4.76 million, growing by 15% since 2001. The NSW Government Department of Planning and Environment project the city's population to grow by an additional one million people over the next 10 years. Over the same period, Sydney's median house price increased by 82%, from A$338,000 in 2001 to A$730,000 in 2014 (APM). Median incomes grew from $988 (2001) to $1444 (2011), a 46% increase.

Use is made of AURIN's access to detailed sales and rental information and also, where feasible, the Portal's current analytical techniques. The case study also identifies a few of the current technical limitations and reflects on challenges relating to data access. Underlying the case study, however, is a simple narrative concerning conceptual debates about housing affordability, noted above, and how these can be tested and assessed as more detailed data becomes available.

One of the central difficulties in constructing any meaningful local index of housing affordability has been a paucity of accessible information. Most urban researchers rely, for better or worse, on forms of Census materials for local level analysis. In the Australian context, the quinquennial Census undertaken by the Australian Bureau of Statistics (ABS) collects information on household, family and individual incomes and provides considerable geographic resolution. Unfortunately, housing cost information that is collected by the Census focuses on weekly rent and weekly mortgage payments. Whilst the former provides a relatively sound proxy for calculating rent affordability, the latter cannot be readily used to consider affordability in the owner occupied sector. Mortgages, of course, decrease over the life of repayments with many locations containing households who have been in residence for considerable periods of time. This serves to produce a substantial mismatch between stated mortgage payments and the *actual* current market values of properties.

Firstly, the following capitalises on the local level household incomes data from the Census and local house sales data (provided through the Portal by APM) in order to begin to bridge this gap. Secondly, focusing on the rental market, an example of a workflow to identify local changes in rental affordability is set out. Finally, a more detailed assessment of affordability is set out. In this final example, some of the current limitations of the Portal, both in terms of data availability and analytical capacity are discussed.

4.1 Housing Affordability: Sales

One of the most widely cited affordability indexes is the Demographia affordability index (Performance Urban Planning 2015). Under this methodology Sydney has consistently been identified as one of the top ten most unaffordable cities in the world. As of 2015, Sydney sits third in the rankings behind Hong Kong and Vancouver.

Underneath these macro city-wide scale assessments, the index itself is very simple in construction; comparing city wide median incomes to city wide median sales prices to define a ratio. Typically data is sourced from national statistical agencies in aggregate form. Although lacking any theoretical justification, ratios of 3 or below are classified as 'Affordable' and 5.1 or over as 'Severely Unaffordable' (see Table 1). Sydney's current median multiple is 9.8.

Despite its limitations (Phibbs and Gurran 2008), for broad comparisons of different cities the Demographia approach is both robust and replicable. However,

Table 1 Demographia affordability thresholds

Demographia international housing affordability survey	
Housing affordability rating categories	
Rating	Median multiple
Severely unaffordable	5.1 and over
Seriously unaffordable	4.1–5.0
Moderately unaffordable	3.1–4.0
Affordable	3.0 and under

Source: Table ES-1 (Demographia 2015)

headline multiples belie underlying spatial complexity. The Portal provides access to prepared APM sales data for the period 1993 through 2015, broken down by percentile ranges. Such provision enables a more spatially disaggregated analysis to be derived through the application of a standard median multiple approach, but in this case, at the local (SA2) level.

Figure 3 presents the median multiple classifications derived across Sydney using local median incomes and reported median sales values for 2011 at the SA2 level. From Fig. 1 it is apparent that not even Sydney's suburban fringe locations are affordable under the Demographia classification: in these more peripheral locations, multiples between 4.3 and 5.9 are common. In the inner city, and particularly the more affluent eastern coastal suburbs, the multiple between median prices and median incomes is well over double (11+) the Severely Unaffordable classification applied by Demographia.

4.2 Housing Affordability: Rents

The case study now shifts focus to rental affordability. Australia has a relatively large and well developed private rental sector in comparison to many other western economies. In 2011 around 32 % of all residential dwellings were rented, a significant increase from the 25 % recorded in 2006 (source: ABS).

Rents are perhaps one of the most dynamic measures of housing affordability. Whilst house prices tend to garner the most attention in debates concerning this area of concern, property sales themselves represent around half of the overall level of housing transactions occurring. This is especially true in Sydney, which is the focus of this case study. In 2011 there were a little over 120,000 sales of residential properties. In comparison, 130,000 new rental contracts were signed. For context, the average rental contract in Sydney lasts for a little over 2 years.

One of the most widely utilised assessments of rental affordability is a strict assessment of median rents to median incomes, with a benchmark of 30 % of household income used as the cut-off threshold for affordability. Such calculations can be carried out utilising a number of resources, most typically, Census data. Ultimately, however, Census data tend not to provide useful contextual information about the scale of housing market activity, a key qualifying component for more

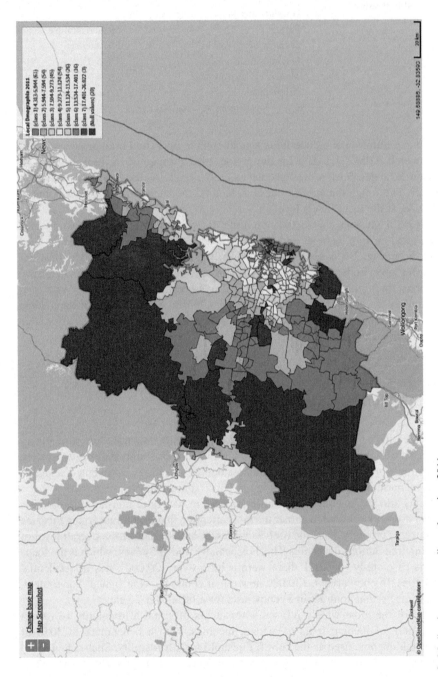

Fig. 3 Median income to median sales prices 2011

rounded research. Realistically, additional information is needed concerning aspects of property availability to support the basic affordability analyses. Affordability assessments based on spatially disaggregated median multiple approaches tell us little about how many sales or lettings in a given time and place were *actually* affordable to target groups. Here, AURIN's provision of both Census and market derived data provides a reliable means through which a range of relevant data can be brought together for a more rounded analysis.

The following worked example compares rental affordability in 2006 to that in 2011 using APM data and Census household income data for SA2 areas. In Sydney, overall, median rents have increased by 40 % between 2006 and 2011 and the percentage of median household income spent on rents has increased from 21 to 25 % over the same period. However, there are significant differences in local rental affordability trajectories.

Utilising the workbench's access to Census time series data and individual rental transactions, the following provides:

- Analysis of localised trajectories in rental affordability over the period 2006–2011;
- Identification of case study locations, and extraction of stock level data to contextualise these trends;

This workflow is made possible through provision of both national and local data in a consistent and comparable format accessible via the Portal. Through such provision, researchers have access to a common research base, thus providing an unprecedented resource from which to develop common research themes. Despite the relative simplicity of the analysis presented in this case study, it is this last point that should be borne in mind.

Figures 4 and 5 set out local rents as a percentage of local median incomes in 2006 and 2011. Within this distribution it is apparent that there has been a substantial decrease in affordability over the time period. There is also a particular geography at work, with affordability rates decreasing considerably in Sydney's middle ring suburbs. This trajectory is further confirmed in Fig. 6, with the class 5 (red) locations witnessing increases of between 5 and 9 % in the local ratio of rents to incomes over the period. This is a useful finding to underline, not only that the ratio between incomes and rents has decreased substantially over the period, but also that it has disproportionally effected these locations the most.

Using the Portal's built-in graphing function it is possible to generate a simple scatter plot to highlight these relationships further. Figure 7 provides this and further confirms the previous observation; locations with the greatest rent to income increases have also witnessed the greatest deterioration in incomes to rent ratios.

To further confirm this spatial pattern, use can be made of the spatial analytic tools provided in the Portal. Figure 8 provides the output of a Geti-Ord (Getis and Ord 1992, Ord and Getis 1995) assessment of changes in median rents to median income. Essentially, this operation enables visually obvious patterns to be assessed statistically to assess whether concentrations are significant. Here the workflow in the Portal moves from basic visualization into statistical confirmation of findings.

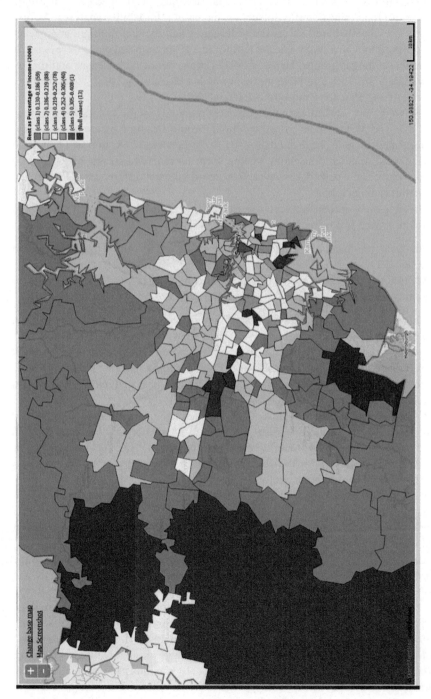

Fig. 4 Rent as percentage of income (2006)

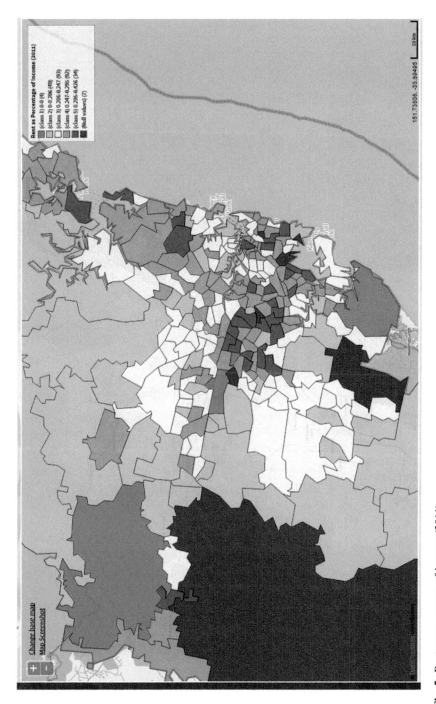

Fig. 5 Rent as percentage of income (2011)

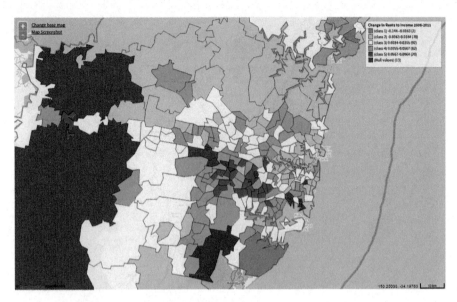

Fig. 6 Change in rents to income 2006–2011

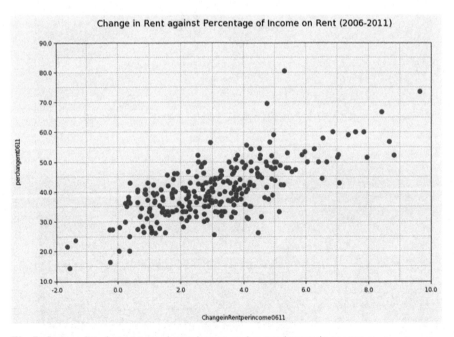

Fig. 7 Scatter plot of percentage change in rents and rent to income increases

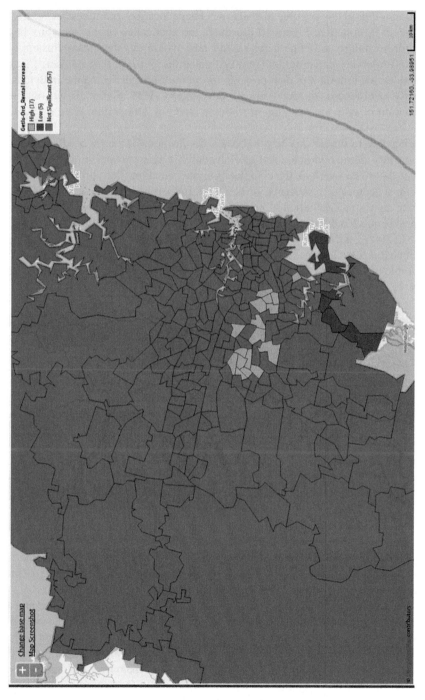

Fig. 8 Output of Getis-Ord classifications on rental affordability clustering

Utilising the Portal's spatial selection tools (currently limited to bounding box selection—Fig. 9) it is possible to drill down inside locations of interest and extract local level information. Figure 10 illustrates the application of the bounding box selection technique to define a case study area within the significant clusters of rental affordability deterioration identified using the Getis-Ord assessment. Using these selection criteria it is then possible to run multiple queries on the individual property level records of rental lettings provided by APM. Figure 10 sets out the distribution of individual lettings for houses made in 2006 within the case study area.

Being able to further quantify and detail not just a basic metric of affordability but how this metric relates to local level availability offers greater sophistication to debates that often overlook the complexity of the situation. The ability to generate realistic schedules of how many, or how few, properties *were* actually affordable (even at an abstract definition of household income) is only made feasible through the provision of individual information. For the case study location, the story is stark. In 2006 a little over 32 % of all rental properties advertised where affordable at the local median income level. In 2011 this had declined to 2.5 %. Whilst in 2006 the profile of affordable rental properties was split equally between one and two bed properties, in 2011 the affordable component only contained one bed properties.

Up to this stage, the case study has utilised some basic concepts of housing affordability that could be readily applied in the current version of the Portal. Being able to replicate these metrics at the local scale has demonstrated that Sydney appears to face severe housing unaffordabilty problems. However, as stated at the outset, the strict median multiple definitions of housing affordability were derived in earlier times, periods where there was a paucity of data available (and *especially* availability of data at a local level). Such calculations are useful at the macro (or city level), but begin to raise questions concerning their relative utility at the local scale.

In the face of these questions, more nuanced calculations have been developed over time. One of these is an assessment of household incomes translated into current borrowing capacity, allowing comparisons to be made between open market sales prices and the amount a typical household could reasonably be expected to afford to pay. Such a metric enables ratios between affordable levels of mortgage repayments (nominally, 30 % of family income) and observed sales prices to be made. For this, a well-established price threshold using the '30:40 rule' is used. Here, comparison is made on the basis of the price affordable at 30 % of income to households earning at the bottom 40th income percentile (Gabriel et al. 2005). This value can then be compared to different sales price positions.

In the following final example, the 30:40 definition of affordability is compared to median recorded sales price from each SA2. This has been made possible through the pre-prepared provision of the APM data set broken down by detailed prices for each percentile (5th through 95th). Whilst there are over 500 ABS data products directly accessible via the AURIN portal that urban researchers can access and interrogate, for more customized data products other software is currently preferred for extraction and manipulation. Subsequently, data on incomes has been extracted

Fig. 9 Case study selection—guided by Getis-Ord analysis

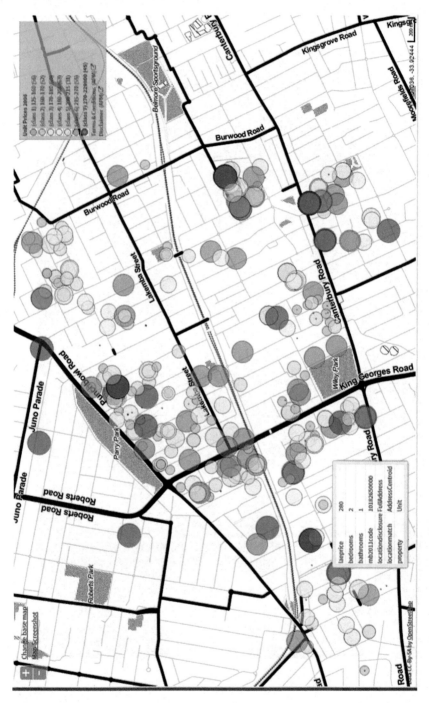

Fig. 10 Distribution of individual rental lettings in case study area, 2006

from the ABS Table Builder product and mortgage calculations have been conducted in Excel prior to uploading to the Portal. One of the strengths of the Portal is users can upload their own data into their cloud based project space. This customized data can then be joined to one of the 1800 AURIN data products to perform further analysis and visualize the results.

As Fig. 11 sets out, the distribution of the ratio between the 30:40 calculated affordability thresholds and local median sales prices presents a different picture to that discussed previously. Values below 1 identify locations where the affordable threshold is considerably below the achieved median prices, values above 1 indicate locations where incomes can comfortably afford monthly repayments. It is evident that more peripheral suburban locations are considerably more affordable for households at the 40th income percentile than inner city locations.

Whilst this more nuanced take on housing affordability has produced a geography broadly comparable to that seen in Fig. 3, it should be noted that the resulting analysis provides a more realistic picture. For example, whilst sales across much of the inner city remains severely unaffordable at the 40th income percentile, key pockets exist where ratios are only marginally above 30:40 definition.

This final stage of the case study has demonstrated how differing aspects of housing affordability can be highlighted through the application of more complex data sets. Critically, due to the nature of the ABS Census derived income variables, only a partial picture could be developed. Income variables are a product of all stated household incomes and therefore is influenced by homeowners, renters, young families and retirees alike: it therefore does not specifically reflect the incomes of those trying to enter home ownership or trade up or down within the housing market. In part, this limitation is due to the manner in which Census data has been traditionally provided, namely through pre-existing flat file tabulations. A more meaningful income profile (or indeed groups of profiles) cannot be readily developed from such resources. Income data for these kinds of disaggregated cohorts derived from individual Census records at the local scale would be more rewarding. However, currently the costs of obtaining such data preclude this more detailed analysis.

This latter point suggests that attempting to harmonise all possible data resources across all possible domains whilst retaining the flexibility to interrogate data fully is a challenging goal. Different data resources have different requirements and to maximise their utility requires a significant level of technical expertise and domain knowledge, not to mention cost-effective access to data. Such approaches suggest hybrid working practices where the format of outputs are harmonised; either within comparable formats or spatial frameworks (and increasingly both). Where the Portal fits within such considerations is in its ability to provide machine to machine access to significant data resources across Australia and the ability to also allow users to import resources derived from other sources in a standardised manner. The Portal does provide access to urban 'big' data but is only exposed to the end user after it has been curated and cleaned. Providing access to such valuable data and a suite of analytical tools via a one-stop-shop interface, whether via a portal or dashboard, is an important step in its utility in big data. Similar observations have been recently raised by various commentators (Wilson 2015, Batty 2015, Rae and Singleton 2015).

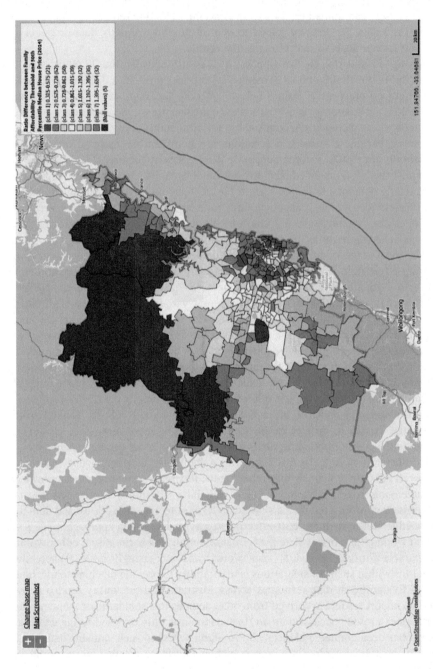

Fig. 11 Ratio between 30:40 mortgage calculations and median house sales prices

4.3 Next Steps

The final case study provided points towards a tantalizing possibility for further development. Applying mortgage calculations to individual income profiles quantifies the level of affordable housing loans that different cohorts can achieve. Assessing these against detailed house price transaction data would allow a quantification of affordability; how many properties (if any) could have been purchased? This possibility could be used to extend affordability debates through adding the context of availability. Instead of classifying groups of areas (or whole cities) as unaffordable, more attuned analysis could be undertaken in order to capture shortfalls in affordable properties, or indeed the financial gaps between prices and different cohorts.

In the context of the Portal, specifically, the basics for such analysis could be provided through the development of a simple tool into which affordability assumptions could be entered (base rates, lifetime of loan, etc.), much in the same manner as are available on many mortgage lender websites currently. This housing affordability tool could be developed to calculate the value of home loans achievable under various assumptions at the local level for different household cohorts. These can be compared to the APM data in order to begin to track not just affordability but levels of *availability* with the housing market overtime. The development of such a tool could be implemented via the Portal's object-oriented modelling framework (Sinnott et al. 2015).

5 Conclusions

Affordable housing is part of a large systems views of affordable living (Kvan and Karakiewicz 2012). The elements for the latter are necessarily more diverse that those that contribute to an analysis of affordable housing alone and would include factors such as accessibility to transportation, employment, education and recreational facilities. Many of these data themes exist within the AURIN data sets. The potential of the workbench is to deliver access to a rich variety of data sets such that the broader questions of affordable living can be examined with a robust model of affordable housing underpinning the analysis.

In this paper we have discussed the workbench in the context of accessing big data from across the AURIN federated data architecture, specifically from the Australian Property Monitors Geoserver web service. We have focused on the issue of housing affordability in the context of Sydney. Specifically, we have shown how the data and the analytical tools within the Portal (the flagship product within the AURIN workbench) can be used to understand the spatial-temporal patterns of housing affordability across a city. However, whilst the Portal is a powerful vehicle for data access, with over 1800 datasets, and analytics with over 100 spatial-statistical tools for analysis, there are limitations in attempting to provide a technical solution that endeavours to integrate geoportal functionality with a suite of advanced analytics. In the authors' experience of undertaking the research outlined in this chapter, it is highlighted that a number of additional software tools were required for data extraction and manipulation in order to complete the analysis. Issues of usability were also encountered. These considerations warrant further evaluation to determine how to best deliver both data and analytical tools to a broad church of users across the many domains comprising the urban research community.

Acknowledgements AURIN is a $24 million project developing national networked urban research infrastructure capability in Australia, funded by the Commonwealth Government under the Education Investment Fund and the National Collaborative Research Infrastructure Strategy. The University of Melbourne is the lead agent for the project. The authors thank the AURIN Office and the Core Technical Team, along with the many contributors to the AURIN project in a number of institutions across Australia, including many government and other agencies providing access to data, through Data Access Agreements and Sub-contracted projects.

References

Badland H, White M, Macaulay G, Eagleson S, Mavoa S, Pettit C, Giles-Corti B (2013) Using simple agent-based modeling to inform and enhance neighborhood walkability. Int J Health Geogr 12:58
Batty M (2015) A perspective on city dashboards. Reg Stud Reg Sci 2:29–32
Castells M (1989) The information city. Blackwell, Oxford
Committee for Sydney (2015) A city for all: five game-changers for affordable housing in Sydney. Sydney issues Paper No. 8
Delaney P, Pettit CJ (2014) Urban data hubs supporting smart cities. In: Proceedings of the Research@Locate'14. 7–9 Apr 2014
Economic Review Committee (2015) Out of reach? The Australian housing affordability challenge. Australian Senate
Gabriel M, Jacobs K, Arthurson K, Burke T, Yates J (2005) Conceptualising and measuring the housing affordability problem. Australian Housing and Urban Research Institute, Sydney
Getis A, Ord JK (1992) The analysis of spatial association by use of distance statistics. Geogr Anal 24:189–206
Heilig GK (2012) World urbanization prospects: the 2011 revision. United Nations, Department of Economic and Social Affairs (DESA), Population Division, Population Estimates and Projections Section, New York
Hulchanski JD (1995) The concept of housing affordability: six contemporary uses of the housing expenditure-to-income ratio. Hous Stud 10:471–491
Klosterman RE (1999) The what if? Collaborative planning support system. Environ Plan B: Plan Des 26:393–408
Kvan T, Karakiewicz J (2012) Affordable living. In: Pearson C (ed) 2020 vision for a sustainable society. Melbourne Sustainable Society Institute, Melbourne, Victoria
Ord JK, Getis A (1995) Local spatial autocorrelation statistics: distributional issues and an application. Geogr Anal 27:286–306
Performance Urban Planning (2015) 11th Annual demographia international housing affordability survey: 2015. Christchurch
Pettit CJ, Stimson R, Tomko M, Sinnott R (2013a) Building an e-infrastructure to support urban and built environment research in Australia: a Lens-centric view. In: Surveying & spatial sciences conference, 2013

Pettit CJ, Klosterman RE, Nino-Ruiz M, Widjaja I, Russo P, Tomko M, Sinnott R, Stimson R (2013b) The online what if? Planning support system. In: Planning support systems for sustainable urban development. Springer

Pettit CJ, Klosterman RE, Delaney P, Whitehead AL, Kujala H, Bromage A, Nino-Ruiz M (2015a) The online what if? Planning support system: a land suitability application in Western Australia. Appl Spatial Anal Policy 8:93–112

Pettit CJ, Barton J, Goldie X, Sinnott R, Stimson R, Kvan T (2015b) The Australian urban intelligence network supporting smart cities. In: Geertman S, Ferreira JJ, Goodspeed R, Stillwell J (eds) Planning support systems and smart cities. Springer International Publishing

Phibbs P, Gurran N (2008) Demographia housing affordability surveys: an assessment of the methodology/Peter Phibbs and Nicole Gurran. Shelter NSW, Sydney

Rae A, Singleton A (2015) Putting big data in its place: a Regional Studies and Regional Science perspective. Reg Stud Reg Sci 2:1–5

Sinnott RO, Bayliss C, Bromage A, Galang G, Grazioli G, Greenwood P, Macaulay A, Morandini L, Nogoorani G, Nino-Ruiz M, Tomko M, Pettit CJ, Sarwar M, Stimson R, Voorsluys W, Widjaja I (2015) The Australia urban research gateway. Concur Comput Pract Exp 27:358–375

Stimson R, Tomko M, Sinnott R (2011) The Australian Urban Research Infrastructure Network (AURIN) initiative: a platform offering data and tools for urban and built environment researchers across Australia. In: State of Australian Cities (SOAC) conference 2011

Stone ME (2006) What is housing affordability? The case for the residual income approach. Hous Policy Debate 17:151–184

Ting I (2015) The Sydney Suburbs where minimum wage workers can afford to rent. Sydney Morning Herald, June 2015

Townsend AM (2013) Smart cities: big data, civic hackers, and the quest for a new utopia. WW Norton & Company, New York

Wilson MW (2015) Flashing lights in the quantified self-city-nation. Reg Stud Reg Sci 2:39–42

Yates J, Milligan V, Berry M, Burke T, Gabriel M, Phibbs P, Pinnegar S, Randolph B (2013) Housing affordability: a 21st century problem, national research venture 3: housing affordability for lower income Australians Final Report No. 105. Australian Housing and Urban Research Institute, Melbourne

A Big Data Mashing Tool for Measuring Transit System Performance

Gregory D. Erhardt, Oliver Lock, Elsa Arcaute, and Michael Batty

Abstract This research aims to develop software tools to support the fusion and analysis of large, passively collected data sources for the purpose of measuring and monitoring transit system performance. This study uses San Francisco as a case study, taking advantage of the automated vehicle location (AVL) and automated passenger count (APC) data available on the city transit system. Because the AVL-APC data are only available on a sample of buses, a method is developed to expand the data to be representative of the transit system as a whole. In the expansion process, the General Transit Feed Specification (GTFS) data are used as a measure of the full set of scheduled transit service.

The data mashing tool reports and tracks transit system performance in these key dimensions:

- Service Provided: vehicle trips, service miles;
- Ridership: boardings, passenger miles; passenger hours, wheelchairs served, bicycles served;
- Level-of-service: speed, dwell time, headway, fare, waiting time;
- Reliability: on-time performance, average delay; and
- Crowding: volume-capacity ratio, vehicles over 85 % of capacity, passenger hours over 85 % of capacity.

An important characteristic of this study is that it provides a tool for analyzing the trends over significant time periods—from 2009 through the present. The tool allows data for any two time periods to be queried and compared at the analyst's

---

G.D. Erhardt (✉)
Department of Civil Engineering, University of Kentucky, 261 Oliver H. Raymond Bldg., 508 Administration Drive, Lexington 40506, KY, USA
e-mail: greg.erhardt@uky.edu

O. Lock
Arup, 1 Nicholson St, East Melbourne, VIC 3002, Australia
e-mail: oliverclock@gmail.com

E. Arcaute • M. Batty
Centre for Advanced Spatial Analysis (CASA), University College London, 90 Tottenham Court Road, London W1T 4TJ, UK
e-mail: e.arcaute@ucl.ac.uk; m.batty@ucl.ac.uk

© Springer International Publishing Switzerland 2017
P. Thakuriah et al. (eds.), *Seeing Cities Through Big Data*, Springer Geography,
DOI 10.1007/978-3-319-40902-3_15

request, and puts the focus specifically on the changes that occur in the system, and not just observing current conditions.

Keywords Big data • Performance-based planning • Transit • Automated vehicle location • Automated passenger count

1 Introduction

Performance-based planning builds upon the traditional transportation planning process by aligning planning goals and objectives with specific performance measures against which projects can be evaluated. The emergence of performance-based planning received a boost from the current U.S. federal transportation legislation (*MAP-21* 2012) which makes it more central to the overall planning process. In recent years, researchers and practitioners have made significant progress in developing approaches to performance-based planning (Turnbull 2013), including approaches to establishing performance-based planning programs (Lomax et al. 2013; Price et al. 2013), methods for converting data into performance measures (Benson et al. 2013; Winick et al. 2013; Zmud et al. 2013), and experience formulating relevant performance measures from institutional priorities (Lomax et al. 2013; Pack 2013; Price et al. 2013). In spite of this momentum, a number of challenges still remain, including the availability of supporting data, the ability to synthesize those data into meaningful metrics and the resources required for analysis (Grant et al. 2013).

This research aims to meet these challenges by developing software tools to support the fusion and analysis of large, passively collected data sources for the purpose of measuring and monitoring transportation system performance. Because they are continuously collected, Big Data sources provide a unique opportunity to measure the changes that occur in the transportation system. This feature overcomes a major limitation of traditional travel data collection efforts, which are cross-sectional in nature, and allows for a more direct analysis of the changes that occur before-and-after a new transport project opens.

This first phase of work focuses on transit system performance, with future work planned to integrate highway measures. This study uses San Francisco as a case study, taking advantage of the automated vehicle location (AVL) and automated passenger count (APC) data available on the city transit system.

As of the year 2000, automated data collection systems were becoming more common at transit agencies, but data systems were immature, network and geographic analysis methods were in their infancy, and the data were often used for little beyond federal reporting requirements (Furth 2000). Subsequently, TCRP Report 88 provided guidelines for developing transit performance measurement systems, with a focus on identifying appropriate performance measures to correspond to agency goals (Kittelson & Associates et al. 2003). By 2006, TCRP 113 identified a wider range of AVL-APC applications, but still a dichotomy between APC data which was used in its archived form and AVL data which was often designed for real-time analysis and not archived or analyzed retrospectively (Furth et al. 2006). More complete data systems have since been developed that encapsulate the data processing and reporting (Liao 2011; Liao and Liu 2010), apply data mining methods in an effort to improve operational performance (Cevallos and Wang 2008), and examine bus bunching (Byon et al. 2011; Feng and Figliozzi 2011). Initial attempts have been made to visualize the data at a network level (Berkow et al. 2009; Mesbah et al. 2012).

Two important characteristics distinguish this study from previous work.

First, it operates on a sample of AVL-APC data, and a methodology is established to expand the data to the schedule as a whole and weight the data to represent total ridership. This is in contrast to the examples given above which generally assume full data coverage. Establishing expansion and weighting methods is important because it allows Big Data analysis to be applied in a wider range of locations with lower expenditure on data collection equipment.

Second, this study develops a tool to analyze the trends over a significant time periods—from 2009 through the present—as opposed to many applications which focus on using the data to understand a snapshot of current operations (Liao and Liu 2010; Feng and Figliozzi 2011; Wang et al. 2013; Chen and Chen 2009). The tool allows data for any two time periods to be queried and compared at the analyst's request, and puts the focus specifically on the changes that occur in the system, and not just observing current conditions. For example, changes that occur in a specific portion of the city may be traceable to housing developments or roadway projects at that location, trends that may go unnoticed given only aggregate measures or cross-sectional totals.

The remainder of this chapter is structured as follows: Section 2 describes the data sources used in this study. Section 3 covers the methodology for data processing, including the approach used to expand and weight the data to be representative of the system as a whole. Section 4 presents example outputs to demonstrate the types of performance reports that the data mashing tool can produce. Section 5 is conclusions and expected future work.

2 Data Sources

This research uses two primary data sources provided by the San Francisco Municipal Transportation Agency (SFMTA): automated vehicle location/automated passenger count (AVL-APC) data, and archived General Transit Feed Specification (GTFS) data. A third data set, from the Clipper transit smartcard system, has recently been released and work is currently underway to validate and incorporate these data.

The AVL-APC data is formatted with one record each time a transit vehicle makes a stop. At each stop, the following information is recorded:

- Vehicle location;
- Arrival time;

- Departure time;
- Time with door open;
- Time required to pullout after the door closes;
- Maximum speed since last stop;
- Distance from last stop;
- Passengers boarding;
- Passengers alighting;
- Rear door boardings;
- Wheelchair movements; and
- Bicycle rack usage.

In addition, identifiers are included to track the route, direction, trip, stop, sequence of stops, and vehicle number. The vehicle locations reflect some noise, both due to GPS measurement error and due to variation in the exact location at which the vehicle stops. However, because the stop is identified, those locations can be mapped to the physical stop location, providing consistency across trips. The count data become less reliable as the vehicle becomes more crowded, but the data are biased in a systematic way, and SFMTA makes an adjustment in the data set to compensate for this bias. The data are not currently available on rail or cable car, only on the buses. Equipment is installed on about 25 % of the bus fleet, and those buses are allocated randomly to routes and drivers each day at the depot. These data are available from 2008 to the present.

Because the AVL-APC data are available for only a sample of bus trips, the GTFS data are used to measure the scheduled universe of bus trips. GTFS is a data specification that allows transit agencies to publish their schedule information in a standard format. It was initially used to feed the Google Maps transit routing, and is now used by a wide range of applications. The data are in a hierarchical format and provide the scheduled time at which each vehicle is to make each stop. The full specification is available from ("General Transit Feed Specification Reference - Transit — Google Developers," n.d.). The data used in this study were obtained from the GTFS archive ("GTFS Data Exchange - San Francisco Municipal Transportation Agency," n.d.), from 2009 to present.

In addition, data from the Bay Area's transit smartcard system, Clipper Card, has recently been made available. These data provide on fare transactions made with the cards. Clipper Card was introduced in 2010, and currently has a penetration rate of approximately 50 % of riders on SFMTA buses. The data provide value over the above sources because they allow transfers to be identified. The data are subject to California's laws governing personally identifiable information (Harris 2014), making data privacy and protection issues of particular importance. Therefore, they have been released with a multi-step anonymization and data obfuscation process (Ory 2015).

3 Methodology

This section describes the methodology used to generate transit performance reports from the raw data. To ensure the performance measures are a valid representation of the transit system, the data area cleaned, expanded and weighted as outlined in Fig. 1. The transit smartcard data is included in the figure to demonstrate how it fits with the process, although its incorporation is left to future work.

First, each individual data set is cleaned and converted into a common format. For the AVL-APC data, this involves filtering out non-revenue service, records without a valid route ID, stop ID or trip ID, duplicate records, and those that do not meet quality control requirements. A number of derived fields are added, including the arriving and departing passenger load, the schedule deviation, flags for on-time arrival, time period groupings, and end-of-line flags. All date and time fields are converted from string format to a native Datetime format that allows for easy sorting and calculation of differences. As part of this Datetime conversion, special care is taken to handle the wrap-around effects of trips occurring between midnight and 3 am, which continue the schedule of the day prior, and whose ridership is counted with the day prior. An equivalency file is read to attach route IDs consistent with the GTFS data so the two files can later be joined. As part of this cleaning and processing the data are converted from their raw text file format to an HDF datastore format, as described later in this section.

The raw GTFS data are read and converted to a record-based format such that they are directly comparable to the AVL-APC data. This format has one record for each stop made by each vehicle trip on each transit route. These data are written separately for each day, making the identification of weekday, Saturday or Sunday/holiday service explicit. The process makes time periods, trip IDs, direction IDs and

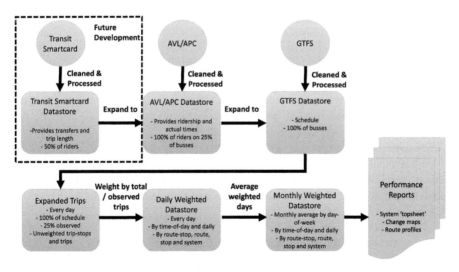

Fig. 1 Data processing flow

route IDs consistent with the equivalency used for the AVL-APC data. It calculates the scheduled headway of each trip, the scheduled runtime from the previous stop, and the distance traveled from the last stop, and along the route shape as a whole.

After the initial cleaning and conversion, the data are joined to create an expanded data store. The goal of this expansion is to identify exactly what is missing from the sampled data, so they can be factored up to be representative of the universe as a whole. The relationship between the data sets is that transit smartcard data provides a sample of about 50 % of riders, the AVL-APC data provides 100 % of riders on a sample of about 25 % of vehicle trips, and the GTFS data identifies 100 % of vehicle trips. Therefore, the expansion chain allows the more information-rich data sets to be combined with the more complete, but less rich data, much like a household travel survey would be expanded to match Census control totals. In this case, the expansion is a left join of the AVL-APC data records onto the GTFS records. Rail does not have AVL-APC equipment installed, so is excluded from the GTFS records as well. Note that this process is not able to account for scheduled trips that are not run, due to driver or equipment availability or other operational issues. The resulting datastore has the full enumeration of service, but ridership and actual time information attached to only a portion of records. Without this step, it would not be possible to differentiate between trips that are missing because of a service change or those that are missing because they were simply not sampled. In a setting where we are explicitly interested in examining service changes, this distinction is important.

The output of this expansion process is a datastore whose structure is shown in Table 1. The table also shows the source and data type of each field. These disaggregate records are referred to as trip-stop records because there is one record for each time a bus trip makes a stop (even if that stop is bypassed due to no passengers boarding or alighting). More specifically, records are defined by a unique combination of values in those fields identified with as an index in the source column. A related set of trip records is generated that aggregates across the SEQ field such that there is a single record for each time a bus makes a trip.

While only about one in five will be observed, the datastore at this root level is suitable for making comparisons of individual trips or trip-stops. However, summing values across trips to generate time-of-day, daily, or system totals would result in an under-estimate of the total ridership because of the missing values. Therefore, a set of weights is developed to factor up the records to estimate the totals at these more aggregate levels for each day.

Because an entire trip is observed together, weights are calculated for trips and then broadcast to all stops in that trip. At the root level, the TRIP_WEIGHT is set equal to 1 if the trip is observed and 0 otherwise. While not entirely necessary, including this TRIP_WEIGHT allows the remaining calculations to be performed consistently at all levels of aggregation.

The weights are calculated by grouping the trips to the level of aggregation of interest, and within the group, applying the formula:

A Big Data Mashing Tool for Measuring Transit System Performance

Table 1 Data dictionary for expanded and weighted trip-stop datastore

Category	Field	Description	Type	Source
Time and date	MONTH	Month and year	Datetime	Index
	DATE	Date	Datetime	Index
	DOW	Day of week (1 = Weekday, 2 = Saturday, 3 = Sunday/Holiday)	Integer	Index
	TOD	Time of day	String	Index
Index fields	AGENCY_ID	Agency ID (i.e. SFMTA)	String	Index
	ROUTE_SHORT_NAME	Route short name (i.e. 38)	String	Index
	ROUTE_LONG_NAME	Route long name (i.e. GEARY)	String	Index
	DIR	Direction (0 = outbound, 1 = inbound)	Integer	Index
	TRIP	Trip ID, as HHMM_SEQ of first stop on trip	Integer	Index
	SEQ	Stop sequence within route	Integer	Index
	ROUTE_TYPE	Type of route (0 = tram, 3 = bus, 5 = cable car)	Integer	GTFS
Route attributes	TRIP_HEADSIGN	Headsign on bus indicating destination (i.e. Ocean Beach)	String	GTFS
	HEADWAY_S	Scheduled headway (min)	Float	Calculated
	FARE	Full fare ($)	Float	GTFS
	PATTERN	Pattern identifier calculated for all trips	String	GTFS
	PATTCODE	Pattern code (i.e. 38OB3)	String	AVL/APC
Stop attributes	STOPNAME	Name of stop (i.e. Geary Blvd & Divisadero St)	String	GTFS
	STOPNAME_AVL	Name of stop in AVL/APC data	String	AVL/APC
	STOP_LAT	Latitude of stop location	Float	GTFS
	STOP_LON	Longitude of stop location	Float	GTFS
	SOL	Start of line flag (1 = start of line, 0 = not)	Integer	GTFS
	EOL	End of line flag (1 = end of line, 0 = not)	Integer	GTFS
	TIMEPOINT	Timepoint flag (1 = stop is a timepoint in schedule, 0 = not)	Integer	AVL/APC

(continued)

Table 1 (continued)

Category	Field	Description	Type	Source
Times	ARRIVAL_TIME_S	Scheduled arrival time	Datetime	GTFS
	ARRIVAL_TIME	Actual arrival time	Datetime	AVL/APC
	ARRIVAL_TIME_DEV	Deviation from arrival schedule (min)	Float	Calculated
	DEPARTURE_TIME_S	Scheduled departure time	Datetime	GTFS
	DEPARTURE_TIME	Actual departure time	Datetime	AVL/APC
	DEPARTURE_TIME_DEV	Deviation from departure schedule (min)	Float	Calculated
	DWELL_S	Scheduled dwell time (min)	Float	GTFS
	DWELL	Actual dwell time (min)	Float	AVL/APC
	RUNTIME_S	Scheduled running time (min), excludes dwell time	Float	GTFS
	RUNTIME	Actual running time (min), excludes dwell time	Float	AVL/APC
	TOTTIME_S	Scheduled total time (min), runtime + dwell time	Float	GTFS
	TOTTIME	Actual total time (min), runtime + dwell time	Float	AVL/APC
	SERVMILES_S	Scheduled service miles	Float	GTFS
	SERVMILES	Service miles from AVL/APC data	Float	AVL/APC
	RUNSPEED_S	Scheduled running speed (mph), excludes dwell time	Float	Calculated
	RUNSPEED	Actual running speed (mph), excludes dwell time	Float	Calculated
	ONTIME5	Vehicle within -1 to $+5$ min of schedule ($1 =$ yes, $0 =$ no)	Float	Calculated

(continued)

Table 1 (continued)

Category	Field	Description	Type	Source
Ridership	ON	Boardings	Float	AVL/APC
	OFF	Alightins	Float	AVL/APC
	LOAD_ARR	Passenger load upon arrival	Float	AVL/APC
	LOAD_DEP	Passenger load upon departure	Float	AVL/APC
	PASSMILES	Passenger miles	Float	Calculated
	PASSHOURS	Passenger hours, including both runtime and dwell time	Float	Calculated
	WAITHOURS	Passenger waiting hours, with wait as 1/2 headway	Float	Calculated
	PASSDELAY_DEP	Delay to passengers boarding at this stop	Float	Calculated
	PASSDELAY_ARR	Delay to passengers alighting at this stop	Float	Calculated
	RDBRDNGS	Rear door boardings	Float	AVL/APC
	CAPACITY	Vehicle capacity	Float	AVL/APC
	DOORCYCLES	Number of times door opens and closes at this stop	Float	AVL/APC
	WHEELCHAIR	Number of wheelchairs boarding at this stop	Float	AVL/APC
	BIKERACK	Bikerack used at this stop	Float	AVL/APC
Crowding	VC	Volume-capacity ratio	Float	Calculated
	CROWDED	Volume > 0.85 * capacity	Float	Calculated
	CROWDHOURS	Passenger hours when volume > 0.85 * capacity	Float	Calculated
Additional ID fields	ROUTE_ID	Route ID in GTFS	Integer	GTFS
	ROUTE_AVL	Route ID in AVL/APC	Integer	AVL/APC
	TRIP_ID	Trip ID in GTFS	Integer	GTFS
	STOP_ID	Stop ID in GTFS	Integer	GTFS
	STOP_AVL	Stop ID in AVL/APC	Float	AVL/APC
	BLOCK_ID	Block ID in GTFS	Integer	GTFS
	SHAPE_ID	Shape ID in GTFS	Integer	GTFS
	SHAPE_DIST	Distance along shape (m)	Float	GTFS
	VEHNO	Vehicle number	Float	AVL/APC
	SCHED_DATES	Dates when this schedule is in operation	String	GTFS

(continued)

Table 1 (continued)

Category	Field	Description	Type	Source
Weights	TRIP_WEIGHT	Weight applied when summarizing data at trip level	Float	Calculated
	TOD_WEIGHT	Weight applied when calculating time-of-day totals	Float	Calculated
	DAY_WEIGHT	Weight applied when calculating daily totals	Float	Calculated
	SYSTEM_WEIGHT	Weight applied when calculating system totals	Float	Calculated

$$W_t = \frac{N}{\sum_t w_t} w_t$$

where:

W_t is the weight for trip t,

N is the number of trips in the group, and

w_t is the base weight for trip t.

These weights are built hierarchically, such that the higher-level weights incorporate the lower-level weights.

The first calculation is for the time-of-day weight, TOD_WEIGHT, in which the trips are grouped by DATE, TOD, AGENCY_ID, ROUTE_SHORT_NAME and DIR, and the TRIP_WEIGHT serves as the base weight. Because the TRIP_WEIGHT is a mask for the observed trips, the formula simply gives the ratio of the total trips to observed trips in the group. So if a particular route makes ten trips in the inbound direction during the AM peak and two of those trips are observed, the resulting weight of five is used to scale up the observations to represent the total ridership in the AM peak.

One level up is the DAY_WEIGHT, used for expanding to daily totals by route. The trips are grouped by DATE, AGENCY_ID, ROUTE_SHORT_NAME and DIR, and the base weight is the TOD_WEIGHT. If a route has some observations in each time period for which there is service, the DAY_WEIGHT will be equal to the TOD_WEIGHT because the base weight already factors up to the total trips during the day. The difference occurs when there are some time periods for which there is service, but zero trips are observed. In this situation, the data from the remaining time periods are scaled up to account for this missing period. This weight should not be applied when summarizing data at the time period level because doing so would over-state the ridership in non-missing time periods, but it provides a better estimate of the daily total by route.

Moving another level up, it is also possible that some routes will not be observed at all during a day, so adding up the total system ridership using the DAY_WEIGHT would miss the ridership on those unobserved routes. To account for this problem, the SYSTEM_WEIGHT is calculated using the DAY_WEIGHT as a base weight and grouping by DATE, TOD and AGENCY_ID. The time-of-day grouping ensures the result is representative of the total number of trips in each time period. As long as some trips are observed in each time period, which is true for all days that have been inspected, another higher-level weight is not needed.

After calculating these weights, they are assigned to the disaggregate records, and the data are aggregated with the weights applied to calculate route-stop, route, stop and system totals by time-of-day and for the daily total. This is done separately for each day, providing an estimate of the state of the system on each day for which data are available. The weighted data are then aggregated by month used to calculate conditions for an average weekday, an average Saturday and an average Sunday/holiday in each month. These monthly average datastores are the primary source of information for the system performance reports, discussed in the next section, although the daily data remain available for more detailed analysis.

The estimates resulting from this process will be more reliable if there is reasonably good coverage of observations across routes. To examine the route coverage, Table 2 shows the percent of trips observed on each route for each weekday in July 2010. Twenty-two percent of trips are observed, although this varies somewhat by route. The weighting process should do a good job of accounting for these varying penetration rates. More limiting are the cases where zero trips are observed on a route, which are highlighted with red cells. In these cases, the weighting process scales up the ridership on other routes to account for the missing values on that route. The missing values tend to occur on the routes that make fewer trips. Overall, 93 % of routes are observed at least once during the month, with those routes covering 96 % of trips.

One of the challenges in this effort is that the sampling of trips is not entirely random. There are operational constraints, such as certain types of buses (motor bus versus trolley bus, and articulated versus standard length) being needed on certain routes, and the fact that once a bus is assigned it tends to drive the same route back and forth. The result is that the data will not be as reliable as could be achieved with a well-designed sampling plan, but with a good overall coverage can be expected to provide good estimates of the state of the system.

To evaluate the magnitude of the error that can be expected from the sampling and weighting process, the number of service miles is used as an indicator. Service miles serves as a useful indicator because it is calculated from the GTFS data, so the enumerated value for the system as a whole is known. For comparison, the service miles are also calculated from the subset of observed records, with the weights applied to scale up those observed records to the system total. These calculations reveal that for months from 2009 through 2013, the average magnitude of the weighting error at the system level is 1.0 %, and the maximum magnitude is 3.3 %.

The software was developed in an open-source framework in the Python environment. It is available under the GNU General Public License Version 3 for

Table 2 Percent of trips observed on each route on weekdays in July 2010

Route		Scheduled Trips	6/1	6/2	6/3	6/4	6/7	6/8	6/9	6/10	6/11	6/14	6/15	6/16	6/17	6/18	6/21	6/22	6/23	6/24	6/25	6/28	6/29	Average
1	CALIFORNIA	389	24%	20%	33%	13%	32%	15%	27%	35%	14%	30%	18%	23%	28%	19%	30%	30%	17%	25%	10%	20%	19%	23%
1AX	CALIFORNIA A EXPRESS	22	9%	32%	14%	23%	5%	9%	9%	18%	18%	27%	9%	18%	23%	32%	18%	0%	27%	27%	9%	18%	5%	17%
1BX	CALIFORNIA B EXPRESS	27	11%	11%	37%	22%	7%	7%	26%	11%	37%	11%	11%	7%	22%	7%	26%	26%	7%	22%	30%	26%	19%	18%
2	CLEMENT	118	15%	31%	26%	4%	37%	55%	23%	14%	29%	19%	27%	17%	31%	36%	28%	14%	15%	21%	21%	26%	25%	25%
3	JACKSON	128	18%	27%	11%	35%	0%	20%	59%	3%	47%	20%	0%	16%	45%	34%	6%	20%	16%	35%	48%	17%	38%	24%
5	FULTON	324	24%	26%	22%	31%	35%	20%	19%	25%	16%	14%	31%	24%	13%	19%	24%	34%	10%	26%	34%	20%	18%	23%
6	PARNASSUS	183	16%	17%	26%	21%	46%	16%	30%	42%	35%	18%	34%	52%	9%	28%	26%	33%	38%	23%	19%	26%	11%	27%
8AX	BAYSHORE A EXPRESS	42	24%	24%	26%	26%	17%	12%	33%	29%	12%	14%	26%	24%	29%	33%	5%	24%	33%	21%	29%	5%	26%	22%
8BX	BAYSHORE B EXPRESS	42	29%	33%	10%	12%	29%	36%	14%	14%	2%	21%	24%	29%	7%	36%	26%	24%	33%	19%	26%	12%	24%	22%
8X	BAYSHORE EXPRESS	175	25%	32%	5%	10%	33%	35%	12%	13%	0%	23%	26%	30%	5%	43%	21%	27%	29%	14%	29%	14%	24%	21%
9	SAN BRUNO	170	2%	13%	22%	26%	18%	26%	8%	20%	24%	9%	33%	25%	33%	11%	9%	24%	5%	14%	16%	22%	16%	18%
9L	SAN BRUNO LIMITED	119	5%	11%	6%	8%	0%	22%	11%	13%	10%	16%	10%	6%	10%	6%	23%	5%	10%	12%	10%	0%	34%	10%
10	TOWNSEND	87	7%	0%	0%	13%	6%	7%	0%	0%	20%	0%	20%	0%	10%	26%	7%	7%	6%	17%	7%	0%	14%	8%
12	FOLSOM/PACIFIC	102	13%	58%	36%	32%	15%	48%	19%	23%	0%	32%	7%	30%	13%	0%	8%	18%	13%	0%	35%	28%	0%	20%
14	MISSION	322	19%	20%	30%	19%	9%	25%	23%	25%	20%	15%	18%	24%	16%	14%	16%	21%	25%	24%	22%	22%	23%	20%
14L	MISSION LIMITED	124	34%	0%	29%	9%	8%	10%	23%	27%	19%	10%	8%	10%	15%	10%	40%	27%	19%	19%	31%	28%	18%	19%
14X	MISSION EXPRESS	33	12%	18%	30%	15%	24%	12%	33%	27%	39%	3%	15%	24%	33%	0%	9%	27%	18%	15%	24%	6%	6%	20%
16X	NORIEGA EXPRESS	29	31%	0%	41%	3%	38%	38%	7%	21%	7%	31%	38%	28%	14%	21%	21%	34%	3%	24%	10%	24%	34%	22%
17	PARK MERCED	64	0%	0%	0%	50%	50%	50%	50%	50%	0%	0%	50%	0%	50%	50%	0%	50%	0%	100%	0%	0%	0%	24%
18	46TH AVENUE	94	45%	29%	24%	20%	34%	29%	33%	56%	0%	6%	44%	47%	18%	26%	21%	41%	18%	83%	38%	39%	24%	32%
19	POLK	136	14%	27%	35%	18%	12%	47%	44%	24%	12%	51%	21%	35%	32%	31%	27%	17%	28%	33%	37%	30%	23%	28%
21	HAYES	174	5%	40%	32%	24%	13%	38%	29%	0%	52%	36%	27%	23%	52%	37%	28%	38%	19%	14%	2%	39%	26%	27%
22	FILLMORE	270	29%	23%	41%	26%	45%	32%	34%	19%	38%	31%	27%	24%	37%	42%	30%	22%	33%	22%	31%	31%	39%	31%
23	MONTEREY	108	21%	21%	19%	48%	13%	26%	20%	29%	6%	34%	13%	19%	16%	21%	24%	0%	15%	0%	31%	6%	13%	18%
24	DIVISADERO	204	15%	18%	30%	23%	19%	35%	23%	37%	30%	42%	46%	19%	35%	30%	24%	36%	26%	27%	46%	30%	50%	31%
27	BRYANT	139	0%	43%	9%	15%	17%	9%	29%	22%	19%	34%	6%	0%	5%	21%	0%	10%	19%	6%	14%	21%	30%	16%
28	19TH AVENUE	175	47%	24%	20%	13%	27%	18%	9%	44%	27%	13%	22%	57%	25%	38%	15%	32%	33%	26%	19%	30%	22%	27%
28L	19TH AVENUE LIMITED	49	51%	35%	10%	14%	33%	18%	8%	59%	29%	10%	27%	63%	12%	39%	18%	35%	24%	16%	16%	27%	29%	27%
29	SUNSET	177	11%	19%	29%	16%	12%	23%	33%	21%	30%	21%	12%	18%	18%	11%	29%	31%	18%	16%	33%	20%	17%	21%
30	STOCKTON	423	33%	22%	27%	39%	36%	19%	32%	30%	33%	42%	26%	39%	35%	46%	34%	42%	23%	46%	27%	37%	22%	33%
30X	MARINA EXPRESS	48	23%	13%	33%	13%	21%	29%	27%	17%	33%	40%	21%	25%	40%	21%	29%	21%	31%	19%	8%	27%	33%	25%
31	BALBOA	164	13%	26%	0%	13%	10%	43%	26%	23%	18%	20%	12%	37%	6%	12%	0%	0%	33%	39%	29%	34%	41%	25%
31AX	BALBOA A EXPRESS	21	5%	5%	29%	10%	10%	10%	5%	19%	24%	5%	10%	24%	19%	14%	19%	19%	10%	10%	5%	29%	5%	13%
31BX	BALBOA B EXPRESS	22	14%	23%	9%	14%	18%	23%	27%	18%	18%	14%	18%	18%	18%	23%	14%	18%	14%	14%	27%	18%	9%	18%
33	STANYAN	118	35%	29%	31%	17%	17%	15%	30%	18%	43%	17%	30%	52%	52%	35%	51%	17%	31%	66%	0%	63%	0%	31%

Table 2 (continued)

Route		Scheduled Trips	Percent of Trips Observed on Date																			
			6/1	6/2	6/3	6/4	6/7	6/8	6/9	6/10	6/11	6/14	6/15	6/16	6/17	6/18	6/21	6/22	6/23	6/24	6/25	6/28	6/29	Average
35	EUREKA	60	0%	53%	0%	0%	0%	0%	0%	0%	0%	0%	0%	0%	0%	0%	0%	0%	75%	0%	0%	0%	98%	27%
36	TERESITA	64	33%	19%	0%	0%	0%	33%	0%	64%	63%	33%	8%	55%	33%	0%	34%	53%	58%	0%	0%	0%	31%	24%
37	CORBETT	91	0%	5%	19%	0%	19%	27%	0%	0%	22%	55%	51%	0%	0%	0%	0%	27%	27%	30%	22%	22%	43%	17%
38	GEARY	312	8%	8%	2%	31%	2%	11%	17%	16%	0%	34%	23%	12%	16%	26%	40%	13%	13%	38%	25%	14%	24%	19%
38AX	GEARY A EXPRESS	24	13%	4%	8%	13%	23%	4%	0%	13%	27%	13%	8%	25%	4%	9%	0%	4%	17%	21%	17%	0%	8%	11%
38BX	GEARY B EXPRESS	23	13%	26%	13%	13%	17%	22%	17%	17%	17%	13%	17%	13%	17%	4%	9%	13%	17%	22%	22%	13%	17%	16%
38L	GEARY LIMITED	276	17%	35%	12%	13%	9%	10%	8%	24%	13%	13%	12%	25%	14%	9%	12%	14%	21%	12%	8%	16%	15%	17%
39	COIT	62	0%	0%	52%	13%	18%	52%	22%	0%	12%	20%	48%	52%	30%	30%	52%	0%	48%	0%	48%	18%	39%	29%
41	UNION	96	40%	30%	0%	29%	48%	0%	13%	0%	98%	50%	0%	0%	0%	0%	0%	0%	0%	0%	33%	40%	33%	35%
43	MASONIC	160	31%	46%	45%	41%	40%	35%	15%	29%	36%	23%	43%	39%	41%	45%	32%	39%	28%	32%	33%	40%	29%	30%
44	O'SHAUGHNESSY	167	38%	29%	32%	17%	16%	16%	36%	18%	34%	18%	19%	23%	38%	32%	42%	28%	19%	33%	51%	19%	11%	22%
45	UNION-STOCKTON	200	17%	45%	1%	4%	32%	19%	16%	19%	24%	13%	32%	14%	19%	45%	3%	26%	26%	25%	21%	6%	5%	17%
47	VAN NESS	203	15%	5%	37%	27%	22%	16%	4%	35%	4%	15%	9%	33%	14%	31%	25%	6%	35%	25%	21%	11%	26%	25%
48	QUINTARA – 24TH STREET	147	33%	9%	28%	14%	29%	8%	12%	11%	7%	64%	33%	7%	21%	31%	0%	45%	40%	36%	10%	26%	5%	20%
49	MISSION-VAN NESS	289	28%	17%	17%	20%	24%	2%	18%	24%	20%	39%	14%	24%	18%	40%	25%	31%	22%	12%	7%	41%	10%	19%
52	EXCELSIOR	68	9%	0%	44%	44%	22%	25%	21%	20%	20%	25%	12%	18%	19%	12%	20%	23%	9%	20%	26%	17%	5%	18%
54	FELTON	104	38%	11%	16%	0%	10%	9%	0%	31%	37%	44%	0%	9%	0%	15%	0%	43%	44%	0%	44%	0%	0%	17%
56	RUTLAND	58	0%	69%	0%	0%	0%	0%	5%	13%	23%	0%	16%	0%	16%	54%	3%	15%	0%	23%	0%	24%	31%	20%
66	QUINTARA	93	0%	0%	47%	44%	0%	0%	64%	0%	0%	0%	0%	0%	0%	59%	45%	0%	0%	21%	100%	66%	0%	9%
67	BERNAL HEIGHTS	96	0%	21%	17%	0%	0%	0%	45%	0%	0%	0%	0%	0%	0%	0%	0%	0%	0%	0%	16%	53%	0%	22%
71	HAIGHT-NORIEGA	141	17%	34%	16%	26%	14%	35%	26%	28%	16%	28%	27%	23%	79%	50%	0%	100%	25%	0%	6%	100%	50%	23%
71L	HAIGHT-NORIEGA LIMITED	25	12%	44%	20%	20%	20%	36%	20%	24%	16%	36%	27%	20%	12%	28%	23%	10%	32%	35%	30%	6%	24%	25%
80X	GATEWAY EXPRESS	1	0%	0%	0%	0%	0%	0%	0%	0%	0%	0%	44%	0%	0%	28%	28%	12%	0%	36%	8%	36%	28%	0%
81X	CALTRAIN EXPRESS	6	0%	0%	0%	0%	0%	0%	0%	0%	0%	0%	0%	0%	0%	0%	0%	0%	0%	0%	0%	0%	0%	0%
82X	LEVI PLAZA EXPRESS	22	0%	0%	0%	0%	0%	0%	0%	0%	0%	0%	0%	0%	0%	0%	0%	0%	0%	0%	0%	0%	0%	0%
88	B.A.R.T. SHUTTLE	16	0%	50%	50%	50%	50%	0%	50%	50%	50%	50%	54%	46%	50%	50%	19%	0%	50%	0%	0%	0%	0%	32%
90	OWL	13	0%	0%	0%	0%	0%	54%	0%	0%	46%	0%	0%	100%	0%	0%	0%	0%	46%	100%	47%	46%	0%	19%
91	OWL	15	40%	0%	0%	47%	20%	0%	0%	0%	27%	33%	40%	0%	20%	0%	13%	13%	13%	0%	47%	40%	67%	20%
95	INGLESIDE APTOS	2	0%	0%	0%	0%	0%	0%	0%	0%	0%	0%	0%	0%	0%	0%	0%	0%	0%	0%	0%	0%	0%	0%
108	TREASURE ISLAND	150	21%	28%	25%	33%	55%	16%	31%	63%	23%	31%	21%	28%	24%	29%	35%	0%	20%	7%	0%	7%	7%	24%
KM BUS		262	0%	0%	0%	0%	0%	0%	0%	0%	0%	0%	0%	0%	0%	0%	0%	0%	0%	0%	0%	0%	0%	0%
K-OWL		3	33%	33%	0%	0%	0%	0%	33%	0%	33%	33%	0%	33%	33%	33%	0%	0%	33%	0%	0%	0%	0%	11%
L-OWL		9	56%	56%	0%	0%	0%	44%	44%	0%	44%	44%	0%	44%	44%	44%	0%	0%	44%	0%	0%	44%	0%	18%
N-OWL		12	25%	0%	25%	0%	0%	0%	0%	25%	33%	33%	0%	58%	58%	0%	33%	33%	0%	17%	0%	0%	33%	18%
Total		8,092	19%	22%	23%	21%	22%	21%	21%	23%	21%	24%	21%	23%	22%	25%	23%	23%	21%	23%	21%	23%	21%	22%

distribution (Erhardt 2014). It leverages several open-source packages specifically designed to provide high-performance data storage, access and analysis for extremely large data sets. Specifically:

- *Pandas* is used for in-memory data operations, providing data structures and analysis tools for fast joins, aggregations, and tabulations of the data. Its functionality is similar to what is available in an R dataframe.
- *HDF5* (Hierarchical Data Format 5) is used to store the data on disk. It is designed for the fast and flexible storage of large data sets, allows for any combination of key-value pairs to be written, and allows on-disk indexing of the data.
- *PyTables* is a package for managing hierarchical datasets designed to easily cope with extremely large data sets. PyTables serves as the interface between Pandas operations in memory and the HDF5 storage on disk.

The advantage to using this combination of technology is that it allows datasets too large to be stored in memory to be written to disk, but allows for random access to those data with very fast queries. The development has shown that the converted data are dramatically faster to access than in their raw text format. This workflow also provides much greater flexibility than using a traditional database, which typically perform best with a stable data structure, making them less ideal for exploratory analysis.

4 Sample Results

This section presents sample results from the data mashing tool. The purpose of this section is to illustrate the types of performance measure the tool is capable of reporting, and how those measures might be useful in planning. In all cases, the performance reports seek to report information that is both relevant to the planning process and readily explainable to policy makers. It further seeks to put the focus of the analysis on the changes that occur over time, rather than a single snapshot of the system.

Table 3 shows a sample of the monthly transit performance report. It consolidates the core performance measures onto a single page, and compares them to performance from another period, often the month before. The measures are grouped in the following categories:

- *Input Specification*: Attributes selected by the user to define the scope of the report. The geographic extent can be the bus system as a whole, a route or an individual stop, with some minor differences for the route or stop reports. The day-of-week is weekday, Saturday or Sunday/holiday. Time-of-day can be specified for the daily total, or for individual time periods allowing for evaluation of peak conditions. The report generation date and a comments section are provided. The notes in this case indicate that system-wide service cuts occurred

Table 3 Sample transit performance summary report

SFMTA transit performance report				
Input specification				
Geographic extent:	All buses			
Day-of-week:	Average weekday			
Time-of-day:	Daily			
Report generated on:	2015-07-29 13:48:00			
Comments:	Service cuts in April 2010			
	Periods			
	Jul-2009	Jul-2010	Difference	% difference
Service provided				
Vehicle trips	9183	8092	−1091	−11.9%
Service miles	57,751	52,046	−5705	−9.9%
Ridership				
Boardings	526,423	509,400	−17,023	−3.2%
Rear-door boardings	1688	1436	−253	−15.0%
Passenger miles	1,034,838	1,043,599	8760	0.8%
Passenger hours	123,675	125,622	1947	1.6%
Wheelchairs served	1066	1281	215	20.2%
Bicycles served	1791	1713	−78	−4.4%
Level-of-service				
Average run speed (mph)	10.60	10.29	−0.31	−2.9%
Average dwell time per stop (min)	0.20	0.23	0.03	14.4%
Average scheduled headway (min)	13.84	14.34	0.51	3.7%
Average full fare ($)	$2.00	$2.00	$0.00	0.0%
Average distance traveled per passenger (mi)	1.97	2.05	0.08	4.2%
Average passenger speed (mph)	8.37	8.31	−0.06	−0.7%
Average wait time per passenger (min)	5.34	5.70	0.37	6.9%
Reliability				
Percent of vehicles arriving on-time (−1 to +5 min)	66.8%	63.0%	−0.04	−5.8%
Average waiting delay per passenger (min)	2.74	2.49	−0.24	−8.9%
Average arrival delay per passenger (min)	2.22	2.00	−0.22	−10.0%
Crowding				
Average volume-capacity ratio	0.43	0.47	0.04	10.5%
Percent of trips with V/C > 0.85	7.1%	8.8%	0.02	22.8%
Passenger hours with V/C > 0.85	7143	8411	1268	17.8%
Observations and error				
Number of days	22	21		
Days with observations	22	21		
Percent of trips observed	19.5%	22.1%		
Measurement error (ON/OFF)-1	−0.2%	0.4%		
Weighting error (SERVMILES/ SERVMILES_S)-1	−0.9%	0.2%		

between the two periods, which corresponds to the nearly 10 % reduction in service miles. The report is generated for the same month before and after these cuts to report to avoid reporting seasonality changes.
- *Service Provided*: The service provided metrics measure the total scheduled transit service, as found in the GTFS. Identical values mean that the schedule did not change between those 2 months.
- *Ridership*: Ridership measures provide the total passenger boardings, the distance and time passengers spend onboard, and the number of wheelchairs and bicycles served. In this example, the ridership decreases by 3.2 %, potentially in response to the service cuts.
- *Level-of-Service*: The level-of-service section provides measures of the quality of service provided, as experienced by users. The average run speed, the dwell time per stop and the scheduled headway are measured as a function of the buses themselves. Run speed is defined as the speed between stops, so excludes the dwell time at stops. Scheduled headway is measured at each route-stop based on the time from the previous trip of the same route. That is, it accounts for combined headways for multiple patterns of the same route, but it does not account for combined headways across multiple routes. The fare is reported as the average full cash fare across all routes and stops, as shown in the GTFS. Separate revenue data would be needed to measure the average fare paid accounting for discounts and passes. The average distance traveled, average passenger speed and average passenger wait are measured as a function of the passengers themselves. In contrast to the run speed, the average passenger speed includes dwell time, making it generally slower. Average waiting time is measured as half the scheduled headway, assuming random passenger arrivals. The system-wide average passenger wait tends to be less than half the system-wide average scheduled headway because passengers tend to use more frequent service. In this example, both the average scheduled headway and passenger wait increase, which is logical given less frequent bus service.
- *Reliability*: Reliability measures indicate how well the buses adhere to their schedule. Consistent with the Transit Capacity and Quality of Service Manual (TCQSM) (Kittelson & Associates et al. 2013), a vehicle is considered on-time if it departs from a timepoint no more than 1 min early or arrives more than 5 min late. In addition, two measures of delay are reported which are weighted to passengers instead of buses. The waiting delay is the average time passengers wait at their stop for a bus to arrive after its scheduled arrival time. Arrival delay is the average time passengers arrive at their alighting stop, past the scheduled time.
- *Crowding*: For the purpose of this tool, a vehicle is considered to be crowded if the volume of passengers onboard exceeds 85 % of the capacity. The range of 85–100 % of total capacity corresponds roughly to the range of 125–150 % of the seated load, which is referenced in the TCQSM as the maximum design load for peak-of-the-peak conditions. The crowding statistics report the average volume-capacity ratio, the percent of trips where the vehicle is crowded at some point during the trip, and the number of passenger hours in crowded conditions. These

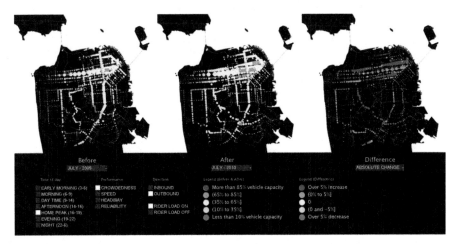

Fig. 2 Sample transit performance change map

performance reports can easily be generated for each time period, allowing for monitoring of crowding during the peak periods.

- *Observations*: The report includes the percent of trips observed, the total number of days and the number of days with observations. At a system level, there will generally be observations on each day, but specific routes or stops may not be observed on some days. The measurement error calculates the percent difference between the total boardings and alightings, providing an indication of the level of error that can be expected from the APC technology. The weighting error calculates the percent difference between the scheduled service miles and the weighted and expanded service miles, giving an indication of the error that can be expected as a result of the sampling and weighting process.

This performance report provides an overview allowing planners to quickly scan a range of indicators for changes that might be occurring.

While the numeric performance measures provide valuable information, their aggregate nature can wash out change that may be occurring in one portion of the city. Therefore, an interactive mapping tool was developed to plot key metrics in their geographic context. Figure 2 shows a screenshot from this tool. The left map shows a before period, the middle map an after period, and the right map shows either the absolute or relative change between the two periods. In this case, the comparison is between July 2009 and July 2010, before and after the service cuts in spring 2010. The user can select which time-of-day, which performance measure and which direction to plot. In this instance, the user has chosen to map the degree of crowdedness in the outbound direction during the 4–7 pm time period. The warm colors on the left two maps indicate more crowding, as measured by the average volume-capacity ratio during the period. The results are logical, with reasonably full buses moving west from the central business district towards residential areas of the city, as well as north-south on Van Ness Avenue. The map on the right shows

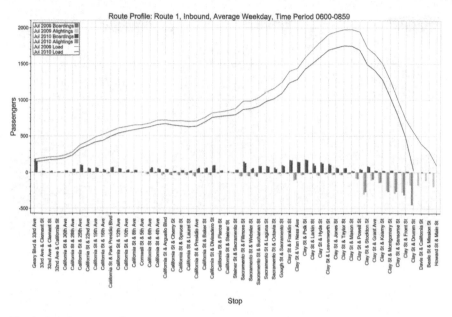

Fig. 3 Sample route profile

the relative change in the metric between the two periods, with the warm colors indicating an increase in crowdedness and the cool colors indicating a decrease. In this instance, the change in concentrated on about three specific routes.

To accommodate further analysis of the changes that occur to specific routes, the software generates route profiles as shown in Fig. 3. In this example, average weekday ridership on the 1-California route is plotted in the inbound direction during the AM peak. The x-axis is the sequence of stops along the route. The line charts show the number of passengers on the bus between each stop. The bar charts show the number of passengers boarding and alighting at each stop, with positive bars indicating boardings and negative bars indicating alightings. In all cases, the blue colors indicate the July 2009 period, and the red colors indicate the July 2010 period. The pattern of ridership remains similar between the two periods, with riders accumulating through the residential portions of the route, and passengers getting off the bus when it reaches the central business district, starting at the Clay Street and Stockton Street stop. The PM peak ridership profile would show the reverse. The route was shortened by the July 2010 period, with service no longer provided to the last three stops. Therefore, in the July 2010 period there are no alightings at these stops, and an increase in alightings at the new end-of-line stop. The overall volume on this route during the AM peak is lower after these changes. These boarding profiles are useful when evaluating service changes made to specific routes, or the ridership resulting from newly opened land developments.

Finally, line plots are output, as in Fig. 4, to show the trends over a longer period of time, rather than just for two periods. This particular example shows the on-time

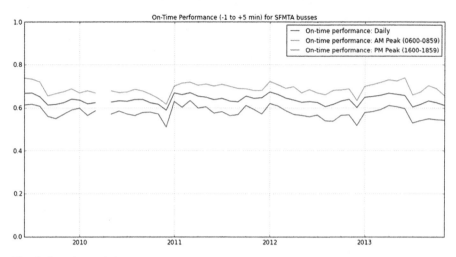

Fig. 4 Sample trend plot

performance, defined as the share of buses arriving no more than 1 min early or 5 min late. This is plotted for the daily totals, the AM peak and the PM peak. The results show the on-time performance is generally 60–70 %, with higher values in the AM peak and lower values in the PM peak. Any of the performance measures can be easily plotted in this way, and doing so is an important step to understanding whether the changes observed are real, or simply within the natural variation of the data.

The software can automatically generate each of the performance reports described above, allowing for core analysis of the most important measures. In addition, the full weighted and imputed datastore is available for advanced users who seek to conduct further in-depth analysis or custom queries.

5 Conclusions and Future Development

The product of this research is a Big Data mashing tool that can be used to measure transit system performance over time. The software is implemented for San Francisco, but can be adapted for use in other regions with similar data.

The paper addressed some of the methodological and mechanical challenges faced in managing these large data sets and translating them into meaningful planning information. One such challenge was the sampled nature of the data, where not all vehicles have AVL-APC equipment installed. To make these data more representative of the system as a whole, the vehicle trips in the AVL-APC data are expanded to match the universe of vehicle trips identified by the GTFS data and weights are developed to scale up to compensate for data that remain unobserved. The expansion process applies strategies from traditional surveys where a small but rich data set is expanded to match a less rich but more complete data set. Such strategies are key to spreading the use of Big Data for urban analysis beyond the first tier of cities that have near-complete data sets to those that are constrained by partial or incomplete data.

The software is available under an open-source license from (Erhardt 2014). For working with these large data sets, it was an important decision to work with libraries that allow fast querying of on-disk data, but also the ability to easily modify the data structure.

The data mashing tool reports and tracks transit system performance in the core dimensions of: service provided, ridership, level-of-service, reliability and crowding. The performance measures are reported for the system, by route and by stop, can also be mapped using an interactive tool. The focus of the tool is on providing the ability to monitor the trends and changes over time, as opposed to simply analyzing current operations. By making performance reports readily available at varying levels of resolution, and the data mashing tool encourages planners to engage in data-driven analysis on an ongoing basis.

Several extensions of this research are currently underway. First, the transit smartcard data has been obtained, and work is underway to evaluate the data set and incorporate it into the current tool. Doing so will provide additional value by allowing transfers and linked trips to be monitored. Second, parallel tools have been developed to monitor highway speeds and plans are in place to incorporate highway performance measures into a combined tool, allowing both to be tracked in concert.

Ultimately, the data mashing tool will be applied to measure the change in performance before and after changes to the transportation system. The study period covers a time with important changes to the transit system, such as the percent service cut in discussed in the test results (Gordon et al. 2010), and several pilot studies aimed at improving the speed and reliability of transit service in specific corridors (City and County of San Francisco Planning Department 2013). Evaluating these changes will provide planners and researchers with greater insight into the effects of transportation planning decisions.

Acknowledgement The authors would like to thank the San Francisco County Transportation Authority (SFCTA) for funding this research, the San Francisco Municipal Transportation Agency (SFMTA) for providing data, and both for providing valuable input and advice.

References

Kittelson & Associates, Urbitran, Inc., LKC Consulting Services, Inc., MORPACE International, Inc., Queensland University of Technology, Yuko Nakanishi (2003) A guidebook for developing a transit performance-measurement system (Transit Cooperative Research Program No. TCRP 88). Transportation Research Board of the National Academies, Washington, DC

Kittelson & Associates, Parsons Brinckerhoff, KFH Group, Inc., Texas A&M Transportation Institute, Arup (2013) Transit capacity and quality of service manual (Transit Cooperative Research Program No. TCRP 165). Transportation Research Board, Washington, DC

Benson JR, Perrin R, Pickrell SM (2013) Measuring transportation system and mode performance. In: Performance measurement of transportation systems: summary of the fourth international conference, 18–20 May 2011, Irvine, California. Transportation Research Board, Irvine, CA

Berkow M, El-Geneidy A, Bertini R, Crout D (2009) Beyond generating transit performance measures: visualization and statistical analysis with historic data. Transp Res Rec J Transp Res Board 2111:158–168. doi:10.3141/2111-18

Byon Y-J, Cortés CE, Martinez C, Javier F, Munizaga M, Zuniga M (2011) Transit performance monitoring and analysis with massive GPS bus probes of transantiago in Santiago, Chile: emphasis on development of indices for bunching and schedule adherence. Presented at the transportation research board 90th annual meeting

Cevallos F, Wang X (2008) Adams: data archiving and mining system for transit service improvements. Transp Res Rec J Transp Res Board 2063:43–51. doi:10.3141/2063-06

Chen W-Y, Chen Z-Y (2009) A simulation model for transit service unreliability prevention based on AVL-APC data. Presented at the international conference on measuring technology and mechatronics automation, 2009, ICMTMA'09, pp 184–188. doi:10.1109/ICMTMA.2009.77

City and County of San Francisco Planning Department (2013) Transit effectiveness project draft environmental impact report (No. Case No. 2011.0558E, State Clearinghouse No. 2011112030)

Erhardt GD (2014) sfdata_wrangler [WWW Document]. GitHub. https://github.com/UCL/sfdata_wrangler. Accessed 15 July 2014

Feng W, Figliozzi M (2011) Using archived AVL/APC bus data to identify spatial-temporal causes of bus bunching. Presented at the 90th annual meeting of the transportation research board, Washington, DC

Furth PG (2000) TCRP synthesis 34: data analysis for bus planning and monitoring, Transit Cooperative Research Program. National Academy Press, Washington, DC

Furth PG, Hemily B, Muller THJ, Strathman JG (2006) TCRP Report 113: using archived AVL-APC data to improve transit performance and management (No. 113), Transit Cooperative Research Program. Transportation Research Board of the National Academies, Washington, DC

General Transit Feed Specification Reference - Transit — Google Developers [WWW Document] (n.d.) https://developers.google.com/transit/gtfs/reference. Accessed 15 July 2014

Gordon R, Cabanatuan M, Chronicle Staff Writers (2010) Muni looks at some of deepest service cuts ever. San Francisco Chronicle, San Francisco

Grant M, D'Ignazio J, Bond A, McKeeman A (2013) Performance-based planning and programming guidebook (No. FHWA-HEP-13-041). United States Department of Transportation Federal Highway Administration

GTFS Data Exchange - San Francisco Municipal Transportation Agency [WWW Document] (n.d.) http://www.gtfs-data-exchange.com/agency/san-francisco-municipal-transportation-agency/. Accessed 15 July 2014

Harris KD (2014) Making your privacy practices public: recommendations on developing a meaningful privacy policy. Attorney General, California Department of Justice

Liao C-F (2011) Data driven support tool for transit data analysis, scheduling and planning. Intelligent Transportation Systems Institute, Center for Transportation Studies, University of Minnesota

Liao C-F, Liu H (2010) Development of data-processing framework for transit performance analysis. Transp Res Rec J Transp Res Board 2143:34–43. doi:10.3141/2143-05

Lomax T, Blankenhorn RS, Watanabe R (2013) Clash of priorities. In: Performance measurement of transportation systems: summary of the fourth international conference, 18–20 May 2011, Irvine, California. Transportation Research Board, Irvine, CA

Mesbah M, Currie G, Lennon C, Northcott T (2012) Spatial and temporal visualization of transit operations performance data at a network level. J Transp Geogr 25:15–26. doi:10.1016/j.jtrangeo.2012.07.005

Moving Ahead for Progress in the 21st Century Act (2012)

Ory D (2015) Lawyers, big data, (more lawyers), and a potential validation source: obtaining smart card and toll tag transaction data. Presented at the 94th transportation research board annual meeting, Washington, DC

Pack M (2013) Asking the right questions: timely advice for emerging tools, better data, and approaches for systems performance measures. In: Performance measurement of transportation systems: summary of the fourth international conference, 18–20 May 2011, Irvine, California. Transportation Research Board, Irvine, CA

Price TJ, Miller D, Fulginiti C, Terabe S (2013) Performance-based decision making: the buck starts here. In: Performance measurement of transportation systems: summary of the fourth international conference, 18–20 May 2011, Irvine, California. Transportation Research Board, Irvine, CA

Turnbull KF (ed) (2013) Performance measurement of transportation systems: summary of the fourth international conference, 18–20 May 2011, Irvine, California. Transportation Research Board, Washington, DC

Wang J, Li Y, Liu J, He K, Wang P (2013) Vulnerability analysis and passenger source prediction in urban rail transit networks. PLoS One 8:e80178. doi:10.1371/journal.pone.0080178

Winick RM, Bachman W, Sekimoto Y, Hu PS (2013) Transforming experiences: from data to measures, measures to information, and information to decisions with data fusion and visualization. In: Performance measurement of transportation systems: summary of the fourth international conference, 18–20 May 2011, Irvine, California. Transportation Research Board, Irvine, CA

Zmud J, Brush AJ, Choudhury MD (2013) Digital breadcrumbs: mobility data capture with social media. In: Performance measurement of transportation systems: summary of the fourth international conference, 18–20 May 2011, Irvine, California. Transportation Research Board, Irvine, CA

Developing a Comprehensive U.S. Transit Accessibility Database

Andrew Owen and David M. Levinson

Abstract This paper discusses the development of a national public transit job accessibility evaluation framework, focusing on lessons learned, data source evaluation and selection, calculation methodology, and examples of accessibility evaluation results. The accessibility evaluation framework described here builds on methods developed in earlier projects, extended for use on a national scale and at the Census block level. Application on a national scale involves assembling and processing a comprehensive national database of public transit network topology and travel times. This database incorporates the computational advancement of calculating accessibility continuously for every minute within a departure time window of interest. This increases computational complexity, but provides a very robust representation of the interaction between transit service frequency and accessibility at multiple departure times.

Keywords Accessibility • Connectivity • Transit

1 Introduction

Accessibility measures the number of opportunities that can be reached in a given travel time—an important metric for assessing the effectiveness of transportation–land use systems. To date, while these metrics have been used locally, there has been no standardized way to compare metropolitan areas systematically. This paper describes the development of an integrated software framework for a nationwide evaluation of the accessibility to jobs provided by public transit systems at the Census block level. Application on a national scale involves assembling and processing a comprehensive national database of public transit network topology and travel times. This database incorporates the computational advancement of calculating accessibility continuously for every minute within a departure time window of interest. Values for contiguous departure time spans can then be

---

A. Owen (✉) • D.M. Levinson
Department of Civil, Environmental, and Geo-Engineering, University of Minnesota, 500 Pillsbury Drive SE, Minneapolis, MN 55408, USA
e-mail: aowen@umn.edu; dlevinson@umn.edu

© Springer International Publishing Switzerland 2017
P. Thakuriah et al. (eds.), *Seeing Cities Through Big Data*, Springer Geography,
DOI 10.1007/978-3-319-40902-3_16

averaged or analyzed for variance over time. This increases computational complexity, but provides a very robust representation of the interaction between transit service frequency and accessibility at multiple departure times.

This project focused on measuring access to jobs, and the output dataset indicates how many jobs can be reached from each Census block within various travel time thresholds, assuming trips made by walking and transit. With minor modifications, this framework can be adapted to provide accessibility metrics for any destination type.

The development of a comprehensive and consistent national public transit accessibility database involved three major components. First, appropriate data sources were identified, collected, and aggregated in a single input geodatabase. Second, a travel time calculation methodology was selected which provides a reasonable and useful representation of expected travel times by public transit. Finally, block-level travel times and the resulting accessibility were calculated in a parallelized, scalable cloud computing environment.

The following sections overview the project's motivation, goals, and implementation and discuss lessons learned and future directions for improving the research and practice of accessibility evaluation.

2 Motivation and Goals

In both practice and in research, accessibility evaluation remains experimental and methodologically fragmented: researchers and planners focusing on different geographical areas often implement different techniques, making it difficult to compare accessibility metrics across different locations. This encourages the development and refinement of improved accessibility evaluation techniques, but heightens the "first mover" risk for agencies seeking to implement accessibility-based planning practices, as they must select a method that might produce results that can only be interpreted locally. Development of a common baseline accessibility metrics advances the use of accessibility-based planning in two ways. First, it provides a stable target for agencies seeking to implement accessibility-based methods in upcoming planning processes. Second, it provides researchers a frame of reference against which new developments in accessibility evaluation can be compared.

In 2012, the Minnesota Department of Transportation (MnDOT) implemented an "Annual Accessibility Measure for the Twin Cities Metropolitan Area" that provides a methodology for calculating accessibility in the Minneapolis–Saint Paul metropolitan area, and that establishes an evaluation methodology for accessibility to jobs by car and transit (Owen and Levinson 2012). Development phases of this project relied on proprietary and custom transit schedule data formats because the GTFS format (described below) had not been adopted by local transit operators (Krizek et al. 2007, 2009).

Simultaneously, the value of consistent, systematic accessibility evaluations across multiple metropolitan areas was demonstrated by the work of Levine et al. (2012), which collected zone-to-zone travel time information from 38 metropolitan planning organizations to implement a cross-metropolitan evaluation of accessibility by car.

The goal of this project is to combine the lessons learned from these earlier works with recent advances in transit schedule data format and availability to produce a new, comprehensive dataset of accessibility to jobs by transit.

3 Data Sources

Detailed digital transit schedules in a consistent format are a critical component of this system, and the availability of such data is a relatively recent phenomenon. The General Transit Feed Specification (GTFS) (Google 2013) was developed by Google and Portland TriMet as a way to provide transit schedules for use in traveler routing and information tools.

Though the initial goal of GTFS was to provide a common format for traveler-focused schedule and routing software, it has also become a key resource for research and analysis of transit systems. Jariyasunant et al. (2011) and Delling et al. (2014) describe recent work in algorithmic approaches to calculating travel times on transit networks that rely on GTFS. Puchalsky et al. (2012) describe how the stop and schedule data contained in GTFS datasets can strengthen regional planning and forecasting processes. Wong (2013) examines how data currently available in GTFS enables network- and agency-level analysis of transit systems, while Catala et al. (2011) identifies ways that the GTFS format could be expanded to support additional uses in transit operations and planning. It would be difficult to overstate the importance of the GTFS data format, and its widespread adoption, in enabling consistent analysis methodology across multiple transit operators.

Despite their importance and digital nature, the collection of GTFS datasets can be frustratingly inconsistent and error-prone. While the format of GTFS data itself is standardized there are no standards for the digital publication of the datasets, and practices vary widely across transit operators. A majority of operators (at least among medium and large metropolitan areas) provide GTFS datasets via a direct web site link. However, even among these variations in URL naming conventions pose challenges for systematic retrieval. Other operators allow GTFS dataset downloads only after users interactively submit a form or agreement. Still others generate GTFS datasets and provided them directly to Google for use in their popular online routing tool, but release them to the public only in response to direct email or hard-copy requests.

These issues are somewhat mitigated by the web site www.gtfs-data-exchange.com, a crowd-sourced archive of GTFS datasets from around the world (This web site has since been discontinued; a similar service is provided today by transitfeeds.com). However, the crowd-sourced nature of this resource poses its own challenges. Most importantly, it is difficult—and in some cases impossible—to validate that a

GTFS dataset obtained from this source was originally published by the actual transit operator, or that it has not been modified in some way. For this project, schedules downloaded from this web site are used only when they cannot be obtained directly from a transit operator.

4 Software

All of the major components of this evaluation system are open source. While this was not a specific goal or requirement, experience from earlier projects suggested some important benefits of using open source tools. First, open source software often provided greater flexibility in input and output data formats. This is an important consideration when a project involves multiple stages of data transformation and processing, each performed with a separate tool. Second, open source software can be rapidly customized to fit the project needs. In this project, local customizations to OpenTripPlanner provided more efficient parallelization and allowed for better data interoperability. Finally, open source approaches reduce barriers to replication and validation. Because the output of this project is itself a dataset designed for use in research and practice, it is important that all parts of the methodology—including those implemented using existing software—are thoroughly transparent and understandable.

This project makes use of the following major software packages:

- **OpenTripPlanner** (OTP), an open-source platform for multi-modal journey planning and travel time calculation.
- **PostgreSQL**, an open-source SQL database engine.
- **PostGIS**, a PostgreSQL extension that allows efficient storage and querying of spatial data.

Additionally, numerous smaller scripts and tools for data collection and processing were developed specifically for this project.

4.1 Data Processing and Organization

Figure 1 illustrates the basic project architecture and workflow, which is described in the following sections.

4.1.1 Inputs

The project inputs are stored primarily in a single SQL database. PostgreSQL is used along with the PostGIS extension; this combination allows spatial and non-spatial data in a single database, automated spatial queries (*e.g.* to select all origins within a given analysis zone), and spatial indexing methods that accelerate

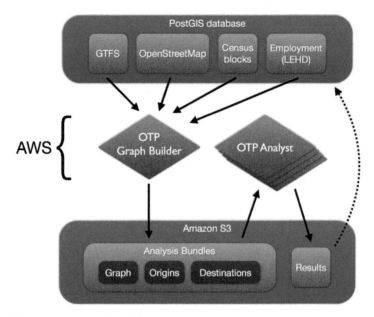

Fig. 1 Project architecture and workflow

these queries. Specifically this database contains an extract of all OpenStreetMap pedestrian data for North America; the full block, county, and core-based statistical area (CBSA) datasets from the U.S. Census Bureau; all 2011 resident area characteristics (RAC) and workplace area characteristics (WAC) data files from LEHD, and spatial bounds information for all collected GTFS datasets (which are stored separately).

4.1.2 Calculation

Travel time calculation is an "embarrassingly parallel" problem—a popular term among computer scientists for computation scenarios that can be easily decomposed into many independent repetitions of the same basic task. Given a suitable data architecture, these tasks can then be performed simultaneously, exponentially increasing the overall calculation speed.

In this case, the calculation of travel times from one origin at one departure time follows exactly the same process as for every other origin and every other departure time. Just under 11.1 million Census blocks (2010) comprise the United States; combined with 1440 min in a day this gives almost 16 billion possible space-time origins. The effective number is less, however, because in blocks with no access to transit service only a single departure time is used—transit travel times vary significantly over the day but walking travel times do not.

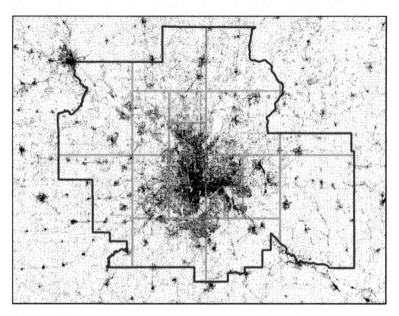

Fig. 2 A metropolitan area divided into analysis zones. Each zone contains a maximum of 5000 Census block centroids

The core unit of work—calculating travel times from a single origin at a single departure time—is provided by existing OpenTripPlanner capabilities. The parameters and assumptions involved in these calculations are described in following sections. OTP is natively multithreaded and can efficiently parallelize its work across multiple processors. To achieve efficient parallelization without requiring dedicated supercomputing techniques, the total computation workload is divided into "analysis bundles" which include all information necessary to compute a defined chunk of the final data. Each analysis bundle includes origin locations and IDs; destination locations, IDs, and opportunity (job) counts; and a unified pedestrian-transit network created by OTP.

The scope of origins included in each bundle is arbitrary; a useful value of 5000 origins per bundle was found through trial and error. Figure 2 illustrates the division of a single county into analysis zones, each containing no more that 5000 census block centroids. Too-small bundles erode overall efficiency by increasing the overhead costs of job tracking and data transfer, while too-big bundles suffer reliability issues: errors do occur, and when they do it is preferable to lose a small amount of completed work rather than a large amount.

Destinations, on the other hand, are selected geographically. Because travel times are by definition not known until the calculations are complete, it is necessary to include in each bundle all destinations that might be reached from any of the included origins within some maximum time threshold. A buffer of 60 km from the border of the origin zone is used, based on 1 h of travel at an estimated 60 km/h upper limit of the average speed of transit trips. This 1-h limit only applies to the

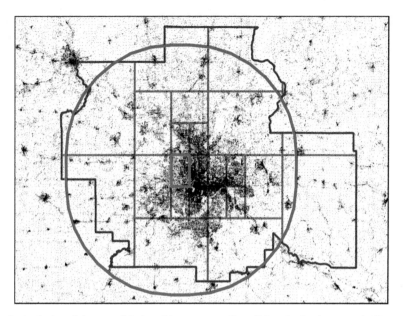

Fig. 3 A single origin zone (*blue*) and its corresponding 60-km destination zone buffer (*red*). Travel times are calculated from each centroid in the origin zone to each centroid in the destination zone

extent of the graph; using such a graph, accessibility metrics can be reported for any time threshold of 1 h or less. Figure 3 illustrates the spatial selection of destinations for a given set of origins.

OTP's Analyst module provides a graph builder function that combines pedestrian and transit network data from the input database into a single graph, and locally-developed software merges the graph into an analysis bundle with the appropriate origins and destinations. The bundle is queued in a cloud storage system making it available for computation.

Computations take place on a variable number of cloud computing nodes that are temporarily leased while calculation is in progress. (Currently, computing nodes are leased from Amazon Web Services (AWS).) Each node is prepared with OTP Analyst software as well as custom software that retrieves available analysis bundles, initiates accessibility calculations, and stores the results.

4.1.3 Outputs

The processing of each analysis bundle results in a single data file that records accessibility values for each origin in the bundle. For each origin, this includes an accessibility value for each departure time and for each travel threshold between 5 and 60 min, in 5-min increments. These values are stored individually and disaggregated to facilitate a wide range of possible analyses. Each result file is tagged with the ID of the analysis zone and range of departure times for which it contains results, and then stored in a compressed format in the cloud storage system.

Because analysis typically takes place at the metropolitan level or smaller, it is rarely necessary to have the entire national result dataset available at once. Instead, custom scripts automate the download of relevant data from the cloud storage system.

4.2 Accessibility Calculations

4.2.1 Transit Travel Time

This analysis makes the assumption that all access portions of the trip—initial, transfer(s), and destination—take place by walking at a speed of 1.38 m/s along designated pedestrian facilities such as sidewalks, trails, etc. On-vehicle travel time is derived directly from published transit timetables, under an assumption of perfect schedule adherence. Transfers are not limited.

Just as there is no upper limit on the number of vehicle boardings, there is no lower limit either. Transit and walking are considered to effectively be a single mode. The practical implication of this is that the shortest path by "transit" is not required to include a transit vehicle. This may seem odd at first, but it allows the most consistent application and interpretation of the travel time calculation methodology. For example, the shortest walking path from an origin to a transit station often passes through destinations where job opportunities exist. In other cases, the shortest walking path from an origin to a destination might pass through a transit access point which provides no trips that would reduce the origin–destination travel time. In these situations, enforcing a minimum number of transit boardings would artificially inflate the shortest-path travel times. To avoid this unrealistic requirement, the transit travel times used in this analysis are allowed to include times achieved only by walking.

Transit accessibility is computed for every minute of the day, as described in Owen and Levinson (2015), which demonstrates that continuous accessibility metrics can provide a better description of the variation in transit commute mode share than do metrics evaluated at a single or optimal departure time.

5 Visualization

This project produces highly detailed accessibility datasets, and some level of aggregation is typically needed to produce easily understandable summary maps. Figures 4, 5, 6 and 7 provide examples of block-level accessibility results mapped at a constant data scale across four major metropolitan areas: Washington, DC;

Washington
Washington-Arlington-Alexandria, DC-VA-MD-WV

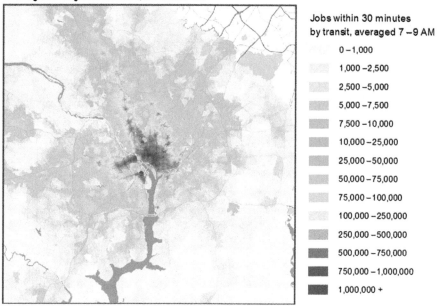

Fig. 4 Map of job accessibility by transit in the Washington, DC metropolitan area

Atlanta
Atlanta-Sandy Springs-Marietta, GA

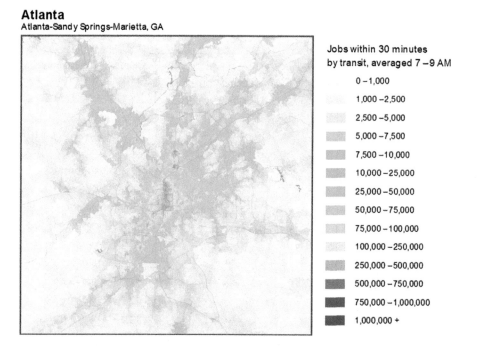

Fig. 5 Map of job accessibility by transit in the Atlanta, GA metropolitan area

Seattle
Seattle-Tacoma-Bellevue, WA

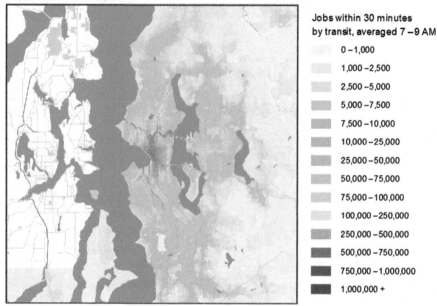

Fig. 6 Map of job accessibility by transit in the Seattle, WA metropolitan area

Minneapolis
Minneapolis-St. Paul-Bloomington, MN-WI

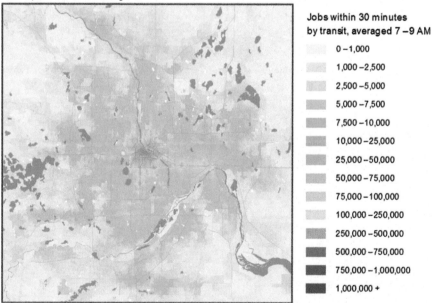

Fig. 7 Map of job accessibility by transit in the Minneapolis–Saint Paul, MN metropolitan area

Atlanta, GA; Seattle, WA, and Minneapolis–Saint Paul, MN. In these maps, accessibility for each Census block has been averaged over the 7–9 AM period. The resulting average accessibility value indicates the number of jobs that a resident of each block could expect to be able to reach given a randomly-selected departure time between 7 and 9 AM.

6 Conclusion

With the framework developed in this project, it is possible to evaluate the accessibility provided by public transit in any area where data is available. Within the United States, the only data limitation is the availability of transit schedules in GTFS format—all other sources are available with full national coverage (with the exception of LEHD data, which is not available for the state of Massachusetts). Also significantly, all data is public or available under an open license.

While this project adopted a specific accessibility metric (cumulative opportunities to jobs) and a set of parameters for implementing it, the framework itself provides flexibility. The core OpenTripPlanner software can calculate weighted accessibility; using a different destination dataset is a trivial modification; various travel time calculation parameters can be easily adjusted. While it is hoped that the accessibility data products described here will be useful for both research and practice, the framework can be used to fit a wide variety of specific accessibility evaluation scenarios. Consistency does not have to mean, "one size fits all."

This project also suggests ways that accessibility evaluation for other transportation modes could be improved. In some ways public transit is the most difficult domain in which to perform this level of evaluation. Accessibility evaluations for car travel, for example, can employ the simplification of using average road speeds to avoid the need to calculate at multiple departure times with fewer consequences; network structure also remains constant over the course of the day. Given appropriate data sources, accessibility by car could be calculated for the same block-level resolution at a fraction of the computation costs.

However, this highlights a critical uniqueness of the transit case: travel time data (in the form of schedules) is publicly available. Outside of loop detector-based systems on urban highways (whose data format varies across cities and states), there exists virtually no equivalent for car travel. Open data initiatives in this realm, such as OpenTraffic.org, though promising, are nascent and lack coverage. Comprehensive data sources for road and highway speeds are effectively limited to commercial datasets; efforts to implement a similar evaluation for accessibility by car will need to confront this reality.

Acknowledgements The project described in this article was sponsored by the University of Minnesota's Center for Transportation Studies. Many of the employed tools and methodological approaches were developed during earlier projects sponsored by the Minnesota Department of Transportation.

References

Catala M, Downing S, Hayward D (2011) Expanding the Google transit feed specification to support operations and planning. Technical Report BDK85 997-15, Florida Department of Transportation

Delling D, Pajor T, Werneck RF (2014) Round-based public transit routing. Transp Sci 49(3):591–604

Google, Inc. (2013) General transit feed specification reference. [Online]. https://developers.google.com/transit/gtfs/reference

Jariyasunant J, Mai E, Sengupta R (2011) Algorithm for finding optimal paths in a public transit network with real-time data. Transp Res Rec J Transp Res Board 2256:34–42

Krizek K, El-Geneidy A, Iacono M, Horning J (2007) Refining methods for calculating non-auto travel times. Technical Report 2007-24, Minnesota Department of Transportation

Krizek K, Iacono M, El-Geneidy A, Liao C-F, Johns R (2009) Application of accessibility measures for non-auto travel modes. Technical Report 2009-24, Minnesota Department of Transportation

Levine J, Grengs J, Shen Q, Shen Q (2012) Does accessibility require density or speed? A comparison of fast versus close in getting where you want to go in US metropolitan regions. J Am Plan Assoc 78(2):157–172

Owen A, Levinson D (2012) Annual accessibility measure for the Twin Cities metropolitan area. Technical Report 2012-34, Minnesota Department of Transportation

Owen A, Levinson DM (2015) Modeling the commute mode share of transit using continuous accessibility to jobs. Transp Res A Policy Pract 74:110–122

Puchalsky CM, Joshi D, Scherr W (2012) Development of a regional forecasting model based on Google transit feed. In: 91st annual meeting of the transportation research board, Washington, DC

Wong J (2013) Leveraging the general transit feed specification for efficient transit analysis. Transp Res Rec J Transp Res Board 2338:11–19

Seeing Chinese Cities Through Big Data and Statistics

Jeremy S. Wu and Rui Zhang

Abstract China has historically been an agricultural nation. China's urbanization rate was reported to be 18 % in 1978 when it began its economic reforms. It has now become the second largest economy in the world. Urbanization in China increased dramatically in support of this economic growth, tripling to 54 % by the end of 2013. At the same time, many major urban problems also surfaced, including environmental degradation, lack of affordable housing, and traffic congestion. Economic growth will continue to be China's central policy in the foreseeable future. Chinese cities are seriously challenged to support continuing economic growth with a high quality of life for their residents, while addressing the existing big city diseases. The term "Smart City" began to appear globally around 2008. Embracing the concept allows China to downscale its previous national approach to a more manageable city level. By the end of 2013, China has designated at least 193 locations to be smart city test sites; a national urbanization plan followed in March 2014. The direction of urban development and major challenges are identified in this paper. Some of them are global in nature, and some unique to China. The nation will undoubtedly continue to build their smarter cities in the coming years. The first integrated public information service platform was implemented for several test sites in 2013. It provides a one-stop center for millions of card-carrying residents to use a secure smart card to perform previously separate city functions and consolidate data collection. The pioneering system is real work in progress and helps to lay the foundation for building urban informatics in China. This paper also discusses the evolving research needs and data limitations, observes a smart city in progress, and makes some comparisons with the U.S. and other nations.

Keywords Urbanization • Smart city • Statistics • Big data • Urban informatics • China

---

J.S. Wu, Ph.D. (✉)
Retired, Census Bureau, Suitland, Maryland, and Department of Statistics, George Washington University, Washington, DC, USA
e-mail: Jeremy.S.Wu@gmail.com

R. Zhang
Public Relations and Communications Department, Digital China Holdings Limited, Beijing, China

1 Introduction

China has historically been an agricultural nation. Its urbanization rate was reported to be about 11 % in 1949 and 18 % in 1978. Subject to differences in definition (Qiu 2012), the U.S. urbanization rate was estimated to be at 74 % in 1980 (U.S. Census Bureau 1990). China began its economic reforms "Socialism with Chinese characteristics" in 1978. It introduced market principles and opened the country to foreign investment, followed by privatization of businesses and loosening of state control in the 1980s.

In the last 36 years, China leapfrogged from the ninth to the second largest economy in the world in gross domestic product (GDP), surpassing all other countries except the U.S. (Wikipedia, "List of countries by GDP (nominal)"). The poverty rate in China dropped from 85 % in 1981 to 13 % in 2008 (World Bank, "Poverty headcount ratio at $1.25 a day").

Through expanding population and land annexation, urbanization in China increased dramatically in support of this economic growth. The objective of this paper is to document the need for a proactive data-driven approach to meet challenges posed by China's urbanization. This strategy will require a number of technological and data-oriented solutions, but also a change in culture towards statistical thinking, quality management, and data integration. New investments in smart cities have the potential to design systems so that the data can lead to much-needed governmental innovations towards impact.

The paper is structured as follows: in Sect. 2, we provide background information on urbanization trends in China and the challenges that have come with it. In Sect. 3, we describe current policy goals and approaches to meet these challenges. The emerging role of statistics and technology, including investments of smart cities and much-needed transformations towards data and their use are discussed in Sect. 4. Section 5 describes progress with a smart city developments with Zhangjiagang in Jiangsu Province, as an example of the types of potential benefits that can be accrued with technology for cities. Conclusions are given in Sect. 6.

2 Background

Table 1 reproduces the State Council of China (2014) report about the growth of Chinese cities from 193 in 1978 to 658 in 2010. Wuhan became the seventh megacity in 2011. No U.S. city qualified to be a megacity in 2012 (U.S. Census Bureau, "City and Town Totals: Vintage 2012"). The urbanization rate in China tripled from 18 % in 1978 to 54 % by the end of 2013 (National Bureau of Statistics of China 2014).

Migration of rural workers to meet the urban labor needs accounted for most of the growth of the Chinese urban population to 711 million. However, under the unique Chinese household registration system known as Hukou (Wikipedia,

Table 1 Number of Chinese cities 1978–2010

Population		1978	2010
Cities		193	658
	≥10 million (Megacity)	0	6
	5–10 million (Extra Large City)	2	10
	3–5 million (Extra Large City)	2	21
	1–3 million (Large City)	25	103
	0.5–1 million (Mid-size City)	35	138
	≤0.5 million (Small City)	129	380
Towns		2173	19,410

Source: The State Council of China (2014)

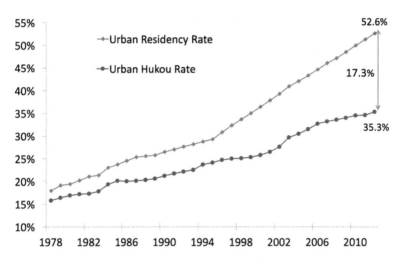

Fig. 1 Urban residency rate and urban hokou rate (1978–2012). (*Source*: The State Council of China (2014))

"Hukou System"), the registered rural residents living in the city are not entitled to the government benefits of the city, such as health care, housing subsidy, education for children, job training, and unemployment insurance. Conversion from the rural to urban registration status has been practically impossible.

This disparity has become a major concern for social discontent in a nation of almost 1.4 billion people. Although 52.6 % of the Chinese population lived in cities in 2012, only 35.3 % were registered urban residents. The gap of 17.3 % is known as the "floating population," amounting to 234 million people and well exceeding the entire U.S. labor force of 156 million people. Figure 1 shows that this gap has been widening since 1978.

In addition to the social inequity caused by the Hukou system, there is an increasing geographical divide. Figure 2 shows that the eastern region of China is more densely populated than the rest of the nation (Beijing City Lab 2014).

Five out of the six megacities in 2010 are located on the east coast. These megacity clusters occupy only 2.8 % of the nation's land area, but contain 18 %

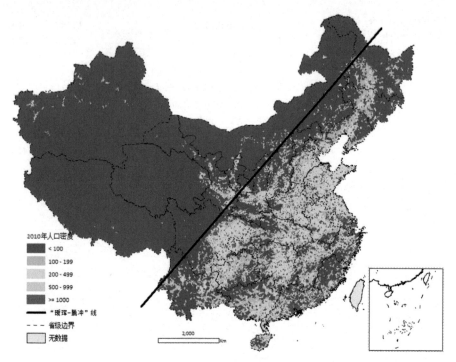

Fig. 2 Population density of China in 2010 (*Source*: Beijing City Lab (2014))

of the population and 36 % of the GDP. While the east coast is increasingly suffocated by people and demand for resources, the central and the western regions lag behind in economic development and income. Reliance on urban land sale and development to generate local revenue during the reform process has led to high real estate and housing prices and conflicts with the preservation of historical and cultural sites in all regions. Conversion of land from agricultural use is also raising concerns about future food supply.

At the same time, "big city diseases" surfaced and became prevalent in China, including environmental degradation, inadequate housing, traffic congestion, treatment of sewage and garbage, food security, and rising demand for energy, water and other resources. Many of these issues have been discussed domestically and internationally (e.g., Henderson 2009; Zhang 2010a, b; United Nations Development Program 2013).

So, where is China heading in terms of economic growth and urbanization? The answer to this question is unambiguous. The urbanization goal of 51.5 % for the 12th Five-Year Plan (2011–2015) has already been exceeded (National People's Congress 2011). The 2014 Report on the Work of the Chinese Government (Li 2014b), which is similar to the annual State of the Union in the U.S., states that "economic growth remains the key to solving all (of China's) problems" and urbanization is "the sure route to modernization and an important basis for integrating the urban and rural structures." On March 16, 2014, the State Council of China (2014) released its first 6-year plan on urbanization for 2014–2020. The comprehensive plan covers 8 parts, 31 chapters, and 27,000 words, providing guiding principles, priorities for development, and numerical and qualitative goals. Under this plan, China sets a goal of 60 % for its urbanization by 2020.

3 Approach and Goals

In the early days of reform, China took the trial-and-error approach of "feeling the rocks to cross the river" when infrastructure and options were lacking. Over time, the original simple economic goals were challenged by conflicting cultural and social values. More scientific evaluations are needed to minimize costly mistakes made by instinctive decisions.

After 30-plus years of reform, Chinese President Xi Jinping (2014) acknowledged that "...the easier reforms that could make everyone happy—have already been completed." Chinese Premier Li Keqiang (2014b) pledged to carry out "people-centered" urbanization and cited three priorities on three groups of 100 million people each:

- Granting official urban Hukou status to 100 million rural people who have already moved to cities;
- Rebuilding rundown, shanty city areas and villages inside cities where 100 million people currently live;
- Guiding the urbanization of 100 million rural residents of the central and western regions into cities in their regions.

Table 2 reproduces the 18 key numerical goals for 2020 along with the 2012 benchmarks under the national urbanization plan.

There are now two key goals with regard to the urban population: raising the level of residents living in cities to 60 % and the level of registered urban residents to 45 %, thereby reducing the floating population from the current 17.3 to 15 % in 6 years. The other key goals promote the assimilation of migrant rural workers into city life, improving urban public service and quality of life, and protecting land use and the environment.

There are less specific qualitative goals in the national urbanization plan. For example, this chapter mandates "Three Districts and Four Lines" in each city. The three districts are defined as areas forbidden from, restricted from, and suitable for construction respectively. Four types of zones will be drawn by color lines: green line for ecological land control; blue line for protection of water resources and swamps; purple line for preservation of historical and cultural sites; and yellow line for urban planning and development. Yet how these districts and zones will be created and sustained has not been specified.

China is pressing forward with concurrent modernization in agriculture, industrialization, information technology, and urbanization. Under the urbanization plan,

Table 2 Numerical goals of 2014–2020 national urbanization plan

Indicator	2012	2020
Urbanization level		
1. Permanent residency urbanization rate (%)	52.6	60
2. City registration urbanization rate (%)	35.3	45
Basic public service		
3. Children of rural migrant workers receiving education (%)		≥99
4. Urban unemployed, rural migrant workers, and new workers receiving free job training (%)		≥95
5. Social security for permanent urban residents (%)	66.9	≥90
6. Health insurance for permanent urban residents (%)	95	98
7. Housing security for permanent urban residents (%)	12.5	≥23
Infrastructure		
8. Public transportation in cities of over 1 million residents (%)	45*	60
9. Urban public water supply (%)	81.7	90
10. Urban sewage treatment (%)	87.3	95
11. Urban garbage treatment (%)	84.8	95
12. Urban family broadband coverage (megabit per second)	4	≥50
13. Integrated urban community services (%)	72.5	100
Resources and environment		
14. Per capita urban construction land use (square meter)		≤100
15. Urban renewable energy consumption (%)	8.7	13
16. Urban green buildings as share of new buildings (%)	2	50
17. Urban green area in developed areas (%)	35.7	38.9
18. County-level or above city meeting national air quality standards (%)	40.9	60

Notes:
① *2011 data
② Social Security Coverage: Permanent resident does not include under 16 years old and students
③ Urban Housing Security: Includes public (subsidized), policy-dictated commercial, and renovated housing
④ Per Capita Land Use National rule: standard urban use 65.0–115.0 sq.m, new cities 85.1–105.0 sq.m
⑤ National Urban Air Quality Standards: On top of 1996 standards, add PM2.5 and ozone 8-h average concentration limits; adjust PM10, nitrogen dioxide, lead concentration limits
Source: The State Council of China (2014)

the central government is responsible for strategic planning and guidance. Authority is delegated to the provincial and municipal levels through political reform. Local administrators are encouraged to innovate, build coalitions, undertake pilot tests, formulate action plans, and implement orderly modern urbanization under local conditions.

Conversion from rural to urban Hukou registration is now officially allowed and encouraged, but the process will be defined by individual cities, under the general rule that the conversion will be more restrictive as the population of the city increases.

4 Emerging Role of Statistics and Technology

The national urbanization plan provides an unprecedented opportunity for the role of statistics and technology to support and monitor the implementation of policies in China. Chapter 31 prescribes the role of defining metrics, standards, and methods to establish a sound statistical system, monitoring the activities dynamically, and performing longitudinal analysis and continuing assessment of the progress of urbanization according to the development trends.

The specification of dynamic monitoring and longitudinal analysis reflects advanced thinking, compared to the current static, cross-sectional reports. Yet how the statistical monitoring system will be implemented also remains unclear at this stage.

Many developed nations have been using a data-driven approach to manage knowledge for their businesses (e.g., Sain and Wilde 2014) and it is assumed that China will also take up this approach for governance in this paper. Figure 4 shows a Data-Information-Knowledge-Wisdom (DIKW) hierarchy model for this process. The foundation of scientific knowledge and wisdom is to observe facts and collect data. However, data in their raw form have little or no meaning by themselves. Not all data have adequate information value or are useful for effective decision making. Statistics, both as a branch of science for knowledge discovery and a set of measurements, provides context and value by converting useful data into relevant information. Knowledge is gained and accumulated from information, and used as the basis for making wise decisions. The decisions will not be correct all the time, but the scientific process promotes efficiency and minimizes errors, especially when conducted with integrity, objectivity, and continuous improvement (Fig. 3).

Although technology is not explicitly shown in the DIKW model, today the base of the pyramid is greatly expanded by technology, and the transformation of data into information has been accelerated. However, the process is also contaminated by hype, useless data, and misinformation (Harford 2014; Wu 2014). Traditional sources of data such as the Census have been used for governance of nations for centuries. Random surveys were later introduced based on probability theory to produce scientifically reliable information with proper design and a relatively small amount of data.

Together censuses and random surveys form the statistical foundation based on structured data (Webopedia, "Structured data") in the twentieth century. Developed

Fig. 3 The DIKW hierarchy model (*Source*: Wikipedia)

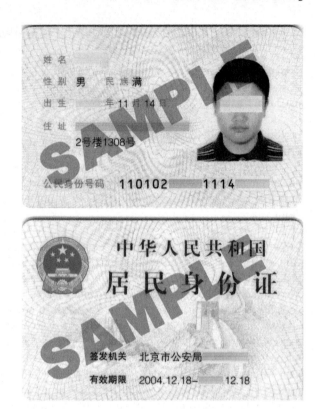

Fig. 4 Sample of Chinese citizen identification card. (*Source*: Wikipedia "Resident Card System")

nations have used them effectively for policy and decision making, with design and purpose, over the past 100 years.

At the turn of this century, massive amounts of data began to appear in or were converted from analog to digital form, allowing direct machine processing (Hilbert and Lopez 2011), a lot of which are unstructured text, map, image, sound, and multimedia data. Big Data was not a well-known term in China until Tu (2012) published the first Chinese-language book on the topic. Although data mining is commonly mentioned as a promising approach to extract information from such data for commercial purposes, their reliability and value can be suspect, especially for the purpose of governance (e.g., Marcus and Davis 2014; Lazer et al. 2014). Few of the key numerical goals in the national urbanization plan can be measured meaningfully or reliably by unstructured data alone.

Other approaches to Big Data arise due to the integration of structured data derived from administrative records to create longitudinal data systems. This approach was the first realized benefit of Big Data for government statistics. For example, the Longitudinal Employer-Household Dynamics (LEHD) program of the U.S. Census Bureau merges unemployment insurance data, social security records, tax filings and other data sources with census and survey data to create a longitudinal frame of jobs. It is designed to track every worker and every business in the nation dynamically through the relationship of a job connecting them, with data updated every quarter, while protecting confidentiality. The data provide insights about patterns and transitions over time, which are not available from the traditional cross-sectional statistics. Similar efforts to build longitudinal data systems for education (Data Quality Campaign n.d.) and health care (Wikipedia, "Health Information Technology for Economic and Clinical Health Act") are underway in the U.S. The 2020 U.S. census will also be supplemented by the integration of administrative records (Morello 2014).

4.1 Chinese National Basic Data Systems and Identification Codes

More than a decade ago, the State Council of China (2002) issued guidance to create four National Basic Data Systems as part of e-Government—longitudinal frames of people, enterprises, and environment/geography respectively with the fourth system as an integration of the first three to form a unified macroeconomic data system. These nationwide data systems possess the desired characteristics of a twenty-first century statistical system (Groves 2012; Wu 2012; Wu and Guo 2013). They help to transition the Chinese government's role from central control to service for citizens and to establish a foundation for data sharing and one-stop integrated service nationwide. Heavy investment followed to establish and implement definitions, identification codes, standards, and related infrastructure.

Identification codes are the keys to unlocking the enormous power in Big Data (Wu and Ding 2013). A well-designed code matches and merges electronic records, offers protection of identity, provides basic description and classification, performs initial quality check, and facilitates the creation of dynamic frames. As early as 1984, China began to build an infrastructure with its citizen identification system (Wikipedia, "Resident Card System"). A sample Chinese citizen card (Fig. 4) displays the citizen identification code, name, gender, ethnicity, birthdate, address, issuing agency, dates of issuance and expiration, and a photograph.

The 18-digit citizen identification code, introduced in 1999, includes a Hukou address code, birthdate, gender, and a check digit. It is issued and administered by the Ministry of Public Security. The citizen code is uniquely and permanently assigned to the cardholder. The card is capable of storing biometric information. It is increasingly required for multiple purposes, such as the purchase of a train ticket for travel. In contrast, the U.S. does not have a comparable national citizen card system. Recent renewed discussions about adding an image of the cardholder to the Social Security card was met again with controversy (e.g., Bream 2014; Eilperin and Tumulty 2014).

China has also established a system of National Organization Codes under the National Administration for Code Allocation to Organizations. The nine-digit organization code includes a check digit and is a unique identification and linking variable to store and retrieve information about companies, institutions, communities, government agencies and other registered organizations.

China has therefore laid a sound foundation for building dynamic frames through these initiatives. However, by 2008, the national approach to create Basic Data Systems was becoming too complex with too many structural, legal, and practical obstacles to overcome. Shen (2008) reported that the Basic Data System on environment and geography was essentially complete but lacked real application. The Basic Data System on population was burdened by the inclusion of over 100 variables, each with a different degree of sensitivity. The Basic Data System on enterprises faced the strongest resistance to data sharing by various agencies with overlapping responsibilities. The Basic Data System on macroeconomics was stalled without the first three data systems in place. It is interesting to note that the LEHD program in the U.S. went through comparable experiences. The national approach faced resistance to data sharing so that the approach had to be strategically adjusted to the state level before data can be re-assembled to the national level.

The Basic Data Systems were relegated to long-term development until the release of the national urbanization plan in 2014. Mandates are now revived and issued for their accelerated development and implementation. For example, the Basic Data System on population is expected to link to cross-agency and cross-regional information systems for employment, education, income, social security, housing, credit services, family planning, and taxation by 2020. The citizen identification code is also mandated to be the only legal standard for recording, inquiring, and measuring population characteristics in China the same year.

4.2 The Rise of Smart City

The term "Smart City" began to appear globally around 2008 as an extension to previous development of e-Government and digital cities. Data collection, processing, integration, analysis and application are at the core of constructing smart cities. In practical terms, embracing the concept of smart city will allow China to downscale the original national approach to the more manageable city level, while protecting past investments and permitting aggregation to the provincial or regional level.

Table 3 describes the direction to develop smart cities as outlined in the national urbanization plan. At the end of 2013, the Chinese Ministry of Housing and Urban-Rural Development has designated 193 locations to be smart city test sites (baidu. com, "National Smart City Test Sites"). They are expected to undergo 3–5 years of experimental development. The Chinese Ministry of Science and Technology has also named 20 smart city test sites (Xinhuanet.com 2013). They are expected to spend 3 years to develop templates of cloud computing, mobile networks, and related technologies for broad implementation.

The issuance of a City Resident Card is a concrete first step for aspiring smart cities to provide one-stop service and to consolidate data collection. The multi-

Table 3 Direction of smart city development

01	**Broadband information network**
	Replace copper by fiber-optics. Implement fiber-optic network covering practically all urban families at connection speed of 50Mbps, 50 % families reach 100Mbps, and some families reach 1Gbps in well-developed cities. Develop 4G network and accelerate public hot spots and WiFi coverage
02	**Information technology for planning and management**
	Develop digital city management, promote platform development and expand functions, establish a unified city geospatial information platform and building (structure) database, build public information platform, coordinate the digitization and refinement of urban planning, land use, distribution network, landscaping, environmental protection and other municipal infrastructure management
03	**Intelligent infrastructure**
	Develop intelligent transportation to guide traffic, command and control, manage adjustments and emergencies. Develop smart grid to support distributed access to energy and intelligent use of electricity by residents and businesses. Develop intelligent water services to cover the entire process from quality and safety of supply to drainage and sewage. Develop intelligent information network to manage urban underground space and pipes. Develop intelligent buildings to manage facilities, equipment, energy consumption, and security
04	**Public service streamlining**
	Establish cross-agency, cross-regional business collaboration, sharing of public service information service system. Use of information technology and innovation to develop urban education, employment, social security, pension, medical and cultural service model
05	**Modernization of industrial development**
	Accelerate the transformation of traditional industries, promote use of information technology, digitization, and networking to transition to intelligent service models for manufacturing. Actively develop and integrate information services, e-commerce and logistics to nurture innovation and new formats
06	**Refinement of social governance**
	Strengthen the application of information to monitor market regulations and the environment, credit services, emergency protection, crime prevention and control, public safety and other areas of governance, establish and improve relevant information service system, innovate to create new model of social governance

Source: The State Council of China (2014)

functional card may be used for social security or medical insurance purposes, as well as a debit card for banking and small purchases. Depending on the city, the City Resident Card may also be used for transportation, public library, bicycle rentals, and other governmental and commercial functions yet to be developed. During the application process, the citizen identification code is collected along with identification codes for social security and medical insurance, residence address, demographic data, and family contact information, facilitating linkage to other data systems and records. The current smart resident cards in use in China (Fig. 5) vary from city to city, but they typically contain two chips and a magnetic memory strip. One example of smart city applications utilizing this card system is a one-stop service platform by Digital China (2013). This is an additional channel of service for millions of card-carrying residents, who can use the secured smart card to perform previously separate functions.

Fig. 5 Sample of smart city resident card. (*Source*: Baike.baidu.com, "Resident Card")

4.3 Urban Informatics

While the aforementioned activities are modest, they have the potential to lay the foundation for urban informatics in China, particularly as they represent the very early results of China's total investment into smart city development, which is estimated to exceed ¥2 trillion ($322 billion) by 2025 (Yuan 2014). Urban informatics, meaning the scientific use of data and technology to study the status, needs, challenges, and opportunities for cities, is presently not a well-known concept in China. It uses both unstructured and structured data, collected with and without design or purpose. The defining characteristics of urban informatics will be the sophisticated application of massive longitudinal data, integration of multiple data sources, and rapid and simple delivery of results, while strictly protecting confidentiality and data security and assuring accuracy and reliability.

However, there are many challenges and needs in establishing urban informatics as a mature field of study in China. These are discussed next.

4.3.1 Need for Change in Culture

There is no assurance that internal resistance to data sharing and standards can be overcome in China despite mandates, political reforms, downscaling, and cloud computing (e.g., UPnews.cn 2014). A major risk of a de-centralized approach is the formation of incompatible "information silos" such that the systems cannot interoperate within or between cities. This challenge is not unique to China. The U.S. had more than 7000 data centers in the federal government alone in 2013; about 6000 of them were considered "noncore." Many of them do not communicate with each other and are costly to maintain. Although a major consolidation initiative was started by the White House in 2010, progress has been slow (CIO.Gov 2014; Konkel 2014).

However, open data-based governance and research are relatively new concepts in China. Although their value is recognized and advocated in the central plans, thus far there has been little support of open data policy and data sharing, and neither has a full awareness of modern statistical or environmental issues demonstrated. Much is also needed regarding statistical quality control (Shewhart 1924; Deming 1994) and quality standards (e.g., International Organization for Standardization 9000). While these statistical principles and thinking originated in the context of industrial production, they are equally applicable to governance.

The National Bureau of Statistics of China relies heavily on data supplied by provincial and local governments. Intervention and data falsification by local authorities are occasionally reported in China (e.g., Wang 2013), including the famed GDP. For example, the incomplete 2013 GDP of 28 out of 31 provinces and cities already exceeded the preliminary 2013 total national GDP by ¥2 trillion or 3.6 % (e.g., Li 2014a). Due to these issues, the credibility and public confidence in China's statistics are not high. Tu (2014) observed that China has not yet developed a culture of understanding and respect for data. Changing this culture is a challenge without historical precedent in China.

4.3.2 Need for Statistical Thinking, Design and Innovation

The Chinese statistical infrastructure is relatively new and fragile. The first Chinese decennial census on population began in 1990 while the U.S. started 200 years earlier; the first Chinese consolidated economic census was conducted only 10 years ago in 2004. Random surveys seldom include detailed documentation on methodology.

Requiring dynamic monitoring and longitudinal analysis by the Chinese government is refreshing in the national urbanization plan. Its implementation faces many statistical and technological issues, including record linkage and integration, treatment of missing or erroneous data, ensuring data quality and integrity, retrieval and extraction of data, scope for inference, and rapid delivery of results. Some of the terms in use, such as "talented persons," "green buildings," and "information level," do not have commonly accepted definition or standard meaning.

In the collection of data about a person, some of the characteristics such as gender and ethnicity remain constant over time; some change infrequently or in a predictable manner such as age, Hukou, and family status; some change more frequently such as education level, income level, employment, and locations of home and work; and others change rapidly such as nutritional intake, use of water and electricity, or opinion about service rendered. Measurement of these characteristics must be made with appropriate frequency, completeness, and quality so that reliable data can be collected to easily and rapidly describe and infer about the population. The definitions must be consistently applied across locations and time so that the results can be compared and meaningful temporal or spatial patterns can be discovered and studied. The base unit may extend from a person to a family or a household. These considerations also apply to an enterprise or a defined geo-space.

Nie et al. (2012) reported the disastrous consequences of mismatched records, outlying data, large variations, and unclear definitions in a Chinese national longitudinal data system on enterprises. Without proper statistical design and implementation of quality control, the data system does not support credible analysis or reliable conclusions despite the high cost of its creation and maintenance. There are few discussions of statistical design or need for quality control of data systems in China. In general, large-scale longitudinal data systems or reliable longitudinal analysis are currently lacking. Drawing conclusions from the Basic Data Systems, created in multiple phases with data from various linked sources, is an example of multi-phase and multi-source inference.

Recent calls for exploring and understanding "Scientific Big Data" (Qi 2014; Jiang 2014) are a promising sign that China may be prepared to make better use of data in scientific disciplines, in addition to the current commercial and marketing environments.

4.3.3 Need for Integration of Technology and Statistics

Yuan (2014) quoted the research firm IDC that "roughly 70 % of government investments went to hardware installation in China, way higher than the global average of 16 %." While China may be strong at hardware, service and software tend to lag behind. Technology and statistics are in many cases disconnected.

For effective administration and rapid information delivery, the underlying data need to be representative and quality-assured. This would facilitate easy extraction, transformation and loading (ETL), as well as dynamic visualization and longitudinal reporting of the status and assessment of progress on the urbanization plan, and overall system performance and customer satisfaction.

Online services based on smart resident cards and one-stop centers have already led to relief in labor-intensive administrative functions and reduction of long queues, but the current static monitoring reports are not connected to data collected from online services in concept or in operation. Although statistical yearbooks are beginning to appear online, interactive queries and dynamic visualization similar to the American Factfinder (U.S. Census Bureau n.d.a) are not yet available. Intelligent mapping applications similar to OnTheMap (Wu and Graham 2009; U.S. Census Bureau n.d.c) have also not been introduced to deliver custom maps and statistical reports based on the most recent data in real time. Such fragmentation may well hamper data-driven impact even though considerable investment may have gone into systems and hardware.

5 Zhangjiagang: A Developing Chinese Smart City

In this section, we discuss current smart city developing by considering the case of Zhangjiagang, a port city of 1.5 million population located along the Yangtze River in eastern China Jiangsu Province (Fig. 6). The urbanization rate for Zhangjiagang was 63 % in 2010, which is higher than the current average in China, exerting high pressure on its city administrators to manage its population, environment, and economic development.

Zhangjiagang'S 12th 5-year plan (2011–2015) emphasizes the support of information technology for e-Government by raising the level of government applications; accelerating the construction of the Basic Data Systems; focusing on public and government data sharing and exchange systems; and further improving the level of inter-departmental applications.

The Zhangjiagang public website was launched in October 2013 with the above goals in mind. The front page of the website (Fig. 7) contains three channels—My Service, My Voice, and My Space. My Service provides government and public services; My Voice connects the government and the resident through an online survey and microblogging; My Space contains the user's "digital life footprint" such as personal information and record of use.

The public website combines online and offline services through the use of the smart resident card, desktop and mobile devices, and government and community service centers, offering 621 types of services by 31 collaborating government and community organizations.

Fig. 6 Zhangjiagang, Suzhou City, Jiangsu Province (*Source*: Google Earth)

Fig. 7 Entry page of Zhangjiagang Public Website (*Source*: Digital China)

The services vary by type of access device. Desktop computers offer the most comprehensive services, including queries, more than 240 online applications, and over 130 online transactions. Mobile device users may check on the progress of their applications, using General Packet Radio Service (GPRS) positioning and speech recognition technology to obtain 56 types of efficiency services such as travel and transportation.

The one-stop service platform attempts to provide a unified, people-centric, complete portal, eliminating the issues of territoriality of various government agencies to build their own websites and service stations and consolidating separate developments such as smart transportation (e.g., showing the location and availability of rental bicycles) and smart health care. The website also combines a variety of existing and future smart city proposals and services. Developers will be able to link to the platform to provide their services with lower operating costs. The city government wants to have a platform to showcase information technology, introduce business services, and assist economic development, especially in e-Commerce.

The Zhangjiagang public website is designed to be an open platform for progressive development. All the applications will be dynamically loaded and flexible to expand or contract. Existing services will be continuously improved, and new functionalities added. It aims to improve public satisfaction of government service and broaden agency participation to facilitate future data sharing and data mining.

Participation of the residents in the platform will determine whether its goals will be achieved or not. In the 6-month period since its launch in October 2013, there have been 15,518 total users through real-name system certification and online registration, 31,956 visitors, and 198,227 page views. The average visit time was 11 min and 7 s. Among all the users, real-name registrants accounted for 67 %, and mobile end users accounted for 44 %.

Online booking of sports venues, event tickets, and long-distance travel are the most popular services to date. They show the value of convenience to the residents, who had to make personal visits in the past.

Although there is no current example of data sharing between government departments, the public website is beginning to integrate information for its residents. A user can view his/her records in a secured My Space.

It is already possible in the Zhangjiagang platform to create a consolidated bill of natural gas, water, electricity, and other living expenses to provide a simple analysis of household spending. Although this is elementary data analysis, it foretells the delivery of more precise future services as online activities and records expand and accumulate over time.

6 Summary

China is in the early stage of its 6-year national urbanization plan, extending its economic development to also address rising social and environmental concerns. There is a defined role for statistics and urban informatics to establish norms and conduct dynamic monitoring and longitudinal analysis. Small steps have been taken to begin data consolidation in some smart city test sites, and modest progress is beginning to appear.

In the next 6 years, cultural changes towards an objective data-driven approach, integration of statistical design and thinking into the data systems, and innovative statistical theories and methods to fully deploy meaningful Big Data will be needed to grow urban informatics in China and to achieve balanced success in its urbanization efforts. China will undoubtedly continue to advance towards building smarter cities with Chinese characteristics, and we will be able to understand more of Chinese cities through statistics and Big Data.

Acknowledgements This research was supported in part by Digital China Holdings Limited and East China Normal University. Correspondence concerning this article should be addressed to Jeremy S. Wu, 1200 Windrock Drive, McLean Virginia 22012. We wish to thank Dr. Carson Eoyang and Dr. Xiao-Li Meng for their valuable comments, corrections, and edits.

References

Baidu.com (n.d.) National smart city test sites. http://bit.ly/1lqGOqq
Baike.baidu.com (n.d.) Resident card. http://bit.ly/1xP8j2u
Beijing City Lab (2014) Spatiotemporal changes of population density and urbanization pattern in China: 2000–2010. Working paper #36. http://bit.ly/UHOTxF

Bream S (2014) Proposal to add photos to Social Security cards meets resistance. Fox News. http://fxn.ws/1lnumHU
U.S. Census Bureau (1990) 1990 census of population and housing, Table 4. http://1.usa.gov/1lnA24E
CIO.Gov (2014) Data center consolidation. http://1.usa.gov/1xKcUTN
Data Quality Campaign (n.d.) Why education data? http://bit.ly/111iQnb
Deming WE (1994) The new economics for industry, government, education. The Massachusetts Institute of Technology, Cambridge
Digital China (2013) Release of first Chinese city public information service platform. http://bit.ly/1q47SPx
Eilperin J, Tumulty K (2014) Democrats embrace adding photos to Social Security cards'. Washington Post. http://wapo.st/1l8TOfM
Groves RM (2012) National Statistical Offices: Independent, identical, simultaneous actions thousands of miles apart. U.S. Census Bureau. http://1.usa.gov/1xJbEQL
Harford T (2014) Big data: are we making a big mistake? FT Magazine. http://on.ft.com/1xJbnx0
Henderson JV (2009) Urbanization in China: policy issues and options. Brown University and NBER. http://bit.ly/1laGvkP
Hilbert M, Lopez P (2011) The world's technological capacity to store, communicate, and compute information. Sci Mag 332(6025):60–65. doi:10.1126/science.1200970. http://bit.ly/1oUvAKm
International Organization for Standardization (n.d.) About ISO. http://bit.ly/1jRdtSs
Jiang C (2014) How to wake the slumbering "Scientific Big Data." Digital Paper of China. http://bit.ly/1p54gcn
Konkel F (2014) Is data center consolidation losing steam? FCW Magazine. http://bit.ly/1q4TdDU
Lazer D, Kennedy R, King G, Vespignani (2014) The parable of Google flu: traps in big data analysis. Sci Mag 343(6176):1203–1205. doi:10.1126/science.1248506. http://bit.ly/1s5RW1h
Li D (2014a). Sum of GDP from 28 provinces already exceed the national GDP by two trillion yuan. The Beijing News. http://bit.ly/1lqfR7Y
Li K (2014b) 2014 Report on the Work of the Government, delivered at the Second Session of the Twelfth National People's Congress. http://bit.ly/1jkDKIr
Marcus G, Davis E (2014) Eight (No, Nine!) problems with big data. The opinion pages, The New York Times. http://nyti.ms/1q4a9KK
Ministry of Science and Technology (2013) Science and technology results to support our nation's smart city development. http://bit.ly/1l9BROg
Morello C (2014) Census hopes to save $5 billion by moving 2020 surveys online. The Washington Post. http://wapo.st/TCJvKP
National Bureau of Statistics of China (2014) Statistical Communiqué of the People's Republic of China on the 2013 national economic and social development. http://bit.ly/1oeqVXM
National People's Congress (2011) The twelfth five-year plan. http://bit.ly/1lHHcm9
Nie H, Jiang T, Yang R (2012) State of China's industrial enterprises database usage and potential problems. World Econ 2012(5). http://bit.ly/1oXtLwp
Qi F (2014) How to wake up the slumbering "Scientific Big Data." Guangming Daily. http://bit.ly/1p54gcn
Qiu A (2012) How to understand the urbanisation rate in China? China Center for Urban Development, National Development and Reform Commission, PRC. http://bit.ly/1lnwsaQ
Sain S, Wilde S (2014) Customer knowledge management. Springer Science + Business Media. http://bit.ly/1qsiyqf
Shen Y (2008) Analysis of the bottlenecks in constructing the four basic data system. Cooper Econ Sci. http://bit.ly/1oUwyq4
The State Council of China (2002) Guidance on development of e-government (17), national information automation leading group. http://bit.ly/1lnx2Wa
The State Council of China (2014) The national new-type urbanization plan (2014–2020). http://bit.ly/1lo1K1g

Tu Z (2012) The big data revolution. Guangxi Normal University, Guilin
Tu Z (2014) Big data: history, reality and future. China CITIC Press, Beijing
U.S. Census Bureau (n.d.a) American Factfinder. http://1.usa.gov/TYyFj5
U.S. Census Bureau (n.d.b) Longitudinal employer-household dynamics program. http://1.usa.gov/1oUB2ge
U.S. Census Bureau (n.d.c) OnTheMap. http://1.usa.gov/1osOX1i
U.S. Census Bureau (n.d.d) City and town totals: vintage 2012, population estimates. http://1.usa.gov/1pkiqKu
United Nations Development Program (2013) China national human development report 2013: sustainable and livable cities: toward ecological civilization. http://bit.ly/1n6bknW
UPnews.cn (2014) Shanghai ponders setting up Big Data agency to promote open data and information sharing. http://bit.ly/1nuw2jR
Wang S (2013) Government forces enterprises to lie, whose responsibility is it? http://bit.ly/1lrVJ5b
Webopedia (n.d.) Structured data. http://bit.ly/1iohxZM
Wikipedia (n.d.c) Health Information Technology for Economic and Clinical Health Act. http://bit.ly/1hORDUf
Wikipedia (n.d.d) Hukou system. http://bit.ly/1uYgAyS
Wikipedia (n.d.e) List of countries by GDP (nominal). http://bit.ly/1lnxJyz
World Bank (n.d.) Poverty headcount ratio at $1.25 a day. http://bit.ly/1pHn2rx
Wu JS (2012) 21st century statistical systems. http://bit.ly/147tSya
Wu JS (2014) Lying with big data. http://bit.ly/1txnOsM
Wu JS, Ding H (2013) The essentials of identification codes. http://bit.ly/15XuaZE
Wu JS, Graham MR (2009) OnTheMap: an innovative mapping and reporting tool. The United Nations Statistical Commission Seminar on Innovations in Official Statistics (February 2009). http://bit.ly/1wP3pCh
Wu JS, Guo J (2013) Statistics 2.0: dynamic frames. http://bit.ly/PaSGR0
Xi J (2014) Further economic reforms a tough task, warns Xi Jinping. Interview with Sino-US.com. http://bit.ly/1oOaC1C
Yuan G (2014) Smart city investment set to top 2 trillion Yuan by 2025. China Daily. http://bit.ly/1odVEnK
Zhang C (2010a) China's urban diseases (1). ChinaDialogue. http://bit.ly/1hWRfD4
Zhang C (2010b) China's urban diseases (2). ChinaDialogue. http://bit.ly/1sfRvBq

Part V
Urban Knowledge Discovery Applied to Different Urban Contexts

Planning for the Change: Mapping Sea Level Rise and Storm Inundation in Sherman Island Using 3Di Hydrodynamic Model and LiDAR

Yang Ju, Wei-Chen Hsu, John D. Radke, William Fourt,
Wei Lang, Olivier Hoes, Howard Foster, Gregory S. Biging,
Martine Schmidt-Poolman, Rosanna Neuhausler, Amna Alruheil,
and William Maier

Abstract In California, one of the greatest concerns of global climate change is sea level rise (SLR) associated with extreme storm events. Several studies were conducted to statically map SLR and storm inundation, while its dynamic was

---

The original version of this book was revised. An erratum to the book can be found at DOI: 10.1007/978-3-319-40902-3_30

Y. Ju (✉) • W.-C. Hsu • W. Fourt • A. Alruheil
Department of Landscape Architecture and Environmental Planning, University of California, Berkeley, CA 94720-1820, USA
e-mail: yangju90@berkeley.edu

J.D. Radke
Department of Landscape Architecture and Environmental Planning, University of California, Berkeley, CA 94720-1820, USA

Department of City and Regional Planning, University of California, Berkeley, CA 94720-1820, USA

W. Lang
Department of Building and Real Estate, The Hong Kong Polytechnic University, Hung Hom, Kowloon, Hong Kong, China

O. Hoes
Faculty of Civil Engineering and Geosciences, Delft University of Technology, Stevinweg 1, 2628 CN, Delft, Netherlands

H. Foster • M. Schmidt-Poolman
The Center for Catastrophic Risk Management, University of California, Berkeley, CA 94720, USA

G.S. Biging • W. Maier
Department of Environmental Science, Policy, and Management, University of California, Berkeley, CA 94720, USA

R. Neuhausler
Department of City and Regional Planning, University of California, Berkeley, CA 94720-1820, USA

Department of Environmental Science, Policy, and Management, University of California, Berkeley, CA 94720, USA

less studied. This study argues it is important to conduct dynamic simulation with high resolution data, and employs a 3Di hydrodynamic model to simulate the inundation of Sherman Island, California. The big data, high resolution digital surface model (DSM) from Light Detection and Ranging (LiDAR), was used to model the ground surface. The results include a series of simulated inundation, which show that when the sea level rises more than 1 m, there are major impacts on Sherman Island. In all, this study serves as a fine database for better planning, management, and governance to understand future scenarios.

Keywords Sea level rise • Mapping • 3Di hydrodynamic model • LiDAR • Sherman island

1 Introduction

In California's coastal areas, one of the great concerns of global climate change is sea level rise (SLR) associated with extreme high tides (Heberger et al. 2009). By 2100, mean sea level (MSL) will rise between 1.2 and 1.6 m (Bromirski et al. 2012), and this will cause a series of impacts along coastal areas, such as inundation and flooding of coastal land, salt water intrusion, increased erosion, and the decline of coastal wetlands, etc. (Nicholls and Cazenave 2010; Titus et al. 1991). Among all the impacts, flood risk is likely the most immediate concern for coastal regions.

This threat can be more severe in the Sacramento-San Joaquin Delta (the Delta), as many of its islands are 3–8 m below sea level (Ingebritsen et al. 2000). These islands are protected by more than 1700 km of levees (Mount and Twiss 2005), with standard cross sections at a height of 0.3 m above the estimated 100-year flood elevation (Ingebritsen et al. 2000). However, with a projected SLR between 1.2 and 1.6 m, these current levees can be easily overtopped, and the islands can be flooded.

Several efforts were made in the adjacent San Francisco Bay area (the Bay Area) to measure and understand the impact of SLR and storm inundation (Biging et al. 2012; Heberger et al. 2009; Knowles 2009, 2010). By using computer models, these studies intersected a water surface with a ground surface to identify inundated areas. The water surface could be interpolated from measured water level data at existing gauges (Biging et al. 2012), while the ground surface was usually obtained from LiDAR that provided fine resolution from 1 to 5 m. It should be noted that the interpolated water surface is static since it only describes the water surface condition at a particular water level, such as MSL or mean higher high water (MHHW) level. However, real tides and storm events are dynamic processes. The Bay Area and the Delta are characterized by semi-diurnal tides each day, meaning there are two uneven heights of high tide and low tide, and should be modeled dynamically to simulate all stages in the tidal cycle and the movements of tides during a storm event.

Therefore, a 3Di hydrodynamic model (Stelling 2012), was used in this study to better simulate the dynamics of tidal interaction during an extreme storm event. In addition, a 1 m resolution digital surface model (DSM) was generated from LiDAR in order to accurately describe the ground surface and to indicate the water flow pathway for 3Di simulations. A near 100-year storm with various scenarios of SLR was simulated in the Delta's Sherman Island, where significant critical infrastructure existed. Inundation extent, frequency, and average depth were mapped and analyzed based on the model outputs. Finally, a spatial resolution sensitivity analysis of DSM was conducted. Through this entire exercise, this study hopes to build a fine database for better planning, management, and governance to understand future scenarios.

2 Study Area

The study area, Sherman Island and its adjacent regions, is located at the confluence of Sacramento River and San Joaquin River (Fig. 1). Sherman Island is one of the major islands in the Delta, and is located at the transition from an estuarine system

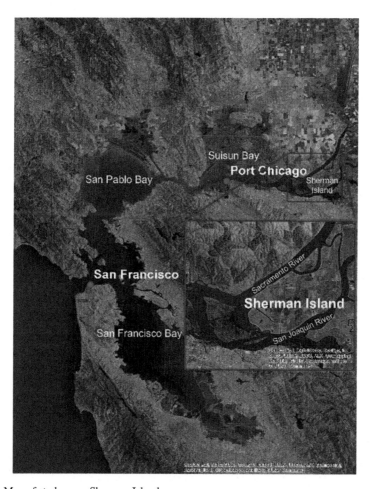

Fig. 1 Map of study area, Sherman Island

to a freshwater system. This island has significant infrastructure, including Highway 160, electric high-power transmission lines and natural gas pipelines. It is also the place where the infrastructure passes the Delta from north to south. According to NOAA, MSL (1983–2001 epoch) measured at a nearby NOAA Port Chicago Gauge is 1.116 m, and MHHW (1983–2001 epoch) is 1.833 m (NOAA Tide and Currents 2010), based on North American Vertical Datum of 1988 (NAVD 88), a commonly used datum in North America for vertical survey. The average elevation of Sherman Island is below MSL. Therefore the island is surrounded by extensive levees to protect it from inundation. Even with this levee system, the island still suffered from flooding due to levee failures. For example, the most recent levee failure and flooding in 1969 costed the Army Corps of Engineers approximately $600,000 for repairing, resloping, and regrading the levee break area (Hanson 2009). Even without a levee failure, the island is still at risk in the face of SLR. The lowest point of the levees in the study area is 2.11 m above NAVD 88. If the SLR by year 2100 is 1.4 m, then the MSL will rise to 2.25 m and the levees in the study area will be easily overtopped and the entire Sherman Island will be flooded. Considering the importance of Sherman Island's infrastructure and its potential exposure to SLR, it is a critical region and needs to be studied.

3 Data and Methods

3.1 Overview

To understand the impact of SLR inundation, a water surface and a ground surface are required to identify the spatial extent of inundated areas. A hydrodynamic model, 3Di (Stelling 2012), was employed in this study to simulate the water surface from a 72-h, near 100-year storm associated with 0, 0.5, 1.0, 1.41 m SLR scenarios. The ground surface, or a DSM, was generated from airborne LiDAR to capture the terrain and important ground objects, e.g. levees. The model output is a time-series of inundations, providing spatial extent, inundation depth, and water level. The workflow is shown in Fig. 2. All elevation data use NAVD 88.

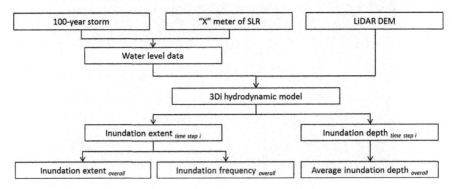

Fig. 2 Work flow of this study

3.2 3Di Hydrodynamic Model

The 3Di hydrodynamic model (Stelling 2012), developed by TU-Delft, Netherlands, dynamically simulates the movement of water through a digital ground surface. It employs an advanced approach for flooding simulation, allowing higher resolution and faster speed compared to existing hydrodynamic models (Van Leeuwen and Schuurmans 2012). The model is unique as it combines four specific methods: a sub-grid method where fine grids are clustered into coarser grids to calculate water levels and velocities (Dahm et al. 2014), a quadtree method which hierarchically decomposes the ground surface into a coarse grid and reduces the number of grid cells (Dahm et al. 2014), a bottom friction technique which accounts for the spatial variation of the roughness in the fine grid when calculating by coarse grid (Dahm et al. 2014; Stelling 2012), and a 'finite-volume staggered grid method for shallow water equations with rapidly varying flows, including semi-implicit time integration' (Stelling 2012).

The inputs of the model include time-series water level data and ground surface data. The output of the model is a time-series of simulated inundation that provides extent, depth, and water level. The time interval of the output is defined by the user. By further processing, the user can generate inundation frequency and average inundation depth by combining results from each time step. With an additional dimension of time, this model simulates the dynamics of flood and identifies the most exposed locations from SLR and storm inundation. In addition, with the time-series output, an inundation animation can be created to provide visual communication for the general public, making it a great educational tool.

3.3 Water Level Data

The first input of the 3Di model is water level data. To estimate the impact of a worst case scenario, a near 100-year storm event was used as the baseline, and SLR increments were added to the baseline. Being a dynamic model, 3Di requires time-series water level data for the entire storm event as input. However, existing 100-year storm calculation methods and studies (Zervas 2013) only provide estimates of water levels for a single stage such as MSL, MHHW, and mean lower low water (MLLW) level. Therefore, a historic storm whose peak water level was close to the 100-year storm was used as the water level input. Shown in Table 1, two storms that occurred in 1983 exceed the estimated 100-year storm at San Francisco NOAA tide station (NOAA ID: 9414290), and a third highest storm occurred on Feb. 6, 1998 that had a peak water level close to the estimated 100-year storm (Zervas 2013). All of these three extreme storms occurred during El Niño years.

Considering the storm's peak water level and availability of data, the Feb. 6, 1998 storm was selected as the storm to be simulated. More specifically, this study simulated this storm event over 72-h, from Feb 5 to Feb 7, 1998, to allow the

Table 1 Estimated and historic extreme storms at San Francisco and Port Chicago NOAA gauge

Station name	Date	Estimated 100-year storm (m)	Peak water level (m)
San Francisco	01/27/1983	2.64	2.707
	12/03/1983		2.674
	02/06/1998		2.587
Port Chicago	02/06/1998	Not available	2.729

model simulation to capture the complete storm movements through the study area. The water level data used for the 3Di simulation was retrieved from the nearby NOAA Port Chicago gauge (NOAA ID: 9415144), which provided measured water levels with 6-min intervals during the storm event.

As for the SLR scenarios, Cayan et al. (2009) and Cloern et al. (2011) studied the projected water level at Golden Gate in the Bay Area, and found that the time that sea levels exceed the 99.99th historical percentile of water elevation would increase to 15,000 h per decade by year 2100. And the 99.99th historical percentile is 1.41 m above year 2000's sea level. This study assumed that 1.41 m would be the maximum SLR by year 2100. This study also analyzed scenarios of 0, 0.5, and 1.0 m SLR to show how the impact changed with rising sea level. SLR was added on top of the baseline water level to simulate each scenario.

3.4 Ground Surface Data

The second input for the 3Di model is a fine spatial resolution DSM. The DSM was constructed based on LiDAR data, which used active remote sensing technology to measure the distance to target by illuminating the target with light pulses from a laser (Wehr and Lohr 1999). The density of the LiDAR data used in this study is 1 point per 0.7 m^2, and there are approximately 140 million points covering the study area. The DSM obtained from LiDAR in this study was originally 1 m resolution, and was resampled to 4 m resolution by the maximum aggregation method (ESRI 2015) to meet the computing limitations. In this method, a coarse grid cell obtains the maximum value of the fine grid cells in the coarse grid cell's spatial extent. Even though the DSM was aggregated to 4 m, this spatial resolution still accurately described the actual ground surface by precisely delineating objects such as levees, ditches, buildings, and the pathways that water moved through.

As aforementioned, the 3Di model has a limitation in the total number of grid cells that it can process, and it uses the quad-tree approach to reduce the total number of grid cells for model computation. The quad-tree is a data structure that is based on the regular decomposition of a square region into quadrants and sub-quadrants (Mark et al. 1989). 3Di draws finer quadrants/grid cells when elevation changes greatly within a short x, y distance, which then preserves detailed information while reducing the amount of data. Considering Sherman Island is

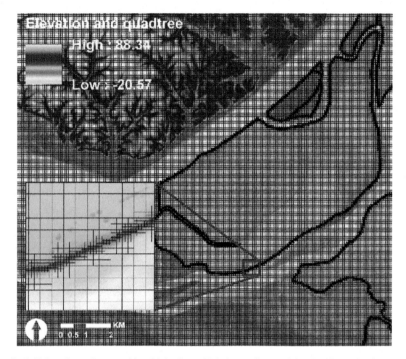

Fig. 3 DSM and guad-tree grid, which show 3Di draws finer grid cells along the levees and coarser grid cells for the other areas

relatively flat and the only abrupt change in topography is due to the levees, only levee data were used in the model to create finer grid cells, and coarser grid cells were generated for the rest of the study area that was more homogenous. The DSM and the quad-tree grid for Sherman Island are shown in Fig. 3.

4 Results

The 3Di simulation output is a time-series of inundated areas with an output time interval that is defined by the user. Each output provides the spatial extent and depth of inundation. In this study, the time interval was set as 1 h, and a total of 72 outputs were generated from the model. With the time-series outputs, this study analyzed inundation frequency and average inundation depth. Figure 4 shows the inundation extent and depth in hour 1, 24, 48, 72 for the simulated near 100-year storm associated with 0, 0.5, 1.0, and 1.41 m SLR.

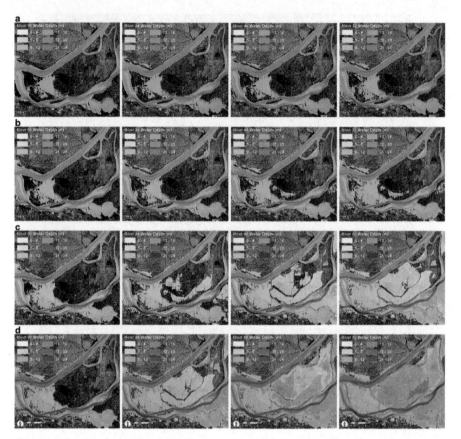

Fig. 4 Examples of simulated inundation from the 72-h, 100-year storm associated with 0 m (**a**), 0.5 m (**b**), 1.0 m (**c**), and 1.41 m (**d**) SLR, showing inudation extent and depth in hour 1, 24, 48, 72, respectively

4.1 Inundation Extent

The results show that during a 100-year storm, a total of 14.68 km^2 of land is inundated in the study area. With 0.5 m SLR, a total 20.67 km^2 of land is inundated, with 1.0 m SLR, a total of 57.87 km^2 of land is inundated, and with 1.41 m SLR, a total of 72.43 km^2 of land is inundated (Table 2). The inundation extent for different SLR scenarios is mapped in Fig. 5. The western end of Sherman Island is constantly underwater in all the modeled SLR scenarios as it is not protected by levees. In the 0.5 m SLR scenario, only minor inundation occurred in the rest of the island. In the 1.0 m SLR scenario, over half of the remaining Sherman Island is inundated and in the 1.41 m SLR scenario, the entire Sherman Island is inundated. This progress shows that when the sea level rises above 1.0 m, it will cause major flood impacts on Sherman Island.

Table 2 Statistical summary of inundation by a 100-year storm with different levels of SLR

SLR (m)	Inundated area (km²)	Area by inundation frequency (km²)			Area by average inundation depth (km²)		
		Low (0.00–0.21)	Medium (0.22–0.64)	High (0.65–1.00)	Low (0.00–1.98 m)	Medium (1.99–4.01 m)	High (4.02–13.22 m)
0	14.68	4.40	4.99	5.30	14.64	0.04	0.00
0.5	20.67	3.26	7.33	10.08	20.51	0.15	0.00
1.0	57.87	6.82	23.93	27.11	49.55	8.31	0.01
1.41	72.43	2.37	15.52	54.54	23.91	32.67	15.86

Fig. 5 Inundated area by a near 100-year storm associated with 0, 0.5, 1.0, 1.41 m SLR

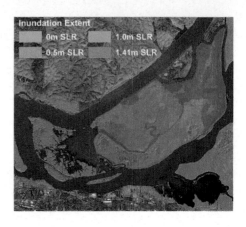

4.2 Inundation Frequency

Storm is a dynamic process, and impacted areas are not permanently under water during the entire storm event. Thus, this study analyzed inundation frequency by using (1) and (2):

$$I_{x,y,i} = \begin{cases} 1, & inundated \\ 0, & not\ inundated \end{cases} \quad (1)$$

$$F_{x,y} = \frac{\sum_{i=1}^{n} I_{x,y,i}}{n} \quad (2)$$

where $I_{x,y,i}$ is whether grid cell in column x, row y gets inundated at hour i, $F_{x,y}$ is inundation frequency for grid cell in column x, row y, and n is total number of outputs, which equals 72 in this study since a 72-h event was simulated.

The inundation frequency calculated here is the proportion of hours each piece of land (i.e. a 4 m × 4 m grid cell) gets inundated in the entire 72-h storm event. This study then classified the inundation frequency in the 1.41 m SLR scenario using a natural breaks method, which minimizes the variance within classes and maximizes the variance between classes (ESRI 2015). From this classification, low frequency is 0.00–0.21, medium frequency is 0.22–0.64, and high frequency is 0.65–1.00. The results from other scenarios were classified using the 1.41 m SLR scenario's classification in order to compare between the scenarios. The inundation frequency is shown in Fig. 6 for the four scenarios, and a statistical summary is shown in Table 2. From the results, it is observed that when the sea level rises, low frequency areas decrease while high frequency areas increase, therefore showing that more land will be permanently inundated in the future with such rises.

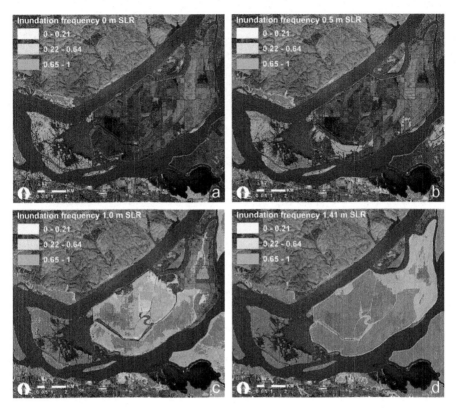

Fig. 6 Inundation frequency during the 3-day 100-year storm associated with 0 m (**a**), 0.5 m (**b**), 1.0 m (**c**), and 1.41 m (**d**) SLR

4.3 Average Inundation Depth

This study also analyzed average inundation depth, as the inundation depth on top of each piece of land varied in a storm event. Average inundation depth was calculated by (3):

$$D_{x,y} = \frac{\sum_{i=1}^{n} d_{x,y,i}}{n} \qquad (3)$$

where $D_{x,y}$ is average inundation depth (m) at grid cell in column x, row y, $d_{x,y,\,i}$ is inundation depth at grid cell in column x, row y, at hour i, and n is total number of outputs, which equals 72 in this study.

Similarly, this study classified average inundation depth in the 1.41 m SLR scenario by the natural breaks method. From this classification, low depth is 0–1.98 m, medium depth is 1.99–4.01 m, and high depth is 4.02–13.22 m. The

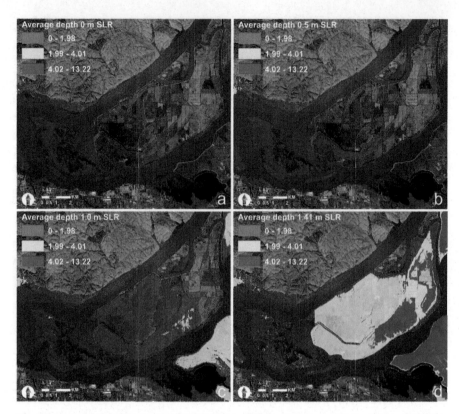

Fig. 7 Average inundation depth during the 3-day 100-year storm associated with 0 m (**a**), 0.5 m (**b**), 1.0 m (**c**), and 1.41 m (**d**) SLR

results from other scenarios were classified using the 1.41 m SLR scenario classification, in order to compare between the scenarios. The average inundation depth is shown in Fig. 7, and a statistical summary is shown in Table 2. The results show that in the 0.5 and 1.0 m SLR scenarios, the majority of inundated areas are under low inundation depth, and when it comes to the 1.41 m SLR, more areas are under medium and even high inundation depth.

4.4 DSM's Spatial Resolution Sensitivity Analysis

While 4 m resolution was used in the simulation as it was the finest resolution possible, it is important to test spatial resolution sensitivity of DSM and the effect on extent, frequency, and average depth. Surfaces with other resolution (5, 6, 10, 20, and 30 m) were used for the sensitivity analysis, while other model parameters remained the same. The same maximum aggregation method was used here for generating those surfaces from the original, 1 m resolution surface, and the results from the 4 m simulation were used as the baseline for comparison. To quantify the sensitivity, percentage difference in area (4) was calculated for extent, and Root Mean Square Difference (RMSD) and Coefficient of Variation (CV) were calculated for depth and frequency (5) and (6). When calculating RMSD, coarse resolution's results were first resampled to 4 m resolution by the nearest neighbor method (ESRI 2015), and RMSD was calculated using the 4 m resolution grid cells.

$$Diff = \frac{A_i - A_4}{A_4} \times 100\% \quad (4)$$

where A_i is the inundated area with other resolution i (5, 6, 10, 20, and 30 m), A_4 is the inundated area with 4 m resolution.

$$RMSD = \sqrt{\frac{\sum_{i=1}^{n}(y_i - y_4)^2}{n}} \quad (5)$$

where y_i is a grid cell's value simulated with other resolution i, y_4 is the grid cell's value simulated with 4 m resolution, n is the total number of grid cells compared.

$$CV = \frac{RMSD}{\bar{y}_4} \quad (6)$$

where $\bar{y}_4$ is the mean value of grid cells simulated with 4 m resolution.

The analysis shows the model is sensitive to resolution (Tables 3, 4, and 5), emphasizing the importance for simulating with fine resolution data. Coarser spatial resolution data leads to greater differences compared to the baseline, as more low elevation areas are diminished by the maximum aggregation method, resulting in an more elevated ground surface, less inundated area, smaller inundation depth, and less frequent inundation. Furthermore, differences are generally greater in lower SLR scenarios (e.g. 0 and 0.5 m) than those in higher SLR scenarios (e.g. 1.0 and 1.41 m). This is because when sea level rises, those elevated areas due to aggregation start to be inundated, and their effect on the results starts to decrease.

Table 3 Percentage difference in inundated area under other resolutions

SLR (m)	Resolution (m)				
	5	6	10	20	30
0	2.51 %	5.25 %	11.46 %	23.19 %	31.24 %
0.5	2.15 %	3.67 %	9.43 %	21.60 %	30.10 %
1.0	0.36 %	−0.86 %	2.42 %	10.20 %	14.00 %
1.41	0.54 %	0.69 %	1.47 %	3.52 %	5.55 %

Table 4 RMSD and CV of average inundation depth under other resolutions

	SLR (m)	Resolution (m)					Mean depth (m) in 4 m simulation
		5	6	10	20	30	
RMSD	0	0.08	0.11	0.12	0.19	0.23	0.37
	0.5	0.11	0.12	0.16	0.25	0.30	0.50
	1	0.17	0.17	0.31	0.43	0.54	1.08
	1.41	0.22	0.26	0.32	0.48	0.80	2.80
CV	0	22.10 %	28.89 %	34.01 %	51.01 %	63.11 %	
	0.5	21.28 %	23.21 %	32.39 %	49.42 %	60.36 %	
	1	15.44 %	16.16 %	29.06 %	40.18 %	50.08 %	
	1.41	8.01 %	9.44 %	11.36 %	17.06 %	28.42 %	

Table 5 RMSD and CV of inundation frequency under other resolutions

	SLR (m)	Resolution (m)					Mean frequency in 4 m simulation
		5	6	10	20	30	
RMSD	0	0.12	0.13	0.20	0.28	0.32	0.05
	0.5	0.20	0.12	0.18	0.27	0.33	0.09
	1	0.07	0.08	0.12	0.19	0.23	0.24
	1.41	0.06	0.07	0.09	0.14	0.18	0.38
CV	0	238.33 %	264.60 %	395.08 %	560.44 %	654.42 %	
	0.5	222.09 %	133.04 %	203.42 %	303.57 %	363.82 %	
	1	30.29 %	33.10 %	49.24 %	76.42 %	92.92 %	
	1.41	15.21 %	17.31 %	23.81 %	36.10 %	46.19 %	

5 Discussion and Conclusions

5.1 Implication for Planning

This study creates a SLR and storm inundation dataset for Sherman Island and its adjacent areas. This is an important and initial step for policy makers, planners, and the public to understand the magnitude and spatial distribution of SLR and storm inundation. The big data, high resolution DSM, was used to accurately model the ground surface. The results show that with more than 0.5 m SLR, the levees protecting Sherman Island start to be overtopped. With 1.0 m SLR, nearly half of Sherman Island is inundated, and with 1.41 m SLR, the entire island is inundated. Based on this study, SLR impacts are significant, especially when SLR is greater than 1.0 m. Local governments can use the inundation water level to improve the levee system and construct new levees to protect areas with high inundation frequency. This dataset can also be employed in a suitability analysis for Sherman Island to identify areas with higher inundation risks, and to improve infrastructure planning and/or adopt different planning strategies for the rising sea level. Finally, the DSM's spatial resolution sensitivity analysis shows the importance of fine resolution data. Local governments should collect the best quality data possible to inform more accurate decision making.

Our future work further studies the SLR impacts on infrastructures, such as pipelines and roads. These infrastructures are designed to allow certain level of inundation, but this tolerance is limited. To better understand SLR and storm inundation impacts on these critical infrastructures, it is beneficial to know the duration and the depth of water sitting on top of any infrastructure. Static models have limitations as they only depict one stage of inundation, where the information on duration and flood dynamic is lost. The dynamic model implemented here provides the additional dimension of time. Subsequent studies can intersect the inundation dataset with infrastructure datasets to calculate the duration of impact, as well as the amount of water sitting on top of impacted infrastructure. To conclude, future researchers should identify when the infrastructure gets impacted, estimate the cost, and provide more detailed planning suggestions.

5.2 Limitations

This study has several limitations. First, the Sacramento-San Joaquin Delta region has a complex hydrologic system which is influenced by both the ocean and the rivers, making it difficult to conduct hydrologic modeling. Considering that Sherman Island is close to the mouth of the Delta, this study simplified the actual process and assumed that the island is only affected by tidal surges from the ocean. With the discharge from the Sacramento and San Joaquin River, the simulated process could be different. Second, the model did not incorporate other factors, such as subsidence, sediment deposition, wind, and rainfall. Being an "artificial" system, the Sacramento-San Joaquin Delta region has limited sedimentation and significant subsidence issues that would further exacerbate the impact of SLR inundation. As a result, the 3Di model might underestimate the SLR impacts on Sherman Island. Third, the water level data used in this study could be inaccurate, as there is no gauge currently available in the immediate region of the study area. Finally, the 3Di model has computing limitations that limit the number of grid cells processed to approximately 125,000 for each simulation. Our study lowered the DSM resolution, from the original 1 m resolution to 4 m, to accommodate the computing limitations. As a result, some topographic information, such as smaller ditches and roads, might not be reflected in the model.

5.3 Conclusions

No GIS model perfectly represents reality (Fazal 2008), and inundation models are usually a simple but effective method that identifies inundated areas (Tian et al. 2010). They provide the possibility to incorporate different datasets and generate models for planners, policy makers and the public to clearly see potential impacts. Compared to previous studies, our study provides a more detailed level of information, and serves as a basis for future analysis in Sherman Island. We continues to generate similar datasets for the entire Bay and Delta, and intersect the datasets with objects of interest. With such efforts, we hope to get a better understanding about SLR, its impacts and possible countermeasures.

Acknowledgements This chapter was funded by the California Energy Commission – PIER, California Climate Change Impacts Program (UCB 500-11-016).

References

Biging GS, Radke JD, Lee JH (2012) Impacts of predicted sea-level rise and extreme storm events on the transportation in the San Francisco Bay Region. California Energy Commission, Sacramento

Bromirski PD et al (2012) Coastal flooding-potential projections 2000–2100. California Energy Commission, Sacramento

Cayan D et al (2009) Climate change scenarios and sea level rise estimates for the California 2008 Climate Change Scenarios Assessment. California Energy Commission, Sacramento

Cloern JE et al (2011) Projected evolution of California's San Francisco bay-delta-river system in a century of climate change. PLoS One 6(9):e24465

NOAA Tide and Currents (2010) Datums for 9415144, Port Chicago CA. http://tidesandcurrents.noaa.gov/datums.html?units=1&epoch=0&id=9415144&name=Port+Chicago&state=CA. Accessed 3 Aug 2015

Dahm R et al (2014) Next generation flood modelling using 3Di: a case study in Taiwan. In: DSD international conference 2014, sustainable stormwater and wastewater management, Hong Kong

ESRI (2015) Nearest neighbor resampling—GIS dictionary. http://support.esri.com/en/knowledgebase/GISDictionary/term/nearest%20neighbor%20resampling. Accessed 4 Aug 2015

Fazal S (2008) GIS basics. New Age International, New Delhi

Hanson JC (2009) Reclamation District 341, Sherman Island Five Year Plan 2009. http://ccrm.berkeley.edu/resin/pdfs_and_other_docs/background-lit/hanson_5yr-plan.pdf

Heberger M et al (2009) The impacts of sea-level rise on the California coast. California Energy Commission, Sacramento

Ingebritsen SE et al (2000) Delta subsidence in California; the sinking heart of the state. U.S. Geological Survey. http://pubs.usgs.gov/fs/2000/fs00500/. Accessed 3 Aug 2015

Knowles N (2009) Potential inundation due to rising sea levels in the San Francisco Bay Region. California Climate Change Center. http://www.energy.ca.gov/2009publications/CEC-500-2009-023/CEC-500-2009-023-D.PDF

Knowles N (2010) Potential inundation due to rising sea levels in the San Francisco Bay Region. San Francisco Estuar Watershed Sci 8(1). http://escholarship.org/uc/item/8ck5h3qn. Accessed 18 Dec 2014

Mark DM, Lauzon JP, Cebrian JA (1989) A review of quadtree-based strategies for interfacing coverage data with digital elevation models in grid form. Int J Geogr Inf Syst 3(1):3–14

Mount J, Twiss R (2005) Subsidence, sea level rise, and seismicity in the Sacramento–San Joaquin Delta. San Francisco Estuar Watershed Sci 3(1). http://escholarship.org/uc/item/4k44725p. Accessed 3 Sept 2014

Nicholls RJ, Cazenave A (2010) Sea-level rise and its impact on coastal zones. Science 328 (5985):1517–1520

Stelling GS (2012) Quadtree flood simulations with sub-grid digital elevation models. Proce ICE Water Manag 165(10):567–580

Tian B et al (2010) Forecasting the effects of sea-level rise at Chongming Dongtan Nature Reserve in the Yangtze Delta, Shanghai, China. Ecol Eng 36(10):1383–1388

Titus JG et al (1991) Greenhouse effect and sea level rise: the cost of holding back the sea. Coast Manag 19(2):171–204

Van Leeuwen E, Schuurmans W (2012) 10 questions to professor Guus Stelling about 3Di water managment. Hydrolink 3:80–82

Wehr A, Lohr U (1999) Airborne laser scanning—an introduction and overview. ISPRS J Photogramm Remote Sens 54(2–3):68–82

Zervas C (2013) Extreme water levels of the United States 1893–2010. Center for Operational Oceanographic Products and Services, National Ocean Service, National Oceanic and Atmospheric Administration

The Impact of Land-Use Variables on Free-Floating Carsharing Vehicle Rental Choice and Parking Duration

Mubassira Khan and Randy Machemehl

Abstract Carsharing is an innovative transportation mobility solution, which offers the benefits of a personal vehicle without the burden of ownership. Free-floating carsharing service remains a relatively new concept and has gained popularity because it offers a flexible one-way auto rental option that charges usage by the minute. Traditionally, carsharing services require returning the rented vehicle to the same location where rented within a specified rental duration. Since free-floating service is a very new addition in the overall transportation system, the empirical research is still very limited. This study focuses on identifying the impact of land-use variables on free-floating carsharing vehicle rental choice and parking duration of car2go services in Austin, Texas on a typical weekday between the 9:00 AM and 12:00 PM off-hour period. Two different methodological approaches, namely a logistic regression model approach and a duration model technique are used for this purpose. The results of this study indicate that demographic variables, the carsharing parking policy, and the number of transit stops all affect the usage of free-floating carsharing vehicles.

Keywords Free-floating carsharing • Land-use and carsharing • Carsharing rental choice • Duration model

---

M. Khan (✉)
HDR, 504 Lavaca St #1175, Austin, TX 78701, USA

Department of Civil, Architectural and Environmental Engineering, The University of Texas at Austin, 301 E. Dean Keeton St. Stop C1761, Austin, TX 78712-1172, USA
e-mail: mubassira@utexas.edu

R. Machemehl
Department of Civil, Architectural and Environmental Engineering, The University of Texas at Austin, 301 E. Dean Keeton St. Stop C1761, Austin, TX 78712-1172, USA
e-mail: rbm@mail.utexas.edu

© Springer International Publishing Switzerland 2017
P. Thakuriah et al. (eds.), *Seeing Cities Through Big Data*, Springer Geography,
DOI 10.1007/978-3-319-40902-3_19

1 Introduction

The urban development in the United States depends highly on the automobile based mobility system. This automobile dependency has led to many transportation and environmental problems, including traffic congestion, greenhouse gas emissions, air and noise pollution, and foreign oil dependency (Kortum and Machemehl 2012). In addition, personal vehicle ownership costs its owners more than $9000 per year (American Automobile Association 2013). Carsharing programs are a novel substitute to personal vehicle ownership in urban areas. Carsharing provides the mobility of using a private car without the burden of car ownership costs and responsibilities (Shaheen et al. 1998). Such innovative programs facilitate reductions in household vehicle ownership by motivating road users' behavior towards personal transportation decisions as on-demand mobility as opposed to an owned asset. The carsharing concept serves as a sustainable mobility solution because recent research studies have shown its positive environmental impact at least in three-ways (Firnkorn and Müller 2011): first, a reduction of total carbon dioxide emissions (Firnkorn and Müller 2011; Haefeli et al. 2006); second, reduction in household vehicle holdings (Shaheen and Cohen 2012); and third, reduction of vehicle miles traveled (Shaheen et al. 2010).

Carsharing service providers allow members on-demand vehicle rental with a basic "pay-as-you-drive". Members of the carsharing program have access to a fleet of vehicles in the service network and pay per use. Currently, traditional/station-based carsharing systems and free-floating/one-way carsharing systems are the two programs in practice. Generally, traditional carsharing programs (such as ZipCar) offer short-term rental with an hourly pricing option and require users to return vehicles to the original location of renting. On the other hand, free-floating carsharing programs (such as car2go and DriveNow) allow users one-way car rental where cars can be rented and dropped-off at any location within a specified service area. The main advantage of free-floating carsharing programs over traditional station-based carsharing programs is flexibility because it overcomes the limiting requirement of dropping off at the same station where it was rented.

German carmaker Daimler's car2go program was the first major initiative that allowed users one-way rental within a city's operating area. In 2008, car2go was first launched in Ulm, Germany and later expanded its service to 28 cities in 8 different countries across Europe and North America. Recently the German carmaker company BMW started offering one-way free-floating carsharing program DriveNow in five cities in Germany and in one city in the U.S. In the U.S., car2go was first launched in Austin, Texas in November, 2009 as a pilot carsharing project for city of Austin employees and later the service was opened to the general public in May, 2010 (Kortum and Machemehl 2012). Presently, over 16,000 car2go members use 300 identical car2go vehicles in Austin and all vehicles can be parked for free at any City of Austin or State of Texas controlled meter, parking space, or designated parking space for car2go vehicles within its operating area (car2go 2014; Greater Austin 2013 Economic Development Guide 2013).

Research studies on carsharing vehicles have routinely claimed that a connection between land-use development patterns and the use of carsharing vehicles exists. However, no rigorous attempt has occurred to explain the causal mechanism that generates the association between land-use and demand for carsharing vehicles. For example, car2go vehicles in different cities exercise free on-street parking at any meter in their operating area. Therefore, it is worthwhile to investigate if the parking cost has any effect on carsharing vehicle use. Again, easy access to alternative transportation modes, emphasized in the carsharing literature, likely affects the use of carsharing vehicles. However, no studies exist that have investigated such relationships for free-floating carsharing programs empirically. One of the factors that limit such investigation is the availability of data about the use of free-floating carsharing vehicles.

This study focuses on identifying impacts of land-use variables on free-floating carsharing usage, namely car2go vehicle rentals and parking duration (unused duration). Carsharing vehicle rental and unused durations are likely to depend on various land-use characteristics such as parking cost, auto-ownership, dominant age of the population in the location, employment density, transit facilities, and land-use mix diversity level. We use a logistic regression model approach to identify factors that affect car2go vehicle rentals. We also use different hazard-based duration model approaches to identify factors that affect their unused duration in Austin, Texas on a typical weekday between 9:00 AM and 12:00 PM. Free-floating vehicle availability at any location within the service area is random. Therefore, during the AM peak periods, a lack of availability equates to a lack of reliability of free-floating carsharing vehicles and limits their desirability as a regular commuting mode. Alternative travel modes such as personal automobile, ridesharing, bike, or transit are perceived as being more available/reliable. Therefore this study did not consider the morning peak/rush hours to investigate factors affecting car2go vehicle rentals and parking duration. Again, the data used for this research was not from a survey and does not have trip purpose information for trips using the car2go vehicles. Therefore, the time periods considered were chosen to homogenize trip purpose. Avoiding trips during rush commuting hours helps to identify trips with discretionary activities as a purpose. Survey data obtained from actual carsharing members' can be helpful to understand those factors; however the costs associated with such surveys are non-trivial. In this study, real-time free-floating car2go vehicle location and condition data over time (in 5 min intervals) are obtained from a car2go application programming interface (car2go API) at no cost. The data provides robust information on vehicle rentals, movement, and availability across the Austin, TX area. Since car2go uses remote control technologies to activate and monitor vehicle locations, the data collection infrastructure is already integrated within the system and can be accessed by the public in real time. This is the primary set of data that enables one to identify the usage of free-floating carsharing vehicles. The vehicle location dataset is supplemented by aggregate level land-use and transit-stop data obtained from the Capital Area Metropolitan Planning Organization (CAMPO) and Capital Metro. A parking survey conducted by CAMPO for the

Austin area is also used as a data source for the study. The organization of this paper is explained below.

The following section presents a brief review of the literature on the topic of this paper. The third section presents the modeling methodology, while the fourth section presents a description of the data set used. The fifth section presents model estimation results and the sixth section offers concluding thoughts.

2 Background and Literature Review

2.1 History of Carsharing

Originating in Switzerland during the mid-nineteenth century, the carsharing program has been popular in Europe for decades. During the early 1980s, this concept was first implemented in the US as pilot research projects to evaluate its feasibility (Fricker and Cochran 1982; Crain & Associates 1984). Commercial operation of station-based carsharing was first established in 1998 in Portland, Oregon (Katzev 2003). Carsharing service has evolved from initial neighborhood residential services where shared-use vehicles parked in designated areas within a neighborhood to robust, more-targeted focus customers in spatially denser areas having easier access to alternative transportation modes (Shaheen et al. 2009). This sector has been emerging and growing over time, and the number of U.S. users has grown from 12,000 in 2002 to over 890,000 in January of 2013 (Shaheen and Cohen 2013).

A relatively new approach of carsharing services is the free-floating operation, which is unique because its flexibility over traditional services. Traditional services offer a short-term round-trip rental option that require the user to return vehicles to the original location of renting, whereas free-floating carsharing services allow users one-way rental within a specified service boundary that facilitates discretionary activities. Free-floating carsharing service may be more appealing to consumers because it is free from fixed costs such as minimum rental duration, a booking fee, and a minimum monthly usage associated with traditional carsharing services as it only charges users usage by minutes. Free-floating carsharing vehicles are described as being close to private automobiles in terms of convenience because they permit one-way rentals. The service convenience is actually inferior to the personal automobile because there is out of vehicle travel time associated with accessing the mode. The convenience of such services may increase as autonomous free-floating carsharing vehicles (or autonomous taxis) increase in availability to consumers.

Since car2go was first established in 2010 in Austin, Texas, the fleet size and the operating area of the service has expanded over time. As of April 2014, a total of 300 identical Smart-for-Two vehicles have provided access to over 16,000 car2go members in the Austin area. Real-time information is provided to the consumers about available vehicle location, internal and external condition, and fuel level or electric charge level from either the Internet, a hotline, or the car2go smartphone application. Users can rent (41 cents per minute, as of April 2014) an available vehicle through the mobile app or from the car2go website in advance. The rent includes costs of parking, fuel, insurance, maintenance, cleaning, GPS navigation, 24/7 customer support, and roadside assistance (car2go 2014).

2.2 Carsharing Vehicle Usages

In consumer research, carsharing is viewed as an access-based consumer product. Historically access-based consumption was considered as an inferior consumption alternative over ownership because of the attachment associated with owned belongings (Bardhi and Eckhardt 2012). However, the confluence of shifting demographics and changing dynamics in the economy facilitated carsharing to gain popularity due to perceived environmental benefits and cost savings (Bardhi and Eckhardt 2012; Shaheen et al. 2010). In fact, ownership is often perceived as burdensome while the short-term access mode allows flexibility and economic savings (Bardhi et al. 2012). Technological advancement, increased use of smartphones, and availability of real-time route information has helped road users reconsider their travel mode choice alternatives.

From the transportation and land-use perspective, carsharing is considered a service providing improved mobility for travelers without a vehicle of their own, who use transit, share rides, or move on bicycle or by foot, but still require access to a personal vehicle for a trip segment. It is an access-based consumer service characterized as "pay-as-you-drive" and often an economical alternative for individuals who may not require daily auto access and drive less than 10,000 miles per year (Shaheen et al. 2010). It is also believed that college and university students will benefit from the service because they can gain access to a fleet of vehicles without bearing the burden of ownership (Shaheen et al. 2004). Therefore it is observed that carsharing programs in the US, both traditional and free-floating, have developed close relationships with universities around the country (Kortum and Machemehl 2012). Recent research on the demographics of carsharing members characterized them as mostly young professionals, single and aged between 21 and 38 years (Bardhi and Eckhardt 2012).

A number of US cities supporting smart growth recognize the environmental and social benefit of carsharing and have developed parking policies allocating on-street parking spots for carsharing vehicles. Shaheen et al. (2010) has a good review about the carsharing parking policy in US cities. For example in Austin, car2go vehicles can be parked on-street at any City of Austin or State of Texas controlled meter or pay station parking space free of charge (car2go 2014). Such a carsharing friendly parking policy is likely to have positive effects on the usage of carsharing vehicles, especially in university towns that have parking challenges.

Overall, it can be seen that there are a host of land-use level socio-economic and demographic factors, as well as parking policies, and attitudinal variables that can affect the usage of carsharing vehicles. The number of studies investigating the effect of land-use variables on carsharing vehicle usage is very limited because most of the studies are based on station-based carsharing services. Moreover, despite the recognition that parking cost and transit service can affect the usage of carsharing services, most of the earlier studies on free-floating carsharing did not consider their effects on carsharing vehicle usage. In this study, we identify the impact of land-use variables on free-floating carsharing vehicle rental choice and parking duration (vehicle unused duration). Two different methodological approaches are used to investigate free-floating carsharing vehicle usage. The first approach is a logistic regression model approach where the dependent variable is a binary outcome that indicates whether or not renting of an available carsharing vehicle occurred, and the second is a duration model approach where the dependent variable is a continuous variable representing the unused duration of carsharing vehicles' fleet time.

3 Methodology

Two different methodologies are adopted to identify land-use factors affecting the usage of car2go carsharing vehicles. The first approach is a logistic regression model approach where the dependent variable is a binary outcome representing the choice of rental of an available free-floating vehicle within a given time period. This approach will be helpful to identify land-use factors that affect the usage of available carsharing vehicles in a 3-h time window. The second approach is a duration model where each available vehicle observed at a specific time period is observed for 3 h or until it is rented again. The methodologies are presented in the following subsections.

3.1 Logistic Regression Model

In the logistic regression model, the response variable (y_j) is binary or dichotomous in nature where y_j takes the value of 1 for a given car2go available vehicle j if the vehicle becomes unavailable (because of renting) during the time of observation and takes the value of 0 if the vehicle remains available to rent at the end of observation period. The equation for the standard logistic regression model is:

$$y_j^* = \mathbf{x_j}\beta + \xi_j, \qquad y_j = 1 \text{ if } y_j^* > 0$$

where, y_j^* corresponds to the latent propensity for vehicle j, $\mathbf{x_j}$ is a vector of explanatory variables, β is a corresponding vector of coefficients that will be estimated and ξ_j is an idiosyncratic random error term assumed to be identically and independently standard logistic distributed. In the latent variable model, we do not observe the latent propensity y_j^*, however, we observe in the sample data whether an available car2go vehicle is still available or not at the end of an observation period. The probability that car2go vehicle j would become unavailable (rented) is

$$\text{Prob}(y_j = 1|\mathbf{x_j}) = \text{Prob}(y_j^* > 0|\mathbf{x_j}) = \mathbf{G}(\mathbf{x_j}\boldsymbol{\beta}).$$

where, $\mathbf{G}(.)$ is the cumulative density function for the error term, which is assumed to be logistically distributed as

$$G(x_j\beta) = \frac{e^{x_j\beta}}{1+e^{x_j\beta}}.$$

Similarly, the probability of remaining available (not rented) is

$$\text{Prob}(y_j = 0|\mathbf{x_j}) = \text{Prob}(y_j^* < 0|\mathbf{x_j}) = \mathbf{G}(-\mathbf{x_j}\boldsymbol{\beta}) = 1 - \mathbf{G}(\mathbf{x_j}\boldsymbol{\beta}).$$

To estimate the model by maximum likelihood, we use the likelihood function for each j. The parameters to be estimated in the logistic regression model are the $\boldsymbol{\beta}$ parameters.

3.2 Duration Model

We let $T \geq 0$ be a continuous random variable that denotes the time measured in minutes until an available car2go vehicle becomes unavailable (rented) again. Then t denotes a particular value of T and the cumulative distribution function (cdf) of T is defined as

$$F(t) = P(T \leq t), \quad t \geq 0.$$

The instantaneous renting rate per unit of time is $\lambda(t)$, commonly referred to as the instantaneous hazard rate in duration model literature. The mathematical definition for the hazard in terms of probabilities is

$$\lambda(t) = \lim_{h \to 0^+} \frac{P(t \leq T, t+h|T \geq t)}{h}.$$

The hazard function can be easily expressed in terms of the density and cdf very simply.

$$\lambda(t) = \frac{f(t)}{1 - F(t)} = \frac{f(t)}{S(t)} = \frac{dF/dt}{S(t)} = \frac{-dS/dt}{S(t)} = -\frac{d\ln S(t)}{dt},$$

where $S(t)$ is referred as a "survivor function" in the duration literature and in the reliability literature it is referred to as a "reliability function". In this study the authors prefer the term "availability function".

The shape of the hazard function (instantaneous renting rate) has important implications for duration analysis. Two parametric shapes for instantaneous renting rate $\lambda(t)$ are considered here. In the first case, the instantaneous renting rate is assumed to be constant, implying that there is no duration dependence or duration dynamics, mathematically,

$$\lambda(t) = \lambda, \ \forall \ t \geq 0.$$

The conditional probability of being rented does not depend on the elapsed time since it has become available to rent. The constant-hazard assumption corresponds to an exponential distribution for the duration distribution. The instantaneous renting rate function with covariates is

$$\lambda(t; \mathbf{x}) = \exp(\mathbf{x}\boldsymbol{\beta}),$$

where $\mathbf{x}$ is a vector of explanatory variables and $\boldsymbol{\beta}$ is a corresponding vector of coefficients to be estimated.

In the second case, the instantaneous renting rate function is generalized to accommodate duration dependency resulting in a Weibull distribution for the duration data. The hazard rate in this case allows for monotonically increasing or decreasing duration dependence and is given by

$$\lambda(t) = \gamma \alpha t^{\alpha - 1}, \ \gamma > 0, \ \alpha > 0.$$

The instantaneous renting rate function with covariates is

$$\lambda(t; \mathbf{x}) = \exp(\mathbf{x}\boldsymbol{\beta}) \alpha t^{\alpha - 1},$$

where $\mathbf{x}$ is a vector of explanatory variables and $\boldsymbol{\beta}$ is a corresponding vector of coefficients to be estimated. When $\alpha = 1$, the Weibull distribution reduces to the exponential with $\lambda = \gamma$, implying no duration dependence. If $\alpha > 1$, there is a positive duration dependence and the instantaneous renting rate is monotonically increasing. If $\alpha < 1$, there is a negative duration dependence and the instantaneous renting rate is monotonically decreasing.

4 Data Description

4.1 Data Sources

Five different data sources are used in this study. The first dataset, as discussed earlier is the real-time 24-h car2go vehicle location and condition data in 5-min intervals obtained from car2go. The data provides robust information on vehicle rentals, movement, and availability across the Austin, TX area. This primary data enables one to identify the usage of free-floating carsharing vehicles. The second dataset is the transportation analysis zone (TAZ) level land-use data provided by CAMPO for year 2005. The CAMPO data provided a host of land-use level socio-demographic information including population, number of households, household size, 2005 median household income, autos owned, total employment etc. The third dataset is land-use level demographic data based on the 2010 census obtained from the Capital Area Council of Governments (CAPCOG). Total population, race/ethnicity, as well as other items, at block level are obtained from the CAPCOG data source. The fourth dataset is a parking survey conducted in 2010 provided by CAMPO that encompasses the study area. The fifth dataset describes transit stops of the study area as an open data source created by Capital Metro, the public transit service provider (Capital Metropolitan Transportation Authority 2012).

4.2 Formulation of Dependent Variables/Car2go Vehicle Usage Data

Car2go vehicle location data was collected for a typical weekday (Tuesday) on February 25th, 2014 from 12:00 AM to 12:00 PM in 5-min intervals. The data collection effort was automated by setting up a Windows Task Scheduler operation that collected and stored data thereby reducing possibilities of human error. Car2go API data provides information about available vehicles only. Therefore, if a vehicle (each car2go vehicle in service has a unique vehicle identification number) is observed at time t and after some time h the vehicle is no longer found in the dataset, then it implies that the vehicle was rented within the t and t + h time intervals. In this study, each of car2go vehicles observed available at 9:00 AM is tracked every 5 min until 12:00 PM to see if they were rented. For instance if one car2go vehicle (say, vehicle 1) is available at 9:00 AM and 9:05 AM but it does not appear in the dataset at 9:10 AM, then it is assumed that the vehicle became unavailable at 9:10 AM because it was rented sometime between 9:05 AM and 9:10 AM. The availability status (rented vs. not rented) of each observed vehicle is used as the dependent variable of the logistic regression model.

The dependent variable for the duration model analysis is continuous. If a vehicle was found to become available at 7:00 AM and then was rented again at

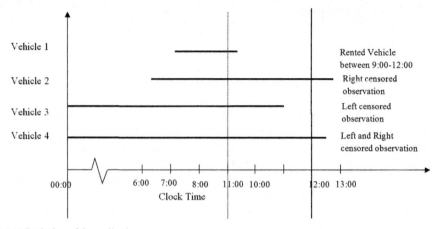

00:00 Beginning of data collection
9:00 Arbitrary time between 00:00 and 12:00
12:00 End of data collection

Fig. 1 Duration data for the sample vehicles

9:10 AM, then the total unused duration is 130 min.[1] The availability status and the duration unused (available duration prior to next rental) of each vehicle was recorded for further analysis.

Figure 1 shows the characterization of car2go vehicles unused duration. Vehicle 1 corresponds to the sample observation described earlier. The dataset also contains observations represented by Vehicle 2, 3, and 4, for which the beginning or ending times are synonymous with the sampling beginning or ending times. Such observations are referred to as right or left censored observations. The duration models used in this study are capable of accounting for the right-censored observations. Vehicle-type 3 illustrates the case for which the beginning time when it became available is unknown and is referred to as a left censored observation. Vehicle-type 4 represents those cases having both left and right censoring. By following the history of the vehicles represented by Types 3 and 4 one can trace the start time of the unused duration. However, this paper focuses only on the unused duration in a given weekday and therefore limits the analysis to midnight to noon on a typical weekday.

4.3 Data Assembly

The final sample was assembled in a number of steps. First, a total of 240 available vehicles observed at 9:00 AM are mapped to one of the 172 TAZs within the car2go

---

[1] Actually the unused duration for the vehicle is somewhat different because the vehicle observed available at 7:00 AM can become available anytime between 6:55 AM and 7:00 AM.

service boundary using a geographic information system (GIS). This operation appended vehicle availability status and unused duration information with the TAZ level land use data for each vehicle. Second, land-use level demographic data obtained from CAPCOG are aggregated into the TAZ level. Third, the number of transit stops located in each TAZ is calculated using the GIS platform. Fourth, parking survey data is also aggregated into TAZ level in order to identify TAZs that impose parking charges. Finally, TAZ level aggregated data on demographics, number of transit stops, and parking cost are appended with the car2go vehicle location data.

4.4 Sample Characteristics

A total of 240 car2go vehicles were observed in the 172 TAZs encompassing the car2go operating area covering 50 mile^2 in Austin, TX at 9:00 AM on February 25th, 2014. At the time of observation the 240 available vehicles were observed in 100 TAZs. The CAMPO data divides all TAZs into five area types namely, CBD, Urban Intense, Urban, Suburban, and Rural. The car2go operating area comprises the first four area types. Total number of zones in each area type, distribution of the vehicle locations in the four area types, average population and employment in the corresponding area types are presented in Table 1. Although higher numbers of available vehicles are observed in the area types with higher population (i.e. urban intense and urban area types), the average number of available vehicles per zone is highest in the CBD area.

Summary statistics of land-use variables for the car2go service area weighted by the number of available cars are presented in Table 2. As one can see from Table 2, the locations of car2go vehicles at 9:00 AM within the service area shows on average more than 11 % of the households have no car. Again, locations of available vehicles are mostly in neighborhoods where the mean percentage of adult population is about 87 %. This is not surprising because the central part of the city, including the downtown area and the neighborhoods (CBD and Urban Intense area Types) around the University of Texas at Austin, have higher concentrations of car2go members (Kortum and Machemehl 2012). The average

Table 1 Distribution of car2go vehicles by area type

Area type	Number of zones	Number (%) of available vehicles	Avg. no of vehicles per TAZ	2005 Population	2005 Employment
CBD	11	37 (15.4 %)	3.36	117	3826
Urban intense	37	102 (42.5 %)	2.76	2059	2267
Urban	46	87 (36.3 %)	1.89	1660	720
Suburban	6	14 (5.8 %)	2.33	1193	360

Table 2 Summary statistics of the land use variables used in the models

Land use variables	Mean	Min.	Max.
Percent of household having no car (year 2010)	11.01	0.87	46.75
Percent of over 18 population (year 2010)	86.91	65.65	100.00
Average household size (year 2005)	2.11	0.00	3.66
Number of transit stops (year 2010)	11.02	1.00	31.00
Employment density (year 2005: # of total employment/acre)	24.35	0.07	199.47

household size is 2.11, which is slightly lower than the 2005 average of 2.40 in Austin (City of Austin 2009).

All 100 TAZs are divided into two categories based on the median income of the TAZs. A total of 33 (14 %) vehicles are located in high income TAZs (Income > $60,000). Based on parking policy, all 100 TAZs are divided into two groups and only seven are found to charge parking fees. A total of 29 (12 %) of available vehicles were observed in TAZs where parking is not free.

A total of 110 (46 %) of the observed vehicles were rented between 9:00 AM and 12:00 PM. The total unused duration of the rented vehicles ranges from 5 to 705 min with an average unused duration of 331 min. The unused duration of the 130 unrented vehicles during the period of observation (12:00 AM to 12:00 PM) ranges from 180 to 720 min with an average of 480 min.

5 Model Estimation Results

This section presents a discussion of the logistic regression model estimation results for carsharing vehicle rental choice and the duration model estimation results for carsharing vehicle unused duration (Tables 3 and 4, respectively).

5.1 Logistic Regression Model Results

A number of socio-demographic attributes affect the choice of whether an available vehicle will be rented or not. As expected, the propensity of an available car2go vehicle to be rented increases as the percentage of households having no car increases. The availability of free-floating carsharing vehicles increases the mobility options for households having no car and therefore increasing the availability of vehicles in those areas increases the likelihood of those vehicles being rented. Households having their own car may perceive the walk-time to access a free-floating vehicle more burdensome compared to those having no personal automobile choice. As the percentage of the population aged over 18 increases in a TAZ, the propensity to rent an available car2go vehicle from that TAZ also increases. Car2go vehicles can accommodate only two adults in the vehicle and therefore may

Table 3 Logistic regression model result of rental choice

Explanatory variable	Parameter estimate	z-Stat
Constant	−12.07	3.32
TAZ land-use measures		
Socio-demographics		
Percent of household having no car (×0.1)	0.71	3.01
Percent of over 18 population (×0.1)	1.02	2.96
High-income TAZ (vs. low-income TAZ)	1.48	2.46
Average household size	0.84	2.10
Employment density	−0.03	3.31
Parking		
Parking cost (vs. doesn't have parking cost)	2.39	2.50
Transit availability		
Number of transit stops	0.05	2.39
Log-likelihood value	−141.15	

Table 4 Free-floating carsharing vehicle unused duration model results

	Weibull distribution		Exponential distribution	
Explanatory variable	Parameter estimate	z-Stat	Parameter estimate	z-Stat
Constant	−12.53	−8.61	−10.10	−8.27
TAZ land-use measures				
Socio-demographics				
Percent of household having no car (×0.1)	0.52	4.41	0.46	4.03
Percent of over 18 population (×0.1)	0.29	2.03	0.26	1.81
Parking				
Parking cost (vs. doesn't have parking cost)	1.58	4.05	1.26	3.37
Transit availability				
Number of transit stops	0.03	2.22	0.03	2.16
Time dependency parameter				
Alpha	1.33	3.02	1.00	–
Log-likelihood value	−238.71		−244.08	

be more attractive where concentration of adult individuals is relatively higher (as opposed to suburban areas where single family homes with young children are predominant). Again, as the number of transit stops increases in a TAZ, the propensity to rent an available car2go vehicle from that TAZ increases. Vehicles located in high median income TAZs (Income > $60,000) are more likely to be rented compared to vehicles located in low-income TAZs. Higher income may be working as a proxy variable representing a higher level of education and environmental consciousness. Individuals with such behavioral attitudes are likely to consciously make an effort to have a smaller carbon footprint on the environment and choose an environmentally-friendly free-floating vehicle as a modal option.

Increasing household size in a TAZ increases the propensity of the vehicles located in those TAZs to be rented. On the other hand, as the employment density increases in a TAZ, the propensity to rent an available car2go vehicle from that TAZ decreases. This result is not surprising because free-floating carsharing vehicles are less likely to be used on a daily basis as a commuting solution.

As expected, there is a higher renting propensity of a car2go vehicle in a TAZ with parking charges compared to a TAZ with free parking. This finding has important policy implications in that parking policy directly affects the usage of such services. To manage higher parking demand, the City of Austin enforces metered parking. The result suggests that allowing free parking at the metered parking spaces for carsharing vehicles can positively affect the usage of carsharing vehicles. The result is also in line with the earlier research findings showing that metropolitan areas with parking policies favoring carsharing have stronger carsharing services (Kortum 2012). The result may also imply that availability and easy access of carsharing vehicles help their potential usage. The result associated with the number of parking spots variable shows that as the number of transit stops increases, the propensity to rent an available car2go vehicle from that TAZ increases. This validates the assumption of carsharing services in areas with good transit services making intermodal trips easier.

The CAMPO area type definition is also used as an explanatory variable in the carsharing vehicle rental choice model, however, the effect was found to be non-significant. Perhaps the disaggregated level land-use variables included in the model explain the data variability more precisely than the aggregated land-use classification.

5.2 Duration Model Results

Table 4 shows the estimated covariate effects for the two parametric duration model specifications. The estimated value of the alpha parameter is $1.33 > 1.0$ for the Weibull parametric distribution of hazard function (instantaneous renting rate) and is statistically significant at the 0.05 level. This alpha parameter >1 indicates a monotonically increasing hazard (the probability to be rented) over time and therefore the results of the Weibull parametric distribution of hazard function appropriately describe the data.

The effect of land-use level socio-demographic characteristics indicates that locations with higher percentages of households having no car have a higher hazard (i.e., a smaller unused duration) than locations with lower percentages of households having no car, probably because of increased mobility options availability to these locations. A 1 % increase in 0-car households in a TAZ increases the probability of the vehicle(s) located in that TAZ to be rented by 5.2 %. Vehicles located in areas with higher percentages of adult population (age >18 years) have a higher hazard (i.e., a smaller unused duration). A 1 % increase in the percentage of adult population increases the renting probability of a vehicle located in that location by 2.9 %.

As expected, parking cost has a very significant impact on the usage of carsharing vehicles (Millard-Ball 2005; Kortum and Machemehl 2012). At any point in time, locations that charge parking fees have about $100 \times \left(e^{1.58} - 1\right) = 385\,\%$ greater hazard (probability of an available vehicle being rented) than the locations where parking is free-of charge. Since car2go vehicles exercises free on-street parking, those vehicles are likely to be chosen for making trips in those locations to gain the benefit of free-parking. Again, vehicles available in those areas where parking is challenging are likely to be rented because of higher trip production and attraction rates in those areas.

The effect of transit indicates that locations with greater numbers of transit stops have a higher hazard (i.e., a smaller unused duration) than locations with smaller numbers of transit stops, probably because of increased intermodal connectivity allowed by better transit availability. An additional transit stop in a TAZ increases the hazard (probability to be rented) of the vehicle(s) located in that TAZ by 3 %.

6 Conclusion

Free-floating carsharing service is gaining popularity because it permits users one-way auto rental and charges users by the minute. Such additional flexibility over traditional station-based carsharing services allows users to select this mode to perform discretionary activities. This study focuses on identifying the impact of land-use variables on free-floating carsharing vehicle rental choice and parking duration of car2go services in Austin, Texas on a typical weekday between 9:00 AM and 12:00 PM. Two different methodological approaches, namely a logistic regression model approach and a duration model are used for this purpose.

The results of this paper identify the importance of land-use level sociodemographic attributes, support from local governments by facilitating carsharing parking policy and the availability of transit facilitating intermodal transportation. The usage of free-floating carsharing vehicles is higher in neighborhoods having lower auto-ownership, greater numbers of adult population (age >18 years), larger household size, and higher household income levels. The study results confirm the impact of these factors with observed real-time data. Support from local government plays a vital role that promotes carsharing. The results indicate that parking policy in favor of carsharing services plays a very important role in carsharing vehicle usage. Since carsharing vehicle usage has significant environmental benefit, supporting these services with additional designated parking spots may help increase usage, and thereby reduce travel by personal automobile. Better transit service availability also positively affects the usage of carsharing vehicles.

The vehicle usage data employed in the study was collected without cost from the car2go API. The land-use data are typical of information maintained by metropolitan planning organizations for travel demand models which can also be accessible in most cases. Often the expense in data collection imposes difficulty in research. The innovative dataset used in this research can shed light on alternative available data sources in transportation planning research.

Admittedly, there are several limitations associated with this study that require further research. First, the data used in this study does not provide information about the trip purposes of the carsharing vehicle trips. Therefore, the study considers 3 h time periods during the AM off-peak periods, primarily to focus on trips made for discretionary activities. However, future research should investigate other time periods to find the effect of time of day on carsharing vehicle usage and to compare the research findings with the current study. Second, the study suggests that vehicles located in high median income TAZs are more likely to be rented compared to vehicles located in low-income TAZs. Now, this result may have two different implications. The availability of the carsharing alternative mode may replace an automobile trip, a walk/bike trip; or a transit trip. From a transportation planning perspective, it would be desirable that carsharing vehicles replace the personal auto trips. However, availability of the carsharing alternative mode may replace transit trips or non-motorized trips and thereby could increase overall auto trips. Future research is required to carefully investigate the effect of income on the choice decisions considering carsharing as an alternative travel mode. Third, the absence of trip purpose data also limits the analysis effort for considering different trip distribution patterns at different times of day. Future research should develop other data sources to investigate the variation in carsharing vehicle usage by trip purpose and time of day. Finally, the real locations of the carsharing vehicles are used to study their usage in terms of rental choice and parking duration. The locations of the available car2go vehicles represent the actual location of the observed vehicles. These locations may reflect (1) where the last users left them; (2) where the car2go staff relocated them within the study area. Future studies should explore the destination choice of carsharing vehicles during different times of the day.

Acknowledgements The authors acknowledge the helpful comments of two anonymous reviewers on an earlier version of the paper. The authors would like to acknowledge Dr. Daniel Yang of the Capital Area Metropolitan Planning Organization (CAMPO) in Austin, Texas for his help with the CAMPO land use and parking data used in the analysis. Brice Nichols of Puget Sound Regional Council provided the python code to download the car2go API Data. The authors are grateful to Lisa Smith for her help in formatting this document.

References

American Automobile Association (2013) Your driving costs 2012. http://newsroom.aaa.com/wp-content/uploads/2013/04/YourDrivingCosts2013.pdf
Bardhi F, Eckhardt GM (2012) Access-based consumption: the case of car sharing. J Consum Res 39(4):881–898

Bardhi F, Eckhardt GM, Arnould EJ (2012) Liquid relationship to possessions. J Consum Res 39 (3):510–529

Capital Metropolitan Transportation Authority (2012) https://data.texas.gov/capital-metro

Car2go Austin Parking FAQs (2014) https://www.car2go.com/common/data/locations/usa/austin/Austin_Parking_FAQ.pdf

City of Austin (2009) Community inventory demographics: demographic & household trends. ftp://ftp.ci.austin.tx.us/GIS-Data/planning/compplan/community_inventory_Demographcs_v1.pdf. Accessed 23 Apr 2014

Crain & Associates (1984) STAR project: report of first 250 days. U.S. Department of Transportation, Urban Mass Transportation Administration, Washington, DC

Firnkorn J, Müller M (2011) What will be the environmental effects of new free-floating carsharing systems? The case of car2go in Ulm. Ecol Econ 70(8):1519–1528

Fricker JD, Cochran JK (1982) A queuing demand model to optimize economic policy-making for a shared fleet enterprise. In: Proceedings of the October ORSA/TIMS conference in San Diego, CA

Greater Austin 2013 Economic Development Guide (2013) Car2go. http://www.businessintexas.org/austin-economic-infrastructure/car2go/

Haefeli U, Matti D, Schreyer C, Maibach M (2006) Evaluation car-sharing. Federal Department of the Environment, Transport, Energy and Communications, Bern, Switzerland

Katzev R (2003) Car sharing: a new approach to urban transportation problems. Anal Soc Issues Public Policy 3(1):65–86

Kortum (2012) Free-floating carsharing systems: innovations in membership prediction, mode share, and vehicle allocation optimization methodologies. Ph.D. dissertation, The University of Texas at Austin

Kortum K, Machemehl RB (2012) Free-floating carsharing systems: innovations in membership prediction, mode share, and vehicle allocation optimization methodologies. Report SWUTC/12/476660-00079-1. Project 476660-00079

Millard-Ball A (2005) Car-sharing: where and how it succeeds, vol 108. Transportation Research Board, Washington

Shaheen SA, Cohen AP (2012) Carsharing and personal vehicle services: worldwide market developments and emerging trends. Int J Sustain Transp 7(1):5–34

Shaheen SA, Cohen AP (2013) Innovative mobility carsharing outlook. Transportation Sustainability Research Center, University of California at Berkeley

Shaheen S, Sperling D, Wagner C (1998) Carsharing in Europe and North America: past, present, and future. Transp Q 52(3):35–52

Shaheen SA, Schwartz A, Wipyewski K (2004) Policy considerations for carsharing and station cars: monitoring growth, trends, and overall impacts. Transp Res Rec 1887:128–136

Shaheen SA, Cohen AP, Chung MS (2009) North American carsharing: 10-year retrospective. Transp Res Rec 2110:35–44

Shaheen SA, Cohen AP, Martin E (2010) Carsharing parking policy. Transp Res Rec 2187:146–156

Dynamic Agent Based Simulation of an Urban Disaster Using Synthetic Big Data

A. Yair Grinberger, Michal Lichter, and Daniel Felsenstein

Abstract This paper illustrates how synthetic big data can be generated from standard administrative small data. Small areal statistical units are decomposed into households and individuals using a GIS buildings data layer. Households and individuals are then profiled with socio-economic attributes and combined with an agent based simulation model in order to create dynamics. The resultant data is 'big' in terms of volume, variety and versatility. It allows for different layers of spatial information to be populated and embellished with synthetic attributes. The data decomposition process involves moving from a database describing only hundreds or thousands of spatial units to one containing records of millions of buildings and individuals over time. The method is illustrated in the context of a hypothetical earthquake in downtown Jerusalem. Agents interact with each other and their built environment. Buildings are characterized in terms of land-use, floor-space and value. Agents are characterized in terms of income and socio-demographic attributes and are allocated to buildings. Simple behavioral rules and a dynamic house pricing system inform residential location preferences and land use change, yielding a detailed account of urban spatial and temporal dynamics. These techniques allow for the bottom-up formulation of the behavior of an entire urban system. Outputs relate to land use change, change in capital stock and socio-economic vulnerability.

Keywords Agent based simulation • Earthquake • Synthetic big data • Socio-economic profiling

1 Introduction

The routine management of cities requires information regarding population characteristics, infrastructure, land-use, house prices, commercial activity, and so on. In the advent of a hazardous event (both natural and man-made) these different, yet

A.Y. Grinberger (✉) • M. Lichter • D. Felsenstein
Department of Geography, Hebrew University of Jerusalem, Mount Scopus, Jerusalem 91900, Israel
e-mail: asherya.grinberger@mail.huji.ac.il; mlichter@gmail.com; daniel.felsenstein@mail.huji.ac.il

interrelated sub-systems demand an immediate response. Invariably, the data requirements for this are only available at a coarse spatial resolution, such as traffic areas zones (TAZs) or statistical units. Different data may be available at varying scales and administrative divisions. Moreover, the spatial extent of the disaster will probably not neatly overlap these divisions. Therefore, urban disaster management requires data which is detailed and dynamic, spatially and temporally. While high resolution, big data for individuals is becoming increasingly available (such as geo-tagged social media data, mobile phones, GPS tracking etc.) and free of arbitrary spatial configurations, these data often over represent eager data-sharers and under represent the technologically-challenged. Furthermore these data often do not include crucial socio-economic profiling such as that provided by surveys.

In this paper we show how existing 'small' administrative data can be utilized to generate 'synthetic' urban big data. Big data is generated by decomposing small areal units into households and individuals, profiling them with socio-economic attributes and combining this data with an agent based simulation model in order to create dynamics. We use a GIS buildings layer to disaggregate administrative small data into households, individuals and eventually to the level of the synthetic location of individuals within buildings by floors. The resultant data can be considered 'big' in terms of volume, variety and versatility. Potentially, it allows for different layers of spatial information to be populated and embellished with synthetic attributes. This process of decomposition facilitates moving from a database describing hundreds or thousands of spatial units to one containing records of millions of buildings and individuals over time. The result is a comprehensive set of spatial 'big data' in which every single individual in a city is synthetically represented by a set of socio-economic attributes and high frequency dynamics.

A popular approach to handling this synthetic data is to intersect it with hazard maps to create a visual dynamic account of the development of a disaster. We have done this elsewhere using a dynamic web-interface that combines flood-hazard maps with the socio economic attributes of the areas under threat (Lichter and Felsenstein 2012). Alternatively, this data can serve as input for dynamic agent based (AB) modeling of urban disasters. This paper illustrates the latter route. We use data fusion techniques at the level of the individual building to generate the initial starting population data for an agent-based simulation of an urban disaster. These techniques allow for the bottom-up formulation of the behavior of an entire urban system.

The behavioral response of each agent is determined by its socio-demographic profiling. For example, age, income and car ownership may constrain, enable, or affect travel mobility, activity selection, and residential location choice. In this manner, the data animates the population analytics for a dynamic agent based simulation of an earthquake in Jerusalem. The AB simulation is based upon individual citizen agents and their interaction with the built environment and with each other. Feeding off the disaggregated data, individual buildings are characterized in terms of land-use, floor-space and value. Agents are characterized in terms of income and socio-demographic characters and are allocated to residential buildings. Using simple behavioral rules grounded in risk-evasiveness, satisficing behavior, and residential location preferences, along with a dynamic house pricing system that informs land-use dynamics, a detailed description of urban spatial and temporal dynamics is presented.

The paper proceeds as follows. After reviewing AB modeling applications for urban disasters, we outline the modeling framework and context of the study. We then describe how the big data is generated and coupled with the AB model. Simulation results are presented relating to change in land use, capital stock and socio-economic structure of the study area. To embellish the visualization potential of this data, we present some output in the form of dynamic web-based maps. Finally, we speculate on further developments derived from this approach.

2 Literature Review; AB Modeling for Disaster Management

Agent based modeling provides an appropriate framework for disaster management (Fiedrich and Burghardt 2007). In an agent based world, autonomous entities (agents) behave according to a set of pre-programmed and simplistic, decision rules. The activities of multiple agents create a computable system in which the actions of individual agents affect each other and the system as a whole. The result is a complex network of behavior patterns that could not have been predicted by simply aggregating individual agent behavior. More importantly, such a system can be simulated and subjected to various exogenous shocks. Its outputs can be tested for parameter stability and structural validity. It is no wonder therefore that a whole slew of disaster management situations have been subjected to AB simulations especially where human organization and learning patterns can be programmed into the response behavior of agents. These applications range from flooding (Dawson et al. 2011), wildfires (Chen and Zhan 2008), epidemics (Simoes 2012) to hurricanes (Chen et al. 2006) and earthquakes (Crooks and Wise 2013). Much of this effort produces output that highlights the collective behavior of the agents. This can range from simulating the intervention of first responders, predicting human behavior under conditions of stress, anticipating traffic and infrastructure congestion due to human movement patterns and even simulating assistance efforts, post disaster.

Less attention has been paid to simulating the response of this behavior on the built environment. At the micro level of individual buildings and human response patterns, Torrens (2014) has shown how the use of highly granular models can yield rich detail of building collapse, agent-agent interaction and evacuation dynamics in the case of a simulated urban earthquake. This contrasts with the 'dumb, coarse and cursory' (Torrens 2014: 965) nature of other AB models that try to reconcile human and physical processes. The spatial and temporal dynamics of such a situation that are animated by an AB model give rise to huge volumes of information that while not intuitively recognized as 'big data' certainly qualify as such in terms of volume and variety.[1] At the broader, system-wide level of the urban area, Zou et al. (2012) argue that the bottom-up dynamics of AB simulation become homogenized when looking at complex urban processes such as sprawl or densification. They propose a different simulation strategy to that commonly used in AB modeling. This involves 'short-burst experiments' within a meta-simulation framework. It makes for more efficient and accelerated AB simulation and allows for the easy transfer across different spatial and temporal scales. Elsewhere, we have also illustrated that complex macroscopic urban change such as land use rejuvenation and morphology change in the aftermath of an earthquake, can be suitably analyzed in an AB framework (Grinberger and Felsenstein 2014, 2017).

While the literature addressing urban outcomes of disasters and using agent-based modeling is limited, there is a larger ancillary literature that indirectly touches on this issue from a variety of traditions. Chen and Zhan (2008) use the commercial Paramics simulation system to evaluate different evacuation techniques under different road network and population density regimes. Another approach is to couple GIS capabilities to an existing analytic tool such as remote sensing and to identify disaster hot spots in this way (Rashed et al. 2007). Alternatively, GIS can be integrated with network analysis and 3D visualization tools to provide a realtime micro-scale simulation tools for emergency response at the the individual building or city block level (Kwan and Lee 2005). At a more rudimentary level, Chang (2003) has suggested the use of accessibility indicators as a tool for assessing the post disaster effects of earthquakes on transportation systems.

A particular challenge to all forms of simulation modeling comes from the inherent dynamics of AB simulation and the visualization of results that has to capture both spatial and temporal dimensions. In the context of big data, this challenge is amplified as real time processing needs to also deal with large quantities of constantly changing data. These are state of the art challenges that require the judicious use of computational techniques, relational databases and effective visualization. The literature in this area is particularly thin. In a non-AB environment, Keon et al. (2014) provide a rare example of how such integration could be achieved. They illustrate an automated geo-computation system in which tsunami inundation and the resultant human movement in its aftermath are simulated. They couple the simulation model with a web-based mapping capability thereby allowing the users to specify input parameters of their choosing, run the simulation and visualize the results using dynamic mapping via a standard web browser. A mix of server-side and client-side programming is invoked that allows the user all the standard functionality of web-based mapping.

Our approach takes this integration one stage further. In addition to using AB simulation we do not just combine a simulation model with a web-based visualization capacity but also generate the synthetic big data that drives the model. Once derived, this data needs to be spatially allocated. The literature provides various

---

[1] While not calling this 'big data' as such, Torrens (2014) notes that the volume of locations/vectors to resolve for each object moved in the simulation is of the order of 10^{12}–10^{14}.

methods such as population gridding (Linard et al. 2011), areal interpolation which calls for 'creating' locations (Reibel and Bufalino 2005) and dasymetric representation which uses ancillary datasets such as roads or night-time lights imagery to approximate population location (Eicher and Brewer 2001; Mennis 2003). An alternative approach, adopted here, is to distribute the data discretely without resorting to interpolation or dasymetric mapping. This calls for combining different sources of survey, administrative and non-structured data to create population count data. In this spirit, Zhu and Ferreira (2015) have recently illustrated how spatially detailed synthetic data from different sources can be input into a microsimulation framework. In our context, the result is detailed spatially-referenced local population count data that uses an appropriate spatial anchor. We use detailed building-level data in order to accurately allocate populations. This method has been adopted elsewhere, for example by Harper and Mayhew (2012a, b) who geo-reference administrative data using a local source of land and property information. Similarly, Ogawa et al. (2015) use building floorspace data to spatially allocate population for earthquake damage assessment.

3 The Modeling Framework

As behooves big data, the modeling framework used here is data-driven. The process is outlined in Fig. 1. Socio-economic data for coarse administrative units (small data) is disaggregated into buildings and individuals on the basis of a GIS

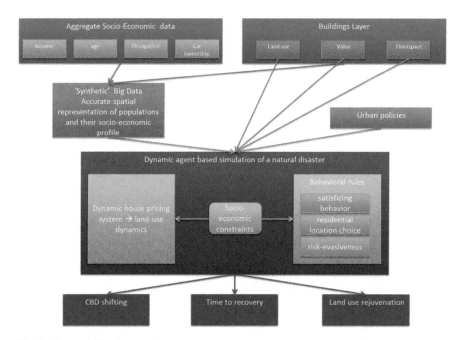

Fig. 1 The modeling framework

building layer and then recombined into households. The resultant synthetic data gives an accurate socio-economic profiling of the population in the study area. Coupling this data with an AB simulation model adds both temporal and spatial dynamics to the data. The result is a multi-dimensional big data set that affords flexibility in transcending conventional administrative boundaries. Outputs relate to socio-cconomic change, change in land use and capital stock in the aftermath of an earthquake. To fully capture the richness of the data generated we use web-based mapping to generate extra visual outputs.

3.1 The Context

We simulate an earthquake in downtown Jerusalem (Fig. 2). While Jerusalem is located 30 km southeast of the active Dead Sea Fault line, the last major earthquake in the city occurred in 1927. The city center lies in a relatively stable seismic area but many of its buildings were constructed prior to the institution of seismic-mitigation building codes making it prone to damage (Salamon et al. 2010).

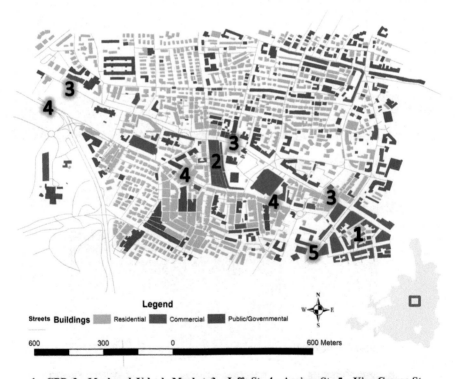

1 – CBD, 2 – Machaneh Yehuda Market, 3 – Jaffa St., 4 – Agripas St., 5 – King George St.

Fig. 2 Study area

The study area houses 22,243 inhabitants, covers 1.45 km^2 and is characterized by low-rise buildings punctuated by high rise structures. A heterogeneous mix of land uses exists, represented by residential buildings (243 thousand squared meters, 717 structures), commercial buildings (505 Th sqm, 119 structures) and government/public use buildings (420 Th sqm, 179 structures). The area encompasses two major commercial spaces: the Machaneh Yehuda enclosed street market and the CBD. Three major transportation arteries roads traverse the area and generate heavy traffic volumes: Agripas and Jaffa Streets (light railway route) run north-west to the south-east and King George Street runs north-south. The area exhibits a heterogeneous mix of residential, commercial, governmental and public land use and high traffic volumes.

4 Big Data for Big Disasters

4.1 Generating Synthetic Big Data

Accurate spatial representation of data is essential for dealing with emergency situations in real time and for preemptive emergency management and training. However, spatial phenomena often do not neatly overlap administrative units of analysis. A need exists for accurately distributing the alpha-numeric socio-economic data for individuals or households within the spatial units used for data collection. We use a GIS-based system that allocates populations into buildings. We then create a spatial database where each inhabitant is represented as a unique entity. Each entity is attached a suite of unique socio-economic properties. Together, all the personal entities in every original spatial unit accurately represent the distribution of all socio-economic variables in the original unit. The disaggregation and reallocation process allows for the accurate spatial distribution of social and economic variables not generally available. This synthetically generated big data drives the AB simulation and underpins the unique characteristics of the agents. These characteristics will determine their behavioral choices.

The model is driven by data at three different resolutions (Fig. 3): buildings, households and individuals. The original data is provided in spatial aggregates called Statistical Areas (SA).[2] This is the smallest spatial unit provided by the Israeli Central Bureau of Statistics (CBS) census.[3] We use a disaggregation procedure whereby spatial data collected at one set of areal units is allocated to a different

---

[2] A statistical area (SA) is a uniform administrative spatial unit defined by the Israeli Central Bureau of Statistics (CBS) corresponding to a census tract. It has a relatively homogenous population of roughly 3000 persons. Municipalities of over 10,000 population are subdivided into SA's.

[3] We also use coarser, regional data on non-residential plant and equipment stock to calculate non-residential building value. The estimating procedure for this data is presented elsewhere (Beenstock et al. 2011).

set of units. This transfer can be effected in a variety of ways such as using spatial algorithms, GIS techniques, weighting systems etc. (Reibel and Bufalino 2005). The GIS layer provides the distribution of all buildings nationally with their aerial footprint, height and land use. We derive the floor-space of each building and populate it with individuals to which we allocate the relevant socio-economic attributes of the SA to which they belong, according to the original distribution of these attributes in the SA. In this way, synthetic big data is created from spatial aggregates. The mechanics of the derivation are described in Appendix 1. The variables used to populate the buildings and drive the model are:

- Building level: land-use, floor-space, number of floors, building value, households
- Household level: inhabitants, earnings, car ownership
- Individuals level: Household membership, disability, participation in the work force, employment sector, age, workplace location.

The variables used in the model, their sources and level of disaggregation appear in Table 1.

Buildings Level disaggregation: The basis of the disaggregation procedure is calculating the floor-space of each building using height and land-use. We assume an average floor height of 5 m for residential building and 7 m for non-residential buildings (see Appendix 1). These figures are the product of comparing total national built floor-space (for each land-use) with total national floor-space as calculated from the building layer.

The entire data disaggregation procedure is automated using SQL and Python code and the results at each stage are stored in a spatial database. The process entails first allocating inhabitants into buildings and then assigning them socio-economic attributes. Later, these inhabitants are grouped into households and a further allocation of households attributes is performed. This necessarily involves loss of data due to dividing whole numbers (integers) such as households and inhabitants by fractions such as building floor-space for density calculations and percentages for socio-economic attribute distributions. In order to avoid loss of data and to meet the national control totals in each calculation, the SQL code is written so that it compensates for data losses (or increases) in the transition from floating points to integer values and ensures that the original control totals are always met. This is done by adjusting the floating point figures rounding threshold for each variable separately, to fit the losses or gain in the count of variables automatically.

Disaggregation at the level of the individual: The disaggregated building level data serves as the basis for the further disaggregation at the level of the individual. The building database includes a total of 1,075,904 buildings. A total of 7,354,200 inhabitants are allocated to 771,226 residential buildings. Disaggregation of the data to the individual begins with assigning each individual in the database a id, so that it is represented as a unique entity tied to a building in the database. Next, each person is allocated a random point location (a lat, lon coordinate) within the building with which it is associated. In each building, demographic attributes

Table 1 Variables used in the model

Variable	Source	Spatial unit	Value	Disaggregation level
Residential building value per m^2	National Tax authority 2008–2013	SA	2008–2013 averaged (2009 real) value in NIS per m^2	Building
Non-residential plant value	Local authorities financial data (CBS)	Local authority	Total value of non-residential building stock in NIS per region	Building
Non-residential machinery and equipment value	Estimation	Beenstock et al. (2011)	Total value of non-residential equipment stock in NIS per region	Building
Number of households	CBS 2008	SA	Total number of households per SA	Building, household
Number of inhabitants	CBS 2008	SA	Total number of inhabitants per SA	Building, household, individuals
Average monthly household earnings	National Insurance Institute, annual data	Local authority	Average household monthly earnings by SA	Building, household
Labor force participation	CBS 2008	SA	Working/not working	Building, household, individuals
Employment by sector	CBS 2008	SA	Commercial/governmental/industrial/home-based/unknown	Building, household, individuals
% Disabled	CBS	SA	% Disabled in SA	Building, household, individuals
Age	CBS	SA	0–18/18–64/65+	Building, household, individuals
Workplace location	GPS survey 2014	Survey of individuals	Inside the region/outside the region	Individuals

(labor force participation, employment sector, disabilities and age group) are allocated to each individual so that they comprise the entire distribution in the building which in turn gets its distribution from the SA in which it is located. In the same way, the distribution of work locations of inhabitants by employment sector is derived from a GPS-based transport survey carried by the Jerusalem Transport Master Plan Team (Oliveira et al. 2011). This is used to determine the distribution of inhabitants working inside or outside the study area according to their sector of employment and to assign the corresponding binary values to individuals.

Household level clustering and attribute allocation: Individuals are clustered into households by size of the household in each building. This creates new unique entities in the database representing households. Households are associated with

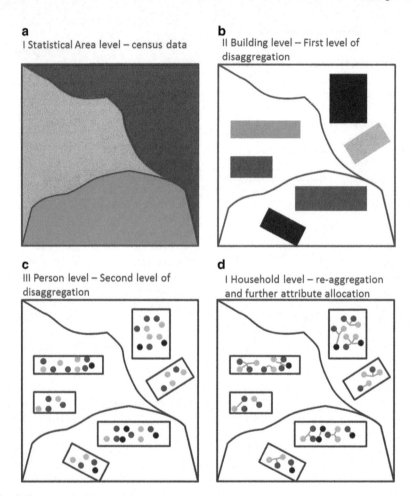

Fig. 3 Data processing stages

buildings, and inhabitants are assigned to them. The clustering introduces heterogeneity in terms of the age distribution in the household to closely represent a "traditional household" containing both adults and children when these are present. This is achieved by an algorithm iterating through inhabitants in each building sorted by age, and assigning them to households. Depending on the age distribution in the building, this algorithm clusters inhabitants in a building into closely age represented but not identical, households. Each household is assigned the SA average household earnings value. Other attributes such as car ownership are assigned to households in the same way.

4.2 Coupling the Data with the Agent Based Model

The high resolution data detailed above is combined with an agent-based model. As agents represent the focal catalysts of change and aggregate patterns are decomposed into actions of individual agents, this further unshackles the constraints imposed by data collected on the basis of arbitrary administrative borders. In the context of the current study this allows us to relate to the specific spatio-temporal nature of the event and its long-term impacts. To do this we characterize the three basic elements of an AB model: agents, their environment and rules governing agent-to-agent and agent-to-environment interactions (Macal and North 2005).

The data provides the first two with individuals and households as agents and buildings as the urban environment. The model itself reflects the dynamics of the urban system via a collection of simplistic behavioral rules. These govern the interactions within the system at each iteration (Fig. 4) which is set to represent one day and also include an exogenous shock in the form of the earthquake. These rules are described in Appendix 2.

A simulation entity is created to represent each individual, household and building along with its spatial and socio-economic characteristics. Identifying the unique workplace for each employed individual in the study area is done on the basis of satisficing behavior. The first building which satisfies randomly generated preferences in terms of distance from home location and floor-space size (assuming

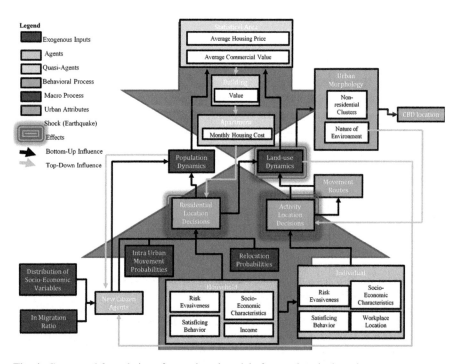

Fig. 4 Conceptual formulation of agent-based model of an earthquake in a city

larger functions attract more employees) and is of the land-use associated with the individual's employment sector is designated as the agent's workplace.

Agent behavior (bottom-up procedures): This simulation characterizes the city as a spatial entity whose organization emerges from the aggregate behavior of all its citizens (individuals). Individuals and households are therefore identified as the agents of the urban system. Their behavior is simplified into two decision sets: the decisions of households about place of residence and the decisions of individuals about choice and sequence of daily activities.

The decision to change place of residence is probabilistic and based on comparing a randomly drawn value to exogenous probabilities of migrating out of the study area (city-level probability), or relocating to another part of the study area (SA specific probability).[4] Choosing a new place of residence location follows two decision rules: a willingness to commit up to one third of monthly household earnings to housing and preferences regarding the socio-economic characteristics of the residential environment. This follows a probabilistic and satisficing procedure similar to that described for selection of work place. If a household fails to find an alternative place of residence after considering 100 possible locations, it leaves the study area. Individual agents that are members of the household relocate/migrate with the household. In-migration is treated in the model by having the number of potential migrating households dependent on the number of available housing units and an exogenous in-migration/out-migration ratio. New households with up to two members are comprised of adults only. Those with more members include at least one non-adult agent and their socioeconomic status reflects the urban average for key socio-economic attributes.

At each iteration individuals conduct a variety of activities within the study area. These are important for land-use dynamics and the mobility paths between land uses (see below). The number of activities undertaken ranges from 0 to 11 and varies by socio-economic characteristics (age, car ownership of household, disability, employment status, location of employment) and randomly generated preferences. The location of each activity (with the exception of work activity for agents employed within the study area) is determined by the attractiveness of different locations. This in turn is dependent on floor-space size, environment, distance from previous activity, and the mobility capability of the individual. This choice criteria is again, probabilistic and satisficing. Movement between each pair of activities is not necessarily shortest-path. A more simplistic aerial-distance-based algorithm is used to reduce computing demands and again reflect the satisficing nature of agents.[5]

---

[4] Calculated from 2012 immigration data in the Statistical Yearbook for Jerusalem 2014 (Jerusalem Institute for Israel Studies).

[5] The algorithm works as follows: at each step, junctions adjacent to the current junction are scanned and the junction with the shortest aerial distance to the destination is flagged as the current junction for the next step. If a loop is encountered, it is deleted from the path. The algorithm ends when agents arrive at the junction closest to the destination or when all junctions accessible from the origin are scanned.

Environmental influences (top-down procedures): AB models typically have a demand side emphasis characterizing the urban system as the outcome of residents' behavior. In normally functioning markets this means that agents will look for cheaper housing. On the supply side however, contractors will tend to build where prices (profits) are high and thus house prices will vary directly with population and inversely with housing stock. In AB models the supply side is often overlooked. We formulate a top-down process in order to get a fuller picture of the operation of urban markets. We conceptualize administrative units, buildings and individual housing units as quasi-agents. These entities are not autonomous or mobile but are nevertheless sensitive to change in their environment according to pre-defined rules of influence (Torrens 2014). Foremost amongst these are sensitivity to the distribution of commercial activity and to values of buildings. In this process, changes on the demand side (population dynamics), and related changes to the land-use system create effects which trickle down from the level of the administrative unit to the individual dwelling unit and ultimately to the individual resident.

We assume that the larger a commercial function, the more activity it will require in order to be profitable. We further assume that the volume of activity is represented by the intensity of traffic in the vicinity of each function. Consequently, traffic volume on roads near commercial functions should be proportional to their floor-space. Hence, any function that is not located near sufficient traffic will not survive and residential uses near large traffic volumes may change their function to commercial, forcing residents to relocate or migrate.

Changes to a dwelling unit's value represent a top-down spatial mechanism based on supply and demand dynamics. Average price per meter is pre-specified at the beginning of the simulation for each SA, for both residential and non-residential uses. As land-use and population dynamics change, supply and demand shift causing these average prices to fluctuate. We assume that non-residential value is affected only by changes in the supply (short supply drives prices up) while residential values are also sensitive to changes in demand (prices drop when population decreases). Changing average prices cause a change in building values according to floor-space. In the case of residential buildings this change is also effected through accessibility to services. Accessibility relates to the share of non-residential uses from total uses accessible from the building. A building whose accessibility is higher than the SA average will be higher priced. These effects further trickle down to the monthly cost of housing per dwelling unit. We assume uniformity of unit prices within a building meaning that the cost of each unit is the average cost of units in the building. This cost affects household residential choice in the next iteration.

The earthquake: this is formalized as a one-time shock, spreading outwards from a focal point with intensity decaying over distance. This earthquake is programmed to occur on the 51th simulation iteration, so the system has a sufficient 'run-in' period in which all processes initiate themselves and stabilize. The epicenter of the quake is determined randomly so that aggregate average results do not represent any place-based bias. As the impact spreads across space it may inflict damage to the environment. The probability a building being demolished as a result of the shock is proportional to both distance from the epicenter and height. A demolished building causes the nearest road to become unavailable for travel until the building is restored to use. The restoration period is proportional to the building's floor-space volume.

5 Simulation and Results

Results are presented relating to three main themes: land use change, change in value of capital stock and socio-economic change. These reflect three dimensions of urban vulnerability to unanticipated shocks: functional, economic and social vulnerability respectively. The simulation is run 25 times each with a duration of 1000 iterations (i.e. days). The earthquake occurs on day 51 in each simulation and is located randomly within the study area. The 50 day run-in time is heuristically derived, comprising of a 30 day period for stochastic oscillation and a further 20 day period for 'settling down'. Like any catastrophic change, the initial impact of the earthquake is an immediate population loss and damage to buildings and infrastructure. Yet as the results illustrate, these lead to a second, indirect round of impacts induced by the way in which agents react to the new conditions created. Sensitivity tests for key model parameters are presented in Appendix 3.

Traffic Patterns and Land Use Change: we present a series of maps that illustrate change in land use (from residential to commercial and vice versa) and the concomitant dispersal of traffic activity at discrete time points (Figs. 5 and 6). As can be seen, in the period following the disaster the main commercial thoroughfares running N-S and E-W across the area lose their prominence induced by changing movement patterns. Within 50 days of the event, major traffic loads shift from the north-west part of the study area into the south and to a lesser extent to the north-east and central sections. Commercial activity responds to this change and a cluster of medium-sized commercial uses appears in the south-west. However, by day 500 we observe a reversal of this pattern. Evidently, the recovery of the traffic network, along with the anchoring effect of large commercial land uses, helps Agripas St. to regain its position causing a new commercial cluster to develop in its vicinity. The immediate, post disaster change in traffic pattern does however leave a permanent footprint. Commercial land use becomes more prevalent in the north-east and CBD centrality seems to be slightly reduced. The buildings containing these activities have larger floor-space area than those located in the new emerging commercial clusters. This implies a potentially large addition of dwelling units to the residential stock in the case that these buildings transform into residences. One year after the shock these patterns hardly change as the new transportation pattern gets locked in, empty buildings become occupied and the potential for land use change is reduced. From the 3D representation of frequency of land use change (Fig. 6) we can identify the time juncture where buildings that

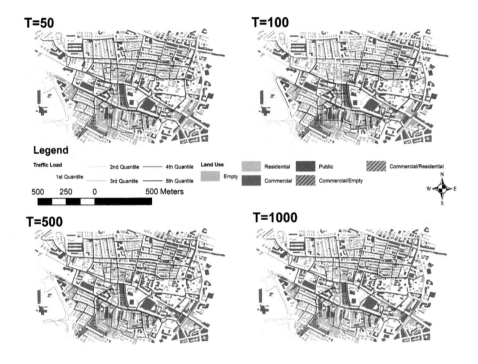

Fig. 5 Average land-use maps at discrete time points

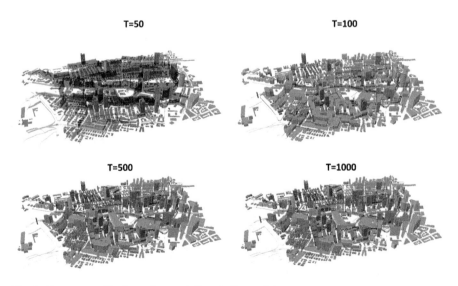

Fig. 6 Frequency of land-use change at discrete time points

were previously empty become occupied. Between day 500 and day 1000 land use tends to become rejuvenated in the sense that unoccupied buildings become populated.

Frequency is represented by building height. Building color represents initial use. Colored section represents the share of total simulations in which the building was in use other than the original use, at that time point. Grey represents the share of times the building was unoccupied. Road height represents absolute traffic volume, while shading represents relative volume within the distribution of all roads.

Change in the Value of Capital Stock: standard urban economic theory suggests that demand for capital stock (residential and non residential) is inversely related to price while the supply of capital stock is positively related to price. In our AB world with dynamic pricing for both residential and non-residential stock, demand changes through population change and supply changes through either building destruction or change in land use as a result of changing traffic loads and accessibility to services. Aggregate simulation results for residential capital stock show that the number of buildings drops to about 600 after about 100 days and in the long run never recovers (Fig. 7). However average residential values tend to rise only after 500 days to about 90 % of their pre-shock values. This lagged recovery may be attributed to supply shortage and increased access to services as suggested by the changing land use patterns noted above along with increasing demand from a growing population. Yet, the fact that a reduced residential stock recovers to almost the same value suggests that this is due to rising average floor-space volumes. This would point to buildings that were initially large commercial spaces becoming residential. Non residential stock behaves rather differently. Long-term increase in stock is accompanied by lower average values. In contrast to residential stock, the turning point in these trends is after roughly one year (Fig. 8). Elsewhere we have identified this with the dispersal of commercial activities from large centers to

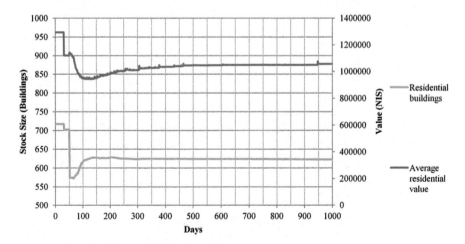

Fig. 7 Change in residential capital stock

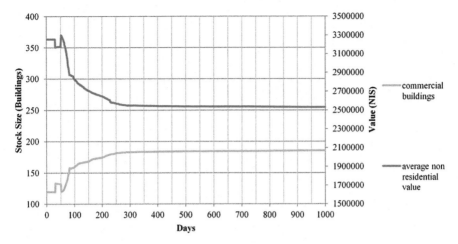

Fig. 8 Change in non-residential capital stock

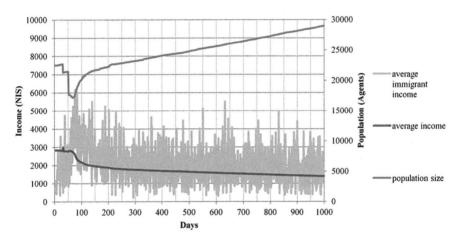

Fig. 9 Population and income change

smaller neighborhood units (Grinberger and Felsenstein 2014). The current results reflect a similar picture. The number of units grows but their average value decreases as the buildings in the south-west and north-east have smaller floor space.

Population Dynamics: The initial impact causes a population loss of about 4000 residents (Fig. 9). After about one year population size recovers and continues to grow to about 29,000 by the end of the simulation period. This increase is the result of in-migration of an heterogeneous population with stochastic earnings. The ability of this extra population to find residence in the area is due to the process of land use change described above. The new, large residential buildings (previously commercial spaces) contain many individual dwelling units. While the

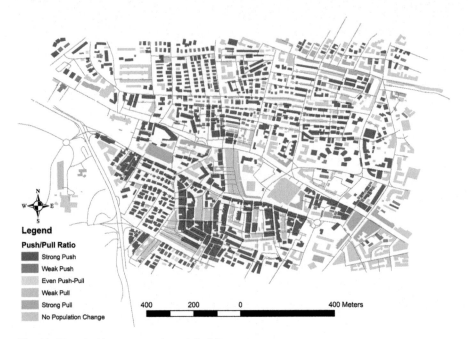

Fig. 10 Household movement through buildings

average price of buildings rises, the rising average floor-space of buildings pushes down the average cost per dwelling unit within the building and consequently the monthly cost of housing services. As a result, lower income households find suitable lower cost residential location making for an increased population that is poorer on average. Since the average income of new in-migrants is slightly higher than the average income of the area, this suggests that the lower average income must result from the out-migration of wealthier households that accompanies in-migration.

A composite indicator of both social, functional and economic vulnerability can be obtained by looking at the flow-through of households through buildings. The simple ratio of in-coming to out-going households per building at each discrete time step, gives an indication of the amount of through-traffic per building and an indication of its population stability. Figure 10 gives a summary account of this ratio. A high 'pull' factor is not necessarily a sign of stability or even attractiveness. It may be an indicator of transience and instability. The overall picture is one of unstable population dynamics. The simulations suggest that none of the buildings that were initially residential are consistently attractive to population. Most of them have difficulty maintaining their population size, post earthquake. For many buildings this is due to either physical damage or change in function for example, from residential to commercial use. It seems that only new potential residential spaces that start initially as commercial uses, consistently succeed in attracting population. The direct and indirect effects of the shock generate much household turnover

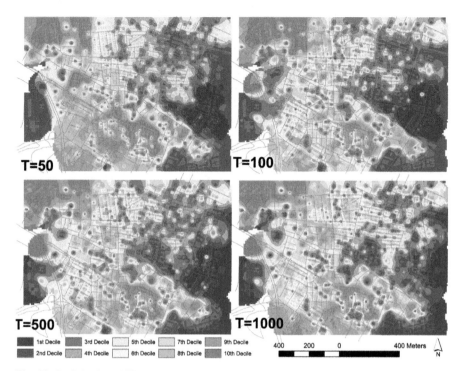

Fig. 11 Social vulnerability heat map

(or 'churning') through buildings but without any indications of location preferences for specific buildings or parts of the study area. Total floor space that registers net positive household movement (strong/weak pull) amounts to only 75 % of the floor space that registers net negative turnover (strong/weak push). This underscores the high state of flux in the study area.

Social Vulnerability: a more in-depth examination of population movement is presented in Fig. 11 where snapshots of the spatial distribution of social vulnerability at different time steps are presented. We follow Lichter and Felsenstein (2012) and use a composite index of social vulnerability.[6] Green indicates less vulnerable population and red more vulnerable. The average index value for each building is calculated, disaggregated into households and used to generate continuous value surfaces using Inverse Distance Weighting (IDW). The parameters used for the interpolation are: pixels of 10×10 m, 100 m search radius and a second order power function.

---

[6] Social vulnerability by household (V_{hh}) is defined as: $V_{hh} = 0.5 * Z_{i_{hh}} - 0.2 * Z_{age_{hh}} - 0.2 * Z_{\%dis_{hh}} + 0.1 * Z_{car}$ where: Z is the normalized value of a variable, i is household income for household hh, age is the average age group of members of household hh, $\%dis$ is the percent of disabled members of all members in household hh, car is car ownership for household hh.

Akin to movement patterns, residence patterns in day 100 present a process of dispersal as relatively less vulnerable households initially clustered in the west move eastward. This population is higher income and consequently has greater availability of resources for relocation. Since agents are characterized by limited tolerance to change it is not surprising to find that the destinations chosen by households are those that were better-off in the pre-event situation (light green areas). Residential patterns re-cluster after day 100, a process similar to the stabilization of traffic and land-use patterns. However, this re-clustering germinates a new spatial pattern with a small cluster of stronger population serving as nuclei for future agglomeration. These clusters attract households of similar characteristics resulting in the emergence of new clusters and a general process of entrenchment. This process occurs at a high spatial resolution that is much more granular than the SA level. Visualizing at this scale prevents the homogenization of patterns. For example, we can detect dispersion over time of the high vulnerability population as the east-west vulnerability patterns in the pre shock era becomes replaced by a much more heterogeneous picture.

6 Web-Based Delivery of Outputs

The richness of simulation outputs is hard to communicate. Displaying spatio-temporal data in a web-based framework is an on-going challenge in the visualization of simulated big data (Keon et al. 2014). This sometimes calls for re-aggregation of the data in order to effectively represent complex dynamics. While it can be claimed that aggregation undermines the efforts to create synthetic big data, in the context of agents there seems to be no alternative to understanding micro-behavior in the absence of disaggregation. Furthermore, disaggregation enables analysis using meaningful spatial units rather than those dictated by administrative convenience (such as SA's that by definition smokescreen inter-zonal variance).

The volume and variety of the big data used as input into the model is manifested in the complexity of these outputs. Not only are these produced for variables at different spatial units but they are also generated temporally for each variable. These dynamics are typically non-linear and therefore cognitively problematic to convey. The nature of the information generated therefore necessitates a visualization technique capable of accommodating large amounts of data and presenting them spatially and temporally. This in turn means sophisticated database construction that will allow not only the dynamic representation of data but also the querying of data in a user-friendly fashion. We view this as a crucial linkage between the simulation outputs generated in a research oriented environment and their use by potential consumers such as planners, decision makers and the informed public.

The results are presented via a web-based application. This allows for communicating outcomes in a non-threatening and intuitive way (see http://ccg.huji.ac.il/AgentBasedUrbanDisaster/index.html). Using a web browser, users can generate

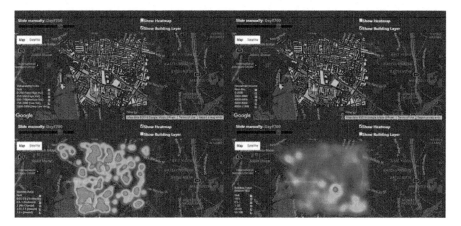

Fig. 12 Time lapse visualization of various household socio-economic attributes on a dynamic web-map (see http://ccg.huji.ac.il/AgentBasedUrbanDisaster/index.html)

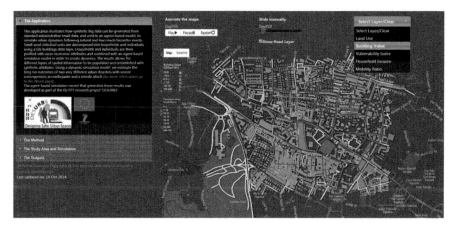

Fig. 13 Time lapse visualization of the change in the number of passengers along roads on a dynamic web-map (see http://ccg.huji.ac.il/AgentBasedUrbanDisaster/index.html)

results for different points in time and for different areas without prior knowledge of handling spatial data or GIS. They can choose a variable of interest, visualize its change over time and space and generate the relevant location specific information they need. To this end, we create a dedicated database for the output results of time series from the model simulation. This needs to be carefully constructed and sometimes does not follow strict DB design but rather contains some flat tables of lateral data in order to be displayed in pop-ups graphs and charts. The visualization includes time lapse representation of human mobility (household level), changes in passengers along roads, changes in buildings' land use and value, household socio-economic attributes change etc. in the study area (Figs. 12 and 13).

We use Google Maps API as the mapping platform and add middleware functionalities. These are functions that are not provided by the mapping API but interact with it to provide ancillary capabilities (Batty et al. 2010) such as time laps animation, sliders, interactive graphs etc. These middleware functionalities are User Interface (UI) features that allow for different ways of data querying and interactive engagement, using a variety of JavaScript libraries and API's. Utilizing this mashup of visualization tools is merely the final stage in the development of the web-map. It is preceded first by extensive data analysis and manipulation of vast model output spatial data and second by creating a dedicated database in order to allow easy, intuitive and sometimes lateral querying of the database.

7 Conclusions

This paper makes both a methodological and substantive contribution to the study of urban dynamics using recent advances in urban informatics and modeling. In terms of method we illustrate how an agent based model can be coupled with a data disaggregation process in order to produce synthetic big data with accurate socio-economic profiling. This fusion adds both temporal and spatial dynamics to the data. The simulation model uniquely treats the built environment as a quasi-agent in urban growth. Consequently, more attention is paid to the supply side dynamics of urban change than generally practiced and the result is a modeling system with dynamic pricing and an active supply side. We also illustrate how outputs can be suitably communicated to practitioners and the informed public. Dynamic web-based mapping is used to enhance civic engagement and public awareness as to the possible implications of a large scale exogenous shock.

On the substantive side the results of the simulation highlight some interesting urban processes at work. We speculate about three of these and their implications for the ability of cities to rejuvenate in the aftermath of an unanticipated event. The first relates to the permanent effects of temporary shocks. Our results have shown that temporary shocks to movement and traffic patterns can generate longer term lock-in effects. In our simulations these have a structural effect on reduction of CBD commercial activity. The issue arising here is the ability to identify when this fossilization takes place and when a temporary shock has passed the point of no return.

The second process relates to the large level of household turnover and 'churning' through the built fabric of the city in the aftermath of an earthquake. Obviously, a traumatic event serves to undermine population stability as housing stock is destroyed and citizens have to find alternative residence. However, high turnover levels of buildings point to a waste of resources, material, human and emotional. In other markets such as the labor market, 'churning' might be considered a positive feature pointing to economic and occupational mobility (Schettkat 1996). However, in the context of a disaster, this would seem to be a process that judicious public policy should attempt to minimize. The welfare costs of the effort needed to search for new accommodation and the dislocation associated with changing place of residence are likely fall hardest on weaker and more vulnerable populations (Felsenstein and Lichter 2014).

Finally, our findings shed new light on the familiar concept of 'urban vulnerability'. Our simulated results show that less vulnerable socio-economic groups 'weather the storm' by dispersing and then re-clustering over time. This points to their higher adaptive capacities. Stronger populations have the resources to accommodate the negative impacts of a disaster. Urban vulnerability is thus as much as an economic welfare issue as it an engineering or morphological concept. From a socioeconomic perspective, it is not the magnitude of the event that is important but the ability to cope with its results. This makes urban vulnerability a relative term: a shock of a given magnitude will affect diverse population groups differentially. Vulnerable populations or communities can be disproportionately affected by unanticipated disasters which are more likely to push them into crisis relative to the general population. Much of this can only be detected at the micro level such as the household. It is often smoke-screened in studies dealing with aggregate city-wide impacts. The use of highly disaggregated and accurately profiled data would seem to be critical in understanding urban vulnerability.

Acknowledgements This research is partially based on work done in the DESURBS (Designing Safer Urban Spaces) research project funded by the European Commission FP7 Program under Grant Agreement # 261652. The authors thank the JTMT for granting access to the HTS database.

Appendix 1: Data Disaggregation Method

This appendix describes the disaggregation procedure of spatially aggregated alpha-numeric real-estate values and populations and their socio-economic attributes into discrete spatial units at the building level. It then proceeds to describe the validation of this allocation method.

The number of floors in residential buildings (F_R), is calculated by dividing building height by average floor height of 5 m:

$$F_R = \frac{H_B}{5}$$

In the case of non-residential buildings, the number of floors (F_N) is estimated as the building height divided by average floor height of 5 m:

$$F_N = \frac{H_B}{7}$$

Floor space for each building (S_B) is then calculated by multiplying the number of floors in each building by its polygon area representing roof space:

$$S_B = S_R \times F$$

where:
S_R = Building polygon footprint
F = Building number of floors

The GIS buildings layer and building type serve as the basis for the calculation of residential building value, non-residential building and equipment value. To create estimates of residential building value we use average house prices per m^2 2008–2013 (in real 2009 prices). In cases where no transactions exist in a specific SA over that period we use a regional estimate for residential property prices.

1. *Value of residential buildings (P_{BR})* is calculated as follows:

$$P_{BR} = P_{SR} \times S_{BR}$$

where:
P_{SR} = Average SA price per m^2
S_{BR} = Residential building floor space.

2. *Value of Non residential buildings* is calculated as follows:
   Non residential value per m^2 by region (P_{RN}):

$$P_{RN} = \frac{V_{RN}}{S_{RN}}$$

where:
S_{RN} = Total regional non residential floor space.
V_{RN} = Total regional non residential building stock

Non residential building value per m^2 for each region is multiplied by the floor space of each non-residential building to produce non-residential building values (P_{BN}):

$$P_{BN} = P_{RN} \times S_{BN}$$

where:
S_{BN} = non residential building floor space.

Regional non-residential stock estimates have been calculated for nine aggregate regions elsewhere (Beenstock et al. 2011).

3. *Value of Equipment and Machinery (P_{RE})* is calculated as follows:

$$P_{RE} = \frac{V_{RE}}{S_{RN}}$$

where:
V_{RE} = Total regional non residential equipment stock

The equipment stock per m^2 for each region is multiplied by the floor space of each non-residential building to produce equipment stock totals by building (P_{BE}):

$$P_{BE} = P_{RE} \times S_{BN}$$

where:
S_{BN} = non residential building floor space.

The source for regional estimates of regional equipment and machinery stock is as above (Beenstock et al. 2011).

The buildings layer also allows for the spatial allocation of aggregated households, and population counts (see Table 1) into building level households and inhabitant totals. Given the existence of these spatial estimates, the distribution of aggregate average monthly earnings, participation in the work force, employment sector, disabilities and age (see Table 1) into a building level distribution is implemented.

4. *Household density by SA (households per m^2)* of residential floor space in a statistical area (H_{SR}) is calculated as follows:

$$H_{SR} = \frac{H_S}{S_{SR}}$$

where:
H_S = Total population per statistical area (IV in Table 1).
S_{SR} = Total statistical area residential floor space.

The number of households per building (H_B) is calculated as follows:

$$H_B = H_{SR} \times S_{BR}$$

5. *Average number of inhabitants per m^2* of residential floor space in a statistical area (I_{SR}) is calculated as follows:

$$I_{SR} = \frac{I_S}{S_{SR}}$$

where:
S_{SR} = Total statistical area residential floor space.
I_S = Total population per statistical area.

Population counts per building (I_B) are then calculated as follows:

$$I_B = I_{SR} \times S_{BR}$$

6. *Total earnings per building (M_B)* is calculated as follows:

$$M_B = M_{SI} \times H_B$$

where:
M_{SI} = Average monthly earnings per household by SA.
H_B = Total number of households in a building.

7. *Number of inhabitants in each building participating in the labor force 2008* (I_W) is calculated by multiplying the number of inhabitants in a building by the labor participation rate in the corresponding SA.

$$I_W = W_S \times I_B$$

where:
$W_S = \%$ of inhabitants participating in the labor force in an SA
$I_B =$ Population count per building

8. *Number of inhabitants per building by employment sector* (I_O) is calculated by multiplying the percentage of inhabitants employed by sector (commercial, governmental, industrial or home-based) per statistical area by the number of inhabitants in each building.

$$I_O = O_S \times I_B$$

where:
$O_S = \%$ of inhabitants employed in an employment category.
$I_B =$ Population counts per building

9. *Number of disabled inhabitants in each building* (I_D) is calculated by multiplying the number of inhabitants in a building by the percentage of disabled in the corresponding SA.

$$I_D = D_S \times I_B$$

where:
$D_S = \%$ of disabled inhabitants in an SA
$I_B =$ Population count per building

10. *Number of inhabitants in each age category* (I_A) is calculated by multiplying the number of inhabitants in a building by the percentage of inhabitants in each age category in the corresponding SA.

$$I_A = A_S \times I_B$$

where:
$A_S = \%$ of inhabitants in each age group category in an SA
$I_B =$ Population count per building.

Validation of Data Disaggregation

To validate the data disaggregation procedure, we compare population counts and demographic attributes of our synthetic distribution with the Household Travel Survey (HTS) conducted in 2010 by the Jerusalem Transport Masterplan Team (JTMT; Oliveira et al. 2011). The idea behind the validation procedure is to re-aggregate the discrete counts yielded by our distribution to the spatial units used by the survey. The survey uses Aggregated Transportation Area Zones (ATAZs) which have a different spatial configuration to those of the SAs on which our distribution is predicated.

Table 2 Validation results for key variables

		Fully sampled ATAZs		ATAZs sampling rate >70%	
Variable		Number of ATAZs	MAPE (%)	Number of ATAZs	MAPE (%)
Population size		39	19.16	65	18.71
Number of households		39	19.74	65	18.89
Population size by age	0–18	39	25.09	65	27.08
	19–64	39	22.65	65	22.83
Number of cars		39	24.14	65	25.59

The HTS is based on representative samples of the population in the Jerusalem metropolitan area. It uses an expansion factor by which the population and attributes of each household are multiplied to replicate their representation in the ATAZs.

ATAZs covered by the HTS but not fully covered by our distribution (for example, ATAZs in East Jerusalem) are excluded from the analysis. This is because the HTS data are aggregates and we cannot account for their distribution within the ATAZ, as we can with our discrete data. Note also that some ATAZs are fully sampled in the HTS and others contain much lower sampling coverage. This can cause misrepresentation of populations and demographics in some ATAZs. This is taken into account in the validation procedure. Validation is performed for the following attributes: number of households, number of persons, persons ages 0–18, 19–64 and number of cars.

Table 2 presents the validation results. MAPE statistics are calculated for two subsets of the original ATAZ dataset. Results are compared between our synthetic distribution with those ATAZs with complete coverage and those with a sampling coverage of 70+%. Mean errors across the two sources are within the range of 18–25 %.

Appendix 2: Behavioral Rules for the ABM

1. *Residential Location Choice* is derived as follows:

$$h_h = b_j \Rightarrow \left[\frac{I_h}{3} > HP_j\right] * [k_h > S(b_j)] = 1$$

where:
h_h is the new residential location for household h,
b_j is the building considered,
[] is a binary expression with value of 1 if true and 0 otherwise,
I_h is household h's monthly income,
HP_j is monthly housing cost of an average apartment in building j,

k_h is a random number between [0,1] indicating tolerance to change in residential environment incurred by relocation,
$S(b_j)$ is a similarity score for building j in relation to current place of residence, calculated as follows:

$$S(b_j) = \frac{\Phi\left(\frac{\bar{I}_j - \bar{I}_h}{I_{\sigma_h}}\right) + \Phi\left(\frac{\bar{A}_j - \bar{A}_h}{A_{\sigma_h}}\right)}{2}$$

where:
Φ is the standard normal cumulative probability function,
$\bar{I}_j, \bar{A}_j$ are the average household income and average age of individuals in building j, respectively
$\bar{I}_h, \bar{A}_h$ are average household income and average age of individuals in residential buildings within 100 m of current residential location of household h,
$I_{\sigma_h}, A_{\sigma_h}$ are standard deviations of household income and of resident age in residential buildings within 100 m from current home location of household h, respectively.

2. *Choice of sequence of activities*: occurs in two stages. First, the number of activities is fixed and then activities are allocated to locations:

$$\#Ac_i = \left\| a * \left(\frac{k_i}{0.5}\right) * (1 + car_h * 0.33) * (1 - dis_i * 0.33) * (1 + [age_i = 2] * 0.33) \right.$$
$$\left. * (1 - [age_i \neq 2] * 0.33) \right\| + employed_i * here_i$$

where:
$\#Ac_i$ is the number of activities for resident i,
k_i is a randomly drawn number between [0, 1] reflecting preferences regarding number of activities,
car_h is a binary variable equal to 1 if the household h owns a car and 0 otherwise,
dis_i is a binary variable equal to 1 if individual i is disabled and 0 otherwise,
age_i is the age group of individual i,
$employed_i$ is a binary variable equal to 1 if i is employed and 0 otherwise,
$here_i$ is a binary variable equal to 1 when i's workplace is located within the study area and 0 otherwise,
$\|x\|$ indicates the nearest integer number to x,
a is the average number of activities based on employment status; equals 2.5 for employed residents and 3 for non-employed.

$$a_{t+1,i} = b_j \Rightarrow \lfloor b_j \neq a_{t,i} \rfloor * \lfloor k_i \geq Att(b_j) \rfloor = 1$$

where:
$a_{t,i}$ is the current location of individual i,
$a_{t+1,i}$ is the next location of activity of individual i,

k_i is a randomly drawn number between [0, 1] reflecting activity location preferences,

$Att(b_j)$ is the attractiveness score for building j, calculated as follows:

$$Att(b_j) = \frac{1 - \Sigma E_j/\Sigma B_j + 1 - D_{ij}/\max D_i * (1 + 0.33*(-car_h + dis_i + [age_i = 3])) + [LU_j = nonRes] * FS_j/\max FS}{2 + [LU_j = nonRes]}$$

where:

ΣE_j is the number of non occupied buildings within a 100 m buffer of building j,
ΣB_j is the number of all buildings within a 100 m buffer of building j,
D_{ij} is the distance of building j from the current location of individual i,
$\max D_i$ is the distance of the building within the study area furthest away from the current location of individual i,
LU_j is the land-use of building j,
$nonRes$ is non-residential use,
FS_j is the floor-space volume of building j,
$maxFS$ is the floor-space volume of the largest non-residential building within the study area.

3. *Choice of workplace location* is calculated similarly to the choice of activity location:

$$WP_i = b_j \Rightarrow [LU_j = ELU_i] * \left[k_i > \frac{D_{ij}/\max D_i + 1 - FS_j/maxFS}{2} \right] = 1$$

where:
WP_i is the workplace location of individual i,
ELU_i is the employment-sector-related land-use for individual i,
k_i is a randomly drawn number between [0, 1] representing workplace location preferences,
D_{ij} is the distance between building j and individual i's place of residence,
$\max D_i$ is the distance of the building within the study area furthest away from individual i's place of residence.

4. *Building values and the monthly cost of living in a dwelling unit* are derived in a 3-stage process. First, daily change in average house price per SA is calculated. Then, values of individual buildings are derived and finally the price of the single, average dwelling unit is calculated. For non-residential buildings, the calculation of individual building values is similar.

$$AHP_{z,t+1} = AHP_{z,t} * \left(1 + \log\left(\frac{pop_{z,t+1}/pop_{z,t} + res_{z,t}/res_{z,t+1} + nRes_{z,t+1}/nRes_{z,t}}{3} \right) \right)$$

$$ANRV_{z,t+1} = ANRV_{z,t} * \left(1 + \log\left(\frac{nRes_{z,t}}{nRes_{z,t+1}} \right) \right)$$

where:
$AHP_{z,t}$ is average housing price per meter in SA z at time t,
$pop_{z,t}$ is population in SA z at time t,
$res_{z,t}$ is the number of residential buildings in SA z at time t,
$nRes_{z,t}$ is the number of non-residential buildings in SA z at time t,
$ANRV_{z,t}$ is the average non-residential value per meter in SA z at time t,

$$HP_{j,t} = AHP_{z,t} * FS_j * {}^{SL_{j,t}}/_{SL_{z,t}}$$
$$V_{j,t} = ANRV_{z,t} * FS_j$$

where:
$HP_{j,t}$ is the house price of a dwelling unit in building j at time t,
$SL_{s,t}$ is the service level within area s at time t—the ratio of non-residential buildings to residential buildings in this perimeter,
$V_{j,t}$ is the non-residential value of building j.

$$P_{du,t} = \frac{\bar{I}_t * \left(1 + \frac{HP_{j,t}/\Sigma Ap_j - \sum_{l=1}^{L_t} HP_{l,t} / \sum_{l=1}^{L_t} \Sigma Ap_l}{P_{\sigma_t}}\right)}{c}$$

where:
$P_{du,t}$ is the monthly cost of living in dwelling unit du at time t,
$\bar{I}_t$ is the average household income in the study area at time t,
ΣAp is the number of dwelling units within a building. If the building is initially of residential use, this is equal to its initial population size, otherwise it is the floor-space volume of the building divided by 90 (assumed to be average dwelling unit size in meters),
L_t is the number of residential buildings in the study area at time t,
P_{σ_t} is the standard deviation of dwelling unit prices within the study area at time t,
c is a constant.

5. *Land-use changes*, from residential to commercial and from commercial to unoccupied are based on the congruence between the building floor-space volume and the average intensity of traffic on roads within a 100 m radius over the preceding 30 days. Both these values are compared with the (assumed) exponential distribution of all values in the study area. This is done by computing the logistic probability of the relative difference in their locations in the distribution:

$$P_{j,t}(\Delta x_{j,t}) = \frac{e^{-\Delta x_{j,t}}}{1 + e^{-\Delta x_{j,t}}}$$

$$\Delta x_{j,t} = \frac{z_{TR_{j,t}} - z_{FS_{t,t}}}{|z_{FS}|}$$

$$z_{y_{t,j}} = \frac{e^{-y_{j,t}/\bar{y}_t} - e^{-y_{med_t}/\bar{y}_t}}{\bar{y}_t}$$

where:
$P_{j,t}$ is the probability of land-use change for building j at time t,
$\Delta x_{j,t}$ is the relative difference in position of traffic load and floor-space for building j at time t,
$z_{yj,t}$ is the position of value y in the exponential distribution, relative to the median for building j at time t,
$\frac{e^{-y_{j,t}/\bar{y}_t}}{\bar{y}_t}$ is the exponential probability density value for $y\left(\frac{1}{\bar{y}_t} = \hat{\lambda}_t\right)$ for building j at time t,
y_{med_t} is the median of y at time t.

If $P > 0.99$ for residential use, it changes to commercial. If the value is in the range $[P(1) - 0.01, P(1)]$ for commercial uses, the building becomes unoccupied. This functional form and criteria values reduce the sensitivity of large commercial uses and small residential uses to traffic volume. Consequently, the process of traffic-related land-use change is not biased by a tendency to inflate initial land uses.

6. *Earthquake impact* is calculated as follows:

$$Im_j = \frac{c * 10^{mag}}{D_j * |\log(D_j)| * F_j}$$

where:
Im_j is the impact building j suffers, c is a constant, mag is the earthquake magnitude (similar to Richter scale), D_j is distance of building j from the earthquake epicenter, F_j is number of floors in building j.

Appendix 3: Sensitivity Analysis

The hard-to-predict nature of model dynamics is a substantive issue in agent-based simulation (Ngo and See 2012). To test for parameter stability, we test two key model parameters:

1. Land-use sensitivity ('lu')—in this scenario, the sensitivity of actual and potential commercial land-use to traffic volumes is set in relation the nearest road and not all roads within 100 m as in the baseline case.
2. Housing budget distribution ('hb')—instead of defining a rigid parameter share for housing service costs out of total monthly income (0.33), variation within the

willingness to pay for housing parameter is allowed. A unique value is drawn for each household from a normal distribution centered on the average expenditure on housing in Israel (23.4 % of total income), see http://www1.cbs.gov.il/reader/newhodaot/hodaa_template_eng.html?hodaa=201415290.

Each scenario is simulated 25 times with extreme results discarded (two cases for the lu scenario and one for the hb scenario). All other parameters remain unchanged. To obtain high resolution results we compute the most frequent land-use and the average residential value, non-residential value and vulnerability index for each building at discrete time points. For land-use we compute the share of times a different land-use is registered in relation to the baseline scenario. For other variables we compute the Median Absolute Percentage Error (MAPE). The results are presented in Fig. 14.

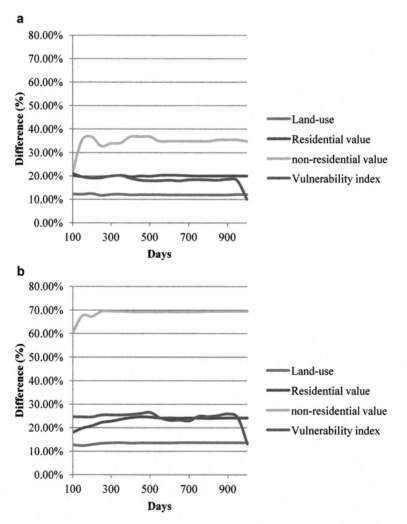

Fig. 14 Sensitivity tests for (**a**) lu scenario and (**b**) hb scenario.

The results illustrate morphological stability with the same land-use over scenario registered for almost 90 % of the buildings at all time points. The other variables also exhibit parameter stability (which is greater in the lu scenario). The only sensitive variable is average non-residential value. This is due to two changes exerting influence on the spatial distribution of commercial functions. The lu scenario constrains agglomeration tendencies and the hb scenario affects residential, and hence traffic, patterns. As commercial stock is small relative to residential stock these changes do not exhibit strong morphological influence but register greater and more fluctuating differences than in the other variables. While general patterns indicate parameter stability, micro-level differences are still observed.

References

Batty M, Hudson-Smith A, Milton R, Crooks A (2010) Map mashups, Web 2.0 and the GIS revolution. Ann GIS 16(1):1–13

Beenstock M, Felsenstein D, Ben Zeev NB (2011) Capital deepening and regional inequality: an empirical analysis. Ann Reg Sci 47:599–617

Chang SE (2003) Transportation planning for disasters: an accessibility approach. Environ Plan A 35(6):1051–1072

Chen X, Zhan FB (2008) Agent-based modeling and simulation of urban evacuation: relative effectiveness of simultaneous and staged evacuation strategies. J Oper Res Soc 59:25–33

Chen X, Meaker JW, Zhan FB (2006) Agent-based modeling and analysis of hurricane evacuation procedures for the Florida Keys. Nat Hazards 38:321–338

Crooks AT, Wise S (2013) GIS and agent based models for humanitarian assistance. Comput Environ Urban Syst 41:100–111

Dawson RJ, Peppe R, Wang M (2011) An agent-based model for risk-based flood incident management. Nat Hazards 59:167–189

Eicher CL, Brewer CA (2001) Dasymetric mapping and areal interpolation: implementation and evaluation. Cartogr Geogr Inf Sci 28(2):125–138

Felsenstein D, Lichter M (2014) Social and economic vulnerability of coastal communities to sea level rise and extreme flooding. Nat Hazards 71:463–491

Fiedrich F, Burghardt P (2007) Agent-based systems for disaster management. Commun ACM 50 (3):41–42

Grinberger AY, Felsenstein D (2014) Bouncing back or bouncing forward? Simulating urban resilience. Urban Des Plan 167(3):115–124

Grinberger AY, Felsenstein D (2017) A tale of two earthquakes: dynamic agent-based simulation of urban resilience. In: Lombard J, Stern E, Clarke G (eds) Applied spatial modeling and planning. Routledge, Abingdon, forthcoming

Harper G, Mayhew L (2012a) Using administrative data to count local populations. J Appl Spat Anal Policy 5(2):97–122

Harper G, Mayhew L (2012b) Re-thinking households—using administrative data to count and classify households with some applications. Actuarial research paper, no. 198. Cass Business School, London

Keon D, Steinberg B, Yeh H, Pancake CM, Wright D (2014) Web-based spatiotemporal simulation modeling and visualization of tsunami inundation and potential human response. Int J Geogr Inf Sci 28(5):987–1009

Kwan MP, Lee J (2005) Emergency response after 9/11: the potential of realtime 3D GIS for quick emergency response in micro-spatial environments. Comput Environ Urban Syst 29:93–113

Lichter M, Felsenstein D (2012) Assessing the costs of sea level rise and extreme flooding at the local level: a GIS-based approach. Ocean Coast Manag 59:47–62

Linard C, Gilbert M, Tatem AJ (2011) Assessing the use of global land cover data for guiding large area population distribution modeling. GeoJournal 76(5):525–538

Macal CM, North MJ (2005) Tutorial on agent-based modeling and simulation. In: Proceedings of the 37th conference on winter simulation, WSC'05. pp 2–15

Mennis J (2003) Generating surface model of population using dasymetric mapping. Prof Geogr 55(1):31–42

Ngo TA, See L (2012) Calibration and validation of agent based models of land cover change. In: Heppenstall AJ, Crooks AT, See LM, Batty M (eds) Agent-based model of geographical systems. Springer, Dordrecht, pp 191–199

Ogawa Y, Akiyama Y, Kanasugi H, Sengoku H, Shibasaki R (2015) Evaluating the damage of great earthquakes in aggregate units based on detailed population distribution for each time frame. Paper presented at the 14th international conference on computers in urban planning and urban management (CUPUM 2015), Boston MA. http://web.mit.edu/cron/project/CUPUM2015/proceedings/Content/index.html

Oliveira MGS, Vovsha P, Wolf J, Birotker Y, Givon D, Paasche J (2011) Global positioning system-assisted prompted recall household travel survey to support development of advanced travel model in Jerusalem, Israel. Transp Res Rec J Transp Res Board 2246(1):16–23

Rashed T, Weeks J, Couclelis H, Herold M (2007) An integrative GIS and remote sensing model for place-based urban vulnerability analysis. In: Mesev V (ed) Integration of GIS and remote sensing. Wiley, New York, pp 199–231

Reibel M, Bufalino ME (2005) Street-weighted interpolation techniques for demographic count estimation in incompatible zone systems. Environ Plan A 37:127–139

Salamon A, Katz O, Crouvi O (2010) Zones of required investigation for earthquake-related hazards in Jerusalem. Nat Hazards 53(2):375–406

Schettkat R (1996) Flows in labor markets: concepts and international comparative results. In: Schettkat R (ed) The flow analysis of labor markets. Routledge, London, pp 14–36

Simoes JA (2012) An agent based/ network approach to spatial epidemics. In: Heppenstall AJ, Crooks AT, See LM, Batty M (eds) Agent-based models of geographical systems. Springer, Dordrecht, pp 591–610

Torrens PM (2014) High-resolution space-time processes for agents at the built-human interface of urban earthquakes. Int J Geogr Inf Sci 28(5):964–986

Zhu Y, Ferreira J Jr (2015) Data integration to create large-scale spatially detailed synthetic populations. In: Geertman S, Ferreira J Jr, Goodspeed R, Stillwell J (eds) Planning support systems and smart cities. Springer, Dordrecht, pp 121–141

Zou Y, Torrens PM, Ghanem RG, Kevrekidis IG (2012) Accelerating agent-based computation of complex urban systems. Int J Geogr Inf Sci 26(10):1917–1937

Estimation of Urban Transport Accessibility at the Spatial Resolution of an Individual Traveler

Itzhak Benenson, Eran Ben-Elia, Yodan Rofe, and Amit Rosental

Abstract Accessibility, particularly for public transport users is an important consideration in sustainable mobility policies. Various accessibility measures have been suggested in the literature, most at coarse aggregate spatial resolution of zones or neighborhoods. Based on recently available Big Urban GIS data our aim is to measure accessibility from the viewpoint of an individual traveler who traverses the transportation network from one building as origin to another at the destination. We estimate transport accessibility by car and by public transport based on mode-specific travel times and corresponding paths, including walking and waiting. A computational application that is based on the intensive querying of relational database management systems is developed to construct high-resolution accessibility maps for an entire metropolitan area. It is tested and implemented in a case study involving the evaluation of a new light rail line in the metropolitan area of Tel Aviv. The results show essential dependence of accessibility estimates on spatial resolution—high-resolution representations of the trip enable unbiased estimates. Specifically, we demonstrate that the contribution of the LRT to accessibility is overrated at low resolutions and for longer journeys. The new approach and fast computational method can be employed for investigating the distributional effects of transportation infrastructure investments and, further, for interactive planning of the urban transport network.

Keywords Accessibility • Public transport • Big data • Graph database • GIS • High-resolution spatial analysis

---

I. Benenson (✉) • A. Rosental
Department of Geography and Human Environment, Tel Aviv University, Tel Aviv, Israel
e-mail: bennya@post.tau.ac.il; amittour@gmail.com

E. Ben-Elia
Department of Geography and Environmental Development, Ben-Gurion University of the Negev, Beersheba, Israel
e-mail: benelia@bgu.ac.il

Y. Rofe
Switzerland Institute for Drylands Environmental and Energy Research, Ben-Gurion University of the Negev, Beersheba, Israel
e-mail: yrofe@bgu.ac.il

1 Introduction

Accessibility is a concept used in a number of scientific fields such as transport planning, urban planning, economics, and geography. In its simplest form accessibility is the ability of people to reach necessary or desired activities using the available transportation modes (Geurs and Ritsema van Eck 2001; Garb and Levine 2002; Handy and Niemeier 1997) and is regarded as a key criterion to assess the quality of transport policy and land use development (Kenyon et al. 2002; Bristow et al. 2009). Accessibility indicators are used to evaluate the contribution of transportation investments to enable the mobility of people (labor) and goods (products) and hence to an efficient functioning of the economy.

Accessibility is an essential yardstick for evaluating the three common pillars comprising sustainable development: economic development, environmental quality and social equity: (Bruinsma et al. 1990; Bertolini et al. 2005; Kwok and Yeh 2004; Feitelson 2002). The social/environmental justice dimension of sustainable development, in turn, draws the attention towards the distribution of accrued benefits and burdens over different members of society. Accessibility can then be used as an indicator of the extent to which all groups can participate in activities considered 'normal' to their society, such as access to employment and essential services (Farrington and Farrington 2005; Martens 2012; Lucas 2012). For all these dimensions, accessibility is thus a key policy indicator, and accessibility measures are a necessary prerequisite for adequate urban planning and transportation policy.

While there is no disagreement as to the importance of addressing accessibility in urban planning goals and policy, accessibility is often misunderstood, poorly defined and poorly measured. The literature presents a wide range of approaches to actually measure accessibility. These vary according to different methods for computing travel times or travel distance, combination of and comparison between transportation modes, spatial scale of analysis, and accounting for network details. Due to these differences, accessibility measures have been broadly categorized as either place-based or person-based. Geurs and van Wee (2004) define four, essentially overlapping, types of accessibility measures: (a) Infrastructure-based—that estimate service level of transport infrastructure (e.g. "the average travel speed on the road network"); Location-based—estimating accessibility of facilities starting/finishing from a certain location (e.g. "the number of jobs within 30 min. travel from origin locations"); Person-based—estimating an individual's advantages of accessibility (e.g., "the number of activities in which an individual can participate at a given time"); and Utility-based—estimating the economic benefits that people derive from access to destinations. A less overlapping categorization is provided by Liu and Zhu (2004): (a) Opportunity-based measures that are based on the number of destinations available within a certain distance/time from origin (e.g. Benenson et al. 2011; Mavoa et al. 2012; O'Sullivan et al. 2000; Witten et al. 2003, 2011; Ferguson et al. 2013; Martin et al. 2008). (b) Gravity-type measures (potential models) that refer to the potential of opportunity between two places (e.g. Alam et al. 2010; Minocha et al. 2008; Grengs et al. 2010). (c) Utility-based measures that relate accessibility to the notion of consumer surplus and net benefits to the users of the transportation system (e.g. Ben-Akiva and Lerman 1979). (d) Measures of the spatio-temporal fit—emphasize the range and frequency of the activities in which a person takes part and whether it is possible to sequence them so that all can be undertaken within given space-time constraints (e.g. Neutens et al. 2010; Miller 1999).

Accessibility is, usually, a relative measure. The growing interest in the interdependence between sustainable development and mobility has emphasized the importance of proper estimation of various dimensions of public transport relative to private car or bike accessibility (Kaplan et al. 2014; Tribby and Zanderbergen 2012; Martin et al. 2008; O'Sullivan et al. 2000). Since the disparity of accessibility between cars and public transport provides important information about the degree of car dependence in urban areas, measuring the relative accessibility of transport versus car has been recently analyzed in many urban regions (Benenson et al. 2011; Ferguson et al. 2013; Grengs et al. 2010; Mao and Nekorchuk 2013; Mavoa et al. 2012; Blumenberg and Ong 2001; Hess 2005; Martin et al. 2008; Kawabata 2009; Salonen and Toivenen 2013). The importance of these studies is straightforward—in a majority of studies, public transport accessibility is lower, often significantly, than that of the private car.

Although different kinds of measures were developed in the literature not much guidance is provided on how to choose or apply these measures in policy assessments. Even more—a key gap exists between how accessibility is primarily addressed in the literature as a physical-financial construct that can be improved by proper design or costing, and how human travelers actually conceive it mentally in their day-to-day experiences. This point was well summarized by Kwan (1999: 210): "the accessibility experience of individuals in their everyday lives is much more complex than that which can be measured with conventional measures of accessibility".

A main hurdle in measuring or modeling accessibility is fitting the spatial resolution of analysis to the scale where real human travelers make travel decisions. An adequate view of accessibility demands analysis at the spatial resolution of human activity—moving from one building as an origin to a destination in another building, by navigating the transportation network, comprised of different modes, lines and stops.

This human viewpoint is difficult to model due to the necessary big data and heavy computation requirements it entails. Consequently, accessibility has been mainly evaluated at a coarse and granular scale of municipalities (Ivan et al. 2013); counties, (Karner and Niemeier 2013); transport analysis zones (Black and Conroy 1977; Bhandari et al. 2009; Ferguson et al. 2013; Foth et al. 2013; Rashidi and Mohammadian 2011; Haas et al. 2008; Burkey 2012; Lao and Liu 2009; Grengs et al. 2010; Kaplan et al. 2014) or neighborhoods (Witten et al. 2011). Aggregate analysis, assumes that the centers of the zones (centroids) are the origin and destination points, and eventually results in a discontinuity when evaluating two adjacent zones. Few studies have attempted to model accessibility at parcel level data and more often, disaggregate data is eventually aggregated for analysis (Mavoa et al. 2012; Tribby and Zanderbergen 2012; Welch and Mishra 2013; Salonen and Toivenen 2013). The exception is Owen and Levinson (2015), who measure accessibility by public transport at the resolution of zipcodes. While often sufficient for car-based accessibility, where the travel between traffic zones takes time comparable to the imprecision of velocity estimate, aggregate estimates tend to either over- or under-estimate public transport accessibility, where the walk or waiting time can be critical for the personal mode choice. Moreover, important components of public transport accessibility, such as walking times to embarkation stops, destinations and between stops when changing lines, and waiting for transfer times, are not usually considered explicitly in the calculations of total travel time between aggregate units.

While existing tools used in transportation practice, (such as TRANSCAD or EMME), are able to calculate detailed public transport-based accessibility, they were not designed for this purpose. High-resolution calculation of accessibility with these and similar general-purpose tools for transport analysis is extremely time-consuming, and it is thus prohibitively expensive to use, in generating accurate estimates of public transport accessibility for an entire metropolitan area.

Being conceptually evident, spatially explicit high-resolution measurement of accessibility raises severe computational problems. A typical metropolitan area with a population of several millions contains 0.5–1 million buildings and demands the processing of hundreds of thousands origins and destinations. Trips themselves cover tens of thousands street segments and hundreds of public transport lines of different kinds (Benenson et al. 2010, 2011). Thus, computations of accessibility at a human traveler resolution involve the processing of huge volumes of raw data. For example, the metropolitan area of Tel Aviv has a population of 2.5 million, c.a. 300,000 buildings and over 300 bus lines. Until recently, attempting such an endeavor seemed impossible, and there was no option but to aggregate. However, recent developments in graph databases and algorithms (Buerli 2012), offer a solution to these problems.

The aim of this paper is the following. First, we describe a new GIS-based computer application that manages to make fast accessibility calculations at the resolution of individual buildings, and exploits high-resolution urban GIS and provides *precise estimates, in space and in time,* of public transport-based accessibility. We are thus able to assess the transportation system at every location and for every hour of the day. Our application implements the ideas proposed by Benenson et al. (2011) and their recent development (Benenson et al. 2014), running on a SQL server engine. Second, we apply the application on a real case study involving the implementation of a new Light Rail (LRT) line in the Metropolitan Area of Tel Aviv. We use this case study to illustrate the importance of high-resolution accessibility estimates to obtain unbiased estimates of the relative accessibility between public transport and car in two scenarios with and without the proposed LRT line.

The rest of the paper is organized in the following manner. Section 2, presents the operational approach we applied to measure accessibility. Section 3 presents the computer application. Section 4 presents the case study analysis. Section 5, concludes and suggests future research directions.

2 Operational Approach to Accessibility Estimation

Accessibility is a function of all three components of an urban system: (a) land use pattern that defines the distribution of people's activities; (b) the residential distribution and socio-economic characteristics of people; (c) the transportation system with its road configuration, modes of travel, and the time, cost and impedance of travel from and to any place within the metropolitan area. All these define the satisfaction or benefits that individuals obtain from traveling from their homes to different destinations. Ideally, the urban system evolves and adapts i.e., a fit between land use and transportation components on the one hand, and individual needs on the other, is maintained. However, even in an ideal setting, different components of the urban system commonly have different the rates of change and adaptation. It takes years and sometimes decades for street configurations, land-use and buildings to change. Therefore, for this research we can safely assume they are constant. Conversely, residential and activity patterns that are related to people, change and adapt quickly to changes in the other subsystems. Consequently, we can safely assume that accessibility is governed by the land-use and transportation subsystems, and measure it as representing people's potential mobility. Moreover, we consider the level of service of the transportation system as more sensitive to policy changes and investment in projects than the land-use system, and concentrate on evaluating the impacts of changes in this subsystem on people's potential accessibility to a fixed set of destinations. At this stage, we avoid entering the quagmire of predicting people's actual spatiotemporal movements.

Our measure of accessibility is itself relational. We measure the accumulated value of a selected characteristic (e.g., jobs, commercial area) available to a person at destinations within a given timeframe (between 30 min to an hour), using a chosen transportation mode, usually comparing between private car and public transport.[1] We then compare between these two values and estimate a relative measure of accessibility.

The new application implements these measures of accessibility at a high spatial resolution approximating a single building. These measures are based on a precise estimate of the *travel time* between (O)rigin and (D)estination and are defined for a given transportation (M)ode. For example, for (B)us and private (C)ar:

- *Bus travel time (BTT):*
    BTT = Walk time from origin to a stop of Bus #1 + Waiting time of Bus #1 + Travel time of Bus #1 + [Transfer walk time to Bus #2 + Waiting time of Bus #2 + Travel time of Bus #2] + [Transfer component related to additional buses] + Walk time from the final stop to destination (square brackets denote optional components).

---

[1] There is no major restriction to account in the future for bicycle and pedestrian movements

- *Car travel time (CTT):*
 CTT = Walk time from origin to the parking place + Car in vehicle time + Walk time from the final parking place to destination.

Estimates of accessibility are based on the notions of Service area and Access area:

Let MTT denotes Mode travel time and S denotes the region of interest

Access area: Given origin O, transportation mode M and travel time τ, let us define Mode Access Area—$MAA_O(\tau)$—as the area of S containing all destinations D that can be reached from O with M during MTT $\leq \tau$.

Service area: Given destination D, transportation mode M and travel time τ, let us define Mode Service Area—$MSA_D(\tau)$—as the area of S containing all origins O from which given destination D can be reached during MTT $\leq \tau$.

It is important to note that accessibility depends on the definition of the region S. The closer is a trip's origin or destination to the region's border, the higher is the potential bias of the estimate: Travelling opportunities and destinations (in case of the access area) or origins (in case of the service area) that are located beyond the S border are ignored. To diminish this effect, S should avoid cutting out parts of a continuously populated area and include essential spatial margins around investigated origins and destinations. In this respect, an entire metropolitan area can be a good initial choice.

By considering accessibility as a relative notion, as the private car usually provides better levels of accessibility throughout a metropolitan area, we assume the transportation system as adequate if it provides public transport accessibility that does not deviate too much behind that provided by the car. Accordingly, we build on this approach to develop an analysis of the gaps in car and public transport accessibility.[2]

Our application includes two main measures of accessibility for a given location, calculated as the *ratio of the service* or *access areas* estimated for a pair of travel modes, for origins O and destinations D of a certain type.

Given an origin O, the *Bus to Car (B/C) Access Areas ratio* is

$$AA_o(\tau) = BAA_o(\tau)/CAA_o(\tau) \tag{1}$$

Given the destination D, the *Bus to Car (B/C) Service area ratio* is

$$SA_D(\tau) = BSA_D(\tau)/CSA_D(\tau) \tag{2}$$

Note: This view of accessibility as relative measure essentially decreases the boundary effect. Namely, the travelling opportunities as well as origins and

---

[2] Benenson et al. (2011) implemented quite similar accessibility measures at a resolution of Traffic Analysis Zones (TAZ).

destinations beyond the region boundary will be excluded from the calculations for *all modes*. This is especially important for the case of high MTT for one or both modes.

Equations (1) and (2) can be easily specified for any particular type (k) of destinations D_k or origins O_k and, further, towards weighting to include the destinations' and origins' capacities, $D_{k,Capacity}$ and $O_{k,Capacity}$. Examples of capacity can be number of jobs at high-tech enterprises as destinations, low wage jobs at industry buildings as destinations, affordable dwellings with origin capacity defined as number of dwellings and so forth.

Equations (1) and (2) can be then generalized to include the ratio of the sums of capacities of the destinations that can be accessed during time τ by Bus and Car:

$$AA_{o,k}(\tau) = \sum\nolimits_{Dk}\{D_{k,Capacity}|D_k \in BAA_o(\tau)\} / \sum\nolimits_{Dk}\{D_{k,Capacity}|D_k \in CAA_o(\tau)\} \quad (3)$$

$$SA_{D,k}(\tau) = \sum\nolimits_{Dk}\{O_{k,Capacity}|O_k \in BSA_D(\tau)\} / \sum\nolimits_{Dk}\{O_{K,Capacity}|O_k \in CSA_D(\tau)\} \quad (4)$$

The numerators of the fractions in (3)–(4) is the overall capacity of the service/access areas estimated for the bus mode, while the denominator is the overall capacity of the service/access areas estimated for the car mode.

3 Computer Application

We implement mode-dependent calculations of access and service areas. Provided the required data is available, accessibility indices for origins or destinations of particular types accounting for their capacities, using (1)–(2) and (3)–(4) can also be estimated.

3.1 Graph-Based View of Public Transport and Car Travel

To enable the use of graph-theory in our calculations, we translate road (R) and public transport (PT) networks into directed graphs. The road network is translated into a graph (RGraph) in a standard fashion: junctions into nodes, street sections into links (two-way segments are translated into two links) and travel time into the impedance. The graph of the public transport network (PTGraph) depends on timetable and, thus, is constructed for a given time interval.

Given the PT lines, stops and timetable, the node N of the public transport network graph is defined by the quadruple:

N = <PT_LINE_ID, TERMINAL_DEPARTURE_TIME, STOP_ID, STOP_ARRIVAL_TIME>

Two nodes N_1 and N_2 of a PT-graph

N_1 = <PT_LINE_ID$_1$, TERMINAL_DEPARTURE_TIME$_1$, STOP_ID$_1$, STOP_ARRIVAL_TIME$_1$ > and
N_2 = <PT_LINE_ID$_2$, TERMINAL_DEPARTURE_TIME$_2$, STOP_ID$_2$, STOP_ARRIVAL_TIME$_2$>

are connected by link L_{12} in two cases:

1. The vehicle of the same PT line drives from STOP_ID$_1$ to STOP_ID$_2$ that is:

    PT_LINE_ID$_1$ = PT_LINE_ID$_2$,
    TERMINAL_DEPARTURE_TIME$_1$ = TERMINAL_DEPARTURE_TIME$_2$
    and
    STOP_ID$_2$ is the next to STOP_ID$_1$ on the bus line PT_LINE_ID$_1$.
    The impedance of the link L_{12} is equal in this case to:

    STOP_ARRIVAL_TIME$_2$ − STOP_ARRIVAL_TIME$_1$.

2. A passenger can transfer between N_1 and N_2—get off a PT vehicle at a stop STOP_ID$_1$ and get on another vehicle at a stop STOP_ID$_2$.

    A transfer can happen if the transfer time (usually walk time) TRASFER$_{12}$ between STOP_ID$_1$ and STOP_ID$_2$ is less or equal to the maximal possible transfer time TRANSFER$_{max}$ that is:

    TRANSFER$_{12}$ ≤ TRANSFER$_{max}$ and
    TRANSFER$_{12}$ ≤ STOP_ARRIVAL_TIME$_2$ − STOP_ARRIVAL_TIME$_1$ ≤ TRANSFER$_{12}$ + PTWAIT$_{max}$, where PTWAIT$_{max}$ is a maximal waiting time at a PT stop

To obtain a full travel graph we extend RGraph and PTGraph to include buildings and connection between buildings and road junctions or public transport stops. In both cases, each building (B) is considered as a node (B-node) and, in RGraph, B-node is connected to the node N representing road junction if the walk time between B and N is less than CARWALK$_{max}$.

The PTGraph represents travel between the PT stops and does not include buildings. To enable this, PTGraph should be extended to the Buildings-PT Graph (BPTGraph). The definition of the BPTGraph is easy: The set of nodes of the BPTGraph is a union of two sets—PTGraph quadruples N, and nodes that represent buildings (denoted below as B). In addition to the links between the nodes N of the PTGraph, BPTGraph contains links that connect between nodes N of the PTGraph and nodes B representing buildings. The N → B and B → N connections are established differently because the impedance of these links is different. Node N is connected to B if the walk time between them is less than PTWALK$_{max}$, while node B is connected to a node N if the walking time between B and N is less than PTWALK$_{max}$ and waiting time for the arriving line is less than PTWAIT$_{max}$.

Formally, in a BPTgraph, a B-node is connected to an N node,

N = <PT_LINE_ID, TERMINAL_DEPARTURE_TIME, STOP_ID, STOP_ARRIVAL_TIME>,

depending on the trip start time T_{start}. Namely, B is connected to N if the walk time between B and N is less than $PTWALK_{max}$, and

$$T_{start} + PTWALK_{max} < STOP\_ARRIVAL\_TIME$$
$$< T_{start} + PTWALK_{max} + PTWAIT_{max}$$

The impedance of a link between the B-node and a node N of a PT-graph is the walking time between B and N plus the waiting time for the arriving line. The impedance of a link between a node N of the PT-graph and B-node is just the walking time from a corresponding stop to the building.

Figure 1 presents a description of the translation of a typical bus trip from origin to destination into a sequence of connected links.

Representation of PT network as a graph enables the application of standard algorithms for building minimal spanning tree (Dijkstra 1959) and, thus, estimation of the service and access areas for a given building. However, this cannot be done with the help of the standard GIS software, as ArcGIS, that start with roads and junctions as GIS layers. Moreover, calculation of the service/access areas for all urban buildings, which number in the hundreds of thousands, is an extremely time-consuming operation. For comparison, calculation of car accessibility for the entire Tel Aviv Metropolitan Area, organized with the help of the ArcGIS Model Builder (Allen 2011) and employing standard ArcGIS Network Analyst procedure, took us several days.

Our goal is to investigate and compare travel opportunities supplied by public transport. That is why car accessibility calculation is done only once and subsequently used as the denominator in all the rest of the calculations of relative

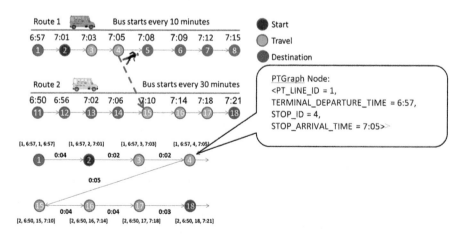

Fig. 1 Conceptualization of translating a public transport network into a graph

accessibility according to (1)–(4). Our approach is thus sensitive to the performance of accessibility calculations for the numerator of these formulas i.e. public transport. Based on the specific properties of the PTGraph, PT accessibility can indeed be estimated in a different form compared to car accessibility, employing the abilities of the modern relational databases.

3.2 Calculation of the Public Transport Accessibility with Relational Database

The use of the Relational Database Management System (RDBMS) for calculating PT accessibility is based on the observation that the degree of most of the quadruple nodes within the PTGraph is exactly 2. That is, most of quadruple nodes of the PTGraph are connected, by just two (directed) links, to previous and next stops of the same PT vehicle traversing the same line.

Let us, for convenience, use the identifiers (ID) of bus lines and stops as their names. Formally, a quadruple node

N1 = <PT_LINE_ID, TERMINAL_DEPARTURE_TIME, **STOP1_ID**, **STOP1_ARRIVAL_TIME**>

of the PTGraph is connected, by the bus line PT_LINE_ID, to two nodes only

N0 = <PT_LINE_ID, TERMINAL_DEPARTURE_TIME, **STOP0_ID**, **STOP0_ARRIVAL_TIME**>, and
N2 = <PT_LINE_ID, TERMINAL_DEPARTURE_TIME, **STOP2_ID**, **STOP2_ARRIVAL_TIME**>.

Where **STOP0_ID** is a stop that is previous to **STOP1_ID** on the PT_LINE_ID for the bus starting the travel at TERMINAL_DEPARTURE_TIME, and **STOP2_ID** is a stop next to **STOP1_ID** on the PT_LINE_ID for this bus.

Conversely, the degree of quadruple nodes at which transfers take place is higher than 2: in addition to being connected to quadruples denoting the previous and next stops of the line, this node is connected to the quadruples that can be reached by foot, as defined in Sect. 3.1.

In what follows, we use BuildingID for the identifier of the building B, and NodeID for the identifier of the quadruple node N. Full RDBMS representation of the PT graph consists of four tables. We present them below as TableName (Meaning of field1, Meaning of field2, etc.):

- Building-PTNode (BuildingID, trip start quadruple NodeID, Walk time between building B and stop of the quadruple N + waiting time at a stop of quadruple N);
- DirectPTTravel (quadruple NodeID, ID of the quadruple N that can be directly reached from the NodeID by the line of quadruple NodeID, travel time between stop of quadruple NodeID and stop of quadruple N);

- Transfer (NodeID of the stop of transfer start quadruple N1, NodeID of the stop of the transfer end quadruple N2, walk time between stops of N1 and N2 + waiting time at a stop of N2 for a line of quadruple N2)
- PTNode-Building (trip final quadruple NodeID, BuildingID, Walk time between the stop of the quadruple N and Building B)

The idea in our approach is expressed by the DirectPTTravel one-to-many table: It presents all possible direct trips from a certain stop with a certain line. Standard SQL queries that join between this and other tables are sufficient for estimating access or service areas.

To illustrate the idea, we build the occurrences of DirectPTTravel and Transfer tables for the example presented in Fig. 1 (number in circle is used as quadruple's ID):

DirectPTTravel		
StartNode_ID	ArrivalNode_ID	TravelTime(mins)
1	1	0
1	2	4
1	3	6
1	4	8
1	5	11
1	6	12
1	7	15
1	8	18
2	2	0
2	3	2
2	4	4
2	5	7
...	...	...
11	11	0
11	12	6
...	...	...
12	12	0
12	13	6
...	...	...

Transfer		
StartNode_ID	ArrivalNode_ID	TransferTime(mins)
4	15	5
...	...	...

To understand the logic of querying: a "Join" between DirectPTTravel and Transfer tables provides trips from every stop to every transfer stop. A further join between the result and second copy of the DirectPTTravel provide all trips between stops with zero or one transfer. The join between Building-PTNode and the result above provides the trips between buildings and final stops of a trip and a further join with the PTNode-Building finally provides trips between buildings. Most of the computation time is spent to two last queries that include GroupBy clause, as we need the minimal time travel between two buildings.

The output table contains three major fields: ID of the origin building, ID of the destination building and total travel time. For further analysis, we also store the components of the trip time: walk times in the beginning and end of a trip and transfer time in case a trip includes transfer, bus travel time(s) and waiting times at the beginning of a trip and transfer. Based on the output table we construct the table that contains number of buildings, total population and total number of jobs that can be accessed during a given trip time, usually 15, 30, 45 and 60 min. This table is further joined with the layer of buildings thus enabling the map presentation of accessibility. Output of the car accessibility computations is organized in the same way and joined with the layer of buildings as well. Relative accessibility—the ratio of the PT-based and car-based accessibility is calculated within ArcGIS. These maps of relative and absolute accessibility are those presented in Sect. 4.

It is important to note that all four tables are built based on the exact timetables of all lines and change whenever the timetable is changed. For the case of ~300 bus lines and several thousand stops in the Tel Aviv Metropolitan Area, the number of rows in the tables, for the bus trips that are less than 1 h, is about several million, within a standard range of modern SQL RDBMS abilities. All our calculations were thus performed with the free version of the Microsoft SQL server. The total computation time for the full PT accessibility map for the Tel Aviv metropolitan is 2 h.

3.3 Inputs and Outputs

The database needs to include:

- Layers of roads and junction, with the attributes sufficient for constructing a network.
- Travel times for cars as based on average velocities at morning peak times from the traffic assignment model, obtained from the metropolitan mass-transit planning agency.
- Layers of all public transport modes: railroads, metros, light rail lines and buses with their stops, with each line related to its stops. Often, the General Transit Feeder Specification (GTFS) that is used by Google and includes all this information is available for the city
- A table of public transport departure and arrival times for each line and stop given as an exact timetable (also included as part of GTFS) or in terms of frequencies

Estimation of Urban Transport Accessibility at the Spatial Resolution...   395

Table 1 Administrator control parameters for managing accessibility analysis

Parameter	Typical value
Max aerial walking distance	400 m
PT boarding time	7:00–7:30
Transfers between lines	Yes
Max walking distance between stops	200 m
Max waiting time at transfer stop	10 min
Max total trip time	60 min
Walking time from origin to parking	0 min
Parking search and walk time at destination	5 min

– Layer of origins and destinations with the use and capacity given. Typically, this is a layer of buildings, while parks and leisure locations may comprise other layers.

The input is defined by the travel start/arrival time and additional parameters as presented in Table 1:

The output is a new layer detailing accessibility indices, at which every origin O or destination D is characterised by the values of index $AA_{O,k}(\tau)$ or $SA_{D,k}(\tau)$ for the given range of τ e.g. maps for $\tau = 15, 30, 45$ and 60 min of travel. These results are captured in shape files that are easily exported to any GIS software.

4 Case Study: The Accessibility Impact of Tel Aviv's "Red" Light Rail Line

We apply the proposed methodology and our application to investigate a planned LRT line (called the "Red" line) connecting the southern suburbs with the eastern suburbs of Tel Aviv. For lack of planning information at this stage, a strong (but problematic) assumption in the subsequent analysis is that the new LRT is introduced without changes to the rest of the running bus lines. Figure 2 shows the spatial resolution with the entire metropolitan area on the left panel and a zoom in to the core on the right. The chosen partition corresponds to squares of 60×60 m—a typical size of a building and its surrounding area in metropolitan Tel Aviv. As noted in Sect. 1 the metropolitan area of Tel Aviv has a population of 2.5 million, c.a. 300,000 buildings and over 300 bus lines.

4.1 Relative accessibility to Jobs in Tel Aviv

Figure 3 shows the access areas to jobs by car (15–45 min) and by bus (45 min) for the morning peak hour. It is clear that the car access area in the metropolitan area is essentially larger than the bus access area, which is limited to the urban core for a

Fig. 2 Tel Aviv metropolitan area (**a**) and its core (**b**) at resolution of 60×60 m cells that correspond to typical distance between centers of two building footprints

45-min trip. In 45 min, any area in the metropolitan region is accessible by car whereas only areas in or near the core are accessible by bus.

Figure 4 illustrates why we consider high resolution an important aspect of accessibility estimation. In the first row is the accessibility index calculated at TAZ level in Benenson et al. (2011), in the second is the relative accessibility at the resolution of buildings aggregated to transportation analysis zones (TAZ). Both maps were calculated based on data of the numbers of jobs obtained from the metropolitan mass transit planning authority at the resolution of statistical areas (approximating TAZ). For the high-resolution calculation, the number of jobs in each statistical area was uniformly distributed over the area.

Despite very close averages over the entire metropolitan area—0.356 for the calculation at resolution of TAZ and 0.336 for the calculation at resolution of buildings and further aggregation, it is easy to verify the contradictory results of computation at the two levels of resolution. On the right, the map built based on the high-resolution results presents seamless changes in accessibility between neighboring zones with higher accessibility in the center compared to lower in the suburbs. Conversely, on the left, low resolution results in the familiar patchwork discontinuity in accessibility levels, with the suburb areas often having higher accessibility relative to the center.

Estimation of Urban Transport Accessibility at the Spatial Resolution... 397

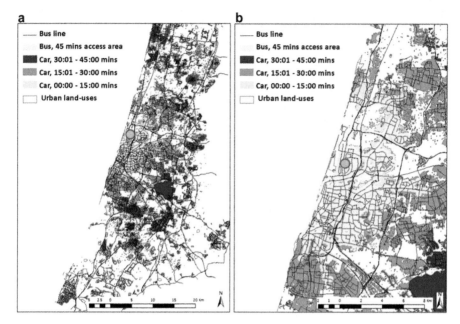

Fig. 3 Access areas by car and bus starting from the Tel Aviv University terminal marked by the arrange circle, Metropolitan built area (**a**), zoom on the city core (**b**)

4.2 The Impact of the LRT on Relative Accessibility Between Car and Public Transport

Figure 5 presents the relative accessibility levels obtained when the Red LRT line is running in the background. The assumption is that the LRT operates in parallel to existing road and bus networks. We did not have information to update to a future configuration at this stage of the analysis.

Figure 6 shows the absolute number of additional accessible jobs due to the introduction of the Red LRT line. The configuration of the LRT is based on an average peak-hour frequency of 5 min in both directions.

As one can see, most of the benefit is around the LRT corridor clearly visible in darker colors and this change is most noticeable for the short trips. For longer trips, the impact of the LRT is less evident. That is, buildings close to the LRT line enjoy an improvement in accessibility to all other areas even within short trips. However, with the bus network that is not adjusted to the LRT, this benefit dissipates as journey time and distance from the LRT line increase. It should also be noted that since the analysis is based on access areas the main LRT function, which is to provide accessibility to job destinations in the core area (i.e. service area), is not viewed in these maps. Naturally, at the morning peak hour most of the traffic on the LRT is bound to the core and not the other way. A complete analysis of the LRT benefit would require looking also at the service area complementary view of accessibility. Moreover, to be effective the LRT introduction should be supported

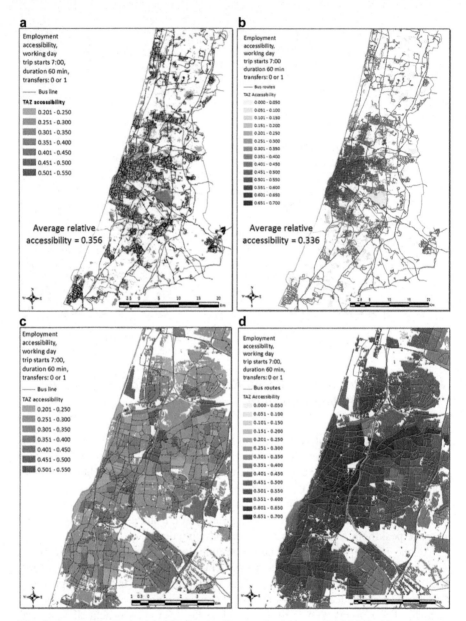

Fig. 4 Access Area Ratio $AA_O(60)$ for the 60 min trip that starts 7:00 AM, with zero or one transfer between the public transport lines. Directly calculated at resolution of TAZ, for the entire metropolitan (**a**) and metropolitan's center (**c**) from Benenson et al. (2011); calculated at resolution of a buildings and aggregated by TAZ for the entire metropolitan (**b**) and metropolitan's center (**d**)

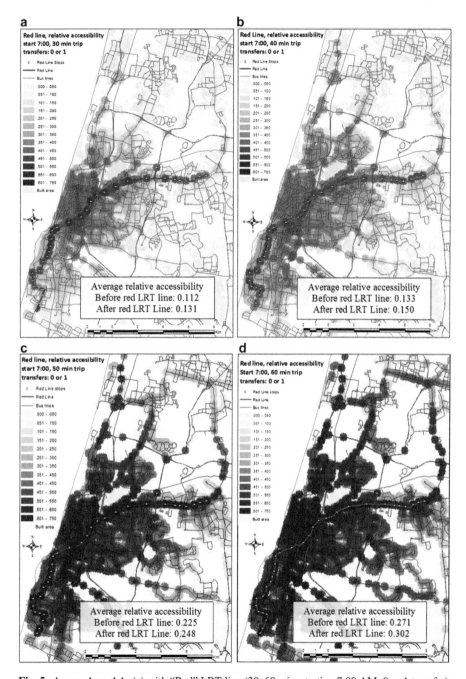

Fig. 5 Access Area $AA_O(\tau)$ with "Red" LRT line (30–60 min, starting 7:00 AM, 0 or 1 transfer)

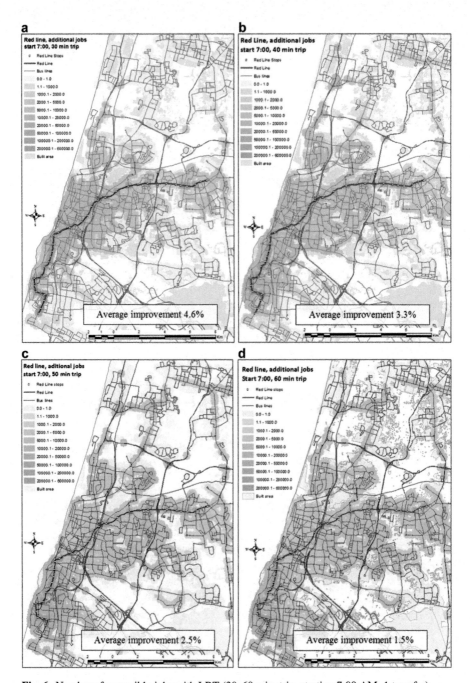

Fig. 6 Number of accessible jobs with LRT (30–60 min. trip, starting 7:00 AM, 1 transfer)

by essential planned changes in the bus network. Otherwise, higher speed of the LRT will not be sufficient to compensate for the transfer time between bus and rail, especially for the longer trips.

5 Conclusions and Further Research

In this chapter, we present an approach and implement a computational application for calculating accessibility measures at high spatial resolution based on "BIG" urban data. While the rationale of the proposed indices is not new, our computations show the importance of high-resolution view of urban travel. Namely, the wide gap between the accessibility estimates done at the resolution of buildings (and then aggregated by TAZ), and the estimates done directly at the aggregate TAZ levels. In particular, the familiar patchwork of accessibility discontinuity when evaluating adjacent zones is eliminated.

A high-resolution view of accessibility demands novel computational abilities and is implemented with the graph database engine. Our focus was on accessibility to jobs (access area analysis). However, the method can be applied to any trip purpose depending on the availability and quality of the data. The case study we present shows how a new LRT line can be evaluated in terms of public transport accessibility improvement. This kind of analysis could be easily incorporated into economic and social equity assessments and feasibility studies of planned transportation infrastructure.

The approach and application we presented needs further developments in several respects. First, the calculation of the car-based and PT-based accessibility should be unified. Second, the performance, especially of the car accessibility calculations should be essentially increased. This would include also adding improved functionalities based on real arrival time at stops for public transport and elements of uncertainty in car travel like distribution of que lengths at intersections and parking cruising times in different areas. Third, the calculation engine for car accessibility should be rebuilt and become independent of the ArcGIS. Fourth, user interface should be developed and the pieces of the software should be organized as unified software system (e.g. Owen and Levinson 2015). All this should be done for turning the system into cloud-based software that enables interactive transportation planning. A planner will upload her road and PT networks and schedules and then perform tasks such as changing public transport lines details, adding/removing stops, showing/hiding GIS layers etc. The backend leverages a sophisticated analytical engine using cutting edge graph theory technologies for real-time (or near real-time) processing.[3]

---

[3] The interested reader who wishes to make use of the application is invited to contact the corresponding author. The application can be used for scientific purposes without restrictions provided that data is given in the requested format and bearing in mind the costs of cluster operation time.

High-resolution calculation of transport accessibility will serve as a basis for human-based evaluation of the urban transportation system. While it is expected that transportation investments will result in increasing accessibility and participation rates to vital activities, experience shows that the distribution of these outcomes is not equal throughout the system. Our further research, which is now in progress, utilizes the high-resolution accessibility indices to evaluate the equity of the transportation system. Namely, we can evaluate the variation of accessibility provided by the transportation system at the level of buildings and compare this variation between several scenarios (e.g. with/without LRT). Moreover, the application can be used to compute service areas, for different scenarios of the public transport network, accessibility to key institutions such as hospitals or universities or terminals and rail stations.

Acknowledgments Parts of this research were funded by the Chief Scientist Office of the Israeli Ministry of Transport. The second author kindly thanks the Returning Scientist fellowship (2014) of the Israeli Ministry of Immigration and Absorption and support of Tel Aviv University. Part of this work is based on the Master thesis of Amit Rosenthal supervised by the corresponding author. The authors would like to thank two anonymous reviewers for constructive and helpful comments on the draft version.

References

Alam BM, Thompson GL, Brown JR (2010) Estimating transit accessibility with an alternative method. Transp Res Rec J Transp Res Board 2144(1):62–71

Allen DW (2011) Getting to know ArcGIS ModelBuilder. ESRI Press, 336 p

Ben-Akiva M, Lerman SR (1979) Disaggregate travel and mobility choice models and measures of accessibility. In: Hensher DA, Stopher PR (eds) Behavioural travel modelling. Croom Helm, London, pp 654–679

Benenson I, Martens K, Rofé Y, Kwartler A (2010) Measuring the gap between car and transit accessibility estimating access using a high-resolution transit network geographic information system. Transp Res Rec J Transp Res Board N2144:28–35

Benenson I, Martens K, Rofé Y, Kwartler A (2011) Public transport versus private car: GIS-based estimation of accessibility applied to the Tel Aviv metropolitan area. Ann Reg Sci 47:499–515

Benenson I, Geyzersky D, et al (2014) Transport accessibility from a human point of view. In: Key presentation at the geoinformatics for intelligent transportation conference, Ostrava, 27 Jan 2014

Bertolini L, le Clercq F, Kapoen L (2005) Sustainable accessibility: a conceptual framework to integrate transport and land use plan-making. Two test-applications in the Netherlands and a reflection on the way forward. Transp Policy 12(3):207–220

Bhandari K, Kato H, Hayashi Y (2009) Economic and equity evaluation of Delhi Metro. Int J Urban Sci 13(2):187–203

Black J, Conroy M (1977) Accessibility measures and the social evaluation of urban structure. Environ Plan A 9(9):1013–1031

Blumenberg EA, Ong P (2001) Cars, buses, and jobs: welfare participants and employment access in Los Angeles. Transp Res Rec J Transp Res Board 1756:22–31

Bristow G, Farrington J, Shaw J, Richardson T (2009) Developing an evaluation framework for crosscutting policy goals: the Accessibility Policy Assessment Tool. Environ Plan A 41(1):48

Bruinsma FR, Nijkamp P et al (1990) Infrastructure and metropolitan development in an international perspective: survey and methodological exploration. Research memoranda. Faculteit der Economische Wetenschappen en Econometrie/Vrije Universiteit Amsterdam, Amsterdam

Buerli M (2012) The current state of graph databases. www.cs.utexas.edu/~cannata/dbms/Class%20Notes/09%20Graph_Databases_Survey.pdf

Burkey ML (2012) Decomposing geographic accessibility into component parts: methods and an application to hospitals. Ann Reg Sci 48(3):783–800

Dijkstra EW (1959) A note on two problems in connexion with graphs. Numerische Mathematik 1 (1):269–271

Farrington J, Farrington C (2005) Rural accessibility, social inclusion and social justice: towards conceptualisation. J Transp Geogr 13(1):1–12

Feitelson E (2002) Introducing environmental equity dimensions into the sustainable transport discourse: issues and pitfalls. Transp Res Part D Transp Environ 7(2):99–118

Ferguson NS, Lamb KE, Wang Y, Ogilvie D, Ellaway A (2013) Access to recreational physical activities by car and bus: an assessment of socio-spatial inequalities in mainland Scotland. PLoS One 8(2):e55638

Foth N, Manaugh K, El-Geneidy AM (2013) Towards equitable transit: examining transit accessibility and social need in Toronto, Canada, 1996–2006. J Transp Geogr 29:1–10

Garb Y, Levine J (2002) Congestion pricing's conditional promise: promotion of accessibility or mobility? Transp Policy 9(3):179–188

Geurs KT, Ritsema van Eck JR (2001) Accessibility measures: review and applications. Evaluation of accessibility impacts of land-use transport scenario's, and related social and economic impacts. RIVM—National Institute of Public Health and the Environment (NL), Bilthoven

Geurs KT, van Wee B (2004) Accessibility evaluation of land-use and transport strategies: review and research directions. J Transp Geogr 12(2):127–140

Grengs J, Levine J, Shen Q, Shen Q (2010) Intermetropolitan comparison of transportation accessibility: sorting out mobility and proximity in San Francisco and Washington, DC. J Plan Educ Res 29(4):427–443

Haas P, Makarewicz C, Benedict A, Bernstein S (2008) Estimating transportation costs by characteristics of neighborhood and household. Transp Res Rec J Transp Res Board 2077:62–70

Handy SL, Niemeier DA (1997) Measuring accessibility: an exploration of issues and alternatives. Environ Plan A 29(7):1175–1194

Hess DB (2005) Access to employment for adults in poverty in the Buffalo–Niagara region. Urban Stud 42(7):1177–1200

Ivan I, Horak J, Fojtik D, Inspektor T (2013) Evaluation of public transport accessibility at municipality level in the Czech Republic. In: GeoConference on informatics, geoinformatics and remote sensing, conference proceedings, vol 1, p 1088

Kaplan S, Popoks D, Prato CG, Ceder AA (2014) Using connectivity for measuring equity in transit provision. J Transp Geogr 37:82–92

Karner A, Niemeier D (2013) Civil rights guidance and equity analysis methods for regional transportation plans: a critical review of literature and practice. J Transp Geogr 33:126–134

Kawabata M (2009) Spatiotemporal dimensions of modal accessibility disparity in Boston and San Francisco. Environ Plan A 41(1):183–198

Kenyon S, Lyons G et al (2002) Transport and social exclusion: investigating the possibility of promoting inclusion through virtual mobility. J Transp Geogr 10(3):207–219

Kwan MP (1999) Gender and individual access to urban opportunities: a study using spacetime measures. Prof Geogr 51:210–227

Kwok RCW, Yeh AGO (2004) The use of modal accessibility gap as an indicator for sustainable transport development. Environ Plan A 36(5):921–936

Lao Y, Liu L (2009) Performance evaluation of bus lines with data envelopment analysis and geographic information systems. Comput Environ Urban Syst 33(4):247–255

Liu S, Zhu X (2004) Accessibility analyst: an integrated GIS tool for accessibility analysis in urban transportation planning. Environ Plan B Plan Des 31(1):105–124. doi:10.1068/b305

Lucas K (2012) Transport and social exclusion: where are we now? Transp Policy 20:105–113

Mao L, Nekorchuk D (2013) Measuring spatial accessibility to healthcare for populations with multiple transportation modes. Health Place 24:115–122

Martens K (2012) Justice in transport as justice in access: applying Walzer's 'Spheres of Justice' to the transport sector. Transportation 39(6):1035–1053, Online 21 February 2012

Martin D, Jordan H, Roderick P (2008) Taking the bus: incorporating public transport timetable data into health care accessibility modelling. Environ Plan A 40(10):2510

Mavoa S, Witten K, McCreanor T, O'Sullivan D (2012) GIS based destination accessibility via public transit and walking in Auckland, New Zealand. J Transp Geogr 20(1):15–22

Miller HJ (1999) Measuring space-time accessibility benefits within transportation networks: basic theory and computational procedures. Geogr Anal 31:187–212

Minocha I, Sriraj PS, Metaxatos P, Thakuriah PV (2008) Analysis of transport quality of service and employment accessibility for the Greater Chicago, Illinois, Region. Transp Res Rec J Transp Res Board 2042(1):20–29

Neutens T et al (2010) Equity of urban service delivery: a comparison of different accessibility measures. Environ Plan A 42(7):1613

O'Sullivan D, Morrison A, Shearer J (2000) Using desktop GIS for the investigation of accessibility by public transport: an isochrone approach. Int J Geogr Inf Sci 14(1):85–104

Owen A, Levinson DM (2015) Modeling the commute mode share of transit using continuous accessibility to jobs. Transp Res Part A Policy Pract 74:110–122

Rashidi TH, Mohammadian AK (2011) A dynamic hazard-based system of equations of vehicle ownership with endogenous long-term decision factors incorporating group decision making. J Transp Geogr 19(6):1072–1080

Salonen M, Toivenen T (2013) Modelling travel time in urban networks: comparable measures for private car and public transport. J Transp Geogr 31:143–153

Tribby CP, Zanderbergen PA (2012) High-resolution spatio-temporal modeling of public transit accessibility. Appl Geogr 34:345–355

Welch TF, Mishra S (2013) A measure of equity for public transit connectivity. J Transp Geogr 33:29–41

Witten K, Exteter D, Field A (2003) The quality of urban environments: mapping variation in access to community resources. Urban Stud 40(1):161–177

Witten K, Pearce J, Day P (2011) Neighbourhood Destination Accessibility Index: a GIS tool for measuring infrastructure support for neighbourhood physical activity. Environ Plan Part A 43(1):205

Modeling Taxi Demand and Supply in New York City Using Large-Scale Taxi GPS Data

Ci Yang and Eric J. Gonzales

Abstract Data from taxicabs equipped with Global Positioning Systems (GPS) are collected by many transportation agencies, including the Taxi and Limousine Commission in New York City. The raw data sets are too large and complex to analyze directly with many conventional tools, but when the big data are appropriately processed and integrated with Geographic Information Systems (GIS), sophisticated demand models and visualizations of vehicle movements can be developed. These models are useful for providing insights about the nature of travel demand as well as the performance of the street network and the fleet of vehicles that use it. This paper demonstrates how big data collected from GPS in taxicabs can be used to model taxi demand and supply, using 10 months of taxi trip records from New York City. The resulting count models are used to identify locations and times of day when there is a mismatch between the availability of taxicabs and the demand for taxi service in the city. The findings are useful for making decisions about how to regulate and manage the fleet of taxicabs and other transportation systems in New York City.

Keywords Big data • Taxi demand modeling • Taxi GPS data • Transit accessibility • Count regression model

1 Introduction

Spatially referenced big data provides opportunities to obtain new and useful insights on transportation markets in large urban areas. One such source is the set of trip records that are collected and logged using in-vehicle Global Positioning

---

C. Yang, Ph.D (✉)
Senior Transportation Data Scientist, DIGITALiBiz, Inc., 55 Broadway, Cambridge, MA 02142, USA
e-mail: Jesse.Yang.CTR@dot.gov

E.J. Gonzales, Ph.D. (✉)
Department of Civil and Environmental Engineering, University of Massachusetts Amherst, 130 Natural Resources Road, Amherst, MA 01003, USA
e-mail: gonzales@umass.edu

© Springer International Publishing Switzerland 2017
P. Thakuriah et al. (eds.), *Seeing Cities Through Big Data*, Springer Geography,
DOI 10.1007/978-3-319-40902-3_22

Systems (GPS) in taxicab fleets. In large cities, tens of thousands of records are collected every day, amounting to data about millions of trips per year. The raw data sets are too large to analyze with conventional tools, and the insights that are gained from looking at descriptive statistics or visualizations of individual vehicle trajectories are limited. A great opportunity exists to improve our understanding of transportation in cities and the specific role of the taxicab market within the transportation system by processing and integrating the data with a Geographic Information System (GIS). Moving beyond simple descriptions and categorizations of the taxi trip data, the development of sophisticated models and visualizations of vehicle movements and demand patterns can provide insights about the nature of urban travel demand, the performance of the street network, and operation of the taxicab fleet that uses it.

Taxicabs are an important mode of public transportation in many urban areas, providing service in the form of a personalized curb-to-curb trip. At times, taxicabs compete with public transit systems including bus, light rail, subway, and commuter trains. At other times, taxis complement transit by carrying passengers from a transit station to their final destination—serving the so-called "last mile." In the United States, the largest fleet of taxis is operated in New York City (NYC), where yellow medallion taxicabs generated approximately $1.8 billion revenue carrying 240 million passengers in 2005 (Schaller 2006). All taxicabs in NYC are regulated by the Taxi and Limousine Commission (TLC), which issues medallions and sets the fare structure. As of 2014, there are 13,437 medallions for licensed taxicabs in NYC (Bloomberg and Yassky 2014), which provide service within the five boroughs but focus primarily on serving demand in Manhattan and at the city's airports; John F. Kennedy International Airport and LaGuardia Airport. Since 2013, a fleet of Street Hail Livery vehicles, known as green cabs, have been issued medallions to serve street hails in northern Manhattan and the Outer Boroughs, not including the airports.

The TLC requires that all yellow taxicabs are equipped with GPS through the Taxicab Passenger Enhancements Project (TPEP), which records trip data and collects information on fare, payment type, and communicates a trace of the route being traveled to passengers via a backseat screen. This paper makes use of a detailed set of data that includes records for all 147 million trips served by taxicabs in NYC in the 10-month period from February 1, 2010 to November 28, 2010. Each record includes the date, time, and location of the pick-up and drop-off as well as information about payment, the driver, medallion, and shift. This dataset provides a rich source of information to conduct analysis of the variation of taxi pick-ups and drop-offs across space and time.

In order to effectively plan and manage the fleet of taxicabs, it is necessary to understand what factors drive demand for taxi service, how the use of taxicabs relates to the availability of public transit, and how these patterns vary across different locations in the city and at different times of day. A trip generation model that relates taxi demand to observable characteristics of a neighborhood (e.g., demographics, employment, and transit accessibility) is developed with high temporal and spatial resolution. This paper demonstrates how GPS data from a large set of taxicab data can be used to model demand and supply and how these models can be used to identify locations and times of day when there is a mismatch between the availability of taxicabs and the demand for taxi service. The models are useful for making decisions about how to manage the transportation systems, including the fleet of taxicabs themselves.

Recent work has been done to identify the factors that influence demand for taxicabs within each census tract in NYC based on observable characteristics of each census tract (Yang and Gonzales 2014). The separate models were developed to estimate the number of taxicab pick-ups and drop-offs within each census tract during each hour of the day, and six important explanatory variables were identified: population, education (percent of population with at least a bachelor's degree), median age, median income per capita, employment by industry sector, and transit accessibility. Yang and Gonzales (2014) specifically developed a technique to measure and map transit accessibility based on the time that it takes to walk to a transit station and wait to board the next departing vehicle. By modeling taxi demand based on spatially specific information about population characteristics, economic activities as indicated by employment, and the availability of public transit services, the models showed how the influence of various relevant factors changes over different times during the day.

This paper builds on existing research by introducing a novel method to quantify the supply of available taxicabs in a neighborhood based on where passengers are dropped off and the vehicle becoming available for hire. Although the total supply of taxicabs is itself of interest to policymakers and regulators, the spatial distribution of this supply has a big effect on where customers are able to hail taxicabs on the street and how long they can expect to wait for an available vehicle. Thus, accounting for the supply of taxis in models of the number of taxicab pick-ups provides additional insights about where taxi demand is being served and where there may be latent or underserved demand for taxicab services. The models that are developed in this paper present additional improvements over previous models by explicitly acknowledging that the number of taxi pick-ups in a census tract is a count process and should be modeled with a count distribution such as a Poisson or negative binomial regression. Both the inclusion of an independent variable for taxicab supply and the use of a count data regression yield detailed models that provide improved insights to the factors that drive taxi demand and affect taxi supply. Furthermore, the visualizations of the modeled results provide greater insights than common techniques that merely plot raw data or show simple aggregations. By developing sophisticated models of supply and demand using the extensive set of data of NYC taxicabs, the underlying patterns in the data reveal how the mode is used and how it may be managed to serve the city better.

2 Literature Review

There are a number of studies of taxicabs in the literature from the fields of policy and economics. Earlier theoretical models developed for taxicab demand are mainly economic models for the taxicab market (Orr 1969). Although classic economy theory states that demand and supply will reach equilibrium in a free market, most taxi markets are not actually free, and the roles of regulations that either support or constrain the taxicab industry need to be considered. Furthermore, it has been argued that the price generated by "competitive equilibrium" may be insufficient to cover the social welfare costs (Douglas 1972) or too high to fully realize certain social benefits (Arnott 1996). Based on a study of taxicabs in London, England, Beesley (1973) argued that five contributing factors account for the number of taxis per head: (1) independent regulations, (2) the proportion of tourists, (3) income per capita (especially in the center of London), (4) a highly developed radially-oriented railway system, and (5) car ownership. Although these classic papers build a theoretical foundation for modeling taxicab demand, they are based only on aggregated citywide data, such as the medallion price by year (Schreiber 1975), occurrence of taxicab monopolies by city (Eckert 1973; Frankena and Pautler 1984), and the total number of taxicabs by city (Gaunt 1995).

More recently, attention has been directed toward identifying the factors that influence the generation of taxicab demand. Schaller (1999) developed an empirical time series regression model of NYC to understand the relationship between taxicab revenue per mile and economic activity in the city (measured by employment at eating and drinking places), taxi supply, taxi fare, and bus fare. However, Schaller's (1999) model is not spatially specific, and is based only on the evolution of citywide totals and averages over time. Other studies compare the supply of taxis in different cities in order to investigate the relationships between taxi demand and factors such as city size, the availability and cost of privately owned cars, the cost of taxi usage, population, and presence of competing modes (Schaller 2005; Maa 2005). These studies provide comparisons across different locations, but they do not account for changes with time.

There have been many technology developments that are beneficial to modeling taxicab demand. Examples include in-vehicle Global Positioning Systems (GPS) implanted in modern taxicabs and analytical tools like Geographic Information Systems (GIS), which facilitate analysis of spatially referenced data (Girardin and Blat 2010; Balan et al. 2011; Bai et al. 2013). As a result, a massive amount of detailed data is recorded automatically for trips served by modern taxicab fleets, such as pick-up locations, drop-off locations, and in some cities a complete track of the route connecting the two (Liang et al. 2013). These large-scale taxicab data make it is possible to build empirical models to understand how taxi trips are generated and distributed across space and time, and how they compete with other transportation modes.

The potential for extracting useful information about taxicab demand and the role of taxis in the broader transportation systems has just begun to be tapped. One recent study considers whether taxicabs operate as a substitute or complement to the public transit system in Boston, Massachusetts (Austin and Zegras 2012). The study makes use of 4 days of GPS data from taxicab trips and demographic information about neighborhoods in Boston to develop a Poisson count model for taxicab trip generation.

The model specification is important for properly representing the trip generation process, and since the number of taxicab trips generated per census tract is a count variable, the model should be based on a count distribution. The most common count model is the Poisson model, which has been applied to many fields, such as property and liability insurance (Ismail and Jemain 2007), counting organisms in ecology (Hoef and Boveng 2007), crime incidents (Piza 2012), and transportation safety (Geedipally and Lord 2008). A critical assumption for the Poisson model is that mean and variance of the response variable are equal. In many cases, the variance of the response variable exceeds the mean, and data with such characteristics are considered overdispersed. There are several count models that may be suitable for overdispersed data, such as quasi-Poisson, zero-inflated Poisson, and negative binomial models (Washington et al. 2003).

In this paper, we show how an extensive data set including 10 months of taxicab trip data from NYC can be used to model taxi demand, and the proposed model differs from previous work in four ways: (1) the data set is large, spanning over 2000 census tracts and including observations from several months; (2) a negative binomial regression is compared to a conventional Poisson model, because a negative binomial distribution is more appropriate for modeling overdispersed count data; (3) transit accessibility is measured in terms of distance and service headway; and (4) taxicab supply is included as an explanatory variable in order to account for the effect that taxicab availability has on realizing demand.

3 Data

The database consists of complete information for all 147 million taxicab trips made in NYC between February 1, 2010 and November 28, 2010. Between 5.5 and 5.8 million taxi trips are made each day in New York City. Each record includes information about when and where a trip was made, the distance traveled, and the fare paid. Specifically, the dataset includes the following fields for each record:

1. Taxi Medallion Number, Shift Number, Trip Number, and Driver Name;
2. Pickup Location (latitude and longitude), Date, and Time;
3. Drop-off Location (latitude and longitude), Date, and Time;
4. Distance Travelled from Pickup to Drop-Off;
5. Number of Passengers;
6. Total Fare Paid, including breakdown by Fare, Tolls, and Tips;
7. Method of Payment (e.g., cash, credit card).

These data are collected by the Taxi & Limousine Commission (TLC) using the GPS and meter devices that are installed in every licensed (yellow medallion) taxi. The data was received in a TXT file format with a magnitude of 40 GB. This dataset is too large to manage using traditional data tools such as EXCEL or ACCESS, therefore a database management system; an SQL server, is used instead. Its primary querying language is T-SQL.

There are three steps to make this large dataset more manageable for regression analysis: first, the raw data is imported to the SQL server; second, less than 2 % original records are obviously false and have been eliminated, e.g., records without valid locational information or records with zero distance traveled or records with zero fare paid; finally, the locations of pick-ups and drop-offs are aggregated by NYC census tract, and the times are aggregated by hour of the day.

The response variable is the number of taxicab pick-up counts in each census tract per hour. Six explanatory variables are included, which have been identified as important in a previous study with the same data set (Yang and Gonzales 2014): population, education (percent of population with at least a bachelor's degree), median age, median income per capita, employment by industry sector, and transit accessibility. The number of taxicab drop-offs in each census tract during each hour is added as an additional explanatory variable representing the immediately available supply of taxicabs at each location and time. The sources of data for the explanatory factors considered in this study include:

- Drop-off taxi demand per hour aggregated by NYC census tract (DrpOff)
- 2010 total population that has been aggregated by NYC census tract (Pop)
- Median age that has been aggregated by NYC census tract (MedAge)
- Percent of education that is higher than bachelors aggregated by NYC census tract (EduBac)
- Transit Access Time (TAT), the combined estimated walking time a person must spend to access the nearest station (transit accessibility) and the estimated time that person will wait for transit service (transit level of service);
- Total jobs aggregated by NYC census tract (TotJob)
- Per capita income aggregated by NYC census tract (CapInc)

Since the model can only be estimated for census tracts with valid data for the response and explanatory variables, the data set is cleaned to eliminate census tracts that do not contain population or employment data. Of the 2143 census tracts in NYC, 17 census tracts are deleted from the data set.

4 Methodology

Linear regression models are inadequate for count data, because the response variable is a count of random events, which cannot be negative. As a result, the models that are developed and compared in this study are count models that are specifically developed to represent count processes. First, the Poisson regression model is introduced, following the reference of Ismail and Jemain (2007). In order to account for the varying effects that each of the explanatory variables have on the response variable at different times of the day, a separate model is estimated for the data in each hour.

Let Y_i be the independent Poisson random variable for the count of taxicab trips in census tract $i = 1 \ldots 2126$. The probability density function of Y_i is defined as:

$$\Pr(Y_i = y_i) = \frac{\exp(-\mu_i)\mu_i^{y_i}}{y_i!} \qquad (1)$$

where y_i is the number of observed counts in each census tract i, and $\mu_i = E(Y_i) = var(Y_i)$. The fitted value from a Poisson regression is defined as:

$$E(Y_i|x_i) = \mu_i = \exp(x_i^T \beta) \qquad (2)$$

where x_i is a $p \times 1$ vector of the p explanatory variables consider for each census tract i and β is a $p \times 1$ vector of regression parameters. The Poisson mean on the left side of the definition represents the non-negative expected number of trips generated. Note that for this model form, the logarithm of the observed count varies linearly with the explanatory variables.

The Poisson process is based on the assumption that the mean of the independent random variables is equal to the variance, so Poisson regression is only appropriate if the model count data has equal mean and variance. If the observed variance for the count data exceeds the mean, the data is said to be overdispersed, and an alternative model specification using either quasi-likelihood estimation or negative binomial regression may be used instead. Both methods use a generalized linear model framework. The approach using quasi-likelihood estimation follows Poisson-like assumptions. The mean and variance of the random variable Y_i are:

$$E(Y_i) = \mu_i \qquad (3)$$
$$Var(Y_i) = \theta\mu_i \qquad (4)$$

where θ is an over-dispersion parameter; $\theta = 1$ corresponds to the Poisson model. The quasi model formulation leaves the parameters in a natural state and allows standard model diagnostics without losing efficient fitting algorithms.

The negative binomial model is characterized by a quadratic relationship between the variance and mean of the response variable (Hoef and Boveng 2007). The density function of negative binomial is defined as in Ismail and Jemain (2007) for the Negative Binomial I model:

$$\Pr(Y_i = y_i) = \frac{\Gamma(y_i + v_i)}{\Gamma(y_i + 1)\Gamma(v_i)} \left(\frac{v_i}{v_i + \mu_i}\right)^{v_i} \left(\frac{\mu_i}{v_i + \mu_i}\right)^{y_i} \qquad (5)$$

where v_i is a parameter of the negative binomial distribution that is equivalent to the inverse of the dispersion parameter α. For the negative binomial model, the mean of Y_i is still μ_i as given by (3), but the variance is:

$$Var(Y_i) = \mu_i + \mu_i^2 v_i^{-1} = \mu_i + \mu_i^2 \alpha \tag{6}$$

The mean and the variance will be equal if $\alpha = 0$, so the Poisson distribution is also a special case of the negative binomial distribution. Values of $\alpha > 0$ indicate that the variance exceeds the mean, and the observed distribution is overdispersed.

4.1 Selecting Quasi-Poisson Distribution or Negative Binomial Distribution

In order to select the most appropriate model specification for the count regression, we must compare the mean and variance of the taxicab pick-up counts, which are the response variable for the proposed model. Table 1 presents a summary by hour of the day of the mean and variance of the total number of taxicab pick-ups per census tract in the 10-month data sample.

The variance of taxicab pick-ups per census tract in the 10-month dataset greatly exceeds the mean, as shown in Table 1, which provides an indication that the data is overdispersed. This pattern holds whether all counts from all hours of the day are considered together or the records are broken down by hour of the day. The implication is that the count model for the regression should be appropriate for overdispersed data. To choose between the quasi-Poisson distribution and the negative binomial distribution, it is necessary to look at how the mean and variance appear to be related. Since a goal of this study is to consider how the effect of explanatory variables changes with the hour of the day, a separate model is estimated for each hour, and the comparison of mean and variance must be considered within each hourly aggregation. Figure 1 presents separate plots comparing count mean and variance for three representative hours: hour 0 is 12:00 A.M.–1:00 A.M. (midnight); hour 8 is 8:00 A.M.–9:00 A.M. (morning peak); and hour 17 is 5:00 P.M.–6:00 P.M. (evening peak).

In order to choose the distribution that most appropriately represents the response variable, the data within each hourly aggregation are divided in 100 subsets using the quantiles of the taxicab pick-up counts. The first category includes taxicab pick-ups for census tracts whose counts fall between the 0 quantile and 0.01 quantile, the second category includes census tract data in the range of the 0.01 quantile and 0.02 quantile, and so on. Within each quantile category, the mean and variance of the included data are calculated, and plotted in Fig. 1. A linear function of the form shown in (4) is fitted to estimate θ and see how well the data matches the assumed relationship for a quasi-Poisson regression model. A quadratic function of the form shown in (6) is fitted to estimate α and see how well the data matches the

Table 1 Mean and variance of response variable by hour

Hour of day	Mean value of pickup counts	Variance of pickup counts
0	2446	100,804,416
1	1827	63,070,167
2	1373	41,751,337
3	994	24,875,367
4	711	10,338,265
5	558	4,547,384
6	1202	25,984,332
7	2213	79,611,311
8	2862	128,381,917
9	2948	137,735,882
10	2801	122,508,279
11	2857	131,214,089
12	3063	153,234,156
13	3030	149,228,639
14	3133	161,516,405
15	2976	143,190,381
16	2586	106,107,666
17	3115	151,429,667
18	3759	224,925,568
19	3913	244,880,173
20	3615	209,472,196
21	3469	195,600,486
22	3360	186,065,870
23	3017	147,131,917
All hours	61,828	57,071,324,807

assumed relationship for a negative binomial regression model. The goodness of fit parameter, R^2, is used to identify which specification fits the data better. A value of R^2 closer to 1 indicates a better fit. It can be seen in the examples for hours 0, 8, and 17 (Fig. 1) that the quadratic function provides a better fit for relating the variance and mean, indicating that the negative binomial distribution is more appropriate for the counts of taxicab pick-ups.

4.2 Selecting Poisson Regression or Negative Binomial Regression

Although the overdispersed taxicab pick-up data appear to show that a negative binomial regression is a more appropriate model than a Poisson regression, it is also necessary to compare the fit of the models with the explanatory variables that have been identified in the Data section. Several methods can be used to compare the fit

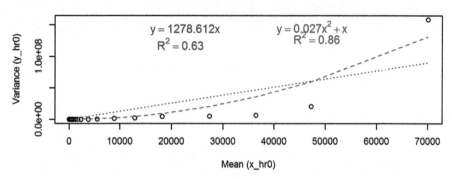

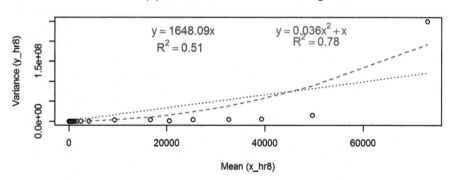

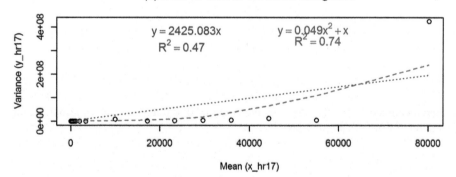

Fig. 1 Plot of the variance vs. mean of the aggregated hourly taxicab pick-up counts for (**a**) hour 0 (midnight), (**b**) hour 8 (morning peak), and (**c**) hour 17 (evening peak). Data are grouped into 100 categories by quantile. The linear equation for the quasi-Poisson model is shown with the *blue dotted line*. The quadratic equation for the negative binomial model is shown with the *red dashed line*

of a Poisson regression and a negative binomial regression. Each of the following statistics provides a different type of measure of how well the regression model fits the data set, and these are used to compare the models and select the most appropriate model to relate the explanatory variables to the response variable, the number of taxicab pick-ups per hour.

1. *Akaike Information Criterion*

   The Akaike Information Criterion (AIC) is a measure of the relative quality of statistical models by trading off the complexity and goodness of fit. This measure is defined as a function of the Log Likelihood (Fox 2008; Yan and Su 2009):

   $$AIC = 2p - 2LL \qquad (7)$$

   where p is the number of parameters in the model and LL is the Log Likelihood of the model. A smaller AIC value represents a better model, and the measure is used to ensure that the model is not overfitted to the data.

2. *Goodness-of-Fit Test*

   The goodness-of-fit test is an analysis of variance (ANOVA) test based on calculating the Pearson's residuals. The Pearson test statistics is (Cameron and Windmeijer 1996):

   $$\chi^2 = \sum_{i=1}^{n} e_i^2 \qquad (8)$$

   where the definition of the Pearson residual, e_i, depends on whether the regression is a Poisson model of a negative binomial model:

   Poisson Model $\qquad e_i = \dfrac{Y_i - \hat{\mu}_i}{\sqrt{Var(Y_i)}} = \dfrac{Y_i - \hat{\mu}_i}{\sqrt{\hat{\mu}_i}} \qquad (9)$

   Negative Binomial Model $\qquad e_i = \dfrac{Y_i - \hat{\mu}_i}{\sqrt{Var(Y_i)}} = \dfrac{Y_i - \hat{\mu}_i}{\sqrt{\hat{\mu}_i + \hat{\mu}_i^2 \alpha}} \qquad (10)$

   In both of these cases, Y_i is the observed count of taxicab pick-ups in census tract i, $\hat{\mu}_i$ is the modeled count, and there are a total of n census tracts included in the dataset. The Pearson statistic, χ^2, is approximately distributed as chi-square with $n-p$ degrees of freedom, where n is the number of observations and p is the number of predicted parameters (i.e., one parameter per explanatory variable included in the model). If $\chi^2 > \chi^2_{n-p, 0.05}$ or P-value $(\chi^2 \, test) < 0.05$ then the model is statistically different from the observed data at the 95 % confidence level. Therefore, we seek a model with P-value $(\chi^2 \, test) > 0.05$.

3. *Sum of Model Deviances*

The sum of squared deviance residuals, G^2, is a measure of model fit which is used for Poisson regressions. The sum of model deviances is calculated as (Washington et al. 2003):

$$G^2 = 2\sum_{i=1}^{n} y_i \ln \frac{y_i}{\hat{\mu}_i} \tag{11}$$

If the model fits the data perfectly, then $G^2 = 0$, because $\hat{\mu}_i = y_i$ for every census tract i. For a count model, such as the Poisson or negative binomial, the observed values are always integers, but the model produces values of $\hat{\mu}_i$ that are continuous, so it is very unlikely to achieve zero sum of squared deviance residuals. Nevertheless, the value of G^2 provides a useful measure of the error in the model.

4. *Likelihood Ratio Test*

A second goodness-of-fit test is based on a comparison of the fits of competing models. The likelihood ratio is calculated using the Log Likelihood, LL, of each model (Washington et al. 2003).

$$\chi^2_{LL} = -2[LL(\text{model}\,1) - LL(\text{model}\,2)] \tag{12}$$

where the χ^2_{LL} statistics is chi-squared distributed with degrees of freedom equal to the difference of the number of parameters estimated in model 1 and model 2. If χ^2_{LL} is larger than the critical value for the 95 % confidence level, then model 1 is said to be statistically different from model 2.

5 Results

Having identified that the negative binomial distribution is more appropriate for the taxicab pick-up data than the quasi-Poisson distribution (as illustrated in Fig. 1), it useful to compare the results using a conventional Poisson regression model with the results of negative binomial regression in order to demonstrate the effect of acknowledging the overdispersed response variable. In order to make a comparison between the Poisson and negative binomial regressions, a separate model has been estimated for each hour of the day. The coefficients of the explanatory variables for both models are shown in Table 2 and Fig. 2. In order to show all of the coefficients on the same graph, we have normalized the scale such that a magnitude of 1 is the magnitude of the coefficient at midnight (hour 0). This allows us to see the relative changes in each coefficient value, and the sign for positive and negative values is not changed.

Table 2 Parameters of the Poisson and negative binomial regressions for taxicab trip generation

	Hour	DrpOff	Pop	MedAge	EduBac	TAT	TotJob
Poisson Model Coefficients	0	0.000048	0.000056	0.021619	0.035982	-0.158105	0.000005
	1	0.000081	0.000026	0.026725	0.031175	-0.145743	0.000001
	2	0.000103	0.000014	0.021769	0.035571	-0.115985	-0.000008
	3	0.000194	-0.000039	0.030535	0.032231	-0.095762	-0.000020
	4	0.000160	0.000081	-0.006056	0.039077	-0.120689	-0.000004
	5	0.000075	0.000162	-0.007020	0.047770	-0.129105	0.000005
	6	0.000031	0.000185	0.007363	0.050621	-0.126587	0.000003
	7	0.000021	0.000183	0.013675	0.054706	-0.107053	0.000000
	8	0.000022	0.000175	0.014264	0.053651	-0.104637	-0.000003
	9	0.000025	0.000180	0.015501	0.048587	-0.118816	-0.000002
	10	0.000029	0.000175	0.015170	0.045607	-0.099687	0.000004
	11	0.000028	0.000146	0.017751	0.045852	-0.094150	0.000007
	12	0.000028	0.000123	0.015003	0.044640	-0.090213	0.000007
	13	0.000027	0.000114	0.021944	0.046272	-0.100170	0.000006
	14	0.000023	0.000102	0.026618	0.048343	-0.114502	0.000008
	15	0.000023	0.000091	0.027366	0.049723	-0.116026	0.000008
	16	0.000026	0.000082	0.027292	0.050917	-0.118006	0.000008
	17	0.000024	0.000073	0.026853	0.049711	-0.114158	0.000008
	18	0.000023	0.000060	0.028121	0.046506	-0.110181	0.000009
	19	0.000022	0.000052	0.029545	0.046116	-0.112516	0.000009
	20	0.000027	0.000039	0.032460	0.044167	-0.127821	0.000009
	21	0.000030	0.000038	0.028421	0.041503	-0.138917	0.000008
	22	0.000033	0.000037	0.027944	0.038780	-0.137978	0.000006
	23	0.000040	0.000041	0.025840	0.036323	-0.142699	0.000003
	Hour	DrpOff	Pop	MedAge	EduBac	TAT	TotJob
Negative Binomial Model Coefficients	0	0.000311	0.000088	-0.018169	0.037879	-0.071524	0.000045
	1	0.000452	0.000071	-0.014504*	0.035677	-0.076790	0.000044
	2	0.000605	0.000063	-0.011550*	0.033998	-0.084690	0.000043
	3	0.000861	0.000042	-0.011454*	0.031256	-0.082681	0.000039
	4	0.001386	0.000035*	-0.016841	0.025647	-0.076768	0.000008*
	5	0.001266	0.000138	-0.026053	0.036176	-0.081259	0.000007*
	6	0.000177	0.000295	-0.015108*	0.058942	-0.080476	0.000036
	7	0.000101	0.000310	0.008864*	0.058380	-0.082836	0.000035
	8	0.000079	0.000331	0.017103*	0.055015	-0.085756	0.000035
	9	0.000070	0.000337	0.017877*	0.058839	-0.088309	0.000053
	10	0.000088	0.000309	0.001142*	0.059663	-0.075448	0.000071
	11	0.000101	0.000295	0.006268*	0.054723	-0.080959	0.000082
	12	0.000106	0.000275	0.000127*	0.051781	-0.076189	0.000091
	13	0.000115	0.000258	-0.005119*	0.050155	-0.075967	0.000079
	14	0.000121	0.000231	-0.007468*	0.046339	-0.075067	0.000076
	15	0.000138	0.000208	-0.016232*	0.042787	-0.072903	0.000076
	16	0.000163	0.000188	-0.022523	0.041864	-0.069031	0.000086
	17	0.000138	0.000186	-0.023651	0.041063	-0.071469	0.000090
	18	0.000110	0.000210	-0.015490*	0.044808	-0.073380	0.000094
	19	0.000104	0.000215	-0.010033*	0.047140	-0.076604	0.000104
	20	0.000122	0.000200	-0.010243*	0.044864	-0.074462	0.000099
	21	0.000138	0.000183	-0.012317*	0.044208	-0.070225	0.000091
	22	0.000161	0.000152	-0.003011*	0.040828	-0.071010	0.000066
	23	0.000215	0.000122	-0.017411*	0.040040	-0.069105	0.000055

'*' and light shade indicates non-significance of the coefficients at 0.05 level

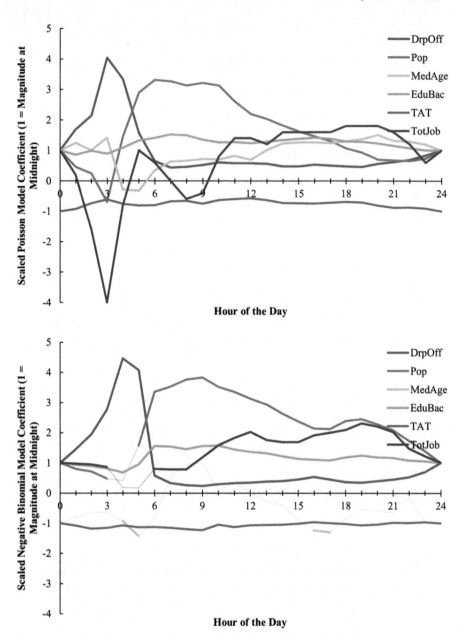

Fig. 2 Scaled coefficients of parameters of the Poisson and negative binomial regressions for taxicab trip generation by hour of the day (the light shade indicates non-significance of the coefficients at 0.05 level)

The six explanatory variables that are significant are the number of drop offs, which represents the available supply of empty taxicabs in a census tract (DrpOff); population (Pop); median age of residents (MedAge); percent of population attaining at least a bachelor's degree (EduBac); transit access time (TAT); and the total number of jobs located in the census tract (TotJob). Although per capita income had been identified as an important explanatory variable in a previous model based on linear regression (Yang and Gonzales 2014), the income is highly correlated with the measure of educational attainment (EduBac) in NYC. Therefore, to avoid problems associated with autocorrelation of the explanatory variable, income has been omitted, and the level of education is kept in the models.

The coefficients for all variables in the Poisson regression are statistically significant at the level of 0.05 (see top half part of Table 2). While most of the coefficients remain significant when the negative binomial regression is used, median age fails to exhibit significance at the 0.05 level for most hours of the day (see lower part of Table 2 and Fig. 2). The magnitude of the model parameters is more stable across models for some explanatory variables than others (Fig. 2). Also, as shown from Fig. 2, the variables TAT, EduBac and MedAge exhibit bigger magnitude and variability throughout the day compared to the other three variables: DrpOff, MedAge, and Pop. The results show that regardless of model specification, the taxi supply (DrpOff), education (EduBac), and transit accessibility (TAT) are always significant determinants of taxicab pick-up demand at all times of the day.

In order to determine whether a Poisson regression or a negative binomial regression fits better with the observed hourly taxicab pick-ups, the fours goodness of fit tests introduced previously are used to assess the fit of the two models. These statistics are summarized in Table 3 for each of the 24 Poisson models and 24 negative binomial models (i.e., one for each hour of the day). The results in Table 3 show through many methods of comparison that the negative binomial regression provides a better fit for the data than the Poisson regression. The interpretations of the statistics are as follows:

1. The *AIC* values for the negative binomial regression models are much lower than for the Poisson regression models.
2. Both models' specifications suffer from low p-values for the χ^2 test, so there is substantial variation in the observed data that is not explained by the models. These errors will be reflected in the residuals, so an analysis of the residuals is valuable and necessary.
3. The sums of squared deviances for the negative binomial regression models are much smaller than for the Poisson regression models.
4. The very low p-values of likelihood ratio test statistics suggest that the negative binomial regression model is very different from the Poisson regression model.

In light of the numerous differences between the models, the better fit and more appropriate model is the negative binomial regression. That said, the model is not perfect, and although a number of statistically significant explanatory variables and parameters have been identified, these are not sufficient to fully explain the

Table 3 Goodness-of-fit statistics for the Poisson (POI) and negative binomial (NB) models

Interpretation	AIC		P-value (X^2 test)		G^2		Likelihood ratio test between POI and NB
	The smaller the better		Significant if P-value is less than 0.05		The smaller the better		Significant if P-value is less than 0.05
Hour	POI	NB	POI	NB	POI	NB	P-value (X^2_{LL})
0	4,139,707	17,535	0	0	4,132,262	2293	0
1	3,095,928	17,005	0	0	3,088,663	2267	0
2	3,196,864	16,254	0	0	3,189,866	2248	0
3	2,600,073	15,684	0	0	2,593,272	2233	0
4	2,165,955	15,695	0	0	2,159,067	2249	0
5	1,398,907	15,302	0	0	1,392,502	2187	0
6	2,866,840	16,003	0	0	2,860,328	2153	0
7	4,510,050	16,852	0	0	4,503,196	2206	0
8	4,949,851	16,851	0	0	4,943,023	2180	0
9	4,703,359	16,006	0	0	4,696,921	2073	0
10	4,188,069	15,906	0	0	4,181,654	2118	0
11	4,357,659	15,656	0	0	4,351,337	2075	0
12	4,609,746	15,734	0	0	4,603,387	2081	0
13	4,620,149	15,732	0	0	4,613,788	2088	0
14	5,142,257	16,043	0	0	5,135,735	2125	0
15	5,072,322	16,034	0	0	5,065,810	2122	0
16	4,448,535	15,912	0	0	4,442,054	2123	0
17	5,147,389	16,414	0	0	5,140,683	2154	0
18	5,958,635	16,625	0	0	5,951,825	2151	0
19	6,493,192	16,740	0	0	6,486,305	2173	0
20	5,795,703	16,793	0	0	5,788,756	2194	0
21	5,569,018	17,090	0	0	5,561,899	2249	0
22	5,591,048	17,466	0	0	5,583,733	2280	0
23	4,965,376	17,625	0	0	4,957,950	2288	0

variation in the number of taxicab trips that are generated in census tracts across NYC.

The interpretation of the model parameters is the same for the negative binomial and Poisson regressions, because both models employ a Log link function. With every unit increase of explanatory variable x, the predictor for the response variable increases by a multiplicative factor $\exp(\beta)$. For example the parameter of population in hour 8 of the negative binomial model is 0.000331, so an increase of population in a census tract by one inhabitant will tend to increase demand by a factor of 1.00033. A positive parameter indicates that the explanatory variable is associated with increased numbers of taxicab pick-ups, and a negative parameter indicates an effect of decreased taxicab pick-ups.

In the negative binomial model, the model parameters all have the expected signs. The number of observed taxicab pick-ups increases with taxi supply (DrpOff), population (Pop), education (EduBac), and the total number of jobs within a census tract (TotJob). The effect of transit access time (TAT) is negative, which means that more taxi pick-ups are made in places that have shorter or faster access to the subway service. There are a couple of possible reasons why a tendency for taxis to be used in the same places that have good transit service should exist. One reason is that people may be likely to take taxis in order to get to or from transit services, so a subway station is a place where a traveler exits the transit system and may look for a taxicab to reach his or her final destination. Another reason is that the types of people and trip purposes that tend to use taxis (e.g., high value of time, unwillingness to search or pay for parking) also tend to occur in parts of the cities that have a lot of transit service. The negative parameter value for TAT is consistent for every hour of the day in the Poisson and negative binomial models, but the precise reason cannot be determined from this regression alone.

One objective of this study is to identify the locations and times of day when there may be a mismatch between taxi demand and supply. One way to investigate this is to look specifically at the Pearson residuals from the models as defined in (10). For a single hour of the day, the residuals for each census tract in the city can be mapped in order to visualize the spatial distribution of the model errors. Maps are presented in Fig. 3 for hour 0 (midnight), hour 8 (morning peak), and hour 17 (evening peak), and the color indicates where the model overestimates taxicab pick-ups (i.e., negative residual shown in green) and where the model underestimates taxicab pick-ups (i.e., positive residuals shown in red). The Pearson residual is calculated by dividing the actual residual by the assumed standard deviation, which for the negative binomial model, increases as a quadratic function of the mean. This manipulation is used to show the magnitude of error in a standardized manner so that busier census tracts don't dominate the figure since larger observed and fitted counts will tend to have errors that are larger in magnitude even if those errors are small relative to the expected variance.

Taxicab supply is included as an explanatory variable in the model, and the availability of cabs is shown to increase the number of realizing taxicab pick-ups (because the parameter value for DrpOff is positive). A negative residual, which represents an overestimate from the model, provides an indication that there are relatively fewer taxicab pick-ups being demanded relative to the supply of empty cabs available, controlling for the characteristics of the neighborhood. Conversely, a positive residual, which represents an underestimate from the model, provides an indication that there are relatively more taxicab pick-ups being demanded relative to the supply of empty taxicabs. Census tracts that fall into this second condition are of interest, because these are the locations during each hour of the day, that appear to have insufficient taxicab service relative to similar neighborhoods in other parts of the city.

In hour 0 (12:00 A.M.–1:00 A.M., midnight), the central part of Manhattan and most of the census tracts in the Outer Boroughs have negative Pearson residuals (colored green in Fig. 3a), and the model overestimates the realizing count of

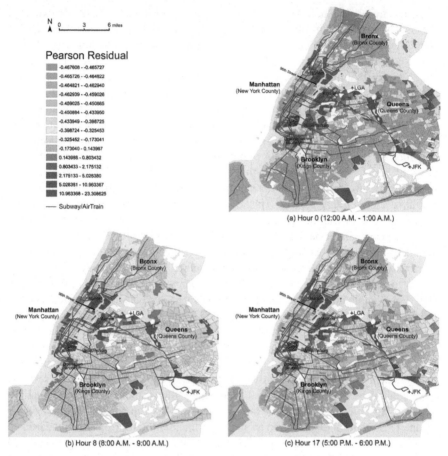

Fig. 3 Pearson residuals of the negative binomial regression models for (**a**) hour 0, (**b**) hour 8, and (**c**) hour 17

taxicab pick-ups. The census tracts with positive Pearson residuals (colored red in Fig. 3a), mean that there are more taxicab pick-ups than our model predicts in northern Manhattan, the Lower East Side, western Queens, and the downtown and Williamsburg parts of Brooklyn. These are neighborhoods where there tends to be more night activity than indicated by the explanatory variables and thus more taxi demand. These are the neighborhoods where there is likely to be the largest mismatch between the supply of available taxicabs and the number of people who seek to be picked up by a taxicab.

It is useful to compare the patterns from hour 0 with other hours of the day, because activity patterns in NYC change over the course of the day. In hour 8 (8:00 A.M.–9:00 A.M., morning peak), the negative Pearson residuals in central Manhattan and much of the Outer Boroughs reduce in magnitude (yellow or light orange in Fig. 3b). This suggests that the binomial regression model provides a better fit during the morning, and taxicab pick-up counts are estimated with less error. One reason for this may be that data associated with the residents of a census tract are most relevant for predicting the number of trips that these residents are likely to make from their homes in the morning. In hour 17 (5:00 P.M.–6:00 P.M., evening peak), the magnitudes of the Pearson residuals become larger again.

Despite the variations, there are consistent patterns in the maps of the Pearson residuals across all hours of the day. The locations where the model underestimates trips at midnight also tend to have underestimated trips in the morning and evening. Many of these neighborhoods, such as Harlem, the Lower East Side, Astoria, Williamsburg, and Downtown Brooklyn, are dense residential neighborhoods with vibrant local businesses but without the same level of large commercial and tourist activities as are concentrated in much of Manhattan. This may be a reason why these inner neighborhoods are associated with high demand for taxicab pick-ups, but the taxicab fleet has a tendency to focus service in more central parts of the Manhattan. Many of the further outlying neighborhoods in the Bronx, Queens, Brooklyn, and Staten Island tend to have demand overestimated by the model. This is likely because the populations in those areas either have lower incomes, which make them less likely to choose to pay for taxicab service, or the neighborhood development is at lower densities, which are more conducive to travel by private car than by hailing a taxicab.

The TLC has already changed policies to address some of the mismatch between taxicab supply and demand in NYC. The green Street Hail Livery vehicles (Boro taxi) are allowed to pick passengers only in Manhattan above 96th Street and in the Outer Boroughs, not including the airports. This coverage area overlaps with many of the underestimated (and potentially underserved) neighborhoods identified in Fig. 3. One part of the city that is consistently underestimated in the models but is not within the green cab's pick-up area is Manhattan's Lower East Side. One reason for this may be the recent growth that has occurred in the neighborhood, which has increased activity but may not be reflected in the service provided by taxis. Nevertheless, the Lower East Side is an example of a neighborhood area that this modeling approach can identify as being in need of additional taxicab service. These models and figures could be useful tools for transportation planners who want to understand where taxicab service is used, and where more taxicab supply is needed.

6 Conclusion

This study made use of a negative binomial regression model to interpret 10 months of overdispersed taxicab demand data in NYC. Negative binomial regressions have been broadly applied to biology, bio-chemistry, insurance, and finance industries, and this paper shows that the model approach is well suited demand modeling for taxicabs. The raw taxicab dataset includes 147 million records, and in order to make sense of the patterns, the records are aggregated by census tract and hour of the day in order to develop meaningful models of the way that taxicab demand varies across space and time. A number of count regression models have been considered to model the number of taxicab pick-ups per census tract within an hour of the day, including the Poisson model, quasi-Poisson model, and negative binomial model. By a series of statistical tests, the negative binomial model is shown to be most appropriate for the overdispersed count data. An analysis of the residuals provides useful insights about where taxicab demand appears to be adequately served by the existing supply of taxicabs, and where there is a need for more taxicab services.

The modeling approach started by using important explanatory variables that were identified in a previous modeling effort that used the same dataset (Yang and Gonzales 2014). An additional explanatory variable was added to represent the taxicab supply, and this is the number of taxicab drop-offs in each census tract during each hour of the day, because each drop-off corresponds to a taxicab becoming available for another customer. The negative binomial regression shows that three explanatory variables are significant during every hour of the day (drop-offs, educational attainment, and transit access time), and two others are significant during most waking hours of the day (population and total number of jobs).

The residual graphs suggest that central Manhattan and most of the Outer Boroughs have at least enough taxi supply for the demand that is observed, controlling for neighborhood characteristics. The northern part of Manhattan, the Lower East Side, and the western parts of the Queens and Brooklyn all have more observed taxicab pick-ups than the model predicts. The fleet of green Street Hail Livery vehicles serves some of these neighborhoods but not the Lower East Side. The maps of residuals provide some useful insights for the transportation planners to understand when and where we need more taxis.

The taxicab data used to create these models has both spatial and temporal dimensions. The effect of time is accounted for by separating the data by hour of the day, and fitting a negative binomial regression for each hour. The effect of the time of the day that each explanatory variable has on the number of taxicab pick-ups can be observed by comparing the parameter values in Table 2. The variability of the coefficients over time is shown in Fig. 2. Additional modeling effort is needed to also account for the spatial correlations in the data set. It is clear from the maps of residuals that adjacent census tracts have correlated performance indicated by the correlated errors. One way to account for these correlations is with a Generalized Linear Mixed Model. Other efforts to improve the model would be to consider additional explanatory variables to account for the activities or popularity of a census tract or to account for the movement of empty taxicabs in search of customers.

Large datasets, such as the records of taxicab trips in NYC, present some challenges, because the raw data are too big to be analyzed directly by conventional methods. By processing the data, and developing models that relate the taxicab data to other sources of information about the characteristics of different parts of the city at different times of day, it is possible to gain useful insights about the role that taxicabs play in the broader transportation system. More importantly, these insights can be used to plan and improve the transportation system to meet the needs of users.

References

Arnott R (1996) Taxi travel should be subsidized. J Urban Econ 40:316–333
Austin D, Zegras C (2012) Taxicabs as public transportation in Boston, Massachusetts. Transp Res Rec 2277:65–74
Bai R, Li J, Atkin JAD, Kendall G (2013) A novel approach to independent taxi scheduling problem based on stable matching. J Oper Res Soc 65(10):1501–1510. doi:10.1057/jors.2013.96
Balan RK, Nguyen KX, Jiang L (2011) Real-time trip information service for a large taxi fleet. In: MobiSys'11—Proceedings of the 9th international conference on mobile systems, applications, and services, Bethesda, Maryland, 28 June–1 July, p 99–112
Beesley ME (1973) Regulation of taxis. Econ J 83(329):150–172
Bloomberg MR, Yassky A (2014) 2014 Taxicab factbook. New York City Taxi & Limousine Commission. http://www.nyc.gov/html/tlc/downloads/pdf/2014_taxicab_fact_book.pdf
Cameron AC, Windmeijer FAG (1996) R-squared measures for count data regression models with applications to health care utilization. J Bus Econ Stat 14(2):209–220
Douglas GW (1972) Price regulation and optimal service standards the taxicab industry. J Transp Econ Policy 6(2):116–127
Eckert RD (1973) On the incentives of regulators: the case of taxicabs. Public Choice 14(1):83–99
Frankena M, Pautler P (1984) An economic analysis of taxicab regulation. Bureau of Economics Staff Report, Federal Trade Commission
Fox J (2008) Applied regression analysis and generalized linear models, 2nd edn. Sage, Thousand Oaks
Gaunt C (1995) The impact of taxi deregulation on small urban areas: some New Zealand evidence. Transp Policy 2(4):257–262
Geedipally SR, Lord D (2008) Effects of varying dispersion parameter of Poisson-gamma models on estimation of confidence intervals of crash prediction models. Transp Res Rec 2061 (1):46–54
Girardin F, Blat J (2010) The co-evolution of taxi drivers and their in-car navigation systems. Pervasive Mob Comput 6:424–434
Hoef JMV, Boveng PL (2007) Quasi-Poisson vs. negative binomial regression: how should we model overdispersed count data? Ecology 88(11):2766–2772
Ismail N, Jemain AA (2007) Handling overdispersion with negative binomial and generalized Poisson regression models. Casualty Actuarial Society Forum, p 103–158
Liang X, Zhao J, Dong L, Xu K (2013) Unraveling the origin of exponential law in intra-urban human mobility. Sci Rep 3:2983. doi:10.1038/srep02983
Maa P (2005) Taxicabs in U.S. cities and how governments act. MMSS senior thesis, Northwestern University
Orr D (1969) The taxicab problem: a proposed solution. J Polit Econ 77(1):141–147
Piza EL (2012) Using Poisson and negative binomial regression models to measure the influence of risk on crime incident counts. Rutgers Center on Public Safety. http://www.rutgerscps.org/docs/CountRegressionModels.pdf
Schaller B (1999) Elasticities for taxicab fares and service availability. Transportation 26:283–297
Schaller B (2005) A regression model of the number of taxicabs in U.S. cities. J Public Trans 8:63–78
Schaller Consulting (2006) The New York City taxicab fact book, 2006. http://www.schallerconsult.com/taxi/taxifb.pdf
Schreiber C (1975) The economic reasons for price and entry regulation of taxicabs. J Transp Econ Policy 9:268–279
Washington SP, Karlaftis MG, Mannering FL (2003) Statistical and econometric methods for transportation data analysis. CRC Press, Boca Raton, FL
Yan X, Su XG (2009) Linear regression analysis: theory and computing, 1st edn. World Scientific, Saddle Brook
Yang C, Gonzales EJ (2014) Modeling taxi trip demand by time of day in New York City. Transp Res Rec 2429:110–120

Detecting Stop Episodes from GPS Trajectories with Gaps

Sungsoon Hwang, Christian Evans, and Timothy Hanke

Abstract Given increased access to a stream of data collected by location acquisition technologies, the potential of GPS trajectory data is waiting to be realized in various application domains relevant to urban informatics—namely in understanding travel behavior, estimating carbon emission from vehicles, and further building healthy and sustainable cities. Partitioning GPS trajectories into meaningful elements is crucial to improve the performance of further analysis. We propose a method for detecting a stay point (where an individual stays for a while) using a density-based spatial clustering algorithm where temporal criterion and gaps are also taken into account. The proposed method fills gaps using linear interpolation, and identifies a stay point that meets two criteria (spatial density and time duration). To evaluate the proposed method, we compare the number of stay points detected from the proposed method to that of stay points identified by manual inspection. Evaluation is performed on 9 weeks of trajectory data. Results show that clustering-based stay point detection combined with gap treatment can reliably detect stop episodes. Further, comparison of performance between using the method with versus without gap treatment indicates that gap treatment improves the performance of the clustering-based stay point detection.

Keywords GPS • Trip detection • DBSCAN • Trajectory data analysis • Uncertainty

1 Introduction

GPS trajectory data, a temporally ordered sequence of GPS track logs with (x, y, t) (or coordinates at a given time interval), have been found useful in many application domains. GPS trajectories have been used to complement personal travel

---

S. Hwang (✉)
Geography, DePaul University, Chicago, IL, USA
e-mail: shwang9@depaul.edu

C. Evans • T. Hanke
Physical Therapy, Midwestern University, Downers Grove, IL, USA
e-mail: CEvans@midwestern.edu; THANKE@midwestern.edu

surveys (Stopher et al. 2008), detect activities of individuals (Eagle and (Sandy) Pentland 2006; Liao et al. 2007; Rodrigues et al. 2014), improve measurement of physical activity and community mobility (Krenn et al. 2011; Hwang et al. 2013), examine effects of transportation infrastructure on physical activity (Duncan et al. 2009), understand the role of environmental factors in occurrence of diseases (Richardson et al. 2013), recommend locations for location-based social network services (Zheng et al. 2011), and detect anomalous traffic patterns in urban settings (Pang et al. 2013).

The utility of GPS trajectory data will only grow as a greater amount of GPS data become readily available from location-aware devices. With increasing population in urban areas, the ability to track human movements in high spatiotemporal resolution offers much potential in urban informatics—"the study, design, and practice of urban experiences across different urban contexts that are created by new opportunities of real-time, ubiquitous technology and the augmentation that mediates the physical and digital layers of people networks and urban infrastructures" (Foth et al. 2011). GPS data integrated with other data such as survey, census, and GIS data can lead to a better understanding of "circulatory and nervous systems" of cities (Marcus 2008). Aided by patterns extracted from GPS trajectories (e.g., patterns of vehicle traffic or human movement), one can examine travel behavior, traffic congestion, community participation, and human impacts on the environment at a fine granularity (Zheng et al. 2014).

Whether this potential can be realized, however, rests on accuracy of algorithms that detect those patterns from GPS trajectory data. Unfortunately, it is difficult to detect patterns reliably from GPS trajectories that are largely characterized as voluminous and noisy. To illustrate, hundreds of thousands of data points are generated from GPS tracking of an individual during the period of a week with 5 s intervals. This renders manual processing of GPS trajectories nearly infeasible especially when dealing with a collection of individual GPS trajectories over an extended period of time. GPS trajectories are often interspersed with *spatial outliers* where coordinates of track logs are deviated from those of temporally neighboring track logs. Further, GPS trajectories contain a fair amount of *gaps* (the time period when track logs are not recorded) caused when GPS satellite signals cannot be received indoors or when the GPS logger battery runs out. While needs for automated procedures for processing GPS data exist, developing algorithms that are robust to uncertainty remains a challenging task.

This paper presents a method for identifying *stay points* from individual GPS trajectories based on spatiotemporal criteria while treating gaps in GPS trajectories. Operationally, stay points are where an individual stays (or does not move) for a minimal time duration (Li et al. 2008). It is important to detect stays points because stay points may indicate where individuals are likely to conduct activities (e.g., work in the office, shop at a store, and gather for social occasions) that are meaningful to individuals. Stay points are fine-grained snapshots of the activity space where individuals work, live, and play. Once stay points or personally important places are identified, one can infer semantics of those places and activities that might have occurred in those places with the aid of analytics and contextual data (like POIs) (Liao et al. 2007; Ye et al. 2009).

Stay point detection is a crucial step in trajectory data analysis from a methodological standpoint. Based on a stay point, trajectories can be divided into a set of homogeneous segments (or *episodes*), which concisely capture semantics of trajectories that are of interest to application domains (Parent et al. 2013). The most commonly used criterion for segmenting trajectories is whether data points are considered to consist of *stop* (a region where activities are conducted) or *move* (a route taken from stop A to stop B). For instance, on such segments it is possible to reliably infer mode of transportation (motorized or not), types of activities, purpose of trips (discretionary or not), and vehicle emissions in the context of formulating policies toward a healthy or sustainable city. Practitioners in disaster management can learn about the most frequent sub-trajectories or trajectories with similar behaviors from collective trajectories (like hurricanes or tornados) for disaster preparedness (Dodge et al. 2012).

This paper intends to highlight stay point detection as a preliminary step to segmentation of individual trajectories. We extend the previous work by Hwang et al. (2013) that proposes a method for segmenting trajectories into stop or move (trip) based on a density-based spatial clustering algorithm (like DBSCAN) adapted to temporal criteria. DBSCAN detects spatial clusters by checking whether the minimum number of data points (*MinPts*) is within a search radius (*eps*) from a randomly chosen *core point* to detect initial clusters, and aggregating those initial clusters that are *density connected* (Ester et al. 1996). Spatial clusters with high density and arbitrary shape can indicate a stay point, and those clusters can be detected reliably by DBSCAN. Finally, those spatial clusters that last for the minimal time duration are identified as a stay point. The previous work, however, does not consider cases when gaps might represent a stop. Satellite signals are lost when one enters a building, and this can result in falling short of *MinPts* necessary to be detected as a spatial cluster. The same problem occurs when one exits a building because the time required for a GPS receiver to acquire satellite signals may take a while. In this paper we present a new method for stay point detection that addresses the limitation above. The proposed method fills gaps such that DBSCAN can detect stay points reliably even in the presence of gaps.

The remaining part of this paper is organized as follows. In Sect. 2, we review related work focusing on stay point detection from GPS trajectories in relation to the field of trajectory data analysis. Section 2 intends to highlight diversity and limitations in methods for detecting a stay point. We describe the proposed method in Sect. 3. The proposal combines density-based clustering with gap treatment to achieve improved results. We implement the proposed method as part of measuring community mobility, more specifically to infer the number of stops made. We compare the performance of methods proposed in the previous and current work in terms of how accurate the number of stops predicted is. These results are presented in Sect. 4. We conclude the study by discussing lessons learned and limitations of the proposed method in Sect. 5.

2 Related Work

The field dedicated to trajectory data analysis is a wide-ranging and well-established area of research (Zheng et al. 2011; Zheng 2015; Parent et al. 2013). Techniques relating to the analysis of trajectory data can be classified into three categories: pre-processing, management, and mining (Zheng 2015). Pre-processing is to makes imperfect raw data clean, accurate, and concise. Management is concerned with efficiently accessing and retrieving data. Mining encompasses various techniques for uncovering patterns and behaviors.

Stay point detection is a part of pre-processing, and provides a means to segment trajectories into *episodes*, which are maximal subsequences of a trajectory that are homogeneous segments with respect to segmentation criteria such as stillness vs. movement (Parent et al. 2013). Identifying stay points or *stop episodes* enhance semantics and reduce complexity on trajectories. A stay point can be detected solely based on GPS raw data that contains spatiotemporal attributes (such as x, y, t, speed, and heading). Other contextual geographic data (such as POIs) can be additionally used to detect a stay point and confirm a stop episode. For instance, spatial clusters concentrated around a shopping mall can be confirmed as a stop episode whereas clusters on the road intersection (waiting for a traffic light) are unlikely to be a stop episode. This paper focuses on techniques for detecting stay points without using contextual information.

If we limit the scope to stay point detection based on raw data, a stay point or stop episode can be mostly detected on the basis of one or more of the following features: gaps, speed, spatial density, and time duration. A position where there is no GPS signal (the GPS is off or the signal is temporarily lost such as when a car enters a tunnel) can be set to a stay point (Ashbrook and Starner 2002). A problem with this approach is that there is no guarantee that those gaps represent a stop episode. Similarly, GPS logs whose speed value is near zero can be identified as a stop episode (Schuessler and Axhausen 2009). This approach is limited because speed values are often inconsistent due to abnormalities present in GPS measurement. It seems that it is necessary to infer semantics of gaps (i.e., do they represent stops or moves?), and reduce variability of inconsistent feature values through pre-processing techniques like noise-filtering or outlier removal.

Trajectories are a set of temporally sequenced location and time values recorded at a given time interval during a monitoring period. A stay point is a subsequence of trajectories that appear to be stationary for some time. Li et al. (2008) propose the algorithm that detects a stay point automatically from raw data. The algorithm identifies sub-trajectories that consist of two or more data points as a stay point if the distance between two successive data points whose interval increments exceeds a distance threshold and the time span among those points exceeds a time threshold. The study uses 200 m as a distance threshold, and 30 min as a time threshold. A limitation of this approach is that spatial outliers after time gaps can be falsely identified as a stay point if outliers are not treated beforehand. In addition, a stop episode that covers a large region with arbitrary shape (e.g., shopping at multiple stores in the mall) cannot be reliably detected as a stay point by the algorithm above.

A density-based spatial clustering algorithm can be used to overcome the limitation above. In other words, spatially connected GPS logs that span beyond a distance threshold can be detected using the clustering algorithm. Yuan et al. (2013) uses the clustering algorithm to recommend where taxi drivers are likely to find customers by detecting parking places where taxi drivers are stationary from collective GPS trajectories of vehicles. Schoier and Borruso (2011) use DBSCAN to detect significant places from GPS trajectories. Spatial clusters detected with these techniques, however, lack temporal dimension (such as time duration between the beginning and end times).

Hwang et al. (2013) detect a stay point by checking whether data points constituting a spatial cluster (detected by DBSCAN) meet temporal criteria. ST-DBSCAN that extends DBSCAN to consider non-spatial attributes (including time) is not well tailored to dealing with time expressed in categorical (binary) scale (i.e., exceed time threshold or not) that are possessed by the unknown number of spatially adjacent data points (Birant and Kut 2007). Finally, additional features like change of direction and sinuosity can be considered in stay point detection given that a stay point can be characterized by higher rate of direction change and higher sinuosity than a move episode (Rocha et al. 2010; Dodge et al. 2012). In the following, we review related work on how gaps are treated.

A gap is usually defined as a stop if the gap duration is above a threshold ranging from 1 to 10 min (depending on study area) (Ashbrook and Starner 2002). Data points can be added for the gap using linear interpolation; that is, location and time of data points added is linearly interpolated between a data point right before the gap, and a data point after the gap (Winter and Yin 2010). Speed for the gap can be estimated based on gap distance and gap time, and if the estimated gap speed matches the speed of the previous trip and of the following trip then the gap is set to a stop (Stopher et al. 2008). This work suggests that the successful stay point detection involves constructing empirically-based rules that consider a combination of features described above in conjunction with reducing uncertainty of GPS trajectories.

3 The Proposed Method

Following on from the previous research, we propose the framework that consists of four modules: data cleaning, gap treatment, stay point detection, and post-processing. The data cleaning module deletes spatial outliers. The gap treatment module fills gaps whose duration is at least 3 min using a linear interpolation. The stay point detection module identifies stay points based on spatial density ($eps = 50$ m; $MinPts = 5$) and the minimal time duration t (3 min) using DBSCAN adapted to temporal criteria. The post-processing module removes noise that remains after the previous module, based on the moving window that consists of five data points. The main difference in this method from previous work (Hwang et al. 2013) is in the gap treatment prior to the stay point detection. By treating gaps, it is possible to address a limitation of DBSCAN that does not detect gaps as a potential stay point. All other parameters remain the same, which allows us to examine effects of gap treatment on results. In the following, we describe the proposed method in detail.

3.1 Data Cleaning

The quality of the raw GPS data was inspected and treated as necessary to minimize effects of uncertainty. GPS track logs were overlaid on data of higher accuracy and independent source (ArcGIS 10.3 Map Service World Imagery). It was shown that most of the GPS data were well aligned with reference data, but a few spatial outliers were found present in trajectories recorded for the 1-week period. Outliers are represented by an abrupt change in position, and can be detected by abrupt changes in speed. More specifically, change in speed between two consecutive track logs was calculated as change in location divided by change in time. A track log is deleted if the elapsed speed for consecutive data points is greater than 130 km per hour. It is possible that data points that are not actually outliers (e.g., as a result of cars speeding) will be deleted if the elapsed speed is greater than or equal to 130 kph, but in any case, those data points would not be used for the purposes of stay point detection.

3.2 Gap Treatment

The gap treatment module adds a given number of data points (k) for a gap whose time duration exceeds a certain threshold (q), where parameters k and q are chosen such that a gap that is likely to represent a stop episode is detected as a stay point, and a gap that is likely to represent a move episode is not detected as a stay point. More specifically, k data points were added just before any gap whose elapsed time from the previous track log is at least q seconds, where $k = MinPts + 1$, and $q = t + r$ where t is the minimum time duration in seconds of a stay point and r is a recording time interval in seconds for GPS logging. In this study, k is 6 and q is 210 s as t is 180 s and r is 30 s. Time and position of those data points added are linearly interpolated.

The idea is that if the spatial distance between a data point before the gap (p_i) and a data point after the gap (p_j) is sufficiently large (i.e., the distance is greater than *eps*, and p_i and p_j are not density connected), then those six data points that were added are unlikely to be detected as a stay point due to low density. Good examples of this are where an individual takes subway, a vehicle enters a tunnel, or a GPS logger runs out of battery. If the spatial distance between p_i and p_j is sufficiently small (i.e., the distance is less than *eps*, and those p_i and p_j are density connected), then those six data points that were added are likely to be detected as a stay point. An example of this would be where GPS satellite signals are lost when one enters a building and the subsequent data point (p_j) after the gap is close to p_i (when one exits a building), which most likely indicates that a stop is made at a place.

3.3 Stay Point Detection

For DBSCAN, *eps* is set to 50 m, and *MinPts* is set to 5 based on observed spatial accuracy of data and extent of spatial clusters. Obviously, *eps* and *MinPts* should be chosen in relation to *t* (the minimum time duration of a stay point) and *r* (recording time interval). Spatial clusters are identified from DBSCAN given parameters *MinPts* and *eps*. With DBSCAN, track logs that are scattered around a stay point are treated as noise, and therefore those track logs (such as beginning of a new trip before or after making a stop) do not form part of a stay point. Staying at home is not included in stay point detection to reduce processing time, where staying at home is identified as track logs within 100 m from the home location provided by study participants.

Three minutes is chosen as *t* because this allows for capturing short duration activities (such as running an errand). Previous studies and observation of the study area indicate that the minimum time duration for stay points or important places is 2–30 min (Ashbrook and Starner 2002; Stopher et al. 2008; Ye et al. 2009). It was then checked to see if track logs constituting a spatial cluster are consecutive for the minimal duration of time *t*. If this condition is met, a spatial cluster is disaggregated into one or more stay points so that a place visited more than once is identified with multiple stay points and different time stamps. Track logs that are identified as a stay point are flagged as "stop", and track logs that are not identified as a stay point are flagged as "move".

3.4 Post-processing

Some track logs can still remain misclassified in the presence of both spatial and non-spatial outliers after the stay point detection module. For instance, anomalies in GPS measurement cause some track logs that are actually part of a stay point to be classified as "move" due to low density around those spatial outliers. Conversely, track logs that are not semantically a stop episode (such as waiting for a traffic light for longer than *t*) can be falsely classified as "stop". What is common in those misclassified track logs is that they are surrounded by track logs that are classified otherwise. Filtering stop/move values in the moving window with a given size can fix this problem. In this study, stop/move values are replaced with the most common value of five consecutive track logs.

For example, a series of track points might have an array of values [stop, move, stop, stop, stop] although they should be classified as [stop, stop, stop, stop, stop]. The program calculates a majority value (the most common value) in the moving window that consists of five consecutive track points, and replaces the binary value of each track log (stop or move) with the majority value. This process is repeated for all track logs. The size of the moving window is calculated as $t/r - 1$ (that is, 5 in this study) to make filtering results balanced between too noisy and too smooth outcomes. That way, an array of five track logs [stop, move, stop, stop, stop] can be classified into a stop as a whole since the second value ("move") will be replaced with "stop". This effectively removes noise that remains after clustering-based stay point detection.

Unique identifiers (IDs) are assigned to stops and moves based on temporal order and a rule that a stop is followed by a trip, and vice versa. In operational terms, a stop is a sequence of track logs that are spatially clustered and temporally continuous; a move is a sequence of track logs that are not spatially clustered but is temporally continuous. Track logs that are marked as a stop with the unique ID are aggregated (or dissolved) into a geographic object called a "stop center" that possesses spatiotemporal attributes, including coordinates of representative location (mean center), time duration, start time and end time.

4 Results and Evaluation

The stay point detection algorithm described above was applied to data collected for the study by Evans et al. (2012). The ultimate goal of this work is to monitor how patients return to the community after rehabilitation treatment following a stroke. Objective measures of community mobility (i.e., ability to get around in the community) allow health care practitioners to infer patients' functional status and monitor how patients respond to clinical intervention. One way to measure community mobility or community participation is to calculate the number of stop episodes where one or more activities are likely to be conducted.

Subjects provided informed consent to participate in the study, involving carrying a GPS logger (GlobalSat DG-100 Data Logger or QStarz Travel recorder XT) during their waking hours for 1 week. One week was chosen as a monitoring period because similar types of activities are typically repeated over a 1-week period, which allows for capturing a minimally distinct set of activities. Recording interval (r) was set to 30 s, and data were collected in a passive and continuous mode. Therefore, one set of GPS trajectory data is comprised of track logs continuously recorded for 1 week. Data was collected from May 2009 to December 2014. We collected data from two sets of subjects (control subjects and patients) from 1 week, 5 weeks, 9 weeks, 6 months, and 1 year after a baseline time (such as time of rehabilitation treatment). We implemented a total of 78 weeks' worth of data from 17 subjects.

Data ID	Trajectory ID	# track logs	begin_time	end_time
1	CS03_W01	5705	4/2/2013	4/9/2013
2	CS03_W05	5012	5/7/2013	5/14/2013
3	CS03_W09	6848	6/4/2013	6/21/2013
4	CS03_W26	12924	10/31/2013	11/5/2013
5	RS01_W01	8720	5/21/2009	6/2/2009
6	RS01_W05	5116	6/23/2009	6/29/2009
7	RS01_W09	8651	7/29/2009	8/3/2009
8	RS01_W26	5101	12/8/2009	12/14/2009
9	RS01_W52	12100	6/1/2010	6/8/2010

Fig. 1 Monitoring periods and sizes of nine trajectory data used for evaluation

To evaluate the performance of the algorithm, a group of research assistants visually inspected track logs of nine trajectories (12 % sample) overlaid on remotely sensed images on Google Earth in November 2014, to determine a series of alternating sequences of stop and move episodes. This resulted in the compilation of the number of validated stop episodes. We then compared the number of stops detected from the algorithm to the number of stops validated in those nine data sets. Figure 1 shows the number of track logs and monitoring periods of the nine trajectories used for evaluation. The trajectory ID identifies the subject and the time elapsed from the baseline time. For example, CS03_W09 is collected from the third control subject 9 weeks after the initiation of study participation. RS01_W52 represents a week of trajectories collected from the first patient 1 year (52 weeks) after rehabilitation. GlobalSat DG-100 Data Logger was used for all subjects listed in Fig. 1.

To evaluate the performance of the current work (i.e., determine how reliably this method detects stop episodes), we compile the number of stops detected by the previous work (Hwang et al. 2013) (denoted by *NoStOld*) and the number of stops detected by the current work (denoted by *NoStNew*), along with the number of stops validated for the nine sets of test data (denoted by *NoStVal*) (Fig. 2). The idea is to determine whether gap treatment contributes to accuracy of the stay point detection. The more closely *NoStOld* or *NoStNew* is aligned with *NoStVal*, the more likely that the respective test method detects stops accurately. It can be seen from Fig. 2 that *NoStNew* is more closely aligned with *NoStVal* than *NoStOld*. The Pearson's correlation coefficient between *NoStVal* and *NoStOld* is $-.249$ (with p-value .519). The Pearson's correlation coefficient between *NoStVal* and *NoStNew* is .692 (with p-value .039). This confirms that the current work outperforms the previous work and therefore would suggest that gap treatment is a worthwhile exercise for improving performance of clustering-based stay point detection.

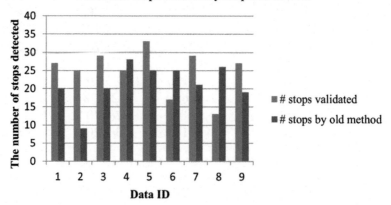

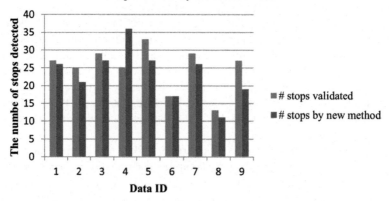

Fig. 2 Evaluation of the proposed method. (**a**) Comparison of # stops detected by the previous work against # stops validated. (**b**) Comparison of # stops detected by the current work against # stops validated

5 Conclusion

The proposed method combines gap treatment with DBSCAN adapted to temporal criteria to detect stay points from GPS trajectories. DBSCAN does not detect stay points when gaps are present and the number of data points near gaps is less than *MinPts*. For temporal DBSCAN to detect a stay point even in the presence of gaps, qualified (i.e., $>t$) gaps are filled first by adding k (or $MinPts + 1$) data points whose position and time are linearly interpolated. Then track logs that meet spatiotemporal constraints (i.e., high density, minimal time duration) are detected as a stay point. If two data points between qualified gaps are far apart, those gaps will be detected as a move episode. If two data points between qualified gaps are close to each other (less than *eps*), those gaps will be detected as a stop episode. Finally, the majority the filtering was performed on the moving window that consists of $t/r - 1$ (five in this study) consecutive data points to smooth out any misclassified values marked (stop or move) for track logs.

We evaluate performance of the proposed method on nine trajectory data. Performance was measured by how closely the number of stops detected by the proposed method is aligned with the number of stops validated. When we compare the performance between the previous work (Hwang et al. 2013) and the current work, it was found that the current work outperforms the previous work. This implies that gap treatment (which is new in the current work compared to the previous work) contributes to improved performance of the stay point detection method based on spatiotemporal criteria. Further, the statistically significant correlation between the number of stops validated through visual inspection and the number of stops detected by the proposed method indicates that the proposed method detects stay points to a satisfactory level of accuracy. The contribution of this study is that the stay point detection is also assessed with respect to gaps in data. This study also explores the effect of uncertainty handling (such as gap treatment) on performance, which is often overlooked in trajectory data analysis.

Several limitations of this research should be acknowledged. The performance of the proposed method can be better evaluated. One way to improve evaluation is to investigate the degree of match between processed data and validation data at the unit of track logs (rather than at the unit of stops). Another way to improve evaluation is to ask participants to annotate spatiotemporal attributes of sample stops, and use those annotated data as baseline data for determining the performance of the proposed method. Although a density-based clustering algorithm has an advantage that differentiates stops from noises (moves) reliably, appropriate values of parameters (such as search radius, the minimum number of data points, and the minimum time duration) should be empirically determined in relation to recording time interval, logging mode, and spatial accuracy of GPS trajectories. This can be done in future research.

The accuracy of stay point detection is crucial to segmenting trajectories into meaningful elements that are of value to a given application. For instance, semantics of stop episodes (e.g., type of location, time of visits, and characteristics of their surroundings) can be inferred with further analysis, which provides clues to understanding activity patterns of individuals. Similarly, inferring semantics of move episodes (e.g., mode of transportation, level of physical activities, and routes taken) will be useful in understanding transportation mode choice, determining features of the built environment that promote physical activity, and estimating transportation-related carbon footprints of individuals. It is hoped that this paper demonstrates the importance of, and various strategies for, segmenting trajectories as a preliminary to furthering knowledge and discovery in Big Data.

References

Ashbrook D, Starner T (2002) Learning significant locations and predicting user movement with GPS. In: Proceedings of the sixth international symposium on wearable computers, 2002 (ISWC 2002). pp 101–108

Birant D, Kut A (2007) ST-DBSCAN: an algorithm for clustering spatial–temporal data. Data Knowl Eng 60(1):208–221, http://www.sciencedirect.com/science/article/pii/S0169023X06000218. Accessed 10 July 2015

Dodge S, Laube P, Weibel R (2012) Movement similarity assessment using symbolic representation of trajectories. Int J Geogr Inf Sci 26(9):1563–1588, http://dx.doi.org/10.1080/13658816.2011.630003. Accessed 10 July 2015

Duncan MJ, Badland HM, Mummery WK (2009) Applying GPS to enhance understanding of transport-related physical activity. J Sci Med Sport 12(5):549–556, http://www.sciencedirect.com/science/article/pii/S1440244008002107. Accessed 10 July 2015

Eagle N, (Sandy) Pentland A (2006) Reality mining: sensing complex social systems. Pers Ubiquitous Comput 10(4):255–268, http://dx.doi.org/10.1007/s00779-005-0046-3. Accessed 10 July 2015

Ester M, et al (1996) A density-based algorithm for discovering clusters in large spatial databases with noise. AAAI Press, pp 226–231

Evans CC et al (2012) Monitoring community mobility with global positioning system technology after a stroke: a case study. J Neurol Phys Ther 36(2):68–78, http://content.wkhealth.com/linkback/openurl?sid=WKPTLP:landingpage&an=01253086-201206000-00004. Accessed 10 July 2015

Foth M, Choi JH, Satchell C (2011) Urban informatics. In: Proceedings of the ACM 2011 conference on computer supported cooperative work, CSCW'11. ACM, New York, pp 1–8. http://doi.acm.org/10.1145/1958824.1958826. Accessed 10 July 2015

Hwang S, Hanke T, Evans C (2013) Automated extraction of community mobility measures from GPS stream data using temporal DBSCAN. In: Murgante B, et al (eds) Computational science and its applications—ICCSA 2013 (Lecture notes in computer science). Springer, Berlin, pp 86–98. http://link.springer.com/chapter/10.1007/978-3-642-39643-4_7. Accessed 10 July 2015

Krenn PJ et al (2011) Use of global positioning systems to study physical activity and the environment: a systematic review. Am J Prev Med 41(5):508–515, http://www.sciencedirect.com/science/article/pii/S0749379711005460. Accessed 10 July 2015

Li Q, et al (2008) Mining user similarity based on location history. In: Proceedings of the 16th ACM SIGSPATIAL international conference on advances in geographic information systems, GIS'08. ACM, New York, pp 34:1–34:10. http://doi.acm.org/10.1145/1463434.1463477. Accessed 10 July 2015

Liao L, Fox D, Kautz H (2007) Extracting places and activities from GPS traces using hierarchical conditional random fields. Int J Rob Res 26(1):119–134, http://ijr.sagepub.com/content/26/1/119. Accessed 10 July 2015

Marcus F (2008) Handbook of research on urban informatics: the practice and promise of the real-time city: the practice and promise of the real-time city. IGI Global, Hershey, PA

Pang LX et al (2013) On detection of emerging anomalous traffic patterns using GPS data. Data Knowl Eng 87:357–373, http://www.sciencedirect.com/science/article/pii/S0169023X13000475. Accessed 10 July 2015

Parent C, et al (2013) Semantic trajectories modeling and analysis. ACM Comput Surv 45(4):42:1–42:32. http://doi.acm.org/10.1145/2501654.2501656. Accessed 10 July 2015

Richardson DB et al (2013) Spatial turn in health research. Science 339(6126):1390–1392, http://www.ncbi.nlm.nih.gov/pmc/articles/PMC3757548/. Accessed 10 July 2015

Rocha JAMR, et al (2010) DB-SMoT: a direction-based spatio-temporal clustering method. In: Intelligent systems (IS), 2010 5th IEEE international conference. pp 114–119

Rodrigues A, Damásio C, Cunha JE (2014) Using GPS logs to identify agronomical activities. In: Huerta J, Schade S, Granell C (eds) Connecting a digital Europe through location and place (Lecture notes in geoinformation and cartography). Springer, pp 105–121. http://link.springer.com/chapter/10.1007/978-3-319-03611-3_7. Accessed 10 July 2015

Schoier G, Borruso G (2011) Individual movements and geographical data mining. Clustering algorithms for highlighting hotspots in personal navigation routes. In: Murgante B, et al (eds) Computational science and its applications—ICCSA 2011 (Lecture notes in computer science). Springer, Berlin, pp 454–465. http://link.springer.com/chapter/10.1007/978-3-642-21928-3_32. Accessed 10 July 2015

Schuessler N, Axhausen K (2009) Processing raw data from global positioning systems without additional information. Transp Res Rec J Transp Res Board 2105:28–36, http://trrjournalonline.trb.org/doi/abs/10.3141/2105-04. Accessed 10 July 2015

Stopher P, FitzGerald C, Zhang J (2008) Search for a global positioning system device to measure person travel. Transp Res Part C Emerg Technol 16(3):350–369, http://www.sciencedirect.com/science/article/pii/S0968090X07000836. Accessed 10 July 2015

Winter S, Yin Z-C (2010) Directed movements in probabilistic time geography. Int J Geogr Inf Sci 24(9):1349–1365, http://dx.doi.org/10.1080/13658811003619150. Accessed 10 July 2015

Ye Y, et al (2009) Mining individual life pattern based on location history. In: Tenth international conference on mobile data management: systems, services and middleware, 2009, MDM'09. pp 1–10

Yuan NJ et al (2013) T-finder: a recommender system for finding passengers and vacant taxis. IEEE Trans Knowl Data Eng 25(10):2390–2403

Zheng Y (2015) Trajectory data mining: an overview. ACM Trans Intell Syst Technol 6(3): 29:1–29:41. http://doi.acm.org/10.1145/2743025. Accessed 10 July 2015

Zheng Y, et al (2011) Recommending friends and locations based on individual location history. ACM Trans Web 5(1): 5:1–5:44. http://doi.acm.org/10.1145/1921591.1921596. Accessed 10 July 2015

Zheng Y, et al (2014) Urban computing: concepts, methodologies, and applications. ACM Trans Intell Syst Technol 5(3): 38:1–38:55. http://doi.acm.org/10.1145/2629592. Accessed 10 July 2015

Part VI
Emergencies and Crisis

Part V
Emergencies and Crises

Using Social Media and Satellite Data for Damage Assessment in Urban Areas During Emergencies

Guido Cervone, Emily Schnebele, Nigel Waters, Martina Moccaldi, and Rosa Sicignano

Abstract Environmental hazards pose a significant threat to urban areas due to their potential catastrophic consequences affecting people, property and the environment. Remote sensing has become the de-facto standard for observing the Earth and its environment through the use of air-, space-, and ground-based sensors. Despite the quantity of remote sensing data available, gaps are often present due to the specific limitations of the instruments, their carrier platforms, or as a result of atmospheric interference. Massive amounts of data are generated from social media, and it is possible to mine these data to fill the gaps in remote sensing observations.

A new methodology is described which uses social networks to augment remote sensing imagery of transportation infrastructure conditions during emergencies. The capability is valuable in situations where environmental hazards such as hurricanes or severe weather affect very large areas. This research presents an application of the proposed methodology during the 2013 Colorado floods with a special emphasis in Boulder County and The City of Boulder. Real-time data collected from social media, such as Twitter, are fused with remote sensing data for transportation damage assessment. Data collected from social media can provide information when remote sensing data are lacking or unavailable.

Keywords Remote sensing • Social media • Flood assessment

G. Cervone (✉)
GeoInformatics and Earth Observation Laboratory, Department of Geography and Institute for CyberScience, The Pennsylvania State University, University Park, PA, USA

Research Application Laboratory, National Center for Atmospheric Research, Boulder, CO, USA
e-mail: cervone@psu.edu

E. Schnebele • N. Waters • M. Moccaldi • R. Sicignano
GeoInformatics and Earth Observation Laboratory, Department of Geography and Institute for CyberScience, The Pennsylvania State University, University Park, PA, USA

1 Introduction

Every year natural hazards are responsible for powerful and extensive damage to people, property, and the environment. Drastic population growth, especially along coastal areas or in developing countries, has increased the risk posed by natural hazards to large, vulnerable populations at unprecedented levels (Tate and Frazier 2013). Furthermore, unusually strong and frequent weather events are occurring worldwide, causing floods, landslides, and droughts affecting thousands of people (Smith and Katz 2013). A single catastrophic event can claim thousands of lives, cause billions of dollars of damage, trigger a global economic depression, destroy natural landmarks, render a large territory uninhabitable, and destabilize the military and political balance in a region (Keilis-Borok 2002). Furthermore, the increasing urbanization of human society, including the emergence of megacities, has led to highly interdependent and vulnerable social infrastructure that may lack the resilience of a more agrarian, traditional society (Wenzel et al. 2007). In urban areas, it is crucial to develop new ways of assessing damage in real-time to aid in mitigating the risks posed by hazards. Annually, the identification, assessment, and repair of damage caused by hazards requires thousands of work hours and billions of dollars.

Remote sensing data are of paramount importance during disasters and have become the de-facto standard for providing high resolution imagery for damage assessment and the coordination of disaster relief operations (Cutter 2003; Joyce et al. 2009). First responders rely heavily on remotely sensed imagery for coordination of relief and response efforts as well as the prioritizing of resource allocation.

Determining the location and severity of damage to transportation infrastructure is particularly critical for establishing evacuation and supply routes as well as repair and maintenance agendas. Following the Colorado floods of September 2013 over 1000 bridges required inspection and approximately 200 miles of highway and 50 bridges were destroyed.[1] A variety of assessment techniques were utilized following Hurricane Katrina in 2005 to evaluate transportation infrastructure including visual, non-destructive, and remote sensing. However, the assessment of transportation infrastructure over such a large area could have been accelerated through the use of high resolution imagery and geospatial analysis (Uddin 2011; Schnebele et al. 2015).

Despite the wide availability of large remote sensing datasets from numerous sensors, specific data might not be collected in the time and space most urgently required. Geo-temporal gaps result due to satellite revisit time limitations, atmospheric opacity, or other obstructions. However, aerial platforms, especially Unmanned Aerial Vehicles (UAVs), can be quickly deployed to collect data about specific regions and be used to complement satellite imagery. UAVs are capable of providing high resolution, near real-time imagery often with less expense than manned aerial- or space-borne platforms. Their quick response

---

[1]http://www.denverpost.com/breakingnews, http://www.thedenverchannel.com/news/local-news.

times, high maneuverability and resolution make them important tools for disaster assessment (Tatham 2009).

Contributed data that contain spatial and temporal information can provide valuable Volunteered Geographic Information (VGI), harnessing the power of 'citizens as sensors' to provide a multitude of on-the-ground data, often in real time (Goodchild 2007). Although these volunteered data are often published without scientific intent, and usually carry little scientific merit, it is still possible to mine mission critical information. For example, during hurricane Katrina, geolocated pictures and videos searchable through Google provided early emergency response with ground-view information. These data have been used during major events, with the capture, in near real-time, of the evolution and impact of major hazards (De Longueville et al. 2009; Pultar et al. 2009; Heverin and Zach 2010; Vieweg et al. 2010; Acar and Muraki 2011; Verma et al. 2011; Earle et al. 2012; Tyshchuk et al. 2012).

Volunteered data can be employed to provide timely damage assessment, help in rescue and relief operations, as well as the optimization of engineering reconnaissance (Laituri and Kodrich 2008; Dashti et al. 2014; Schnebele and Cervone 2013; Schnebele et al. 2014a,b). While the quantity and real-time availability of VGI make it a valuable resource for disaster management applications, data volume, as well as its unstructured, heterogeneous nature, make the effective use of VGI challenging. Volunteered data can be diverse, complex, and overwhelming in volume, velocity, and in the variety of viewpoints they offer. Negotiating these overwhelming streams is beyond the capacity of human analysts. Current research offers some novel capabilities to utilize these streams in new, ground-breaking ways, leveraging, fusing and filtering this new generation of air-, space- and ground-based sensor-generated data (Oxendine et al. 2014).

2 Data

Multiple sources of contributed, remote sensing, and open source geospatial data were collected and utilized during this research. All the data are relative to the 2013 Colorado floods, and were collected between Septmber 11 and 17, 2013. Most of the data were collected in real time, as the event was unfolding. A summary of the sources and collection dates of the contributed and remote sensing data is available in Table 1.

Table 1 Sources and collection dates of contributed and remote sensing data

Data	9.11	9.12	9.13	9.14	9.15	9.16	9.17
Contributed							
Twitter	X	X	X	X	X	X	X
Photos	X	X	X	X			
Remote sensing							
Landsat-8							X
Civil Air Patrol				X	X	X	X

2.1 Contributed Data

2.1.1 Twitter

The social networking site Twitter is used by the public to share information about their daily lives through micro-blogging. These micro-blogs, or 'tweets', are limited to 140 characters, so abbreviations and colloquial phrasing are common, making the automation of filtering by content challenging. Different criteria are often applied for filtered and directed searches of Twitter content. For example, a *hashtag*[2] is an identifier unique to Twitter and is frequently used as a search tool. The creation and use of a hashtag can be established by any user and may develop a greater public following if it is viewed as useful, popular, or providing current information. Other search techniques may use keywords or location for filtering.

Twitter data were collected using a geospatial database setup at The Pennsylvania State University (PSU) GeoVISTA center. Although the volume of Twitter data streams is huge, only a small percentage of tweets contain geolocation information. For this particular study, about 2000 geolocated tweets were collected with hashtag #boulderflood.

2.1.2 Photos

In addition, a total of 80 images relating to the period from 11–14 September 2013 were downloaded through the website of the city of Boulder (https://bouldercolorado.gov/flood). These images did not contain geolocation information, but they included a description of when and where they were acquired. The google API was used to convert the spatial description to precise longitude and latitude coordinates.

While in the current research the data were semi-manually georectified, it is possible to use services such as Flickr or Instagram to automatically download geolocated photos. These services can provide additional crucial information during emergencies because photos are easily verifiable and can contain valuable information for transportation assessment.

2.2 Remote Sensing

2.2.1 Satellite

Full-resolution GeoTIFF multispectral Landsat 8 OLI/TIRS images that were collected on May 12, 2013 and on September 17, 2013 provided data of the Boulder County area before and after the flooding, respectively. The data were downloaded

---

[2]A word or an unspaced phrase prefixed with the sign #.

from the USGS Hazards Data Distribution System (HDDS). Landsat 8 consists of nine spectral bands with a resolution of 30 m: Band 1 (coastal aerosol, useful for coastal and aerosol studies, 0.43–0.45 µm); Bands 2–4 (optical, 0.45–0.51, 0.53–0.59, 0.64–0.67 µm), Band 5 (near-IR, 0.85–0.88 µm), Bands 6 and 7 - (shortwave-IR, 1.57–1.65, 2.11–2.29 µm) and Band 9 (cirrus, useful for cirrus cloud detection, 1.36–1.38 µm). In addition, a 15 m panchromatic band (Band 8, 0.50–0.68 µm) and two 100 m thermal IR (Bands 10 and 11, 10.60–11.19, 11.50–12.51 µm) were also collected from Landsat 8 OLI/TIRS.

2.2.2 Aerial

Aerial photos collected by the Civil Air Patrol (CAP), the civilian branch of the US Air Force, captured from 14–17 September 2013 in the areas surrounding Boulder (105.5364−104.9925° W longitude and 40.26031−39.93602° N latitude) provided an additional source of remotely sensed data. The georeferenced Civil Air Patrol RGB composite photos were downloaded from the USGS Hazards Data Distribution System (HDDS).

2.3 Open Source Geospatial Data

Shapefiles defining the extent of the City of Boulder and Boulder County were downloaded from the City of Boulder[3] and the Colorado Department of Local Affairs[4] websites, respectively. In addition, a line/line shapefile of road networks for Boulder County was downloaded from the US Census Bureau.[5]

3 Methods

The proposed methodology is based on the fusion of contributed data with remote sensing imagery for damage assessment of transportation infrastructure.

3.1 Classification of Satellite Images

For the Colorado floods of 2013, supervised machine learning classifications were employed to identify water in each of the satellite images. Water pixels are

---

[3]https://bouldercolorado.gov.
[4]http://www.colorado.gov.
[5]http://www.census.gov.

identified in both Landsat images by using a decision tree induction classifier. Ripley (2008) describes the general rule induction methodology and its implementation in the **R** statistical package used in this study. In the near-IR water is easily distinguished from soil and vegetation due to its strong absorption (Smith 1997). Therefore, imagery caught by Landsat's Band 5 (near IR, 0.85–0.88 μm) was used for the machine learning classification. Control areas of roughly the same size are identified as examples of water pixels, 'water', and over different regions with no water pixels as counter-examples, or 'other'. Landsat data relative to these regions are used as training events by the decision tree classifier. The learned tree is then used to classify the remaining water pixels in the scene. This process is repeated for both the May and September images.

3.2 Spatial Interpolation

Satellite remote sensing data may be insufficient as a function of revisit time or obstructions due to clouds or vegetation. Therefore, data from other sources can be used to provide supplemental information. Aerial remote sensing data as well as contributed data, or VGI, from photos and tweets were used to capture or infer the presence of flooding in a particular area.

Utilizing different types and sources of data, this research aims to extract as much information as possible about damage caused by natural hazards. Environmental data are often based on samples in limited areas, and the tweets analyzed are approximately only 1 % of the total tweets generated during the time period. This is usually referred to as the 'Big Data paradox', where very large amounts of data to be analyzed are only a small sample of the total data, which might not reflect the distribution of the entire population.

In addition, the absence of data in some parts of the region is likely to underestimate the total damage. In order to compensate for the missing data, the spatio-temporal distribution of the data were analyzed by weighting according to the spatial relationships of the points (Tobler 1970, p. 236). This assumes some levels of dependence among spatial data (Waters 2017). For these reasons a punctual representation of data may not be sufficient to provide a complete portrayal of the hazard, therefore a spatial interpolation is employed to estimate flood conditions and damage from point sources.

Spatial interpolation consists of estimating the damage at unsampled locations by using information about the nearest available measured points. For this purpose, interpolation generates a surface crossing sampled points. This process can be implemented by using two different approaches: *deterministic models* and *statistical techniques*. Even if both use a mathematical function to predict unknown values, the first method does not provide an indication of the extent of possible errors, whereas the second method supplies probabilistic estimates. Deterministic models include *IDW*(Inverse Distance Weighted), *Rectangular*, *NN* (Natural

Neighbourhood) and *Spline*. Statistical methods include *Kriging* (Ordinary, Simple and Universal) and *Kernel*. In this project, Kernel interpolation has been used.

Kernel interpolation is the most popular non-parametric density estimator, that is a function $\hat{p} : \Re \times (\Re)^N \to \Re$. In particular it has the following aspect:

$$\hat{p}(x) = \frac{1}{Nh} \sum_{i=1}^{N} K\left(\frac{x - x_i}{h}\right). \quad (1)$$

where $K(u)$ is the *Kernel function* and h is the *bandwidth* (Raykar and Duraiswami 2006). There are different kinds of kernel density estimators such as *Epanechnikov, Triangular, Gaussian, Rectangular*. The density estimator chosen for this work is a Gaussian kernel with zero mean and unit variance having the following form:

$$\hat{p}(x) = \frac{1}{N\sqrt{2\pi h^2}} \sum_{i=1}^{N} e^{-(x-x_i)^2/2h^2}. \quad (2)$$

Kernel interpolation is often preferred because it provides an estimate of error as opposed to methods based on radial basis functions. In addition, it is more effective than a Kriging interpolation in cases of small data sets (for example, the data set of photos in this project) or data with non-stationary behavior (all data sets used in this work) (Mühlenstädt and Kuhnt 2011).

In general, spatial interpolation is introduced to solve the following problems associated with histograms:

- The wider the interval, the greater the information loss;
- Histograms provide estimates of *local* density (points are "local" to each other if they belong to the same bin) so this method does not give prominence to proximity of points;
- The resulting estimate is not smooth.

These problems can be avoided by using a smooth kernel function, rather than a histogram "block" centered over each point, and summing the functions for each location on the scale. However, it is important to note that the results of kernel density interpolations are very dependent on the size of the defined interval or "bandwidth". The bandwidth is a smoothing parameter and it determines the width of the kernel function.

The result of a kernel density estimation will depend on the kernel $K(u)$ and the bandwidth h chosen. The former is linked to the shape, or function, of the curve centered over each point whereas the latter determines the width of the function. The choice of bandwidth will exert greater influence over an interpolation result than the kernel function. Indeed, as the value of h decreases, the local weight of single observations will increase.

Because confidence in data may vary with source characteristics, bandwidth selection can be varied for each data type. Generally, as certainty, or confidence, in a given data source increases so does the chosen bandwidth. For example, aerial images can be verified visually and therefore are considered to be more credible information sources. By contrast some tweets could not be useful since they are subjective; indeed some may only contain users' feelings and not information related to damage caused by the hazard. For this reason a lower bandwidth for tweets has been chosen in this work.

There are different methods for choosing an appropriate bandwidth. For example, it can be identified as the value that minimizes the approximation of the error between the estimate $\hat{p}(x)$ and the actual density $p(x)$ as explained in Raykar and Duraiswami (2006). In this work, a spatial Kernel estimation has been used because all considered data sets consist of points on the Earth's surface. In case of d-dimensional points the form of the Kernel estimate is:

$$\hat{p}(x;H) = n^{-1}\sum_{i=1}^{N} K_H(x - X_i) \tag{3}$$

where:

$$x = (x_1, x_2, \ldots, x_d)^T$$
$$X_i = (X_{i1}, X_{i2}, \ldots, X_{id})^T$$
$$i = 1, 2, \ldots, N$$
$$K_H(x) = |H|^{-1/2} K(H^{-1/2}x)$$

In this case $K(x)$ is the *spatial kernel* and H is the *bandwidth matrix*, which is symmetric and positive-definite. As in the uni-dimensional case, an optimal bandwidth matrix has to be chosen, for example, using the method illustrated in Duong and Hazelton (2005). In this project, data have been interpolated by using the R command *smoothScatter* of the package *graphics* based on the *Fast Fourier transform*. It is a variation of regular Kernel interpolation that reduces the computational complexity from $O(N^2)$ to $O(N)$. The O ('big O') notation is a Computer Science metric to quantify and compare the complexity of algorithms (Knuth 1976). $O(N)$ indicates a linear complexity, whereas $O(N^2)$ indicates a much higher, quadratic, complexity. Generally, the bandwidth is automatically calculated by using the R command *bkde2D* of the R package *KernSmooth*. However, the bandwidth for tweets interpolation has been specified because information related to them has a lower weight.

4 Analysis and Results

4.1 Damage Assessment

Using a supervised machine learning classification as discussed in Sect. 3.1, water pixels were identified in these images. For example, a comparison of the classifications in the Landsat 12 May, 2013 'before' image (Fig. 1a) and the Landsat 17 September, 2013 'after' image (Fig. 1b) illustrates additional water pixels classified in the 'after' image associated with the flood event. One of the challenges associated with classifying remote sensing imagery is illustrated in (Fig. 1b) where clouds over the front range in Colorado are misclassified as water pixels. In addition, because the Landsat 'after' image was collected on 17 September, 7 days after the beginning of the flood event, it is likely that the maximum flood extent is not captured in this scene.

Following the collection of remote sensing data, contributed data were also collected, geolocated, and interpolated following the methods discussed in Sect. 3.2. The interpolated pixels were then overlayed on the remote sensing classification to give an enhanced indication of flood activity in the Boulder area (Fig. 2). The use of supplemental data sources, such as the Civil Air Patrol (CAP) aerial photos, shows flooding in areas that was not captured by the satellite remote sensing. In the case of the Boulder flooding, the cloud cover over the front range and western parts of the City of Boulder, made the identification of water from satellite platforms difficult. The ability of planes to fly below cloud cover as well as to collect data without the revisit limitations common to space-borne platforms, allowed the CAP to capture flooding and damage not visible from satellite images in the western parts of Boulder County (Fig. 2).

4.2 Transportation Classification

Although the identification of water in the near-IR is a standard technique, the supervised classification of the Landsat data did not indicate any pixels as 'water' in the City of Boulder (Fig. 1b). This could be because the image was collected on 17 September, a week after the flood event began, and flood waters could have receded, as well as the presence of obstructing vegetation and cloud cover. However, it is interesting to note that contributed data such as photos and tweets do indicate the presence of flooding in the City of Boulder. Using the geolocated tweets (n=130) containing the hashtag '#boulderflood' geolocated near Boulder ($105.0814° - 105.2972°$ W longitude and $40.09947° - 39.95343°$ N latitude) as well as geolocated photos (n=80), flood activity is indicated by local citizens in the downtown area. While there may be uncertainties associated with information obtained from tweets, the presence of flooding and damage are more easily verified in photos.

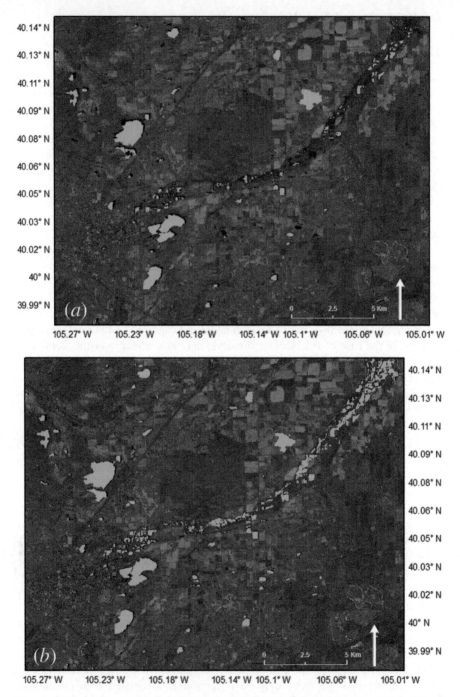

Fig. 1 Water pixel classification using Landsat 8 data collected 12 May 2013 (**a**) and 17 September 2013 (**b**). The background images are the Landsat band 5 for each of the 2 days

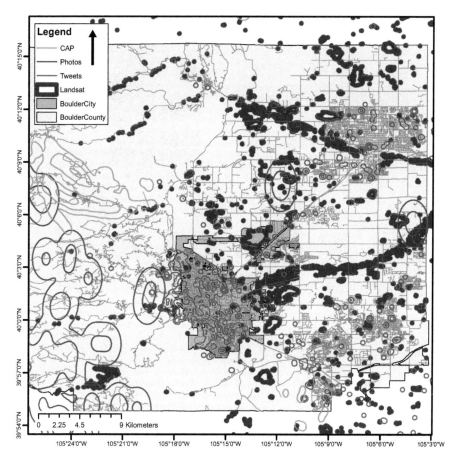

Fig. 2 Classified Landsat 8 image collected 17 September 2013 image and interpolated ancillary data give an indication of flood activity around the Boulder, CO area. While some data sources overlap, others have a very different spatial extent

Using contributed data points (Fig. 3a), a flooding and damage surface is interpolated using a kernel density smoothing application as discussed in Sect. 3.2 for the downtown Boulder area. After an interpolated surface is created from each data set (tweets and photos), they are combined using a weighted sum overlay approach. The tweets layer is assigned a weight of 1 and the photos layer a weight of 2. A higher weight is assigned to the photos layer because information can be more easily verified in photos, therefore there is a higher level of confidence in this data set. The weighted layers are summed yielding a flooding and damage assessment surface created solely from contributed data (Fig. 3b). This surface is then paired with a high resolution road network layer. Roads are identified as potentially compromised or impassable based on the underlying damage assessment (Fig. 3c). In a final step, the classified roads are compared to roads closed by the Boulder Emergency Operations Center (EOC) from 11–15 September 2013 (Fig. 3d).

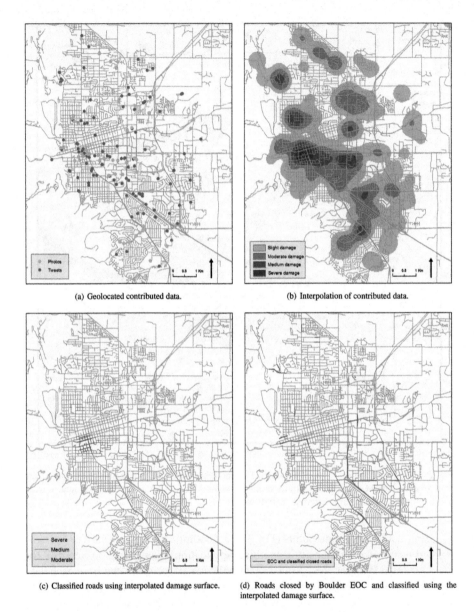

(a) Geolocated contributed data.

(b) Interpolation of contributed data.

(c) Classified roads using interpolated damage surface.

(d) Roads closed by Boulder EOC and classified using the interpolated damage surface.

Fig. 3 Using contributed data geolocated in the downtown Boulder area (**a**), an interpolated damage surface is created (**b**) and when paired with a road network, classifies potentially compromised roads (**c**). Roads which were closed by the Boulder Emergency Operations Center (EOC) that were also classified using this approach (**d**)

5 Conclusions

Big data, such as those generated through social media platforms, provide unprecedented access to real-time, on-the-ground information during emergencies, supplementing traditional, standard data sources such as space- and aerial-based remote sensing. In addition to inherent gaps in remote sensing data due to platform or revisit limitations or atmospheric interference, obtaining data and information for urban areas can be especially challenging because of resolution requirements. Identifying potential damage at the street or block level provides an opportunity for triaging site visits for evaluation. However, utilizing big data efficiently and effectively is challenging owing to its complexity, size, and in the case of social media, heterogeneity (variety). New algorithms and techniques are required to harness the power of contributed data in real-time for emergency applications.

This paper presents a new methodology for locating natural hazards using contributed data, in particular Twitter. Once remote sensing data are collected, they, in combination with contributed data, can be used to provide an assessment of the ensuing damage. While Twitter is effective at identifying 'hot spots' at the city level, at the street level other sources provide a supplemental source of information with a finer detail (e.g. photos). In addition, remote sensing data may be limited by revisit times or cloud cover, so contributed ground data provide an additional source of information.

Challenges associated with utilizing contributed data, such as questions related to producer anonymity and geolocation accuracy as well as differing levels in data confidence make the application of these data during hazard events especially challenging. In addition to identifying a particular hazard, in this case flood waters, by pairing the interpolated damage assessment surface with a road network creates a classified 'road hazards map' which can be used to triage and optimize site inspections or tasks for additional data collection.

Acknowledgements Work performed under this project has been partially funded by the Office of Naval Research (ONR) award #N00014-14-1-0208 (PSU #171570).

References

Acar A, Muraki Y (2011) Twitter for crisis communication: lessons learned from Japan's tsunami disaster. Int J Web Based Communities 7(3):392–402

Cutter SL (2003) Giscience, disasters, and emergency management. Trans GIS 7(4):439–446

Dashti S, Palen L, Heris M, Anderson K, Anderson S, Anderson T (2014) Supporting disaster reconnaissance with social media data: a design-oriented case study of the 2013 Colorado floods. In: Proceedings of the 11th international conference on information systems for crisis response and management. ISCRAM, pp 630–639

De Longueville B, Smith R, Luraschi G (2009) OMG, from here, I can see the flames!: a use case of mining location based social networks to acquire spatio-temporal data on forest fires. In:

Proceedings of the 2009 international workshop on location based social networks. ACM, New York, pp 73–80

Duong T, Hazelton ML (2005) Cross-validation bandwidth matrices for multivariate kernel density estimation. Scand J Stat 32(3):485–506

Earle P, Bowden D, Guy M (2012) Twitter earthquake detection: earthquake monitoring in a social world. Ann Geophys 54(6):708–715

Goodchild M (2007) Citizens as sensors: the world of volunteered geography. GeoJournal 69(4):211–221

Heverin T, Zach L (2010) Microblogging for crisis communication: examination of Twitter use in response to a 2009 violent crisis in the Seattle-Tacoma, Washington, area. In: Proceedings of the 7th international conference on information systems for crisis response and management. ISCRAM, pp 1–5

Joyce KE, Belliss SE, Samsonov SV, McNeill SJ, Glassey PJ (2009) A review of the status of satellite remote sensing and image processing techniques for mapping natural hazards and disasters. Prog Phys Geogr 33(2):183–207

Keilis-Borok V (2002) Earthquake prediction: state-of-the-art and emerging possibilities. Annu Rev Earth Planet Sci 30:1–33

Knuth DE (1976) Big omicron and big omega and big theta. ACM Sigact News 8(2):18–24

Laituri M, Kodrich K (2008) On line disaster response community: people as sensors of high magnitude disasters using internet GIS. Sensors 8(5):3037–3055

Mühlenstädt T, Kuhnt S (2011) Kernel interpolation. Comput Stat Data Anal 55(11):2962–2974

Oxendine C, Schnebele E, Cervone G, Waters N (2014) Fusing non-authoritative data to improve situational awareness in emergencies. In: Proceedings of the 11th international conference on information systems for crisis response and management. ISCRAM, pp 760–764

Pultar E, Raubal M, Cova T, Goodchild M (2009) Dynamic GIS case studies: wildfire evacuation and volunteered geographic information. Trans GIS 13(1):85–104

Raykar VC, Duraiswami R (2006) Fast optimal bandwidth selection for kernel density estimation. In: SDM. SIAM, Philadelphia, pp 524–528

Ripley B (2008) Pattern recognition and neural networks. Cambridge University Press, Cambridge

Schnebele E, Cervone G (2013) Improving remote sensing flood assessment using volunteered geographical data. Nat Hazards Earth Syst Sci 13:669–677

Schnebele E, Cervone G, Kumar S, Waters N (2014a) Real time estimation of the Calgary floods using limited remote sensing data. Water 6:381–398

Schnebele E, Cervone G, Waters N (2014b) Road assessment after flood events using non-authoritative data. Nat Hazards Earth Syst Sci 14(4):1007–1015

Schnebele E, Tanyu B, Cervone G, Waters N (2015) Review of remote sensing methodologies for pavement management and assessment. Eur Transp Res Rev 7(2):1–19

Smith L (1997) Satellite remote sensing of river inundation area, stage, and discharge: a review. Hydrol Process 11(10):1427–1439

Smith AB, Katz RW (2013) US billion-dollar weather and climate disasters: data sources, trends, accuracy and biases. Nat Hazards 67(2):387–410

Tate C, Frazier T (2013) A GIS methodology to assess exposure of coastal infrastructure to storm surge & sea-level rise: a case study of Sarasota County, Florida. J Geogr Nat Disasters 1:2167–0587

Tatham P (2009) An investigation into the suitability of the use of unmanned aerial vehicle systems (UAVs) to support the initial needs assessment process in rapid onset humanitarian disasters. Int J Risk Assess Manage 13(1):60–78

Tobler WR (1970) A computer movie simulating urban growth in the Detroit region. Econ Geogr 46:234–240

Tyshchuk Y, Hui C, Grabowski M, Wallace W (2012) Social media and warning response impacts in extreme events: results from a naturally occurring experiment. In: Proceedings of the 45th annual Hawaii international conference on system sciences (HICSS). IEEE, New York, pp 818–827

Uddin W (2011) Remote sensing laser and imagery data for inventory and condition assessment of road and airport infrastructure and GIS visualization. Int J Roads Airports (IJRA) 1(1):53–67

Verma S, Vieweg S, Corvey W, Palen L, Martin J, Palmer M, Schram A, Anderson K (2011) Natural language processing to the rescue? Extracting "situational awareness" Tweets during mass emergency. In: Proceedings of the 5th international conference on weblogs and social media. AAAI, Palo Alto

Vieweg S, Hughes A, Starbird K, Palen L (2010) Microblogging during two natural hazards events: what twitter may contribute to situational awareness. In: Proceedings of the 28th international conference on human factors in computing systems. ACM, New York, pp 1079–1088

Waters N (2017) Tobler's first law of geography. In: Richardson D, Castree N, Goodchild MF, Kobayashi A, Liu W, Marston R (eds) The international encyclopedia of geography. Wiley, New York, In press

Wenzel F, Bendimerad F, Sinha R (2007) Megacities–megarisks. Nat Hazards 42(3):481–491

Part VII
Health and Well-Being

'Big Data': Pedestrian Volume Using Google Street View Images

Li Yin, Qimin Cheng, Zhenfeng Shao, Zhenxin Wang, and Laiyun Wu

Abstract Responding to the widespread growing interest in walkable, transit-oriented development and healthy communities, many recent studies in planning and public health are concentrating on improving the pedestrian environment. There is, however, inadequate research on pedestrian volume and movement. In addition, the method of data collection for detailed information about pedestrian activity has been insufficient and inefficient. Google Street View provides panoramic views along many streets of the U.S. and around the world. This study introduces an image-based machine learning method to planners for detecting pedestrian activity from Google Street View images. The detection results are shown to resemble the pedestrian counts collected by field work. In addition to recommending an alternative method for collecting pedestrian count data more consistently and subjectively for future research, this study also stimulates discussion of the use of 'big data' for planning and design.

Keywords Google Street View • Pedestrian count • Walkability

L. Yin (✉) • L. Wu
Department of Urban and Regional Planning, University at Buffalo, The State University of New York, Buffalo, NY 14214, USA
e-mail: liyin@buffalo.edu; laiyunwu@buffalo.edu

Q. Cheng
Department of Electronics and Information Engineering, Huazhong University of Science and Technology, Wuhan 430074, China

Z. Shao
The State Key Laboratory of Information Engineering on Surveying Mapping and Remote Sensing, Wuhan University, Wuhan 430079, China

Z. Wang
Department of Urban and Regional Planning, University at Buffalo, The State University of New York, Buffalo, NY 14214, USA

Center for Human-Engaged Computing, Kochi University of Technology, 185 Miyanokuchi, Tosayamada-Cho, Kami-Shi, Kochi 782-8502, Japan

1 Introduction

Findings from recent research have suggested a link between the built environment and physical activity. Responding to the widespread growing interest in walkable, transit-oriented development and healthy communities, many recent studies in planning and public health focused on improving the pedestrian environment (Ewing and Bartholomew 2013; Yin 2014; Yin 2013; Clifton et al. 2007). Pedestrian counts are used as a quantitative measure of pedestrian volume to help evaluate walkability and how it correlates with land use, and other built environment characteristics. It can also be used as baseline data to help inform planning and decision-making.

Pedestrians have been the subject of increasing attention among planners, engineers and public health officials. There is, however, inadequate research on pedestrian volume and movement. In addition, the method of data collection for detailed information about pedestrian activity has been insufficient and inefficient (Hajrasouliha and Yin 2014). Pedestrian count data has been collected by field work, self-reported surveys and automated counting. Field work and self-reported surveys are more subjective than automatic counts using video-taping or sensors. Most pedestrian counts are done manually because of the high cost associated with using technologies such as laser scanners and infrared counters.

With the recent rapid development of internet and cloud computing, we are entering the era of 'big data' with the "Internet of People and Things" and the "Internet of Everything" (O'Leary 2013). The term 'big data' was defined to describe the making of the rapidly expanding amount of digital information analyzable and "the actual use of that data as a means to improve productivity, generate and facilitate innovation and improve decision making" (O'Leary 2013).

Google Street View provides panoramic views along many streets of the U.S. and around the world and is readily available to anyone with access to the internet. Although some parts of the images such as people's faces or automobile license plates are blurred to protect privacy, they are still potentially useful to identify the number of pedestrians on a particular street for generating patterns of walkability across a city. Human detection based on images and videos has had a wide range of applications in a variety of fields including robotics and intelligent transportation for collision prediction, driver assistance, and demographic recognition (Prioletti et al. 2013; Gallagher and Chen 2009). This study introduces an image-based machine learning method to planners for detecting pedestrian activity from Google Street View images, aiming to provide and recommend future research an alternative method for collecting pedestrian counts more consistently and subjectively and to stimulate discussion of the use of 'big data' for planning and design. The results can help researchers to analyze built environment walkability and livability.

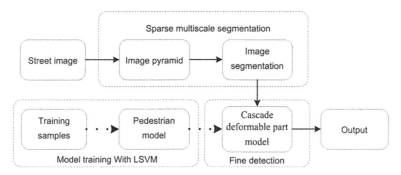

Fig. 1 Framework of proposed algorithm

2 Pedestrian Detection

The current mainstream pedestrian detection algorithms are mainly based on a statistical classification method that classifies features of images and pattern recognition methods that are used to identify pedestrians (Dollár et al. 2010). Dalal and Triggs (2005) proposed one of the most influential pedestrian detection algorithms, characterized by a histogram based on the distribution of the intensity and direction of the gradient of the local image region, referred to as Histogram of Oriented Gradient (HOG). Objects can be characterized with reasonably good detection performance using these distribution features (Andriluka et al. 2008). There have been continuous improvements over the years. A typical example is Zhu et al. (2006) that introduced the Boosted Cascade Face Detection algorithm to the field of pedestrian detection. Felzenszwalb et al. (2008) proposed the Deformable Part Model (DPM), which is a two-layer model with the global and part characteristics that can overcome the partial occlusion to some extent. Felzenszwalb et al. (2010) proposed a Cascade Deformable Part Model (CDPM), which helped to reduce the detection time.

In order to achieve a fast and robust pedestrian detection, this paper adopted a pedestrian detection algorithm based on sparse multi-scale image segmentation and a cascade deformable part model, following Felzenszwalb et al. (2008, 2010). The pedestrian detection framework is illustrated in Fig. 1. First, a model is trained with a latent support vector machine (LSVM) to realize the pedestrian model training offline (Felzenszwalb et al. 2008). Sparse multi-scale image segmentation is then designed to extract the regions of possible pedestrians, to narrow detection range, and to eliminate large areas of background interference in order to achieve the primary detection. Finally, the cascade deformable part model is used to realize finely divided regions for multi-scale fine pedestrian detection.

The training data set of this paper is INRIA, one of the most popular static pedestrian detection datasets (Dalal and Triggs 2005), and LSVM is the training method. The LSVM is used to train the binary classifiers. The LSVM training

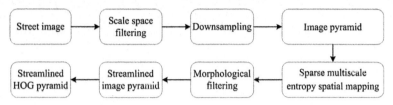

Fig. 2 Multi-scale image segmentation

objective obeys the scoring functions used in Felzenszwalb et al. (2008) in the following form.

$$f_\beta(x) = \max_{z \in Z(x)} \beta \cdot \varnothing(x,z) \quad (1)$$

Here x represents a detection window as input; β is a vector of model parameters; and z contains the values to latent variables such as part placements. The possible latent values for x is set in Z(x) (Yu and Joachims 2008). In analogy to classical SVMs, β is trained from labelled samples of $D = (\langle x_1, y_1 \rangle, \ldots \langle x_n, y_n \rangle)$, where $y_i \in \{-1, 1\}$.

The sparse multi-scale segmentation algorithm adopted is illustrated in Fig. 2. The pedestrian has obvious texture features which are different from those in the background such as the ground and the sky. We used local image entropy to construct a morphological filter and built a streamlined image pyramid. Following Felzenszwalb et al. (2010), the deformation and positional relationships between the parts and the object as a whole can be calculated using the cascade deformable part model (CDPM), which can be applied to detect non-rigid targets such as pedestrians. A sliding window method is used to detect pedestrians in images, step by step, with the CDPM.

2.1 Validation

The detection algorithm discussed above was used on images of over 300 street blocks scattered across the City of Buffalo, where field pedestrian counts were collected by urban planning graduate students at the University at Buffalo, The State University of New York. Depending on the block size, two to five images were used to assemble one image for each street block. Every street block has tabular information to be used to add up the number of pedestrians detected and visualize the pedestrian counts, including block ID, starting and ending coordinates, linked Google Street View image number, etc.

The model results were validated against two sets of data. One is the pedestrian count data collected in the spring and fall seasons from 2011 to 2013 by the students at University at Buffalo. The second is a walkability index assigned to streets, which was collected from WalkScore, a private company that provides walkability services to promote walkable neighborhoods. Pedestrian count data was collected by counting the number of pedestrians on sample street blocks at non-intersection locations, in a 15-min interval. Walk scores were calculated based on distance to the closest amenity such as parks, schools, etc. The Google Street View image data are from static shots for every street. Even though these data sets were collected using different methods, the patterns of walkability represented by these data should be a reasonable reflection of the how streets in Buffalo are really used by pedestrians.

3 Findings

The results of multi-scale image segmentation are shown in Fig. 3, first two columns. The images in the first row have a single pedestrian and the second row images have multiple pedestrians and both showed that pedestrians can be segmented effectively. At the end of this primary stage, the background was removed, some of the environmental interference was reduced, and the effective area for the fine detection stage was retained. This is done to help reduce the false detection rate and to increase accuracy because unnecessary complex calculations are avoided at the fine detection stage. Thus the detection rate was significantly enhanced.

Images in the rightmost column of Fig. 3 show the final detection results. After using sparse multi-scale segmentation to remove the background area, the green rectangles are detected as specific pedestrian areas. As can be seen from the figure, pedestrians can be detected by the algorithm used. Experimental results show that the detection accuracy is increased compared with the traditional pedestrian detection algorithms, and the detection speed is improved.

Figure 4 shows the pedestrian counts based on the pedestrian detection method proposed in this paper (Fig. 4a) in comparison with the counts obtained from the field work (Fig. 4b) and patterns of walkability from WalkScore (Fig. 4c). All three maps show the similar patterns of walkability in the City of Buffalo, with the highest concentration of pedestrians in downtown and Elmwood village areas. Google Street View data has a lower number of pedestrians than the counts from the field work because Google Street View captures only the number of pedestrians at one point in time during one static shot while field work usually captures counts over a period of time.

4 Discussion and Conclusion

This paper used a pedestrian detection algorithm based on sparse multi-scale image segmentation and a cascade deformable model to extract pedestrian counts from the Google Street View images. The experimental results showed that this algorithm performs reasonably fast in comparison to the traditional DPM method. The

Fig. 3 Results from pedestrian detection

detection results were compared and showed to resemble the pedestrian counts collected by field work. The patterns of walkability in the City also resemble the WalkScore data. Future work includes further pedestrian characteristics analysis, combined with a pedestrian tracking algorithm to accelerate the detection efficiency, and robust real-time pedestrian detection.

Google Street View images provide a valuable source for compiling data for design and planning. However, in the form it is currently published and made available to the public, there are two aspects of limitations for pedestrian detection:

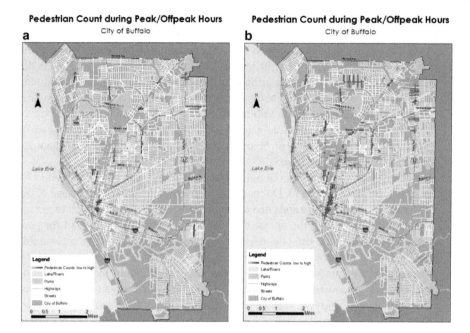
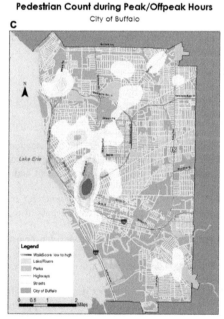

Fig. 4 Pedestrian counts comparison: Google Street View image pedestrian detection (**a**) vs. field work (**b**) vs. WalkScore (**c**)

1. Metadata. Metadata provides information about Google Street View images by Google. Better and more complete metadata would help in expanding the use of these images for research, such as image collection time accurate to a specific day. Currently, the street-view image collection time varies in format and is not accurate to days. This made it difficult to match information extracted from the images to widely available historical weather data for further research into how weather influences pedestrian volume. U.S. city wide weather data is currently accurate to hours and so if the acquisition time of street-view images can be published accurate to days or hours in the future, the validity of the data for research use will be improved.
2. Un-measurable images. The Google Street View data are images acquired by vehicle platform, using a CCD sensor from different viewpoints. Because of the different geometric distortions due to the different angles, it is difficult to get the images registered with the vector data. These images cannot be used for geometric measurements directly because Google does not provide measurable parameters of the street-view currently. If the internal and external orientation elements of the image are released in the future, the stereo image pair can be constructed. Measureable images can also be used to conduct 3D-GIS based calculations and analysis, such as proportion of sky for research on walkability (Yin 2014). Additionally, studies can be done on how to rebuild 3D models from Google Street View panoramic images based on previous research on generating 3D models from regular 2D images. 3D models can be help distinguish pedestrians from poles and street furniture more accurately and effectively to increase pedestrian detection accuracy.

References

Andriluka M, Roth S, Schiele B (2008) People tracking-by-detection and people-detection- by tracking. In: IEEE conference on computer vision and pattern recognition

Clifton KJ, Smith ADL, Rodriguez D (2007) The development and testing of an audit for the pedestrian environment. Landsc Urban Plan 80:95–110

Dalal N, Triggs B (2005) Histograms of oriented gradients for human detection. In: IEEE Computer society conference on computer vision and pattern recognition, 2005, vol 1. pp 886–893

Dollár P, Belongie S, Perona P (2010) The fastest pedestrian detector in the West. BMVC 2 (3):1–11

Ewing R, Bartholomew K (2013) Pedestrian and transit-oriented design. Urban Land Institute and American Planning Association, Washington, DC

Felzenszwalb P, McAllester D, Ramanan D (2008) A discriminatively trained, multiscale, deformable part model. In: IEEE conference on computer vision and pattern recognition, 2008. pp 1–8

Felzenszwalb P, Girshick R, McAllester D (2010) Cascade object detection with deformable part models. In: IEEE conference on computer vision and pattern recognition, 2010. pp 2241–2248

Gallagher A, Chen T (2009) Understanding images of groups of people. In: Proc. CVPR

Hajrasouliha A, Yin L (2014) The impact of street network connectivity on pedestrian movement. Urban Studies Online

O'Leary DE (2013) "Big data', the 'internet of things' and the 'internet of signs'. Intell Syst Account Finance Manage 20:53–65

Prioletti A, Mogelmose A, Grisleri P et al (2013) Part-based pedestrian detection and feature-based tracking for driver assistance: real-time, robust algorithms, and evaluation. IEEE Trans Intell Transp Syst 14(3):1346–1359

Yin L (2013) Assessing walkability in the city of Buffalo: an application of agent-based simulation. J Urban Plan Dev 139(3):166–175

Yin L (2014) Review of Ewing, Reid and Otto Clemente. 2013. Measuring urban design: metrics for livable places. J Plan Literature 29(3): 273e274

Yu CN, Joachims T (2008) Learning structural SVMs with latent variables. In: Neural information processing systems

Zhu Q, Shai A, Chert Y (2006) Fast human detection using a cascade of histograms of oriented gradients. In: Proceedings of IEEE conference on computer vision and pattern recognition, New York. pp 1491–1498

Learning from Outdoor Webcams: Surveillance of Physical Activity Across Environments

J. Aaron Hipp, Deepti Adlakha, Amy A. Eyler, Rebecca Gernes, Agata Kargol, Abigail H. Stylianou, and Robert Pless

Abstract Publicly available, outdoor webcams continuously view the world and share images. These cameras include traffic cams, campus cams, ski-resort cams, etc. The Archive of Many Outdoor Scenes (AMOS) is a project aiming to geolocate, annotate, archive, and visualize these cameras and images to serve as a resource for a wide variety of scientific applications. The AMOS dataset has archived over 750 million images of outdoor environments from 27,000 webcams since 2006. Our goal is to utilize the AMOS image dataset and crowdsourcing to develop reliable and valid tools to improve physical activity assessment via online, outdoor webcam capture of global physical activity patterns and urban built environment characteristics.

This project's grand scale-up of capturing physical activity patterns and built environments is a methodological step forward in advancing a real-time, non-labor intensive assessment using webcams, crowdsourcing, and eventually machine learning. The combined use of webcams capturing outdoor scenes every 30 min and crowdsources providing the labor of annotating the scenes allows for accelerated public health surveillance related to physical activity across numerous built environments. The ultimate goal of this public health and computer vision collaboration is to develop machine learning algorithms that will automatically identify and calculate physical activity patterns.

Keywords Webcams • Physical activity • Built environment • Crowdsourcing • Outdoor environments

---

J.A. Hipp, Ph.D. (✉)
Department of Parks, Recreation, and Tourism Management and Center for Geospatial Analytics, North Carolina State University, 2820 Faucette Dr., Campus Box 8004, Raleigh, NC 27695, USA
e-mail: jahipp@ncsu.edu

D. Adlakha, M.U.D. • A.A. Eyler, Ph.D. • R. Gernes, M.S.W., M.P.H.
Brown School, Washington University in St. Louis, St. Louis, MO, USA

Prevention Research Center, Washington University in St. Louis, St. Louis, MO, USA

A. Kargol • A.H. Stylianou • R. Pless, Ph.D.
Computer Science and Engineering, Washington University in St. Louis, St. Louis, MO, USA

© Springer International Publishing Switzerland 2017
P. Thakuriah et al. (eds.), *Seeing Cities Through Big Data*, Springer Geography,
DOI 10.1007/978-3-319-40902-3_26

1 Introduction

Kevin Lynch's 1960 book, 'The Image of the City', was one of the first to emphasize the importance of social scientists and design professionals in signifying ways that urban design and built environments can be quantitatively measured and improved (Lynch 1960). 'The Image of the City' led to enormous efforts to investigate the structure and function of cities, to characterize perception of neighborhoods (Jacobs 1961; Xu et al. 2012), and promotion of social interactions (Milgram et al. 1992; Oldenburg 1989). To date, large scale studies seeking to understand and quantify how specific features or changes in the built environment impact individuals, their behavior, and interactions have required extensive in-the-field observation and/or expensive and data-intensive technology including accelerometers and GPS (Adams et al. 2015; Sampson and Raudenbush 1999). Such studies only provide a limited view of behaviors, their context, and how each may change as a function of the built environment. These studies are time intensive and expensive, deploying masses of graduate students to conduct interviews about people's daily routines (Milgram et al. 1992) or requiring hand-coding of thousands of hours of video (Whyte 1980) to characterize a few city plazas and parks. Even current state-of-the art technology to investigate associations between behavior and the urban built environment uses multiple expensive devices at the individual level (GPS and accelerometer) and connects this data to Geographic Information System (GIS) layers known to often be unreliable (James et al. 2014; Kerr et al. 2011; Schipperijn et al. 2014).

A key population behavior of interest to our research team is physical activity (Hipp et al. 2013a, b; Adlakha et al. 2014). Physical activity is associated with many health outcomes including obesity, diabetes, heart disease, and some cancer (Office of the Surgeon General 2011). Over 30 % of adults and 17 % of children and adolescents in the US are obese (CDC 2009), with lack of physical activity due to constraints in the built environment being an important influence (Ferdinand et al. 2012). Lack of safe places to walk and bicycle and lack of access to parks and open space can impact the frequency, duration, and quality of physical activity available to residents in urban settings (Jackson 2003; Jackson et al. 2013; Brownson et al. 2009). Physical activity may be purposive such as a jog in a park, or incidental such as a 10 min walk from home to a public transit stop. In both purposive and incidental cases the design of urban built environments influences the decisions and experience of physical activity behaviors. As such, the US Guide to Community Preventive Services (Community Guide) currently recommends the following built environment interventions to increase physical activity behaviors and reduce obesity: (1) community and street-scale urban design and land use policies; (2) creation of, or enhanced access to places for physical activity; and (3) transportation policies and practices (CDC 2011).

Physical Activity Assessment. Physical activity and built environment research has expanded during the past 20 years (Handy et al. 2002; Ferdinand et al. 2012; Harris et al. 2013). The research has followed traditional patterns of growth beginning with ecological studies of association (Ewing et al. 2003), then local validation of associations via retrospective surveys and researcher-present observation (Bedimo-Rung et al. 2006; Mckenzie and Cohen 2006). For example, the System for Observing Physical Activity and Recreation in Communities (SOPARC) (Mckenzie and Cohen 2006) was developed to understand physical activity in context with the environment while being unobtrusive. SOPARC continues to be a popular method of assessing physical activity with pairs of researchers positioning themselves in numerous target areas to scan the environment for the number of people participating in sedentary to vigorous physical activity (Reed et al. 2012; Baran et al. 2013; Cohen et al. 2012). Presently, natural experiments related to physical activity patterns and built environments are growing in popularity (Cohen et al. 2012). These studies have been of great benefit to the field by informing public health and urban design. While there is now a substantial body of evidence to inform local interventions and policies (Ding and Gebel 2012; Saelens and Handy 2008; Kaczynski and Henderson 2007; Feng et al. 2010; Renalds et al. 2010; Sandercock et al. 2010), currently used methodologies and the use of small, local samples limit the external validity and dissemination of many results, interventions, and policies. There is a need for large-scale, evidence-informed evaluations of physical activity to increase external validity as evident in recent calls for more studies across a greater variety of environments (Dyck et al. 2012; Cerin et al. 2009).

Big Data Opportunities. Big data and modern technology has opened up several opportunities to obtain new insights on cities and offer the potential for dramatically more efficient measurement tools (Hipp 2013; Graham and Hipp 2014). The relative ease of capturing large sample data has led to amazing results that highlight how people move through cities based on check-ins (Naaman 2011; Silva et al. 2012) or uploaded photos (Crandall et al. 2009). In addition, GIS, GPS, accelerometers, smart phone applications (apps), and person-point-of-view cameras are each being used in many studies, often in combination (Graham and Hipp 2014; Kerr et al. 2011; Hurvitz et al. 2014). Apps that track running and walking routes are being investigated for where populations move and how parks and other built environment infrastructure may be associated with such movement (Adlakha et al. 2014; Hirsch et al. 2014).

Though these big data sources offer important contributions to the field of physical activity and built environment research, they are each dependent on individuals to upload data, allow access to data, and/or agree to wear multiple devices. This is the epitome of the quantified-self movement (Barrett et al. 2013). A complementary alternative big data source is the pervasive capture of urban environments by traffic cameras and other public, online webcams. This environmental-point-of-view imaging also captures human behavior and physical activity as persons traverse and use urban space.

The Archive of Many Outdoor Scenes (AMOS) has been archiving one image each half hour from most online, publicly available webcams since 2006, creating an open and widely distributed research resource (Pless and Jacobs 2006). AMOS began to collect images from these 27,000 webcams mapped in Fig. 1 to understand

Fig. 1 Map of cameras captured by the Archive of Many Outdoor Scenes (AMOS)

the local effects of climate variations on plants (Jacobs et al. 2007). Other dataset uses include large collections of up-close, on the ground measurements to suggest corrections to standard satellite data products such as NASA's Moderate Resolution Imaging Spectroradiometer (MODIS) estimates of tree growing seasons (Ilushin et al. 2013; Jacobs et al. 2009; Richardson et al. 2011). This global network of existing cameras also captures images of public spaces—plazas, parks, street intersections, waterfronts—creating an archive of how public spaces have changed over time and what behaviors are being performed within these spaces.

With its archive of over 750 million captured images, AMOS not only represents 27,000 unique environments, but is capturing concurrent behaviors in and across the environments. Unique and of significance to public health surveillance, the online, publicly available webcams are non-biased in data collection, consistent and thorough (an image each half hour), and timely (images instantly added to the archive and available to the public). The AMOS project provides an opportunity to virtually annotate changes in the built environment and associated physical activity behaviors. This dataset can provide a census of physical activity patterns within captured environments since 2006 and moving forward. Due to the size of the AMOS dataset, our research team is currently using crowdsources to help annotate the captured scenes.

Use of Crowdsourcing in Public Health Research. Crowdsourcing refers to and utilizes the masses, or crowds, of individuals using the Internet, social media, and smartphone apps. The crowds participating in these websites and applications are the source of data or the source of needed labor (Kamel Boulos et al. 2011). Crowdsourcing data collection in public health is an emerging field, with examples including the collection of tweets and Google searches that detected an increase in influenza before the increase in subsequent influenza-related hospital visits (Kamel Boulos et al. 2011; Ginsberg et al. 2009). Another potential use of crowdsourcing is as the labor in evaluation or assessment of research hypotheses (Bohannon 2011; Buhrmester et al. 2011; Office of the Surgeon General 2011). Our research team was the first to publish on the use of crowdsourcing as physical activity annotators (Hipp et al. 2013a). A crowdsource marketplace, i.e., Amazon Mechanical Turk, can be used to ask workers to complete Human Intelligence Tasks (HITs) such as drawing a box around each pedestrian in a captured image.

Objectives of Current Work. The primary goal of our ongoing collaboration is to use the AMOS dataset and crowdsourcing to develop reliable and valid tools to improve physical activity behavior assessment. This goal will be accomplished by addressing two subsequent aims:

Aim 1: Develop and test the reliability of using publicly available, outdoor webcams to enumerate built environment characteristics and physical activity patterns across thousands of global outdoor environments.

Aim 2: Develop and test the reliability and validity of using crowdsourcing to enumerate built environment characteristics and physical activity patterns across thousands of global outdoor environments.

2 Data Sources

Archive of Many Outdoor Scenes (AMOS). All scenes of human behavior, physical activity, and urban built environments are from the publicly captured AMOS dataset. AMOS is a Washington University project which aims to capture and archive images from every publicly available, online, outdoor webcam (e.g., traffic cams, campus cams, ski-resort cams, etc.—See Fig. 1). This dataset was developed primarily as a basis to research computer vision algorithms for

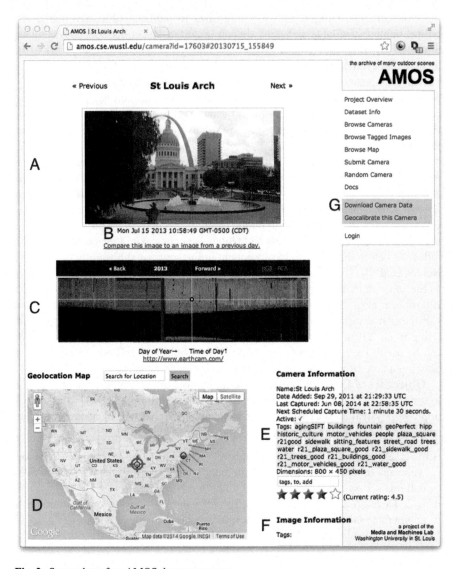

Fig. 2 Screenshot of an AMOS data access page

geolocating and calibrating cameras, and as a demonstration that webcams can be re-purposed as a complement to satellite imaging for large-scale climate measurement (Jacobs et al. 2008, 2009). Images are digitally captured from each camera every 30 min and archived in a searchable dataset (http://amos.cse.wustl.edu/).

Our current work builds on the AMOS model system for working with, sharing, and crowdsourcing big data. Figure 2 shows a screenshot of a main data access page, showing (A) one image and (B) the time this specific image was captured. A yearly summary image, indexed by time of year on the x-axis, and time of day on the y-axis is shown in (C). This summarizes a year of images with each pixel as a representation of the image at that time of year and time of day. Pixels can also be represented using principal component analysis to quickly identify images that differ based on precipitation, snowfall, dusk, dawn, etc. This summary serves several purposes. First, it is a data availability visualization, where dark red highlights when the camera was down and did not capture images. Second, it highlights annual patterns such as the summer nights being shorter than winter nights. Third, it reveals changes in camera angle. Fourth, changes in precipitation, plant growth, and/or shadows are easily recognizable. Finally, data capture errors are quickly visible. This data visualization is "clickable" so that a user can see, by clicking, the image from a particular time of day and time of year.

Each camera also contains extensive metadata as outlined in the Fig. 2: (D) Shows the geolocation of the camera; (E) Shows free form text tags that our research team and other groups use to keep track of and search for cameras with particular properties; (F) Allows the tagging of specific images (instead of cameras), and (G) Is a pointer to zip-files for data from this camera or a python script to allow selective downloading. When exact camera locations are known, the cameras may be geo-oriented and calibrated relative to global coordinates as shown in Fig. 1.

Amazon.com's Mechanical Turk Crowdsource. Virtual audits have emerged as a reliable method to process the growing volume of web-based data on the physical environment (Badland et al. 2010; Clarke et al. 2010; Odgers et al. 2012; Bader et al. 2015). Research has also turned to virtual platforms as a way to recruit study participants and complete simple tasks (Kamel Boulos et al. 2011; Hipp et al. 2013a; Kaczynski et al. 2014). The Amazon.com Mechanical Turk (MTurk) website outsources Human Intelligence Tasks (HITs), or tasks that have not yet been automated by computers. Workers may browse available HITs and are paid for every HIT completed successfully (Buhrmester et al. 2011). MTurk workers are paid a minimum of US$0.01 per HIT, making them a far less expensive option than traditional research assistant annotators (Berinsky et al. 2012). MTurk has been found to be an effective method for survey participant recruitment, with more representative and valid results than the convenience sampling often used for social science research (Bohannon 2011). MTurk has also been used for research task completion such as transcription and annotation. These have generally been small in scale and MTurk reliability for larger scale data analysis has not been established (Hipp et al. 2013a). Within MTurk, our team has designed a unique web-form used with the MTurk HIT that allows workers to annotate images by demarcating each pedestrian, bicyclist, and vehicle located in a photograph.

Trained Research Assistants. Trained undergraduate and graduate Research Assistants from the computer science and public health departments at Washington University in St. Louis have annotated images for physical activity behaviors and built environment attributes. For both behaviors and environments, Research Assistants were provided with example captured scenes. Project Principal Investigators supervised the scene annotation process and provided real-time feedback on uncertain scenes. Difficult or exceptional scenes and images were presented to the research group to ensure that all behaviors and environments were annotated in a consistent manner.

3 Methods

Annotating Physical Activity Behaviors. We have used 12 traffic webcams located in Washington, D.C., to determine initial feasibility of our physical activity behavior research agenda. AMOS has archived a photograph every 30 min from Washington, D.C., Department of Transportation webcams. Since 2007, Washington, D.C., has initiated multiple built environment improvements to increase physical activity behaviors, including a bicycle share program, miles of new bike lanes, and painted crosswalks. For example, a new bicycle lane was added in the middle of Pennsylvania Avenue in spring 2010, and AMOS has an archive of captured images every 30 min for the year prior to the installation of the bike lane, and a year following installation.

The MTurk website was used to crowdsource the image annotation. In a pilot project we uploaded each webcam photograph captured by AMOS at the intersection of Pennsylvania Avenue NW and 9th Street NW between 7 am and 7 pm the first work week of June 2009 and June 2010 to the MTurk webpage (Hipp et al. 2013a). There we designed a HIT that allowed MTurk workers to annotate our images by marking each pedestrian, bicyclist, and vehicle in each captured scene. MTurk workers used their computer mouse to hover over the appropriate behavior, e.g., pedestrian activity, and left-click atop each individual pedestrian. Five unique MTurk workers completed this task for all three transportation behaviors per image. The numbers of each type of annotation were then downloaded to a spreadsheet and imported into SPSS.

In related ongoing work, we have used 12 different AMOS webcams that captured other built environment changes at intersections in Washington, D.C., between 2007 and 2010. We have made improvements to our MTurk task by asking workers to use their cursors to draw polygons, or boxes, around the specified transportation behavior (walking, bicycling, driving). Similar to the first HIT, we used each photograph between 7:00 am and 7:00 pm during the first week of June proceeding and following a built environment change. Finally, we posted to MTurk every photograph from two of the above intersections between 6:00 am and 9:00 pm for 19 consecutive months (5 months prior to a crosswalk being introduced to the intersections and 14 months post).

MTurk workers were paid US$0.01 or US$0.02 per scene to mark each pedestrian, bicyclist, and vehicle in an image and took on average 71 s to complete each task. Each image was annotated five unique times. Two trained Research Assistants completed the same task, annotating each image twice. Training took place in two sessions. In the first session, Research Assistants received the same instructions as MTurk participants and completed a practice set of 100 images. In the second session, Research Assistants compared their practice results and discussed differences in analysis. Research Assistants completed the camera annotations in separate forms, and their results were averaged.

Annotating Built Environments. Selecting the appropriate built environment image tags per camera has been an iterative process. First, we selected two commonly used built environment audit tools to establish a list of potential built environment tags. These were the Environmental Assessment of Public Recreation Spaces (EAPRS) (Saelens et al. 2006) and the Irvine-Minnesota Inventory (Day et al. 2006). From an initial list of 73 built environment items that we believed could be annotated using captured images we narrowed the final list down to 21 built environment tags. Following the combination of similar terms, we further reduced the potential list of tags based on the inclusion criteria that the tag must be theoretically related to human behaviors.

To establish which of the 27,000 AMOS webcams are at an appropriate urban built environment scale, i.e., those with the potential of capturing physical activity, our team designed an interface that selects a set of camera IDs, and displays 25 cameras per screen. This internal (available only to research staff) HIT was created to populate a webpage with the 25 unique camera images. Below each image were a green checkmark and a red x-mark. If physical activity behaviors could be captured in the scene, the green checkmark was selected and this tag automatically added to a dataset of physical activity behavior cameras. This process was repeated with trained Research Assistants for reliability and resulted in a set of 1906 cameras at an appropriate built environment and physical activity scale. In addition to the above inclusion criteria, selected cameras must have captured scenes from at least 12 consecutive months. The final 21 built environment tags are presented in Table 1.

To tag each camera, Research Assistants were provided a one-page written and photographic example (from AMOS dataset) of each built environment tag. For example, a written description for a crosswalk was provided along with captured images of different styles of crosswalks from across the globe. A second internal HIT was created similar to the above that populated a webpage with 20 unique camera images, each marked with a green checkmark and a red x-mark. If the provided built environment tag (e.g., crosswalk) was present in the image then the green checkmark was selected and this tag automatically added to the camera annotation. If a Research Assistant was unsure they could click on the image to review other images captured by the same camera or could request the assistance of other Research Assistants or Principal Investigators to verify their selection. This process was completed for all 21 built environment tags across all 1906 cameras in the AMOS physical activity and built environment dataset. To date, the built environment tags have only been annotated by trained Research Assistants. Reliability and validity of tags is a future step of this research agenda. This initial step provided the team a workable set of publicly available webcams to address our two study aims.

Table 1 List of built environment tags used to annotate AMOS webcam images

No.	Built environment tag	Number of cameras with built environment tag present
1.	Open space	769
2.	Sidewalk	825
3.	Plaza/square	174
4.	Residential/homes	97
5.	Trees	1058
6.	Buildings	1245
7.	Street, intersection	621
8.	Bike lane	71
9.	Athletic fields	60
10.	Speed control	185
11.	Trail path	154
12.	Street, road	1029
13.	Signage	59
14.	Commerce retail	382
15.	Play features	42
16.	Sitting features	166
17.	Motor vehicles	1048
18.	Crosswalk	576
19.	Bike racks	27
20.	Water	326
21.	Snow/Ski	169

4 Data Analysis

Physical Activity Behaviors. In the pilot project we used t-tests and logistic regressions to analyze the difference in physical activity behaviors before and after the addition of the bike lane along Pennsylvania Avenue. T-tests were used for pedestrians and vehicles, where the data was along a continuous scale from 0 to 20 (20 being the most captured in any one scene). Logistic regression was used for the presence or absence of a bicyclist in each image.

Reliability and Validity. Inter-rater reliability (IRR) and validity statistics (Pearson's R, Inter-Class Correlations, and Cohen's Kappa) were calculated within and between the five MTurk workers and between the two trained Research Assistants. The appropriate statistic was calculated for two, three, four, or five MTurk workers to determine the optimal number of workers necessary to capture a reliable and valid count of pedestrians, bicyclists, and vehicles in a scene. Due to each scene being annotated by five unique MTurk workers we were able to test the reliability of ten unique combinations of workers; that is, Worker 1 and Worker 2, Worker 1 and Worker 3, Worker 1 and Worker 4, etc. Similar combinations were used with three workers (ten unique combinations) and four workers (five unique combinations). Each combination was compared to the trained Research Assistants results to measure validity. For all tests we used Landis and Koch's magnitudes of agreement: <0.19 (poor agreement), 0.20–0.39 (fair), 0.40–0.59 (moderate), 0.60–0.79 (substantial) and >0.80 (near perfect agreement) (Landis and Koch 1977).

5 Results

Pilot Project. Previously published results reveal that publicly available, online webcams are capable of capturing physical activity behavior and are capable of capturing changes in these behaviors pre and post built environment changes (Hipp et al. 2013a). Of the 240 images captured by the camera located at Pennsylvania Avenue NW and 9th Street NW (7 am–7 pm for 1 week prior to built environment change and 7 am–7 pm for 1 week post change), 237 (99 %) had at least one pedestrian annotated, and 38 (16 %) had at least one bicyclist annotated. Table 2 presents the number and percent of images with pedestrian, bicyclist and motor vehicle annotation by at least one MTurk worker, by intersection.

The odds of the traffic webcam at Pennsylvania Avenue NW and 9th Street NW capturing a bicyclist present in the scene in 2010 increased 3.5 times, compared to 2009 (OR = 3.57, p < 0.001). The number of bicyclists present per scene increased four-fold between 2009 (mean = 0.03; SD = 0.20) and 2010 (0.14; 0.90; F = 36.72, 1198; p = 0.002). Both results are associated with the addition of the new bike lane. There was no associated increase in the number of pedestrians at the street intersection following the addition of the bike lane, as may be theoretically expected with a bicycle-related built environment change, not a pedestrian-related change.

Table 2 Number and percent of images with at least one bicycle, pedestrian, or motor vehicle annotated by at least one MTurk worker, by intersection/camera ID

Camera ID	N images collected	N images with pedestrians (%)[a]	N images with bicyclists (%)[a]	N images with motor vehicles (%)[a]
919	244	42(17 %)	7(3 %)	162(66 %)
920	241	119(49 %)	18(7 %)	237(98 %)
923	241	78(33 %)	9(4 %)	217(90 %)
925	240	86(36 %)	24(10 %)	215(90 %)
929	236	204(86 %)	40(17 %)	236 (100 %)
930	245	186(76 %)	27(11 %)	242(99 %)
935	240	133(55 %)	15(6 %)	193(80 %)
942	242	225(93 %)	32(13 %)	236(98 %)
950	240	213(89 %)	21(9 %)	235(98 %)
968	240	133(55 %)	8(3 %)	209(87 %)
984	240	237(99 %)	38(16 %)	240 (100 %)
996	246	131(53 %)	28(11 %)	245(100 %)
Total	2895	1787(62 %)	267(9.2 %)	2667(92 %)

[a]Percent rounded to nearest whole number

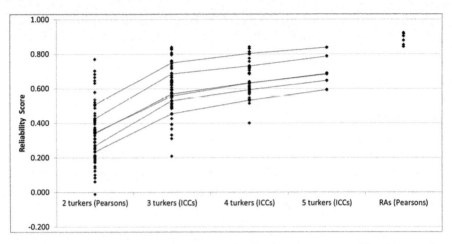

Fig. 3 Reliability results for annotation of pedestrians in 720 webcam scenes. *Lines* represent the average reliability score across five unique cameras

Reliability Assessment. Next, we tested reliability and validity of using publicly available webcams and MTurk HITs to annotate captured scenes for physical activity and transportation behaviors. Reliability statistics varied across MTurk workers based on the number annotating each scene and the annotation task (pedestrians compared to bicyclists).

For pedestrians (n = 720 images), pairs of MTurk workers had an agreement average and a Pearson's R-score of 0.562 (range: 0.122–0.866). The Inter-Class Correlation (ICC) for three MTurk workers averaged 0.767 (0.330–0.944) and four workers averaged 0.814 (0.534–0.954). The average for all five workers across the 720 captured scenes was 0.848 (0.687–0.941). The ICCs for four and five workers represented near-perfect agreement. The pair of trained Research Assistants averaged a Pearson's R-score 0.850 (0.781–0.925), also representing near perfect agreement (Fig. 3).

The averages and ranges of annotator agreement for presence of bicyclists in 2007 were as follows (Table 3): two workers (Cohen's Kappa: 0.333; Range: 0.000–0.764), three workers (0.553; 0.000–0.897), four workers (0.607; 0.000–0.882), five workers (0.645; 0.000–0.874), and Research Assistants (0.329; 0.000–0.602). Annotator agreement with four and five MTurk workers showed substantial agreement. For the pilot project presented above, we used the average of five MTurk workers. When analyzing presence versus absence of a bicyclist, majority ruled; if three or more of the five MTurk workers annotated a bicyclist, then this scene received a 1/present. If two or fewer annotated a cyclist, the scene received a 0/not present.

The averages and ranges for number of vehicles were as follows: two workers (0.354; 0.000–0.769), three workers (0.590; 0.208–0.830), four workers (0.653; 0.398–0.830), five workers (0.705; 0.592–0.837), and Research Assistants (0.885;

Table 3 Inter-rater reliability coefficients for trained research assistants and Mechanical Turk workers

Camera ID	Counts	Reliability coefficients							
		RAs		2 MTurkers		Correlation average	3 MTurkers	4 MTurkers	5 MTurkers
		N	Correlation	Range	Correlation average		ICC range	ICC range	ICC
919	Pedestrian[a]	74	.779**	.122–.769**	.344		.330***–.825***	.534***–.728***	.687***
	Bicyclist[b]	74	n/a[c]	n/a[c]	n/a[c]		n/a[c]	.000–(−.019)	.000***
	Motor Vehicle[a]	74	.916**	.122–.769**	.344		.330***–.663***	.534***–.721***	.687***
920	Pedestrian[a]	121	.925**	.382**–.800**	.604		.724***–.881***	.819***–.874***	.896***
	Bicyclist[b]	121	n/a[c]	n/a[c]	n/a[c]		.267*–.773***	.507***–.737***	.665***
	Motor Vehicle[a]	121	.852**	−.013–.502**	.235		.208***–.630***	.398***–.614***	.592***
929	Pedestrian[a]	120	.822**	.342**–.812**	.577		.706***–.897***	.812***–.882***	.874***
	Bicyclist[b]	120	.524***	.425***–.635***	.566		.706***–.897***	.882***–.857***	.874***
	Motor Vehicle[a]	120	.902**	.257–.702**	.508		.605***–.838***	.730***–.841***	.837***
930	Pedestrian[a]	125	.900**	.659**–.860**	.784		.875***–.944***	.917***–.954***	.941***
	Bicyclist[b]	125	.547**	.317***–.663***	.426		.477***–.685***	.578***–.756***	.704***
	Motor Vehicle[a]	125	.878**	.177*–.577**	.340		.390***–.730***	.575***–.685***	.683***
942	Pedestrian[a]	126	.781**	.353**–.534**	.450		.679***–.749***	.765***–.780***	.803***
	Bicyclist[b]	126	.602***	.281**–.764***	.552		.680***–.868***	.783***–.871***	.852***
	Motor Vehicle[a]	126	.841**	.099–.411**	.269		.392***–.617***	.544***–.634***	.646***
996	Pedestrian[a]	128	.893**	.526**–.709**	.615		.790***–.856***	.851***–.877***	.889***
	Bicyclist[b]	128	.301***	.169–.659***	.317		.437***–.788***	.645***–.792***	.774***
	Motor Vehicle[a]	128	.922**	.291**–.682**	.381		.619***–.742***	.690***–.774***	.786***

Significance levels: $*p < .05$, $**p < .01$, $***p < .001$
RA research assistant, ICC intraclass correlation coefficients
[a] Pearson correlations were used in this calculation
[b] Kappas were used in this calculation
[c] Items could not be calculated due to insufficient variance between raters

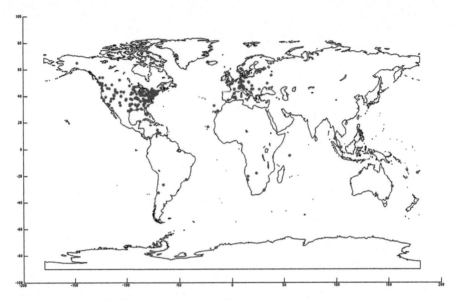

Fig. 4 Location of AMOS webcams tagged with 'open space'

0.841–0.922). The reliability statistics for four and five MTurk workers again showed substantial rater/annotator agreement, and near perfect agreement between the two Research Assistants.

Validity Assessment. From reliability estimates, we concluded that using four MTurk workers was the most reliable and cost-efficient method. Next, validity statistics were calculated for four MTurk workers and two trained RAs. Validity statistics (Pearson's R) for pedestrians (0.846–0.901) and vehicles (0.753–0.857) showed substantial to near perfect agreement. Validity (Cohen's kappa) for cyclists (0.361–0.494) were in the fair-moderate agreement range.

Built Environment Tags. As provided in Table 1, our final list of built environment tags includes 21 unique items. The number of cameras with the tag present is also presented. 'Buildings' were found the most frequently; at a scale to capture human behavior across 1245 webcams. 'Bike racks' was annotated the fewest times, only occurring in 27 scenes. Figure 4 shows an example map of where each of the cameras with the built environment tag of 'open space' is located.

6 Discussion

The use of public, captured imagery to annotate built environments for public health research is an emerging field. To date the captured imagery has been static and only available via Google Street View and Google Satellite imagery (Charreire et al. 2014; Edwards et al. 2013; Wilson et al. 2012; Taylor and Lovell 2012;

Odgers et al. 2012; Kelly et al. 2013, 2014; Wilson and Kelly 2011; Taylor et al. 2011; Rundle et al. 2011). There have been no attempts to crowdsource this image annotation, nor combine annotation of built environments and images capturing physical activity behaviors. Using an 8-year archive of captured webcam images and crowdsources, we have demonstrated that improvements in urban built environments are associated with subsequent and significant increases in physical activity behaviors. Webcams are able to capture a variety of built environment attributes and our study shows webcams are a reliable and valid source of built environment information. As such, the emerging technology of publicly available webcams facilitates both consistent uptake and potentially timely dissemination of physical activity and built environment behaviors across a variety of outdoor environments. The AMOS webcams have the potential to serve as an important and cost-effective part of urban environment and public health surveillance to evaluate patterns and trends of population-level physical activity behavior in diverse built environments.

In addition to presenting a new way to study physical activity and the built environment, our findings contribute to novel research methodologies. The use of crowdsources (Amazon's Mturk) proved to be a reliable, valid, inexpensive, and quick method for annotating street scenes captured by public, online webcams. While MTurk workers have previously been found to be a valid and reliable source of participant recruitment for experimental research, this is the first research agenda that has found MTurk to be a valid and reliable method for content analysis (Buhrmester et al. 2011; Berinsky et al. 2012; Hipp et al. 2013a, b). Our results indicate taking the average annotation of four unique MTurk workers appears to be the optimal threshold. Our results also show that across each mode of transportation assessed, the average reliability score with four unique workers was 0.691, which is considered substantial agreement (Landis and Koch 1977).

In addition to substantial agreement between the MTurk workers, the trained RAs yielded substantial agreement with vehicles, near perfect agreement with pedestrians, but only fair agreement with cyclists. The bicyclist statistics were the least reliable, primarily due to the low number of images (only 10 % of captured scenes) with a cyclist present. Similar to reliability statistics, validity was near perfect for pedestrians and vehicles, but only fair to moderate for cyclists. These results suggest MTurk workers are a quick, cheap annotation resource for commonly captured image artifacts. However, MTurk is not yet primed to capture rare events in captured scenes without additional instruction or limitations to the type of workers allowed to complete tasks.

Our current big data and urban informatics research agenda shows that publicly available, online webcams offer a reliable and valid source for measuring physical activity behavior in urban settings. Our findings lay the foundation for studying physical activity and built environment characteristics using the magnitude of globally-available recorded images as measurements. The research agenda is innovative in: (1) its potential to characterize physical activity patterns over the timescale of years and with orders of magnitude more measurements than would be feasible by standard methods, (2) the ability to use the increase in data to characterize complex interactions between physical activity patterns, seasons, and weather, and (3) its capacity to be an ongoing, systematic public health surveillance system. In addition to increasing the capacity of physical activity research, the methodologies described here are of novel interest to computer vision researchers. Automating algorithms to detect and quantify behavioral transformations due to changes in urban policy and built infrastructure can transform aspects of this research as well.

These findings have several implications related to cost and timeliness for the use of MTurks in content analysis. The total cost for all MTurk analysis to date has been $320.01 for 32,001 images, compared to $1333.33 for a trained Research Assistant paid at US$10 per hour. MTurk workers are paid for each successful completion of a HIT, compared to hourly wages for a Research Assistant. This allows MTurk to pay multiple workers at a time for each HIT, at a cost substantially lower than the same number of trained Research Assistants. The higher speed and lower cost of crowdsourced analysis is especially suitable for annotating AMOS data, to which thousands of images are added daily. Reliable and rapid image annotation using MTurk could allow for large-scale and more robust analysis of results that would be too costly to complete with traditional analysis. Thus far our team has looked at 12 cameras in one metro area. Future studies could increase the number of cameras annotated for a specific area or time, compare results across metro regions, or analyze environmental effects such as weather, season, and day of the week on mode of transportation.

There are several limitations inherent to the use of webcams to study pedestrian behavior. Many publicly available outdoor webcams have limited availability for physical activity and built environment use (roughly 90 % of the 27,000 AMOS cameras did not meet our inclusion criteria of 1 year of data, at least one of the 21 built environment tags, and at a scene scale where physical activity could be annotated). Therefore, current research is restricted to locations in which webcams are present and researchers can access publicly available webcam image data (~2000 webcams). Webcam image occlusion does occur, including glare, inclement weather (precipitation on lens), shifting camera angles, vehicles obstructing the captured image, and other obstacles. In the studies mentioned in this chapter, images were captured once every 30 min. As such, certain annotated objects— such as bicycles—may not be captured reliably by the webcams due to bicycles still being a rare event and due to the size and speed of bicyclists causing motion blur in some images. All of these can restrict the accuracy of image-based annotation. The design of the Human Intelligence Task may also impact the reliability of resulting data. Despite these limitations, the availability of webcams and the objective nature of the data they provide present an opportunity to understand physical activity trends in a variety of environments in an easy, cost-effective, and accurate manner.

There are several ethical and human subjects concerns related to publicly available, online webcams and the use of MTurk. With our initial research projects, we have received exempt status for the use of both AMOS and MTurk. The webcams were exempt because our research is not collecting individual identifiable private information; this activity is not considered to meet federal definitions under the jurisdiction of an Institutional Review Board and therefore falls outside the purview of the Human Research Protection Office. AMOS is an archival dataset of publicly available photos. The photos are being used for counts and annotation of physical activity patterns and built environment attributes and are not concerned with individual or identifiable information. To date, no camera has been identified that is at an angle and height so as to distinguish an individual's face. The use of publicly available webcams fits with the 'Big Sister' approach to the use of cameras for human-centered design and social values (Stauffer and Grimson 2000). Related, recent research utilizing Google Street View and Google Earth images have also been HRPO-exempt (Sequeira et al. 2013; Sadanand and Corso 2012; National Center for Safe Routes to School 2010; Saelens et al. 2006).

Finally, AMOS is quite literally "Seeing Cities Through Big Data" with applications for research methods and urban informatics. With thoughtful psychometrics and application of this half-billion image dataset, and growing, we believe pervasive webcams can assist researchers and urban practitioners alike in better understanding how we use place and how the shape and context of urban places influence our movement and behavior.

Acknowledgements This work was funded by the Washington University in St. Louis University Research Strategic Alliance pilot award and the National Cancer Institute of the National Institutes of Health under award number 1R21CA186481. The content is solely the responsibility of the authors and does not necessarily represent the official views of the National Institutes of Health.

References

Adams MA, Todd M, Kurka J, Conway TL, Cain KL, Frank LD, Sallis JF (2015) Patterns of walkability, transit, and recreation environment for physical activity. Am J Prev Med 49 (6):878–887

Adlakha D, Budd EL, Gernes R, Sequeira S, Hipp JA (2014) Use of emerging technologies to assess differences in outdoor physical activity in St. Louis, Missouri. Front Public Health 2:41

Bader MDM, Mooney SJ, Lee YJ, Sheehan D, Neckerman KM, Rundle AG, Teitler JO (2015) Development and deployment of the Computer Assisted Neighborhood Visual Assessment System (CANVAS) to measure health-related neighborhood conditions. Health Place 31:163–172

Badland HM, Opit S, Witten K, Kearns RA, Mavoa S (2010) Can virtual streetscape audits reliably replace physical streetscape audits? J Urban Health 87:1007–1016

Baran PK, Smith WR, Moore RC, Floyd MF, Bocarro JN, Cosco NG, Danninger TM (2013) Park use among youth and adults: examination of individual, social, and urban form factors. Environ Behav. doi:10.1177/0013916512470134

Barrett MA, Humblet O, Hiatt RA, Adler NE (2013) Big data and disease prevention: from quantified self to quantified communities. Big Data 1:168–175

Bedimo-Rung A, Gustat J, Tompkins BJ, Rice J, Thomson J (2006) Development of a direct observation instrument to measure environmental characteristics of parks for physical activity. J Phys Act Health 3:S176–S189

Berinsky AJ, Huber GA, Lenz GS (2012) Evaluating online labor markets for experimental research: Amazon.com's Mechanical Turk. Polit Anal 20:351–368

Bohannon J (2011) Social science for pennies. Science 334:307

Brownson RC, Hoehner CM, Day K, Forsyth A, Sallis JF (2009) Measuring the built environment for physical activity: state of the science. Am J Prev Med 36: S99–S123.e12

Buhrmester M, Kwang T, Gosling SD (2011) Amazon's Mechanical Turk: a new source of inexpensive, yet high-quality, data? Perspect Psychol Sci 6:3–5

CDC (2009) Division of Nutrition, Physical Activity and Obesity. http://www.cdc.gov/nccdphp/dnpa/index.htm

CDC (2011) Guide to community preventive services. Epidemiology Program Office, CDC, Atlanta, GA

Cerin E, Conway TL, Saelens BE, Frank LD, Sallis JF (2009) Cross-validation of the factorial structure of the Neighborhood Environment Walkability Scale (NEWS) and its abbreviated form (NEWS-A). Int J Behav Nutr Phys Act 6:32

Charreire H, Mackenbach JD, Ouasti M, Lakerveld J, Compernolle S, Ben-Rebah M, McKee M, Brug J, Rutter H, Oppert JM (2014) Using remote sensing to define environmental characteristics related to physical activity and dietary behaviours: a systematic review (the SPOTLIGHT project). Health Place 25:1–9

Clarke P, Ailshire J, Melendez R, Bader M, Morenoff J (2010) Using Google Earth to conduct a neighborhood audit: reliability of a virtual audit instrument. Health Place 16:1224–1229

Cohen DA, Marsh T, Williamson S, Golinelli D, McKenzie TL (2012) Impact and cost-effectiveness of family Fitness Zones: a natural experiment in urban public parks. Health Place 18:39–45

Crandall DJ, Backstrom L, Huttenlocher D, Kleinberg J (2009) Mapping the world's photos. In: Proceedings of the 18th international conference on World Wide Web

Day K, Boarnet M, Alfonzo M, Forsyth A (2006) The Irvine-Minnesota inventory to measure built environments: development. Am J Prev Med 30:144–152

Ding D, Gebel K (2012) Built environment, physical activity, and obesity: what have we learned from reviewing the literature? Health Place 18:100–105

Dyck DV, Cerin E, Conway TL, Bourdeaudhuij ID, Owen N, Kerr J, Cardon G, Frank LD, Saelens BE, Sallis JF (2012) Perceived neighborhood environmental attributes associated with adults' transport-related walking and cycling: findings from the USA, Australia and Belgium. Int J Behav Nutr Phys Act 9:70

Edwards N, Hooper P, Trapp GSA, Bull F, Boruff B, Giles-Corti B (2013) Development of a Public Open Space Desktop Auditing Tool (POSDAT): a remote sensing approach. Appl Geogr 38:22–30

Ewing R, Meakins G, Hamidi S, Nelson A (2003) Relationship between urban sprawl and physical activity, obesity, and morbidity. Am J Health Promot 18:47–57

Feng J, Glass TA, Curriero FC, Stewart WF, Schwartz BS (2010) The built environment and obesity: a systematic review of the epidemiologic evidence. Health Place 16:175–190

Ferdinand AO, Sen B, Rahurkar S, Engler S, Menachemi N (2012) The relationship between built environments and physical activity: a systematic review. Am J Public Health 102:e7–e13

Ginsberg J, Mohebbi MH, Patel RS, Brammer L, Smolinski MS, Brilliant L (2009) Detecting influenza epidemics using search engine query data. Nature 457:1012–1014

Graham DJ, Hipp JA (2014) Emerging technologies to promote and evaluate physical activity: cutting-edge research and future directions. Front Public Health 2:66

Handy S, Boarnet M, Ewing R, Killingsworth R (2002) How the built environment affects physical activity: views from urban planning. Am J Prev Med 23:64–73

Harris JK, Lecy J, Parra DC, Hipp A, Brownson RC (2013) Mapping the development of research on physical activity and the built environment. Prev Med 57:533–540

Hipp JA (2013) Physical activity surveillance and emerging technologies. Braz J Phys Act Health 18:2–4

Hipp JA, Adlakha D, Eyler AA, Chang B, Pless R (2013a) Emerging technologies: webcams and crowd-sourcing to identify active transportation. Am J Prev Med 44:96–97

Hipp JA, Adlakha D, Gernes R, Kargol A, Pless R (2013b) Do you see what I see: crowdsource annotation of captured scenes. In: Proceedings of the 4th international SenseCam & pervasive imaging conference. San Diego, CA: ACM

Hirsch JA, James P, Robinson JR, Eastman KM, Conley KD, Evenson KR, Laden F (2014) Using MapMyFitness to place physical activity into neighborhood context. Front Public Health 2:19

Hurvitz PM, Moudon AV, Kang B, Saelens BE, Duncan GE (2014) Emerging technologies for assessing physical activity behaviors in space and time. Front Public Health 2:2

Ilushin D, Richardson A, Toomey M, Pless R, Shapiro A (2013) Comparing the effects of different remote sensing techniques for extracting deciduous broadleaf phenology. In: AGU Fall Meeting abstracts

Jackson RJ (2003) The impact of the built environment on health: an emerging field. Am J Public Health 93:1382–1384

Jackson RJ, Dannenberg AL, Frumkin H (2013) Health and the built environment: 10 years after. Am J Public Health 103:1542–1544

Jacobs J (1961) The death and life of great American cities. Random House LLC, New York

Jacobs N, Roman N, Pless R (2007) Consistent temporal variations in many outdoor scenes. In: IEEE conference on computer vision and pattern recognition, 2007, CVPR '07, 17–22 June 2007. pp 1–6

Jacobs N, Roman N, Pless R (2008) Toward fully automatic geo-location and geo-orientation of static outdoor cameras. In: Proc. IEEE workshop on video/image sensor networks

Jacobs N, Burgin W, Fridrich N, Abrams A, Miskell K, Braswell BH, Richardson AD, Pless R (2009) The global network of outdoor webcams: properties and applications. In: ACM international conference on advances in geographic information systems (SIGSPATIAL GIS)

James P, Berrigan D, Hart JE, Hipp JA, Hoehner CM, Kerr J, Major JM, Oka M, Laden F (2014) Effects of buffer size and shape on associations between the built environment and energy balance. Health Place 27:162–170

Kaczynski AT, Henderson KA (2007) Environmental correlates of physical activity: a review of evidence about parks and recreation. Leis Sci 29:315–354

Kaczynski AT, Wilhelm Stanis SA, Hipp JA (2014) Point-of-decision prompts for increasing park-based physical activity: a crowdsource analysis. Prev Med 69:87–89

Kamel Boulos MN, Resch B, Crowley DN, Breslin JG, Sohn G, Burtner R, Pike WA, Jezierski E, Chuang KY (2011) Crowdsourcing, citizen sensing and sensor web technologies for public and environmental health surveillance and crisis management: trends, OGC standards and application examples. Int J Health Geogr 10:67

Kelly C, Wilson J, Baker E, Miller D, Schootman M (2013) Using Google Street View to audit the built environment: inter-rater reliability results. Ann Behav Med 45(Suppl 1):S108–S112

Kelly C, Wilson JS, Schootman M, Clennin M, Baker EA, Miller DK (2014) The built environment predicts observed physical activity. Front Public Health 2:52

Kerr J, Duncan S, Schipperijn J (2011) Using global positioning systems in health research: a practical approach to data collection and processing. Am J Prev Med 41:532–540

Landis JR, Koch GG (1977) The measurement of observer agreement for categorical data. Biometrics 33:159–174

Lynch K (1960) The image of the city. MIT Press, Cambridge

Mckenzie TL, Cohen DA (2006) System for observing play and recreation in communities (SOPARC). In: Center for Population Health and Health Disparities (ed) RAND

Milgram S, Sabini JE, Silver ME (1992) The individual in a social world: essays and experiments. McGraw-Hill Book Company, New York

Naaman M (2011) Geographic information from georeferenced social media data. SIGSPATIAL Special 3:54–61

National Center for Safe Routes to School (2010) Retrieved from http://www.saferoutesinfo.org/.

Odgers CL, Caspi A, Bates CJ, Sampson RJ, Moffitt TE (2012) Systematic social observation of children's neighborhoods using Google Street View: a reliable and cost-effective method. J Child Psychol Psychiatry 53:1009–1017

Office of the Surgeon General, Overweight and obesity: at a glance (2011) Retrieved from http://www.cdc.gov/nccdphp/sgr/ataglan.htm.

Oldenburg R (1989) The great good place: cafés, coffee shops, community centers, beauty parlors, general stores, bars, hangouts, and how they get you through the day. Paragon House, New York

Pless R, Jacobs N (2006) The archive of many outdoor scenes, media and machines lab, Washington University in St. Louis and University of Kentucky

Reed JA, Price AE, Grost L, Mantinan K (2012) Demographic characteristics and physical activity behaviors in sixteen Michigan parks. J Community Health 37:507–512

Renalds A, Smith TH, Hale PJ (2010) A systematic review of built environment and health. Fam Community Health 33:68–78

Richardson A, Friedl M, Frolking S, Pless R, Collaborators P (2011) PhenoCam: a continental-scale observatory for monitoring the phenology of terrestrial vegetation. In: AGU Fall Meeting abstracts

Rundle AG, Bader MDM, Richards CA, Neckerman KM, Teitler JO (2011) Using Google Street View to audit neighborhood environments. Am J Prev Med 40:94–100

Sadanand S, Corso JJ (2012) Action bank: a high-level representation of activity in video. In: IEEE conference on computer vision and pattern recognition (CVPR), 2012

Saelens BE, Handy S (2008) Built environment correlates of walking: a review. Med Sci Sports Exerc 40:S550–S566

Saelens BE, Frank LD, Auffrey C, Whitaker RC, Burdette HL, Colabianchi N (2006) Measuring physical environments of parks and playgrounds: EAPRS instrument development and inter-rater reliability. J Phys Act Health 3:S190–S207

Sampson RJ, Raudenbush SW (1999) Systematic social observation of public spaces: a new look at disorder in urban neighborhoods. Am J Sociol 105:603

Sandercock G, Angus C, Barton J (2010) Physical activity levels of children living in different built environments. Prev Med 50:193–198

Schipperijn J, Kerr J, Duncan S, Madsen T, Klinker CD, Troelsen J (2014) Dynamic accuracy of GPS receivers for use in health research: a novel method to assess GPS accuracy in real-world settings. Front Public Health 2:21

Sequeira S, Hipp A, Adlakha D, Pless R (2013) Effectiveness of built environment interventions by season using web cameras. In: 141st APHA annual meeting, 2–6 Nov 2013

Silva TH, Melo PO, Almeida JM, Salles J, Loureiro AA (2012) Visualizing the invisible image of cities. In: IEEE international conference on green computing and communications (GreenCom), 2012

Stauffer C, Grimson WEL (2000) Learning patterns of activity using real-time tracking. IEEE Trans Pattern Anal Mach Intell 22:747–757

Taylor JR, Lovell ST (2012) Mapping public and private spaces of urban agriculture in Chicago through the analysis of high-resolution aerial images in Google Earth. Landsc Urban Plan 108:57–70

Taylor BT, Peter F, Adrian EB, Anna W, Jonathan CC, Sally R (2011) Measuring the quality of public open space using Google Earth. Am J Prev Med 40:105–112

Whyte WH (1980) The social life of small urban spaces. The Conservation Foundation, Washington, DC

Wilson JS, Kelly CM (2011) Measuring the quality of public open space using Google Earth: a commentary. Am J Prev Med 40:276–277

Wilson JS, Kelly CM, Schootman M, Baker EA, Banerjee A, Clennin M, Miller DK (2012) Assessing the built environment using omnidirectional imagery. Am J Prev Med 42:193–199

Xu Z, Weinberger KQ, Chapelle O (2012) Distance metric learning for kernel machines. arXiv preprint arXiv:1208.3422

Mapping Urban Soundscapes via Citygram

Tae Hong Park

Abstract In this paper we summarize efforts in exploring non-ocular spatio-temporal energies through strategies that focus on the collection, analysis, mapping, and visualization of soundscapes. Our research aims to contribute to multimodal geospatial research by embracing the idea of time-variant, poly-sensory cartography to better understand urban ecological questions. In particular, we report on our work on scalable infrastructural technologies critical for capturing urban soundscapes and creating what can be viewed as dynamic soundmaps. The research presented in this paper is developed under the *Citygram* project umbrella (Proceedings of the conference on digital humanities, Hamburg, 2012; International computer music conference proceedings (ICMC), Perth, pp 11–17, 2013; International computer music conference proceedings, Athens, Greece, 2014b; Workshop on mining urban data, 2014c; International computer music conference proceedings (ICMC), Athens, Greece, 2014d; INTER-NOISE and NOISE-CON congress and conference proceedings, Institute of Noise Control Engineering, pp 2634–2640, 2014) and includes a cost-effective prototype sensor network, remote sensing hardware and software, database interaction APIs, soundscape analysis software, and visualization formats. Noise pollution, which is the New Yorkers' number one complaint as quantified by the city's 311 non-emergency hotline, is also discussed as one of the focal research areas.

Keywords Sound-mapping • Soundscape • Noise pollution • Cyber-physical system

1 Introduction: Soundscapes and Soundmaps

The study and practice of environmental cartography has a long and rich human history with early examples found in ancient fifth century Babylonian and ancient Chinese cultures (Horowitz 1988; Berendt 1992). Much more recently, with the

---

T.H. Park (✉)
Department of Music and Performing Arts Professions, Steinhardt School, New York University, 35 West 4th Street, Suite 1077, New York, NY 10012, USA
e-mail: thp1@nyu.edu

advent of precise and cost-effective hardware, global positioning systems, environmental sensors, cloud computing, and various components of geographic information systems (GIS), digital cartography has become ubiquitous and perhaps even a necessity for the modern global citizen. As has been the case throughout history, modern mapping paradigms generally have focused on *fixed* landmarks such as coastlines, buildings, parks, and roads. However, human interaction with the environment is *immersive*, demanding a poly-sensory engagement where all but one of our senses is *visually* oriented. Even in contemporary mapping practices, *invisible* environmental dimensions are severely underrepresented partly due to: (1) importance of the ocular reflected in the so-called *eye culture* (Berendt 1992) and (2) technical complexities in addressing spatio-temporality needed to capture non-ocular environmental dimensions.

One of the most intriguing, yet underexplored areas of digital cartography is soundmapping. That is, digital cartographical platforms that capture, represent, and map *soundscapes*, a term coined by R. Murray Schafer (Schafer 1977). Schafer is one of the earliest researchers to bring awareness to environmental sounds and is also one of the founders of the *World Soundscape Project*[1] (WSP). WSP is concerned with education and research in *acoustic ecology* which include concepts such as *lo-fi* and *hi-fi* (Schafer 1977) sounds as it relates to low and high signal-to-noise ratio (SNR) sound environments. Soundscape research has traditionally focused on recording and capturing environmental sounds while simultaneously bringing attention to disappearing soundscapes largely due to human impact on the environment (Wrightson 2000). The WSP group engaged in numerous projects that primarily entailed recording a variety of soundscapes using portable audio recording devices leading to recordings such as *The Vancouver Soundscapes* (1973), *Five Village Soundscapes* (1975), and *Soundscape Documentation* DVD-ROM (2009). Their efforts, however, have primarily concentrated on historical soundscape preservation through audio recordings. *Citygram*, as summarized in Sect. 2 and 3, builds on the ideas set forth by Schafer et al. and further explores the notion of soundmaps through the application of modern technologies including data communication, data preservation, data analysis, data visualization, data access paradigms, and the Internet of Things (IoT). An area that we have particularly focused on is urban noise pollution as briefly discussed below.

1.1 Urban Noise Pollution

Humans have shown remarkable ability in adjusting to changing environments as evidenced by studies that depict this trait over the last 200,000 years (Cai et al. 2009). In the last 200 years, however, the size of cities, population growth, and accompanying urban infrastructural complexities, along with its multimodal

---

[1] http://www.sfu.ca/~truax/wsp.html

byproducts, has reached astonishing numbers. The industrial revolution, in particular, has been cataclysmic in contributing to the rapid worldwide population growth causing change to the natural environment (Lutz et al. 2008)—including *soundscapes*. Modern city-dwellers are all too familiar with the constant cacophony of urban machinery and the ubiquity of noise pollutants, regardless of time and space. For New Yorkers, the city's soundscape has become second nature. Adapting to noise pollution, however, comes with serious associated health risks, and according to Bronzaft, one of the leading experts in environmental psychology, "It means you've adapted to the noise ... you're using energy to cope with the situation. That's wear and tear on your body" (Anon n.d.a, b, c, d). Studies show that such "wear and tear" does not just contribute to hearing impairment, but also non-auditory health risks, including adverse effects on children's learning skills, hypertension, and sleep deprivation, as well as gastrointestinal, cardiovascular, and other physiological disorders (Lang et al. 1992; Zhao et al. 1991; Ward 1987; Van Dijk et al. 1987; Kryter 1970; Passchier-Vermeer and Passchier 2000; Evans et al. 2001; Bronzaft 2002; Woolner and Hall 2010; Knipschild 1977; Jarup et al. 2008; Barregard et al. 2009). This notion of human "adaptation" is especially concerning if we consider how little time we have had to adapt since the explosion of the modern urban environment. The potential for adverse noise pollution effects in cities is alarmingly more worrisome when considering that three fifths of the global population will live in megacities by 2050.

One of the difficulties that surround noise pollution is its very definition. Noise codes, such as the Portland Noise Code,[2] define noise using a *spatio-temporal* schema and distinguish *sound* from *noise* as a function of sound level, location, and time. New York City (NYC) has been particularly sensitive to its noisy soundscape, and for good reason: since 2003 more than 3.1 million noise complaints have been logged by NYC's 311 city service hotline[3] representing the top category of all the complaints by its community. Other cities nationwide that have implemented 311-style citizen hotlines have figures comparable to NYC: recent consumer ranking of the noisiest cities in the United States include Chicago, Atlanta, Philadelphia, San Francisco, and Houston (Washington n.d.). Noisy urban environments—something that acoustic ecologist Schafer refers to as *lo-fi* soundscapes (Schafer 1977)—is unsurprisingly a global phenomenon and continues to be one of the main environmental problems facing Europe today. For example, studies in the United Kingdom have shown that the general population lives above World Health Organization (WHO) noise level recommendations and an increase of noise has been recorded between 1990 and 2000 (Skinner and Grimwood 2005). Although we have come a long way since recognizing that noise is not just a mere *nuisance* or *irritation* (Skinner and Grimwood 2005) for humans, noise codes as written *and* enforced today are problematic in a number of ways:

---

[2] http://www.portlandonline.com/auditor/index.cfm?c=28714#cid_18511
[3] http://www.amny.com/news/noise-is-city-s-all-time-top-311-complaint-1.7409693

1. The *metrics* by which noise is defined are based on definitions of excessive "volume" that are either severely subjective or, when standard SPL measurements are used, fail to reflect how sound is perceived. For example, soothing ocean waves at 80 dB and the sound of blackboard fingernail scratching at the same level are not perceived in the same way.
2. Cities' capacity to effectively monitor, quantify, and evaluate urban noise is very limited.
3. The mechanism for noise enforcement is impractical as noise is fleeting in nature: even when law enforcement officers do make it to a reported "noise scene," chances are that noise pollution traces will have disappeared completely by the time they arrive.
4. Noise complaints are typically reported via 311 hotlines or directly reported to the police. However, current tools are inadequate for reporting or combating noise. For example, studies show that only 10 % of surveyed residents who were experiencing noise issues bothered to contact authorities: most people directly confront the person responsible (Skinner and Grimwood 2005) which partly may explain the 4.5 annual noise-related homicides affiliated with neighbor disputes (Slapper 1996).

With the recent maturing of cost-effective and sophisticated technologies including wireless communication networks, cloud computing, crowd-sourcing/citizen-science practices, and the explosion of Big Data science, the past few years has provided an opportunity to re-examine many of the issues pertinent to capturing soundscapes. That is, creation of a comprehensive real-time and interactive cyber-physical system (CPS) for collecting, analyzing, mapping, and visualizing soundscapes. Additionally, considering the increasing willingness of cities to provide public access to data[4] and integrate data science techniques and civic participation towards public policy-making decisions (Dickinson et al. 2012), an even more compelling case for developing an adaptive, scalable, and comprehensive CPS system for mapping our hyper-dimensional environment can be made. In the following sections we will provide an overview of our efforts to address many of the issues that exist in creating soundmaps today.

2 Related Work

In recent years, a variety of soundscape-like projects have begun to emerge, many touching upon principles pioneered by Schaefer. In particular, with the maturation of affordable hardware and software technologies, smaller portable devices often in the form of handheld devices, and the popularity of personal wireless telecommunication as a vital accessory of the *homo urbanus*, numerous application workflows

---

[4] https://nycopendata.socrata.com/

have been explored to capture and measure soundscapes and environmental noise. Many of these applications tap into a somewhat new phenomenon of the "citizen-scientist" to help contribute in solving problems hindered by spatio-temporality. In 2008, an application called *NoiseTube* was developed at Sony Computer Science Laboratory to measure noise via smartphones, enabling the development of *crowd-sourced* noise maps of urban areas (Maisonneuve et al. 2009). *NoiseTube* attempted to address the "lack of public involvement in the management of the commons" by empowering the public with smartphones to measure "personal" noise pollution exposure via mean dB SPL levels. *WideNoise*, a 2009 Android/iOS application is also a citizen-science noise metering example that includes an active world map of current noise levels (Anon n.d.a, b, c, d). *WideNoise* includes a social user experience component, which encourages active participation and also includes a sound sample-tagging feature that allows users to annotate sounds by associating it with its source ID and mood labels. Another related project is *Motivity* (2010), which employs a small number of stationary decibel meters at key intersections in the Tenderloin neighborhood of San Francisco.[5] Developed as an acoustic ecology project to demonstrate the efficacy of noise metering in a high-traffic area, the project uses an instrumentation system consisting of *fixed* microphones with embedded computing systems placed at intersections within a 25-block area. As with the other projects, *TenderNoise* and *WideNoise* both use the *one-size-fits-all* SPL metric to evaluate noise. In the area of preservation and capturing dying soundscapes such as rainforests, *Global Soundscape* is a project from Purdue University that offers simple tagging options and additional verbose descriptors inputted via a custom smartphone application. One of the goals of this project is to collect over "one million natural soundscapes" as part of a special *Earth Day* experience on April 22, 2014. Like many of the other software solutions, *Global Soundscape* also provides a mapping interface with "nodes" that represent geo-tagged citizen-scientist[6] contributed audio snapshots—short audio recordings frozen in time and space. The *Locustream SoundMap* is another soundscape-based project and is based on a so-called "networked open mic" streaming concept. In essence, *Locustream* aims to broadcast site-specific, unmodified audio through an Internet mapping interface by participants referred to as "streamers." *Streamers* are persons who deploy custom-made *Locustream* devices, which are provided from the developers to potential users in order to share the "non-spectacular or non-event based quality of the streams." This project was one of the many sources of inspiration for our own *Citygram* project and we have taken many concepts a few steps further by providing means for anyone with a computer and microphone to participate in serving as a "streamer" as further discussed in Sect. 3.1.2. Other examples include remote environmental sensing such as *Sensor City* (Steele et al. 2013), *NoiseMap* (Schweizer et al. 2011) and *Tmote Invent* (So-In et al. 2012). The latter two utilize a periodic record-and-upload SPL level strategy

---

[5] http://tendernoise.movity.com
[6] https://www.globalsoundscapes.org/

for noise monitoring and *Sensor City* is a project from the Netherlands that aims to deploy "hundreds" of fixed sensors equipped with high-end, calibrated acoustic monitoring hardware and its own dedicated fiber-optic network around a small city. A final example is the *Center for Urban Science and Progress* (CUSP) *SONYC* Project (2015–), which is built on *Citygram's* core fixed and citizen-science sensor network designs and technologies to capture urban noise pollution.

3 The Citygram Project: Sensor Network

In 2011, *Citygram* (Park et al. 2012, 2013, 2014b, c, d; Shamoon and Park 2014) was launched by Tae Hong Park in response to addressing inadequacies of past and present digital maps: absence of dynamicity and focus on ocularity. Traditional digital maps typically are built on *static* landscapes characterized by slowly changing landmarks such as buildings, avenues, train tracks, lakes, forests, and other visible objects. Ocular, physical objects, although critical in any mapping model, are not the only elements that define environments, however; various energy types including acoustic energies are also important factors that make up our environments. Noticing the underrepresentation of sound in modern interactive mapping practices, we began to explore and develop concepts that would enable spatio-temporal mapping via real-time capture, streaming, analysis, and human-computer interaction (HCI) technologies. Although it is difficult to pinpoint as to why meaningful soundmaps have not yet been developed, it is not that difficult to observe that (1) our society is visually oriented and (2) technical issues related to real-time sensor networks and spatio-temporal resolution have likely played a role in this phenomenon. Google Maps, for example, updates spatial images every 1–3 years[7] reflecting the low sampling rate needed to capture the nature of slowly changing landscapes. This is clearly inadequate for capturing sound: the standard sampling rate for full spectrum audio is 44.1 kHz. Our first iteration [*Citygram One* (Park et al. 2012)] focuses on exploring spatio-temporal acoustic energy to reveal meaningful information including spatial loudness, noise pollution, and spatial emotion/mood. The backbone of *Citygram* includes a sensor network, server technology, edge-compute models, visualizations, interaction technologies, data archives, and machine learning techniques as further summarized in the following sub-sections.

3.1 Sensor Network and Remote Sensing Devices

Our sensor network design philosophy is based on adopting scalable, robust, cost-effective, and flexible remote sensing device (RSD) technologies that communicate

---

[7] https://sites.google.com/site/earthhowdoi/Home/ageandclarityofimagery

through cloud and edge-computing technologies to create a high-resolution spatio-acoustic sensor network. We approach the dense sensor network problem by rethinking the functionality and adaptation of *Citygram's fixed* and *crowd-sourced* environmental sensing concepts we call *plug-and-sense*. This includes addressing issues associated with traditional spatially *sparse* monitoring practices that cover large areas with a small number of bulky, and often costly, sensors. Although these designs have the advantage of very high-quality sound, they suffer in terms of scalability and spatio-temporal resolution. Our sensor network design also addresses concerns related to an overreliance on consumer handheld devices (e.g. smartphones and tablets) for sensor network creation. These designs may have adverse effects on data quality due to inadequate onboard hardware components as well as issues related to calibration, control, and recording variability in personal crowd-sourced sound capture practices (Dumoulin and Voix 2012). Figure 1 shows the *Citygram* sensor network infrastructure with a server and various forms of RSDs including desktop and laptop computers; handheld devices such as smartphones and tablets, and fixed, calibrated RSDs as further described below.

3.1.1 Fixed RSDs

Fixed RSDs are designed to be permanently installed in "fixed" locations to provide consistent, reliable, secure, and calibrated audio data to our server. These RSDs, which form a distributed and edge computing network, use identical hardware and software components to ensure data consistency and reliability. To date, a number of initial systematic tests have been conducted to select suitable components for RSD development. Tests have included consideration of audio capture capability, processing power, RAM, onboard storage, OS flexibility, wireless connectivity, power consumption, I/O expandability, robustness/sturdiness, cost-effectiveness, and technology transferability. A number of RSD candidates were considered including Alix boards, Raspberry Pi hardware, Arduino microcontrollers, and a variety of consumer handheld devices. We are currently focusing on Odroid, Raspberry Pi 2, Edison, and the *Android mini-PC* platform. In particular, we have found the mini-PC to be a good fit as it includes necessary I/O, built-in WiFi and is approximately the size of a jump-drive as shown in Fig. 2a. Additional measurements and analyses workflows have led to identifying potential microphones by considering frequency response, pickup patterns, dynamic range, size, power consumption, and durability. Our current proof-of-concept RSD currently includes a single custom Micro-Electro-Mechanical Systems (MEMS) microphone board as shown in Fig. 2b (Mydlarz et al. 2014a, b). A number of field tests, spanning several months (and still continuing) since 2012, have been conducted via RSD node deployment in normal outdoor weather conditions around the New York, NY and Valencia, CA areas. These low-cost, fixed RSDs capture, analyze, and transmit consistent soundscape reporting via distributed and cloud computing client-server architectures (Fig. 3).

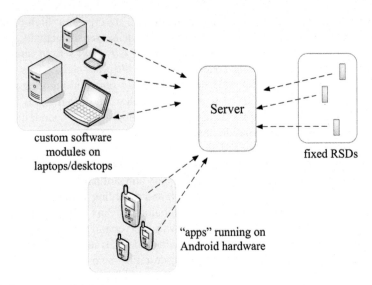

Fig. 1 Sensor network infrastructure

Fig. 2 (**a**) Android mini PC and audio interface and (**b**) single MEMS microphone

3.1.2 Crowd-Sourced RSD

Our crowd-sourced RSDs, in particular, embrace our design philosophy of *plug-and-sense* whereby *any* computing device with a microphone and Internet connection can be rendered into an RSD and allow users to become so-called *streamers*. This includes smartphones, tablets, "phablets" (phone-tablet hybrids), laptops, and desktop computers running "apps" and other software add-ons that run on popular commercial software (Figs. 4 and 5). We believe that our hybrid system exploits the benefits of what fixed and crowd-sourced RSDs have to offer in facilitating the

Fig. 3 Fixed-RSD (*1* mini PC, *2* power supply, *3* audio interface, *4* MEMS microphone board)

creation of a dense sensor network to produce high level of spatial granularity. Spatial granularity is critical in addressing the unpredictability of environments such as urban spaces. Unpredictability, by its very definition, is problematic when aiming to address spatial sensor *scarcity* distributions[8] through modeling strategies in order to fill the spatial "data gaps" (Saukh et al. n.d.; Aberer et al. 2010) [e.g. a commonly used method is the generalized additive model (GAM) (Ramsay et al. 2003) based on statistics]. A number of prototype software have been developed for Android and desktop platforms (Park et al. 2013, 2014a, b, c) and we are currently finalizing a JavaScript solution that will bypass complexities of developing software for various operating systems and hardware which perfectly aligns with our design philosophy of plug-and-sense: insert URL into a web-browser and begin sensing. These crowd-sourced RSDs are designed to capture, analyze, and stream audio data including low-level acoustic descriptions in *addition* to our fixed RSDs to facilitate the creation of a dense network while inviting meaningful community and citizen-science participation. Although *Citygram*'s sensor network infrastructure development is still a work-in-progress, our core designs and technologies are already being adopted and implemented and by other cyber-physical systems including CUSP's SONYC[9] project (Mydlarz et al. 2015).

---

[8] e.g. NABEL and OstLuft have a combined sensor node count of five for the entire city of Zurich.

[9] New York University's Center for Urban Science and Progress (CUSP).

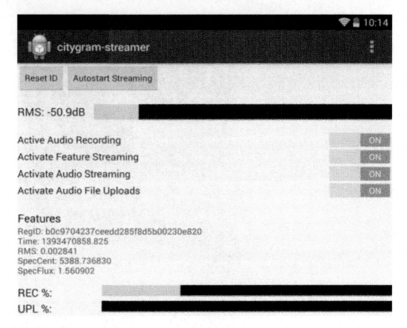

Fig. 4 Android app screenshot

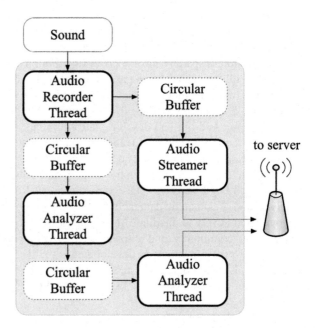

Fig. 5 Block diagram of Android streamer app

3.1.3 Sensor Deployment

Our sensor deployment strategy follows a multi-stage procedure based on incremental deployment steps. A small-scale sensor network step is currently being tested for end-to-end functionality of physical and virtual components. This includes performance validation of fixed RSD nodes, mobile/crowd-sourced RSDs, server-RSD data communication reliability, database query/update efficiency, and data visualization. Our long-term and large-scale deployment plans include activating citizen-scientists and adapting our CPS to existing urban infrastructures. This includes the application of *Citygram* for artistic purposes including real-time music performance and composition (e.g. the *InSeE* project), real-time data-driven visualization, and the development of interactive tools. One key future deployment strategy includes partnering with non-commercial (e.g. NYC currently has 59 unlimited free hotspots) and private sector organizations, which have taken initiatives to provide free and open Wi-Fi to city dwellers. For example, in 2013 Google sponsored the creation of free Wi-Fi to 2000+ residents, 5000+ student populations, and hundreds of workers in Manhattan's Chelsea area. Another example is NYC's initiatives to "reinvent" 11,412 public payphones.[10] These payphones produce approximately $17.5 million annual revenue primarily from advertising, but its function as a public telecommunication station has practically been rendered obsolete—it has been reported that roughly 1/3 of the payphones in Manhattan are inoperable as communication apparatuses.[11] NYC's recent call-for-proposals to "reinvent payphones" aims to install, operate, and maintain up to 10,000 public payphone nodes with free Wi-Fi and other technologies. This has lead to the launching of the LinkNYC project—a project that is being developed by Titan, Control Group, Qualcomm, and Comark.[12] The urban payphone infrastructure could serve as an significant large-scale deployment mechanism for our CPS sensor network as each station will provide uninterrupted power supply, data communication capability, (additional) protection from weather, and as whole, provide an opportunity to repurpose existing—and practically obsolete—urban infrastructures for large-scale fixed RSD deployment. Such a model would be straightforwardly transferable to other cities around the world. NYCLink, in particular, is a natural fit for Citygram's network technology as the NYCLink runs on an Android OS-based core and our software has been redesigned to run via JavaScript which allows it to be run on Internet browsers such as Chrome.

---

[10] http://business.time.com/2013/01/09/google-brings-free-public-wifi-to-its-new-york-city-neighborhood/

[11] http://www.nydailynews.com/new-york/brooklyn/40-public-pay-phones-broken-city-records-article-1.1914079

[12] http://www.link.nyc

3.2 Interaction, Data Archiving, and Visualization

One of the goals of creating a comprehensive cyber-physical system is including mechanisms for interactive exploration. We are thus developing online access and exploration technologies not only for researchers but also for citizen-scientists, students, artists, and the general public. The current prototype web interface is designed to function as an interactive environmental exploration portal and is built on the Google Maps API. Also, a number of visualization prototypes have been realized (Park et al. 2013; 2014) providing real-time visualizations and accompanying interfaces for standard web browsers. An early proof-of-concept heatmap visualization of spatio-temporal acoustic energy is shown in Fig. 6 and a number of additional visualizations are currently being developed in *JavaScript* for the web platform to allow compatibility across operating systems. The interface dynamically visualizes RSD-streamed audio data and also provides the ability to animate historical data stored in the server database. The historical data serves as an archival module, which stores low-level spatio-acoustic feature vectors. To enable users to hear the "texture" and characteristics of spaces without compromising private conversations that may be inadvertently captured in public spaces, we employ a custom voice blurring algorithm based on a *granular synthesis* (Roads 1988). To accomplish these conflicting tasks—blurring the audio while retaining the soundscape's texture—a multi-band signal processing approach was devised as detailed in (Park et al. 2014c).

4 Citygram: Soundscape Information Retrieval (SIR)

The Big Data analytics component of our project involves fundamental understanding of soundscapes through descriptors including semantic, emotive, and acoustic descriptors. In the case of semantic soundscape descriptors, much of the work originated from research in acoustic ecology (Schafer 1977) where the *identity* of the sound source, the notion of *signal* (foreground), *keynote* (background), *soundmarks* (symbolically important), *geophony* (natural), *biophony* (biological), and *anthrophony* (human-generated) play important roles. However, as our immediate focus lies in automatic classification of urban noise polluting agents, a first step towards this goal is the development of agreed-upon urban soundscape taxonomy. That is, (1) determining what important sound classes occupy urban soundscapes, (2) developing an agreed-upon soundscape namespace, and (3) establishing organizational and relational insights of its classes. This research, however, is still underexplored and a standardized taxonomy is yet to be established (Marcell et al. 2000; Guastavino 2007; Brown et al. 2011). In an effort to develop urban environmental *noise* taxonomy, we have prototyped software tools for soundscape annotation that follows an open-ended labeling paradigm similar to work by Marcell et al. (2000). We are also developing an urban soundscape

Fig. 6 *Citygram* dB$_{RMS}$ visualization

taxonomy that reflects the notion of "collective listening" rather than relying on the opinions of a few (Foale and Davies 2012). We believe that our methodology of inviting both researchers *and* the public to define and refine the pool of sono-semantic concepts has the potential to contribute to pluralistic soundscape taxonomy. Of critical importance in this approach is to ensure a sufficiently large collection of data. As an initial proof-of-concept we developed an urban soundscape mining methodology through the open, crowd-sourced database *Freesound*[13], a methodology adopted in (Salamon et al. 2014). Although *Freesound* is a rich resource for "annotated" audio files, initial semantic analysis proved to be less-than-ideal due to the amount of noise of unrelated words present in the data—each soundfile, no matter how long or short includes a set of labels that describe the *entire* recording. After employing a number of de-noising filters including tag normalization, spelling correction, lemmatization (Jurafsky and Martin 2009), and histogram pruning, we were able to substantially clean the original 2203 tags down to 230 tags obtained from 1188 annotated audio files (Park et al. 2014d). An analysis of the filtered tags suggested that, with additional filtering techniques, we could gain insights into hierarchical and taxonomical information in addition to our current conditional probability techniques. In addition to semantic descriptor analysis, we are currently investigating research efficacy in spatio-acoustic *affective*

---

[13] Crowd-sourced online sound repository contains user-specified metadata, including tags and labels.

computing (Barrett 1998). We believe that these supplemental emotive parameters will further provide insights into perceptual qualities of noise pollution, its semantic associations, and soundscape perception in general.

4.1 Machine Learning and Classification

The field of automatic soundscape classification is still in its nascent stages due to a number of factors including: (1) the lack of ground truth datasets (Giannoulis et al. 2013), (2) the underexplored state of soundscape namespace, (3) the overwhelming emphasis on speech recognition (Valero Gonzalez and Alías Pujol 2013; Tur and Stolcke 2007; Gygi 2001), and (4) the sonic complexity/diversity of soundscape classes. A soundscape can literally contain any sound, making the sound classification task fundamentally difficult (Duan et al. 2012). That is not to say that research in this field—something we refer to as Soundscape Information Retrieval (SIR)—is inactive: research publications related to music, speech, and environmental sound as a whole has increased more than four-fold between 2003 and 2010 (Valero Gonzalez and Alías Pujol 2013); and numerous research *subfields* exist today, including projects related to monitoring bird species, traffic, and gunshot detection (Clavel et al. 2005; Cai et al. 2007; Mogi and Kasai 2012; Van der Merwe and Jordaan 2013).

An important research area in audio classification is the engineering of quantifiable audio descriptors (Keim et al. 2004; Lerch 2012; Müller 2007; Peeters 2004) that are extracted every few hundred milliseconds, commonly subjected to statistical summaries (Aucouturier 2006; Cowling and Sitte 2003; Meng et al. 2007), which are fed to a classifier. Standard features for representing audio include centroid (SF), spread (SS), flatness measure (SFM), spread, flux (SF), Mel-Frequency Cepstral Coefficients (MFCC) in the frequency-domain; attack time, amplitude envelope, Linear Predictive Coding (LPC) coefficients, and zero-crossing rate, in the time-domain. What makes automatic classification particularly interesting is that *noise* is not entirely subjective and feature vectors are also not spectro-temporally invariant. In some cases, depending on environmental conditions, a sound source may not be perceived the same way, and a sound's feature vectors may also change for the same sound class. Sound class variance may also be influenced by the notion of *presence* in the form of foreground, middle-ground, and background sound. In a sense, it could be argued that traditional noise measurement practices primarily focus on the foreground characteristic of noise quantified via dB levels—sounds that were produced in the background would yield low noise rankings and thus unnoticeable, although in reality, it may contribute to irritation to listeners nearby.

One of the notable initiatives in soundscape classification research began recently in 2013 with the creation of the *IEEE D-CASE Challenge for Computational Audio Scene Analysis* (CASA) (Giannoulis et al. 2013). Although training and evaluation of SIR systems were primarily focused on indoor office sounds,[14] it is still worthwhile to note some of the SIR techniques presented at the Challenge. In the area of feature extraction, MFCCs were widely used, although in some studies, a case was made for time-domain and computer vision approaches realized via matching pursuit (MP) (Chu et al. 2008; 2009) and a k-NN-based spectrogram image feature (Dennis 2011). The former used a dictionary of atoms for feature presentation and the latter exploited spectrogram images for acoustic event detection (AED) and acoustic event classification (AEC). Both methods were demonstrated as alternative methods to MFCCs and showed robust performance in the presence of background noise. Some of the classifiers that were omnipresent included k-NNs, GMMs, SVMs, HMMs, and SOFMs based on expert-engineered feature vectors also reported in (Duan et al. 2012). The road ahead, however, is still largely uncharted and fundamental topics concerning the taxonomy and soundscape semantics, soundscape dataset availability, and the development of comprehensive and robust classification models that are adaptable to the diversity, dynamicity, and sonic uncertainty of outdoor soundscapes still remains challenging (Peltonen et al. 2002).

4.2 Soundscape Taxonomy, Annotation, and Ground Truth

To address the taxonomical and semantic side of soundscape research, we are currently developing crowd-sourced annotation tools to collect tags, labels, and soundscape descriptions through semantic data mining techniques. This has dual functionality of gaining insights into the soundscape *namespace* and also collecting *ground truth* data for machine learning. The database is projected to contain multiple annotations per sound class. The exact number of sound classes will be determined after careful analysis of tags/labels and community-provided noise complaint reports as further discussed below. The prototype annotation software consists of standard playback controls, zoom functionality, and tagging interfaces, including text entry for open-ended tagging, sound ID, start and end markers for each acoustic event, input for valence/arousal amount, and event fore/middle/background as shown in Fig. 7. Once our initial tagging efforts are complete, we expect the creation of a rich dataset of ground truth data to facilitate urban noise classification research.

One of the key issues in SIR is its sonic, spatial, and temporal diversity. The aim at this stage, however, is not to *solve* the urban SIR problem per se. Rather, the goal is to develop automatic urban sound classification algorithms that can detect and classify the most "popular" (perceptually unpopular) noise pollutants in cities like NYC; benchmark classification performance, and progressively improve and expand soundscape class identification capabilities. And although developing a

---

[14] Dissimilar to music and speech sounds although arguably a type of "environmental sound"

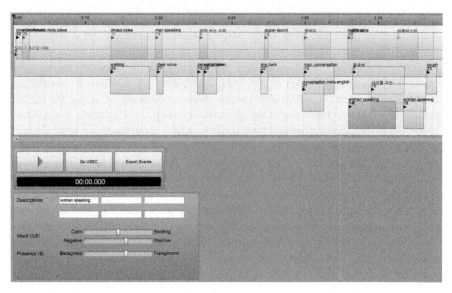

Fig. 7 Soundscape annotation software prototype

comprehensive soundscape classifier is the ultimate goal, if we can identify a small, but significantly impacting subset of noise pollutants as determined by city-dwellers, the automatic classification problem becomes more manageable and an iterative procedure can be applied. Thus, for the classification portion we are focusing on (1) classifying some of most "(un)popular" noise polluting agents while (2) strategically increasing the collection of noise classes guided by a sound class priority scheme based on crowd-sourced *noise agent rankings*. To make a preliminary assessment of this notion of *noise agent ranking*, we have analyzed the NYC 311 noise complaint dataset which shows the following class distributions representing four years of data: 54 % complaints *included* the words *car* or *truck*, 49 % *music*, 20 % *people* or *talking*, 14 % *construction*, and 10 % the word *dog*. In other words, if we focus our attention to a smaller subset of soundscape classes (those that are most "popular") and incrementally expand our algorithms to automatically recognize classes (include less "popular" ones), then the classification task can be divided into a number of manageable iterations. In addition to the 311 dataset we aim to analyze other similar datasets (Anon n.d.a, b, c, d) from cities like Chicago, Atlanta, Philadelphia, San Francisco,[15] Houston[16] and other cities [in the European Union, for example, road traffic noise accounts for 32 % of noise events that are above 55 dB(A) (Barreiro et al. 2005)]. With the amalgamation of fundamental knowledge gained in the analysis research part, we

---

[15] *www.sf11.org* lists barking dog, people talking, and car alarms as top 3 examples for noise complaints.

[16] Houston noise code lists vehicles, amplified sound from vehicles, and noise animals as top noise examples.

aim to transition from a position of questioning, "what is noise?" to a position of enunciation, "this is noise."

4.3 Preliminary Results

As a preliminary step in soundscape classification we have conducted a number of studies including acoustic event detection (AED) (Park et al. 2014a). The overall goal for this initial research was to explore the SIR feature space to inform subsequent classification research strategies. The majority of audio classification systems combine AED and AEC. AEC (background is simply considered to be another class) has been proven especially effective when an audio scene is highly specific, its event classes well-defined, and relatively small in number. In our research, we are exploring an approach where AED is conducted separately from AEC. That is, we first conduct AED and only when an acoustic event is detected, do we classify the sound. This is akin to research done in (Park and Cook 2005) where music instrument family (strings, brasses, and woodwinds) performance is much higher than individual instrument classification using the same radial/elliptical neural network architecture. Separating the AED and AEC task, however, also addresses system efficiency and sensor network transmission bandwidth. For example, in *Citygram* one of the data communication models is to only transmit sound IDs: for 100 sound types we would only need to 8 bits per sound ID to be transmitted. To address computation load, energy loss, and bandwidth usage, data from an RSD to a server is only transmitted when an acoustic event is detected. The latter point considers the fact that soundscapes vary in activity depending on location and time, and continually running AEC 24/7 is therefore inefficient, as acoustic scenes will tend to have significant inactive and uneventful periods at the short, mid, and long-term timescales.

Our current AED algorithm uses 19 feature vectors including RMS, ZCR, SF, SFM, SF, SS, and 13 MFCC coefficients. We designed and tested two AED systems where in one we employ an adaptive noise removal algorithm (Park et al. 2014a). This helps to adaptively attenuate a soundscape's dynamic background noise portion, which is often in flux throughout the day. This in turn improves feature computation due to background noise removal. We also utilized linear discriminant analysis (LDA) to analyze soundscapes into two classes: event or non-event. Both resulted in approximately 73 % correct classification of 176 acoustic events totaling 76 min in duration. The most salient features for our preliminary AED tests were ranked as follows: SS, SFM, and MFCC. Furthermore, we were able to improve performance by approximately 12 % by increasing the statistical window size (and "bag-of-fames") to approximately 1 s (85 % classification). Feature ranking, however, remained similar. Other initial classification research included unsupervised learning techniques, dimensionality reduction (PCA), and k-NN clustering.

A final area of research we are examining is feature learning. This area of machine learning is based on automatic "feature learning" as opposed to engineering features "manually" via expert knowledge. Feature learning requires large quantities of examples. Deep learning and deep networks, as they are also referred to, are based on neural networks, which are particularly well suited for learning complex functions that are difficult to model manually. Preliminary experiments have led to some interesting initial results in (Jacoby 2014). Although, the feature learning method did not outperform traditional discriminative models (SVM), it showed interesting artifacts. For example, for some urban sound classes (those that exhibited clear resonant characteristics such as pitch), it outperformed discriminative models. One way to look at this result is that, as we are using algorithms, methods, and approaches primarily developed for music and speech, they may work reasonably well only for specific soundscape types.

5 Summary and Future Work

We presented a CPS approach for interactive sound mapping and provided an overview of the *Citygram* project; summary of various software and hardware prototypes; strategies for representing spatio-temporal granularity and RSD deployment including discussions related to scalability and cost-effectiveness; soundscape-based machine learning topics including soundscape taxonomy, namespace, and ground truth data; as well as research and development that is currently underway. We discussed the project's scope including its potential for providing public service by embracing an immersive mapping paradigm—especially as it relates to measuring urban noise pollution by approaching it from a "you can't fix what you can't measure" approach. For example, its potential for combating urban noise pollution through mechanisms that can facilitate gaining insights into spatio-temporal noise pollution patterns is an interesting one. Also, its potential for exploring noise producing agents in the context of the *broken windows theory*[17] (Harcourt and Ludwig 2006)—or in our case, "broken sound theory,"—is also a potential avenue for multidisciplinary research and application for urban development: it has been shown that noise has been symptomatic of criminal activity including confrontations between neighbors; spousal/child abuse and abuse of the elderly; and drug dealers who, in NYC for example, have been pushed from the streets into buildings and are often marked by noise such as constant loud music, "customer" visitation noise, and repeated usage of whistles and other "all-clear" signals (Blair 1999; Bronzaft 2000). Another application example is in the context of NYC Department of Transportation's (DOT) *Off-Hour Delivery Program* initiated in 2013 to jumpstart an integrative process, involving cities, private sector, and researchers to develop new freight transportation paradigms that are designed to

---

[17] A crime prevention theory that has seen substantial success in crime control in large cities including New York, Chicago, Los Angeles, Baltimore, Boston, Albuquerque, Massachusetts, and also Holland.

improve quality-of-life of its residents. In Manhattan, for example, just 56 buildings create approximately 4 % of delivery traffic; 6800 restaurants generate more truck traffic than city ports, and 10 zip codes with freight parking require more parking space than all available city street parking spaces combined (Anon n.d.a, b, c, d). Pilot studies conducted by DOT between 2009 and 2010 show potential for significant advantages in implementing a robust OHDP program for high-traffic megacities: up to 50 % reduction of delivery times, reduction of service time by a third of current numbers, potential for 149.86 PM10 (kg) annual environmental pollution reduction, and $100–$200 million/year in travel time and pollution reduction (Anon n.d.a, b, c, d). The off-hour delivery program, however, is a double-edged sword: on one hand it significantly reduces resources and environmental pollution for megacities like NYC, while on the other hand, it contributes to noise pollution and disruption to residents during off-hour delivery periods where background noise is low and SNR high. For the OHDP program we are exploring utilizing *Citygram* to capture more noise patterns, develop noise reduction solutions through enforcement polices based on noise codes already in place, and test new low-noise materials and strategies for reducing noise complaints. We believe that the *Citygram* CPS has the potential to create data detailing off-hour delivery noise sources as a function of time, location, periodicity, and amount. This in turn will create a practical control and enforcement mechanism for the City, which will directly contribute to improving living conditions for its residents, reduce delivery time and reduce overall city traffic congestion and fuel costs. These examples, however, are only one of many ways our research could be utilized. Other examples include: (1) maps for the visually impaired: with enough RSDs density the possibility of soundmaps for the visually impaired becomes practicable; (2) spatio multimodality: folding in spatio-acoustic data in better understanding spaces and places—e.g. exploring its impact on education, crime, businesses, real-estate industry, and urban development; and (3) acoustic ecology: avian ecology and its relationship to soundscapes and urban noise pollution. The research and development outlined in this paper suggests several pathways towards a fundamental understanding of soundscapes with an emphasis on urban noise pollution and its effect on our cities.

Citygram is designed to be an iterative, long-term project. In our first iteration, our focus has been on soundmapping and acoustic ecology. Completion of iteration one will result in a comprehensive cyber-physical system that includes modules for real-time data capture, data streaming, sensor networks, analytics, archiving, and soundscape exploration. These same modules can be repurposed to form a comprehensive CPS framework that will facilitate system expansion and inclusion of additional sensors to allow measurement of other spatio-temporal energies captured via accelerometers, air pollution sensors, electro-magnetic, humidity, and other sensors types. Iteratively focusing on a particular "non-ocular" spatio-temporal energy source is part of our long-term goal. Additionally, we plan to also fold-in and consider other spatial modalities including census data, crime statistics, real-estate data, weather data, as well as social media data feeds. We anticipate that creating a multi-format, multidimensional, and multimodal CPS will allow for the opportunity to provide immersive, interactive, and real-time maps that will improve our understanding of our cyber-physical society. This research represents collaborative efforts of faculty, staff, and students from a number of organizations including the NYU Steinhardt, NYU ITP, NYU CUSP; the California Institute of the Arts; and the NYC DOT and NYC Department of Environmental Protection.

References

Aberer K, Sathe S, Chakraborty D, Martinoli A, Barrenetxea G, Faltings B, Thiele L (2010) OpenSense. In: Proceedings of the ACM SIGSPATIAL international workshop on geostreaming—GIS'10

Anon (n.d.a) Lessons from the off-hour delivery program in New York City

Anon (n.d.b) widenoise @ cs.everyaware.eu [Online]. http://cs.everyaware.eu/event/widenoise

Anon (n.d.c) AMERICA'S NOISIEST CITIES—tags: CITIES & towns—ratings & rankings NOISE pollution [Online]. http://connection.ebscohost.com/c/articles/40111669/americas-noisiest-cities. Accessed 29 May 2014a

Anon (n.d.d) Arline Bronzaft seeks a less noisy New York—NYTimes.com [Online]. http://www.nytimes.com/2013/10/07/nyregion/arline-bronzaft-seeks-a-less-noisy-new-york.html?_r=2&adxnnl=1&ref=health&adxnnlx=1401455249-ggbv5pzX04mLZgCmp6WXpg. Accessed 30 May 2014b

Aucouturier J-J (2006) Dix exp{é}riences sur la mod{é}lisation du timbre polyphonique. PhD thesis, University Paris VI

Barregard L, Bonde E, Öhrström E (2009) Risk of hypertension from exposure to road traffic noise in a population-based sample. Occup Environ Med 66(6):410–415

Barreiro J, Sánchez M, Viladrich-Grau M (2005) How much are people willing to pay for silence? A contingent valuation study. Appl Econ 37(11):1233–1246

Barrett LF (1998) Discrete emotions or dimensions? The role of valence focus and arousal focus. Cognit Emot 12:579–599

Berendt J (1992) The third ear: on listening to the world. Henry Colt, New York

Blair J (1999) Striking drug bosses, not street dealers, pays off, the police say. New York Times 17

Bronzaft AL (2000) Noise: combating a ubiquitous and hazardous pollutant. Noise Health 2(6):1–8

Bronzaft AL (2002) Noise pollution: a hazard to physical and mental well-being. In: Bechtel RB, Churchman A (eds) Handbook of environmental psychology. Wiley, New York, pp 499–510

Brown AL, Kang J, Gjestland T (2011) Towards standardization in soundscape preference assessment. Appl Acoust 72(6):387–392

Cai J, Ee D, Pham B, Roe P, Zhang J (2007) Sensor network for the monitoring of ecosystem: bird species recognition. In: 2007 3rd international conference on intelligent sensors, sensor networks and information, 2007, IEEE. pp 293–298

Cai J, Macpherson J, Sella G, Petrov D (2009) Pervasive hitchhiking at coding and regulatory sites in humans. PLoS Genet 5(1):e1000336

Chu S, Narayanan S, Kuo C-C (2008) Environmental sound recognition using MP-based features. In: IEEE international conference on acoustics, speech and signal processing, ICASSP 2008. pp 1–4

Chu S, Narayanan S, Kuo C-C (2009) Environmental sound recognition with time—frequency audio features. IEEE Trans Audio Speech Lang Process 17(6):1142–1158

Clavel C, Ehrette T, Richard G (2005) Events detection for an audio-based surveillance system. In: 2005 I.E. international conference on multimedia and expo, IEEE. pp 1306–1309

Cowling M, Sitte R (2003) Comparison of techniques for environmental sound recognition. Pattern Recognit Lett 24(15):2895–2907

Dennis JW (2011) Sound event recognition and classification in unstructured environments

Dickinson JL, Shirk J, Bonter D, Bonney R, Crain RL, Martin J, Phillips T, Purcell K (2012) The current state of citizen science as a tool for ecological research and public engagement. Front Ecol Environ 10(6):291–297

Duan S, Zhang J, Roe P, Towsey M (2012) A survey of tagging techniques for music, speech and environmental sound. Artif Intell Rev

Dumoulin R, Voix J (2012) On the use of smartphones for occupational noise monitoring. Canadian Acoustical Week, Canadian ... [Online]. http://my.publications.li/st/jvoix/dumoulin_use_2012. Accessed 15 Sept 2015

Evans GW, Lercher P, Meis M, Ising H, Kofler WW (2001) Community noise exposure and stress in children. J Acoust Soc Am 109(3):1023–1027

Foale K, Davies WJ (2012) A listener-centred approach to soundscape evaluation. In: Acoustics

Giannoulis D, Benetos E, Stowell D, Rossignol M, Lagrange M, Plumbley MD (2013) Detection and classification of acoustic scenes and events: an IEEE AASP challenge. In: 2013 I.E. workshop on applications of signal processing to audio and acoustics, October 2013, IEEE. pp 1–4

Guastavino C (2007) Categorization of environmental sounds. Can J Exp Psychol 61(1): 54–63. http://www.ncbi.nlm.nih.gov/pubmed/17479742. Accessed 23 May 2014

Gygi B (2001) Factors in the identification of environmental sounds [Online]. Indiana University. http://www.ebire.org/speechandhearing/dissall.pdf

Harcourt BE, Ludwig J (2006) Broken windows: new evidence from New York City and a five-city social experiment. Univ Chic Law Rev. pp 271–320

Horowitz W (1988) The Babylonian map of the world. Iraq 50:147–165

Jacoby C (2014) Automatic urban sound classification using feature learning techniques. Master's thesis, New York University

Jarup L, Babisch W, Houthuijs D, Pershagen G, Katsouyanni K, Cadum E, Dudley M, Savigny P, Seiffert I, Swart W, Breugelmans O (2008) Hypertension and exposure to noise near airports: the HYENA study. Environ Health Perspect 116(3):329

Jurafsky D, Martin JH (2009) Speech and language processing: an introduction to natural language processing, computational linguistics, and speech recognition. Prentice Hall, Upper Saddle River, NJ

Keim DA, Panse C, Sips M (2004) Information visualization: scope, techniques and opportunities for geovisualization. Bibliothek der Universität Konstanz, Konstanz

Knipschild P (1977) V. Medical effects of aircraft noise: community cardiovascular survey. Int Arch Occup Environ Health 40(3):185–190

Kryter KD (1970) The effects of noise on man. Academic, New York

Lang T, Fouriaud C, Jacquinet-Salord M-C (1992) Length of occupational noise exposure and blood pressure. Int Arch Occup Environ Health 63(6):369–372

Lerch A (2012) An introduction to audio content analysis: applications in signal processing and music informatics. Wiley, Hoboken

Lutz W, Sanderson W, Scherbov S (2008) The coming acceleration of global population ageing. Nature 451(7179):716–719

Maisonneuve N, Stevens M, Steels L (2009) Measure and map noise pollution with your mobile phone. In: DIY:: HCI—a showcase of methods, communities and values for reuse and customization

Marcell MM, Borella D, Greene M, Kerr E, Rogers S (2000) Confrontation naming of environmental sounds. J Clin Exp Neuropsychol 22(6):830–864

Meng A, Ahrendt P, Larsen J, Hansen LK (2007) Temporal feature integration for music genre classification. IEEE Trans Audio Speech Lang Process 15(5):1654–1664

Mogi R, Kasai H (2012) Noise-robust environmental sound classification method based on combination of ICA and MP features. Artif Intell Res 2(1):107

Müller M (2007) Information retrieval for music and motion. Springer, Berlin

Mydlarz C, Nacach S, Park T, Roginska A (2014a) The design of urban sound monitoring devices. In: Audio Engineering Society convention 137. Audio Engineering Society

Mydlarz C, Nacach S, Roginska A (2014b) The implementation of MEMS microphones for urban sound sensing. In: Audio Engineering Society convention 137. Audio Engineering Society

Mydlarz C, Shamoon C, Baglione M, Pimpinella M (2015) The design and calibration of low cost urban acoustic sensing devices [Online]. http://www.conforg.fr/euronoise2015/proceedings/data/articles/000497.pdf. Accessed 15 Sept 2015

Park TH, Cook P (2005) Radial/elliptical basis function neural networks for timbre classification. In: Proceedings of the Journées d'Informatique Musicale, Paris

Park TH, Miller B, Shrestha A, Lee S, Turner J, Marse A (2012) Citygram one: visualizing urban acoustic ecology. In: Proceedings of the conference on digital humanities 2012, Hamburg

Park TH, Turner J, Jacoby C, Marse A, Music M, Kapur A, He J (2013) Locative sonification: playing the world through citygram. In: International computer music conference proceedings (ICMC), 2013. Perth: ICMA, pp 11–17

Park TH, Lee J, You J, Yoo M-J, Turner J (2014a) Towards soundscape information retrieval (SIR). In: Proceedings of the international computer music conference 2014, Athens, Greece

Park TH, Musick M, Turner J, Mydlarz C, Lee JH, You J, DuBois L (2014b) Citygram: one year later . . . . . In: International computer music conference proceedings, 2014, Athens, Greece

Park TH, Turner J, Musick M, Lee JH, Jacoby C, Mydlarz C, Salamon J (2014c) Sensing urban soundscapes. In: Workshop on mining urban data, 2014

Park TH, Turner J, You J, Lee JH, Musick M (2014d) Towards soundscape information retrieval (SIR). In: International computer music conference proceedings (ICMC), 2014. Athens, Greece: ICMA

Passchier-Vermeer W, Passchier WF (2000) Noise exposure and public health. Environ Health Perspect 108(Suppl 1):123

Peeters G (2004) {A large set of audio features for sound description (similarity and classification) in the CUIDADO project}

Peltonen V, Tuomi J, Klapuri A, Huopaniemi J, Sorsa T (2002) Computational auditory scene recognition. In: 2002 I.E. international conference on acoustics, speech, and signal processing (ICASSP). pp II–1941

Ramsay T, Burnett R, Krewski D (2003) Exploring bias in a generalized additive model for spatial air pollution data. Environ Health Perspect 111(10): 1283–1288. http://www.pubmedcentral.nih.gov/articlerender.fcgi?artid=1241607&tool=pmcentrez&rendertype=abstract. Accessed 22 May 2014

Roads C (1988) Introduction to granular synthesis. Comput Music J 12:11–13

Salamon J, Jacoby C, Bello JP (2014) A dataset and taxonomy for urban sound research. In: Proceedings of 22nd ACM international conference on multimedia, Orlando, USA

Saukh O, Hasenfratz D, Noori A, Ulrich T, Thiele L (n.d.) Demo abstract: route selection of mobile sensors for air quality monitoring

Schafer RM (1977) The soundscape: our sonic environment and the tuning of the world [Online]. http://philpapers.org/rec/SCHTSO-15. Accessed 16 Apr 2014

Schafer RM (1993) The soundscape: our sonic environment and the tuning of the world [Online]. Inner Traditions/Bear. http://books.google.com/books/about/The_Soundscape.html?id=_N56QgAACAAJ&pgis=1.

Schweizer I, Bärtl R, Schulz A, Probst F, Mühlhäuser M (2011) NoiseMap—real-time participatory noise maps. World. pp to appear

Shamoon C, Park T (2014) New York City's new noise code and NYU's Citygram-sound project. In: INTER-NOISE and NOISE-CON congress and conference proceedings, vol 249(5), pp 2634–2640. Institute of Noise Control Engineering

Skinner C, Grimwood C (2005) The UK noise climate 1990–2001: population exposure and attitudes to environmental noise. Appl Acoust 66(2):231–243

Slapper G (1996) Let's try to keep the peace. The Times [Online]. http://scholar.google.com/scholar?q=%22let%27s+try+to+keep+the+peace%22&btnG=&hl=en&as_sdt=0%2C33#0. Accessed 18 July 2014

So-In C, Weeramongkonlert N, Phaudphut C, Waikham B, Khunboa C, Jaikaeo C (2012) Android OS mobile monitoring systems using an efficient transmission technique over Tmote sky WSNs. In: Proceedings of the 2012 8th international symposium on communication systems, networks and digital signal processing, CSNDSP 2012

Steele D, Krijnders D, Guatavino C (2013) The sensor city initiative: cognitive sensors for soundscape transformations. In: Proceedings of GIS Ostrava 2013: geoinformatics for city transformations, 2013, Ostrava, Czech Republic

Tur G, Stolcke A (2007) Unsupervised language model adaptation for meeting recognition. In: 2007 I.E. international conference on acoustics, speech and signal processing—ICASSP'07 [Online], IEEE. pp IV–173–IV–176

Valero Gonzalez X, Alías Pujol F (2013) Automatic classification of road vehicles considering their pass-by acoustic signature. J Acoust Soc Am 133(5):3322

Van der Merwe JF, Jordaan JA (2013) Comparison between general cross correlation and a template-matching scheme in the application of acoustic gunshot detection. In: 2013 Africon, September 2013, IEEE [Online]. pp 1–5

Van Dijk FJH, Souman AM, De Vries FF (1987) Non-auditory effects of noise in industry. Int Arch Occup Environ Health 59(2):133–145

Ward WD (1987) Noise and human efficiency. Ear Hear 8(4):254–255

Washington SE (n.d.) The daily decibel

Woolner P, Hall E (2010) Noise in schools: a holistic approach to the issue. Int J Environ Res Public Health 7(8):3255–3269

Wrightson K (2000) An introduction to acoustic ecology. Soundscape J Acoust Ecol 1(1):10–13

Zhao YM, Zhang SZ, Selvin S, Spear RC (1991) A dose response relation for noise induced hypertension. Br J Ind Med 48(3):179–184

Part VIII
Social Equity and Data Democracy

Big Data and Smart (Equitable) Cities

Mai Thi Nguyen and Emma Boundy

Abstract Elected officials and bureaucrats claim that Big Data is dramatically changing city hall by allowing more efficient and effective decision-making. This has sparked a rise in the number of "Offices of Innovation" that collect, manage, use and share Big Data, in major cities throughout the U.S. This paper seeks to answer two questions. First, is Big Data changing how decisions are made in city hall? Second, is Big Data being used to address social equity and how? This study examines Offices of Innovation that use Big Data in five major American cities: New York, Chicago, Boston, Philadelphia, and Louisville, focusing specifically on three dimensions of Big Data and social equity: data democratization, digital access and literacy, and promoting equitable outcomes. Furthermore, this study highlights innovative practices that address social problems in order to provide directions for future research and practice on the topic of Big Data and social equity.

Keywords Big Data • Smart cities • Equity • Local government

1 Introduction

Elected officials and bureaucrats claim that Big Data is dramatically changing city hall by allowing more efficient and effective decision-making. This has sparked a rise in the number of "Offices of Innovation" that collect, manage, use and share Big Data, in major cities throughout the United States. A watershed moment for Big Data and cities was President Obama's Open Government Initiative announced in January of 2009, which provided a Federal directive to establish deadlines for action on open data (Orzag 2009). Shortly thereafter, a number of local municipalities in the U.S. began making data more accessible, developed policies around open data, and made government services and civic engagement easier through the use of new technologies and Big Data.

M.T. Nguyen, Ph.D. (✉) • E. Boundy, M.A.
Department of City and Regional Planning, University of North Carolina at Chapel Hill, Chapel Hill, NC, USA
e-mail: mai@unc.edu

© Springer International Publishing Switzerland 2017
P. Thakuriah et al. (eds.), *Seeing Cities Through Big Data*, Springer Geography,
DOI 10.1007/978-3-319-40902-3_28

San Francisco launched the first open data portal for a U.S. city in 2009 and opened the first Mayor's Office of Civic Innovation in January 2012 (Appallicious 2014). As of July 2013, at least ten cities had Chief Innovation Officers, and a survey conducted in Spring 2013 found that "44 % of cities of populations of more than 300,000 and 10 % cities of populations between 50,000 and 100,000 had offices of innovation" (Burstein 2013). While San Francisco had the first office, Mayor Bloomberg's dedication to opening data in New York has been heralded by civic innovators as one of the driving forces behind the open data movement and the national trend towards greater civic entrepreneurship (Appallicious 2014).

Offices of Innovation have become popular because they offer the promise of using Big Data for predictive analytics, streamlining local government processes, and reducing costs. Yet, very little research has been conducted on what data is being harnessed, how it is organized and managed, who has access, and how its use affects residents. Even less attention has been paid to the relationship between Big Data and equity.

This paper seeks to answer two questions. First, is Big Data changing how decisions are made in city hall? Second, is Big Data being used to address social equity and how? This paper seeks to answer these questions by examining Offices of Innovation that use Big Data in five major American cities: New York, Chicago, Boston, Philadelphia, and Louisville. In particular, this study examines three dimensions of Big Data and social equity: data democratization, digital access and literacy, and promoting equitable outcomes. Furthermore, this study highlights innovative practices that address social problems in order to provide directions for future research and practice on the topic of Big Data and social equity.

2 Big Data, Governance, and Social Equity

Although the private sector has become highly sophisticated at culling Big Data to shape business practices and planning for some time now, the use of Big Data in the public sector is a relatively new phenomenon. Much of the academic literature to date on Big Data and cities has largely focused on the historical evolution of Big Data and smart cities (Batty 2012, 2013; Kitchin 2014; Batty et al. 2012; Chourabi et al. 2012), the potential impact that Big Data can have on the future of citizens' lives (Domingo et al. 2013; Batty 2013; Chen and Zhang 2014; Wigan and Clarke 2013; Hemerly 2013), and the challenges the public sector faces integrating Big Data into existing processes and strategies (Batty 2012; Joseph and Johnson 2013; Vilajosana et al. 2013; Almirall et al. 2014; Chen and Zhang 2014; Cumbley and Church 2013; Wigan and Clarke 2013; Kim et al. 2014; Hemerly 2013). However, less research has centered on the relationships between Big Data, local governance, and social equity.

Social equity in governance, is defined by the Standing Panel on Social Equity in Governance as "The fair, just and equitable management of all institutions serving the public directly or by contract; the fair, just and equitable distribution of public services and implementation of public policy; and the commitment to promote fairness, justice, and equity in the formation of public policy" (National Academy of Public Administration, www.napawash.org). This definition of social equity focuses on the governance *process* and does not address equitable outcomes. For this study, we consider another dimension of social equity: public policies and government actions that promote greater equitable *outcomes* for the public. When social equity is tied to Big Data, a number of key themes emerge in the literature: digital access, digital literacy, and the use of Big Data to promote more equitable outcomes for the public. Thus, in our study, our definition of social equity considers both processes and outcomes.

2.1 Inequities in Access: The Digital Divide

As cities steadily transition towards a digital governance system, uneven access to digital technology across different groups, known as the "digital divide" may exacerbate social inequality (Light 2001; Gilbert et al. 2008). Existing research on digital access has focused on how and why certain demographic groups have historically been left out of the technological adoption process (Batty et al. 2012; Chourabi et al. 2012; Prieger and Hu 2008; Gilbert et al. 2008; Lee et al. 2015). Research indicates that the groups that have the least access to digital technology include: the poor, unemployed individuals, those with low levels of education, families without children, the elderly, non-whites (especially, Blacks and Hispanics), and those living in rural areas (DiMaggio et al. 2004; Azari and Pick 2005; Gilbert et al. 2008; Gilmour 2007; Hilbert 2011; LaRose et al. 2007; Lee et al. 2015; Prieger and Hu 2008; Prieger 2013; Velaga et al. 2012).

There is a large and growing body of research on the digital divide that focuses largely on broadband adoption, Internet usage, and computer access. To address this divide, federal funding has been used to develop initiatives to provide better access to computing centers and expanding availability of Internet services (Bailey and Ngwenyama 2011; Revenaugh 2000). Furthermore, local governments have attempted to improve computer and Internet access for underserved population groups through partnerships with local schools and community centers and by improving local digital infrastructure in disconnected neighborhoods (Gilbert et al. 2008; Araque et al. 2013). Despite these efforts at eliminating the inequities in digital access, the digital divide still persists (Gilbert et al. 2008; Kvasny and Keil 2006; Correa 2010; Hargittai 2002; Looker and Thiessen 2003; DiMaggio et al. 2004; Light 2001).

2.2 Digital Literacy: The "Participatory Gap"

While scholars have paid attention to the digital divide in relation to digital access over the past 15 years, recent research has emphasized growing concerns over digital literacy—the skills, knowledge, or familiarity with digital technology (Gilbert et al. 2008; Gilmour 2007; Lee et al. 2015; Correa 2010; Hargittai 2002). This form of digital divide has been referred to as the "participatory gap" (Fuentes-Bautista 2013) and signifies that even if individuals have access to computers, smartphones, or the Internet, they may lack the skills, education, or familiarity to take advantage of the opportunities that information and communications technologies (ICTs) can provide (Warren 2007; Gilbert et al. 2008; Kvasny and Keil 2006; Looker and Thiessen 2003; DiMaggio et al. 2004; Light 2001). Differences in levels of accessibility and digital literacy are found to be correlated with typical patterns of social exclusion in society (Warren 2007; Lee et al. 2015; Mossberger et al. 2012; DiMaggio et al. 2004). In particular, socioeconomic status is considered the leading cause of the new digital literacy divide (Guillen and Suarez 2005).

As municipal governments become increasingly reliant on digital technology, the ability to navigate public agency websites, download and submit forms electronically, scan documents, and a host of other digital skills are increasingly becoming important. Digital illiteracy will undoubtedly limit the ability of individuals, organizations, and local businesses to access resources and opportunities. Community-based organizations, for example, often struggle with having low capacity to perform sophisticated studies or evaluations using Big Data and, therefore, do not have the ability to provide quantitative analyses that may be required to apply for and receive government or philanthropic funding. Thus, understanding the barriers to digital literacy and the characteristics of groups and organizations that are persistently illiterate will allow local governments to adopt policies and practices to address it.

2.3 Closing the Digital Divide: What Have We Learned?

The first generation of initiatives developed to address the digital divide proposed that by improving digital accessibility, this would benefit disadvantaged groups and reduce gaps in access and usage (Azari and Pick 2005). The idea behind these initiatives relied on the assumption that closing gaps in technological access would mitigate broader inequalities, including literacy. These initiatives also assumed that providing access to ICTs would improve disadvantaged groups' social statuses (e.g. income) (Light 2001). However, studies found confounding factors associated with digital inequality, including available equipment, autonomy in using ICTs, skills, support (e.g. technical assistance), and variations in purposes (e.g. using ICTS to obtain jobs vs. social networking) (DiMaggio et al. 2004). Increasing digital access does not adequately address these five issues and, therefore may be ineffective at reducing the digital divide (Looker and Thiessen 2003). For example, Kvasny and Keil's (2006) study found that providing computers, Internet access, and basic computer training was helpful, but not sufficient at eliminating the digital divide for low-income families in high-poverty areas. This study pointed to the intersection between digital inequities and other social structural inequities, such as lack of access to high-quality schools, limited public investment, and pervasive poverty. Even when digital divide initiatives do help low-income Americans living in poor neighborhoods to gain digital literacy, these programs often do not mitigate inequities caused by disparities in transportation access or educational status (Light 2001; Tapia et al. 2011). Thus, the literature suggests that in order to be effective in closing the digital gap, digital programs and policies must also be coordinated with other social policies that address the root causes of the digital divide: poverty, poor education, economic residential segregation, and public and private sector disinvestment in poor neighborhoods.

Technological innovations have become an integral part of America's communication, information, and educational culture over the past decade. Access to information and computer technologies is increasingly considered a "necessity" to participate in many daily functions (Light 2001). As ICTs have become more integrated into daily life, populations that have historically not had access to or familiarity with how to use ICTs may become increasingly disadvantaged without improved access and literacy (Tapia et al. 2011). This may also exacerbate other forms of social, economic, and political marginalization for excluded groups (Gilbert et al. 2008; Lee et al. 2015). Furthermore, growing disparities between the digital "haves" and "have-nots" can have lasting negative social and economic consequences to neighborhoods and cities.

Recognizing how data is collected and who provides the data has important implications for democracy and the distribution of government resources. For example, crowdsourcing—a citizen engagement platform—will disproportionately benefit individuals and groups that provide data through mobile applications or web-based applications. Without an understanding of how to use ICTs, disadvantaged groups will not have their voices heard (Jenkins et al. 2009; Bailey and Ngwenyama 2011; Lee et al. 2015). However, being digitally illiterate may be of greater concern for more common procedures, such as job applications or qualifying for federal assistance programs. Today, even some minimum wage jobs require job applications to be filed online. Individuals without access to a computer or the Internet and/or individuals without any familiarity in completing paperwork or forms online may experience significant difficultly completing the application, which may further exacerbate existing economic inequities. Data and digital inequities, in terms of both access and literacy, compound issues that disadvantaged populations face. As a result, it is important to continually address inequity in digital access and literacy in order to prevent populations from becoming increasingly disenfranchised and for inequalities to be exacerbated as technological innovations continue to develop.

3 Case Study Cities: Boston, Chicago, Louisville, New York and Boston

To better understand whether and how Big Data is changing decision-making in city hall and how it is used to address social inequity, we conduct interviews with key informants in five cities that have well-established Offices of Urban Innovation: Boston, Chicago, Louisville, New York and Philadelphia. For more information about each city's office of urban innovations, see Appendix. We used a snowball sampling design, which started with contacting the directors of each of the offices and receiving referrals from individuals we made contact with. We conducted a total of 19 semi-structured phone interviews with staff in the Offices of Innovations and key local stakeholders, such as users of Big Data disseminated by local governments, developers of new mobile applications that collect data, and staff of non-profits. These interviews lasted between 30 min and 1 h and were transcribed. We supplemented our interviews with reports, newspaper articles, and scholarly publications that provided information about the five offices of innovations' mission, goals, organizational and institutional structure, operational priorities, portfolio of programs, and key initiatives.

Offices of Innovation are typically responsible for information technology activities, such as providing Wi-Fi and broadband infrastructure, developing information technology policies, strategies, and benchmarks, and providing access to open data and mapping data through online portals. However, each of these offices have unique initiatives that have garnered nationwide attention.

4 Is Big Data Changing City Hall?

4.1 Big Data and Government Decision-Making

One of the primary way in which Big Data is changing City Halls nationwide is by increasing the number of data sources (e.g. administrative, mobile application data, social media) available to develop data analytics and predictive processes to inform decision-making. These types of systems are being developed to save money, allowing municipal governments to stretch budgets, improve efficiency, and develop new methods of communication and networking internally in order to be more innovative in delivering public services. Two of our case study cities offer insights into how Big Data is used to inform local government decision-making: New York and Chicago. New York has been approaching Big Data from a problem solving approach, whereas Chicago is working to infuse data analytics into their existing governmental structure in a comprehensive way.

4.1.1 Big Data and Predictive Processes: New York and Chicago

New York City's Mayor's Office of Data Analytics (MODA) has been widely recognized for using predictive analytics in government decision-making over the past several years. Since 2012, New York City has approached predictive analytics as a way to "evolve government" to improve the efficient allocation of resources and develop a better response to the real-time needs of citizens (Howard 2011). The City's repository for administrative data is called DataBridge and was designed to perform cross-agency data analysis utilizing data from 45 city agencies simultaneously. According to Nicholas O'Brien, the Acting Chief Analytics Officer of New York, the main challenge the office has had to overcome has been working with Big Data from "45 mayoral agencies spread out in a distributed system...[this has been a challenging and arduous process because] each city department has a different anthology for how they characterize the data, so it's important to understand the overlaps and the exceptions" (Personal Communication, February 24, 2014). Thus, matching the data across administrative units and ensuring the quality of the data is extremely important as the decisions made through the predictive analytics process are only as valid and reliable as the data utilized to predict the outcomes.

Some of MODA's most lauded successes include: "(1) a five-fold return on the time that building inspectors spend on looking for illegal apartments, (2) an increase in the rate of detection for dangerous buildings prone to fire-related issues, (3) more than a doubling of the hit rate for discovering stores selling bootlegged cigarettes, and (4) a five-fold increase in the detection of business licenses being flipped" (Howard 2012). MODA's quantifiable successes using predictive analytics has inspired other cities nationwide to create these types of processes to improve their own internal productivity and decision-making. Although the benefits have been well documented, there is no account of how much it costs to collect, manage, and analyze the data. Therefore, this does not allow for a critical examination of whether Big Data predictive analytics is a more cost-effective problem solving tool than other methods.

In January 2014, Chicago received a $1 million grant from Bloomberg Philanthropies to create the first open-source, predictive analytics platform, called Smart Data (Ash Center Mayors Challenge Research Team 2014). Chicago collects seven million rows of data each day that is automatically populated and gathered in varying formats, through separate systems. The SmartData platform will be able to analyze millions of lines of data in real time to improve Chicago's predictive processes and according to Brenna Berman, will "develop a new method of data-driven decision making that can change how cities across the country operate" (Ash Center Mayors Challenge Research Team 2014).

The SmartData platform will have the power to transform predictive analytics for cities nationwide through its open-source technology. If successful, the development of this replicable model for predictive processes can potentially change decision-making processes for every municipal government nationwide that utilizes this software. The platform designed to be user-friendly and understandable to government employees with varying levels of data familiarity. Brenna Berman states, "...we need to find a way to make analytics become available to the non-data engineer" (Shueh 2014a, b). While Chicago has a team of data engineers, most cities do not have access to those resources and city staffers must make decisions without extensive experience in ICTs or data analytics. In addition, the use of predictive analytics may assist with preventing problems rather than responding to problems, which is how most cities operate. Brett Goldstein, former Chief Data Officer of Chicago, explains that predictive analytics is "having governments think about, 'How do we prevent rather than react?'...Part of my role now is to say, 'How can we use those techniques to do government better and do it smarter?'" (Rich 2012).

4.1.2 Using Big Data for Predictive Policing

One area of predictive analytics adopted by cities is predictive policing, which employs data analytics to assist police in making decisions that can prevent crimes, such as homicides, burglaries, and vehicle thefts (Novotny 2013). Predictive policing systems use software that builds models similar to private sector models of forecasting consumer behavior. Municipal police agencies use this technology to predict and prevent crime. The concept behind predictive policing is that situational awareness can be improved to create strategies that improve public safety and utilize police resources more efficiently and effectively (Hollywood et al. 2012). Employing limited resources more effectively and working proactively can help police departments anticipate human behavior and identify and develop strategies to prevent criminal activity. However, in order for predictive policing to be successful, rigorous evaluations must be conducted to determine if reductions in crime are directly due to predictive policing (Hollywood et al. 2012).

Two of our case study cities, Philadelphia and Boston, have already seen some quantifiable successes utilizing predictive policing. Philadelphia has been training police officers as data scientists in a "smart policing" program since April 2012. Officers completed a 2-week crime science program focused on utilizing technology to map crimes, understanding predictive software, and generating digital surveys to collect information from residents, which is a form of data crowdsourcing (Reyes 2014). This program changes traditional police protocol because police officers, rather than external consultants, are directly trained in these technologies and build upon their skills and knowledge as a result of the program. Violent and property crimes decreased by 5.8 % and residential burglaries decreased by 39 % in one district between 2012 and 2013. Philadelphia's Deputy Commissioner, Nola Joyce, believes that the reductions in crime are due to the smart policing program and as a result of developing this program, the police department is moving from "counting and reporting crime" to "understanding" crime (Reyes 2014).

Boston has established the Problem Properties Task Force, an interdepartmental effort to identify properties with persistent criminal activity and/or blight that have caused problems (Boston's Mayoral Transition: The Problem Properties Task Force 2013). Developed to improve the allocation of the city's limited resources, the Task Force convenes executive staff members from more than 12 departments and uses a data-driven, predictive analytics system that combines data from multiple city agencies. This program has resulted in reductions in property assessment times from days or weeks to seconds (Boston's Mayoral Transition: The Problem Properties Task Force 2013). The Problem Properties Task Force is notable because it is an example of multiple sources of city administrative data (Big Data) grounded in local knowledge from executive departmental staff to conduct predictive analytics and inform decision-making.

Critics of predictive policing raise concerns that the data used to prevent crime, such as race, ethnicity, or neighborhood could result in profiling. However, supporters of predictive policing argue that only data on past crimes, not criminals' characteristics, are used in the data analysis. These controversies raise questions about ethical and legal concerns over using Big Data for predictive analytics.

4.1.3 Balancing Predictive Analytics with Contextual Realities

While Big Data and predictive analytics offers the potential for greater efficiency and cost-savings, it can also do harm. Users of Big Data should be careful to ensure the accuracy and completeness of the data used in predictive models. There is also the potential for human error or misinterpretation of the results, thus it is important to cross check the findings from predictive models with individuals in the field—including staff who work within communities or the public at large. While efficiency is important, accuracy and transparency is equally important when using Big Data for predictive modeling or forecasting.

5 Big Data and Social Equity

Our case studies revealed that Big Data and new technologies have tackled tame problems (Rittel and Webber 1973), such as infrastructure improvements, how to allocate staff time, and making city hall run more efficiently and proactively, rather than focusing on the more intractable problems of inequality, poverty and social equity. Based on our research, we find that there are three ways that cities are addressing social equity with Big Data: democratizing data, improving digital access and literacy, and promoting equitable outcomes using Big Data. We discuss each of these topics in turn below.

5.1 Data Democratization

Data democratization is centered upon the idea of increasing access to typically inaccessible or unpublished data for widespread analysis and consumption. There are many legislative forms in which data democratization can be encouraged or required by municipal governments. According to the Sunlight Foundation, of the 32 cities with open data policies in place by April 2014, two are administrative memos, ten are executive orders, and 20 are "codified in legislation." Codified in legislation is the "strongest" policy form, because "it preserves consistent criteria and implementation steps for opening government data beyond the current administration" (Williams 2014). When open data laws become incorporated into legislation, consistency, enforcement, and management standards become part of the city's legislation and are more difficult to overturn or alter under changing leadership. Regardless of the type of policy passed, each city's relationship with open data is dependent on the municipal government's structure and support for transparency, as well as the city's existing mechanisms and capacity for data tracking and management.

5.1.1 Open Data in New York City

Enacted in March 2012, New York City's landmark Open Data law—Local Law 11—was the first of its kind at the local U.S. municipal level (NYC DoITT 2012). The result of Local Law 11 was that New York City established a plan with yearly milestones to release all of the city's data from city agencies by 2018. When finished, it will become the first U.S. local municipality with a complete comprehensive public agency dataset inventory (Williams, "NYC's Plan to Release All-ish of their data," 2013a). According to Gale Brewer, Manhattan Borough President, New York City's open data law was more significant and transformative than the federal directive because it demonstrated how this type of work could be implemented at the local level (Goodyear 2013).

The Mayor's Office of Data Analytics (MODA) operates New York City's open data portal and works closely with NYC DoITT to populate the data portal and pursue other projects relating to data innovation and analytics (Feuer 2013). MODA has been successful at procuring 1,500 datasets from the city's public agencies thus far. Yet, there remains many challenges to completing this task, including the cost, organizational capacity, data management skills, and ongoing maintenance and upkeep of the data. What is also not clear is who uses the data and for what purpose, which raises questions about data formatting and requisite skills and education of users. Nicholas O'Brien, Acting Director of MODA explains,

> We're also really starting to understand our audience. The customers of our open data portal are primarily non-profits, who we considered mid-tier data users that have some digital and data expertise but aren't necessarily writing code or programming. We also know for our tech-savvy users, we have to direct them to our developer portal for more robust resources,

and we have a third level of users that have very limited skills with data analysis. Understanding what each of these audiences want and need is an ongoing process for us. (Personal Communication, February 24, 2014).

5.1.2 Open Data in Boston, Chicago, Louisville, and Philadelphia

Since 2012, Boston, Chicago, Louisville, and Philadelphia have established open data executive orders. These cities have largely developed open data portals and created new executive positions to manage data initiatives. Philadelphia is the only municipal government in the country that does not "unilaterally control" the city's open data portal (Wink, "What Happens to OpenDataPhilly Now?," 2013). Instead, Philadelphia's portal is managed by a non-profit and contains both municipal and non-municipal data (that users can submit directly).

In December 2012, Mayor Emanuel in Chicago established an open data executive order and created a position for a Chief Data Officer (CDO) to speed up the development of an open data portal. In order to improve transparency and build working relationships between departments with regard to Big Data, the executive order required that an Open Data Advisory Group, which includes representatives from each agency, to convene in order to discuss the portal's ongoing development (Thornton, "How open data is transforming Chicago", 2013a). According to Brenna Berman, Commissioner and Chief Information Officer, "meeting participants prioritize what datasets should be developed and identifies cross agency collaborations for data analytics" (Personal Communication, March 21, 2014). To support Chicago's open data portal, the city established an accompanying data dictionary for information about all data being published (Thornton, "How Chicago's Data Dictionary is Enhancing Open Government", 2013b). The Data Dictionary takes transparency to another level and enhances the open data experience beyond what the other major American cities are doing.

In October 2013, Louisville announced an executive order for creating an open data plan. At that time, Louisville's open data policy was the first U.S. municipal policy that stated open data will be the "default mode" for how government electronic information will be formatted, stored, and made available to the public (Williams, "New Louisville's Open Data Policy Insists Open by Default is the Future", 2013b). The implications are that data that is legally accessible will be proactively disclosed online through the city's open data portal. Since January 2014, Louisville's open data portal has been in development and operated by a small team working within the Louisville Metro Technology Services department (Personal Communication, February 25, 2014).

Among our case study cities, Louisville has the lowest population with almost 600,000 residents and the smallest city government. Currently, the city has a "homegrown portal" that the city staff developed. The current process for this homegrown portal requires a data specialist to determine (with a small team) which datasets should be prioritized based on volume of requests and ease of "cleaning" the data. Louisville hopes to eventually publish between 500 and 1000 datasets in the portal (Personal Communication, February 25, 2014). Unlike New York City's law that mandates that all city agencies make their data accessible, Louisville relies on one staff member to collect, store, and manage the data, thereby making their call for open data to be the "default mode," very challenging.

5.2 Data Democratization and Equity

Developing and maintaining an open data portal is a significant investment in terms of infrastructure and finances. However, to democratize data, open data portals are only the first step. The next step is for open data to be user-friendly to a larger range of the population and to have broad impact. According to Justin Holmes, the Interim Chief Information Officer in Boston,

> We have to figure out how we take that data and make it more relevant. We know that Excel spreadsheets are not relevant to your grandmother. City departments need to be activists and understand why and how data can be impactful and then create a user-friendly platform (March 19, 2014).

While the open data movement has generated excitement and support from municipal governments, civic hackers, and tech-savvy citizens, these innovative applications typically provide benefits or services to those who also already utilize data and technology in their everyday lives. For citizens that have access to and understand these systems, they are able to receive benefits in terms of cost, efficiency, and decision-making.

Despite the publicity surrounding open data, providing data does not mean that every citizen will directly receive or experience a benefit or improve their quality of life. Truly innovative municipal governments should aim to provide widespread access and understanding of data and technologies to their citizenry (McAuley et al. 2011). The following analogy between libraries and open data portals is instructive for how data portals should be conceived: "we didn't build libraries for an already literate citizenry. We built libraries to help citizens become literate. Today we build open data portals not because we have a data or public policy literate citizenry, we build them so citizens may become literate in data, visualization, and public policy" (Eaves 2010). Nigel Jacob of Boston's MONUM echoes these sentiments by saying "open data is a passive role for the government...fine for software development, but it does not actively engaged with citizens." Thus, in order for cities to develop a democratic data system, they need to make the data usable and provide supplementary resources and training to ensure widespread use and impact.

5.3 The New Digital Divide: Digital Literacy

In the last few decades, local governments have been engaged in activities to reduce the digital divide by increasing access to broadband, Wi-Fi, ICTs, and computer centers. Coupled with these programs and the increasing affordability of acquiring technology, digital access is becoming less of a problem.

However, a new digital divide is emerging between individuals who can effectively access and use digital resources and data to improve their well-being and those who can not. For example, being unable to download, fill out, and submit a job application on-line will severely limit their job opportunities. If digital literacy is low among groups that are traditionally disadvantaged, this may exacerbate social inequality.

5.3.1 Chicago and New York: Digital Access and Literacy Initiatives

In 2009, every city in our study, except Louisville received funding from the Broadband Technologies Opportunities Program (BTOP), a federal program designed to expand access to broadband services nationwide. New York City received $42 million from BTOP and developed the NYC Connected Communities program, which focused on broadband adoption and improving access to computer centers in low-income and limited-English neighborhoods throughout five boroughs (Personal Communication, March 10, 2014). Through this program, 100 computing centers were opened at local public libraries, public housing developments, community centers and senior centers. The majority of these centers have remained open as more funding was acquired in 2013 when the BTOP funding expired. NYC Connected Communities included computer training and digital literacy programs designed to meet community needs (Personal Communication, March 10, 2014). Since 2010, the NYC Connected Communities program has hosted more than three million user sessions citywide, approximately 100,000 residents have participated in training classes, and over 4.7 million residents have attended open lab sessions (NYC DoITT, "Technology & Public Service Innovation: Broadband Access" n.d.; City of New York Public Computing Centers 2014).

In Chicago, the Smart Communities program received $7 million of federal funding in 2010 to develop training and outreach initiatives centralized in five low-income communities in Chicago (Tolbert et al. 2012). The Smart Communities program created a master plan that included considerable input from community members to determine program priorities to address challenges specific to their community (Deronne and Walek 2010). Thus, in Chicago, the design of the programs was developed through a "bottom up" participatory process that resulted in unique programmatic components that focused on the idea that the "community knows best." (Personal Communication, February 25, 2014).

Through early 2013, the Smart Communities program has trained approximately 20,000 people in computer and digital literacy skills (City of Chicago Public

Computing Centers 2014). One of the Smart Communities program's most applauded successes is a statistically significant 15 % point increase in Internet usage in Smart Communities neighborhoods compared to other neighborhoods in the city between 2008 and 2011 (Tolbert et al. 2012).

5.3.2 Small-Scale Approaches to Digital Access and Literacy

The programs mentioned above are supported by large sums of federal funds. But, these funds are not available to the vast majority of cities throughout the country. Thus, we offer examples of smaller scale initiatives to improve digital access and literacy found in our case study cities. In Chicago, LISC, a non-profit community-based organization, has built upon the Smart Communities program and developed community-focused initiatives that provide training for residents from a diversity of demographic backgrounds and offers an online presence for low-income neighborhoods. Boston and New York have installed computers in vans to serve as mobile city halls that bring public staff into the field to offer services and to provide access to technology to residents of concentrated poor and minority neighborhoods. Boston operates "Tech Goes Home" (TGH), an initiative that provides digital literacy courses, subsidized computer software, and broadband access to school-age children and their families. Louisville's focus on digital literacy comes from a workforce development perspective. The city has made investments in developing high-level data analysis skills in residents that can improve employment opportunities while simultaneously making the city more attractive to businesses.

The one issue with these smaller-scale approaches is that it is difficult to develop an initiative or program that tackles both digital access and digital literacy on a smaller scale and budget. The initiatives in Chicago have managed to continue this dual emphasis through LISC's partnership with their other organizations. Boston's Tech Goes Home program has managed to expand since its development in 2000 and has evolved into one of the more sustainable models of digital literacy by providing a computer and low-cost access to broadband to those who complete their program.

5.3.3 Digital Access and Literacy Recommendations

All the cities in this study have worked to close the digital divide in terms of access and literacy. However, the innovativeness and diversity of Boston and Chicago's programs demonstrate the significant investment and local resources required, both financially and in terms of the coordination between local stakeholders. Justin Holmes at Boston's Office of Innovation & Technology's describes the complexity of approaching issues of data access and literacy in diverse communities:

> "Our engagement approach is multichannel...we need to be mobile, move beyond call centers and traditional centers, and use social media as a 'value add' to reach people. We're

working to meet people where they are comfortable" (Personal Communication, March 19, 2014).

The main commonality between most cities was the development of public computing centers to improve access. Dan O'Neil, Executive Director of the Smart Chicago Collaborative, believes that "public computing centers are the most essential building block in providing access to technology" (Personal Communication, March 6, 2014). However, the programs with the potential for long-lasting impacts appear to be those with a concentrated effort on providing extensive on-site training through a site-specific curriculum tailored to the wants and needs of the community. Andrew Buss, the Director of Innovation Management in the Philadelphia Office of Innovation & Technology, identified that the key to Philadelphia's KEYSPOT computing center initiative was having an instructor on site:

> "You can't just have a room with a bunch of technology...you need to have a person onsite at each location for assistance on how to use the equipment and to solve minor tech issues [which creates] a guided experience" (Personal Communication, February 28, 2014).

The availability and expertise of on-site instructors was also seen in each of the mobile van initiatives and has proven to be crucial for digital literacy programs. Furthermore, it seems that establishing a high level of trust between the program providers, teachers, and participants is integral to the program's success and to see positive outcomes gained for students.

Improving data literacy is important for a diversity of users, not only user groups that do not have access to new technologies. Brenna Berman, the Chief Information Officer at Chicago's Department of Innovation & Technology, spoke about the importance of non-profits accessing and utilizing data:

> "We've been creating a partnership between commercial organizations and the philanthropic community to make sure non-profits are benefiting from Big Data and using some indirect organizations that have been addressing the gap...we know non-profits were not embracing Big Data and weren't using data to inform decisions. They needed representatives from communities to teach them how to do this, so we've run education workshops to collaborate and educate...like that saying, a rising tide raises all ships" (Personal Communication, March 21, 2014).

Therefore, closing the digital divide gap may not be simply a matter of providing access and training to individuals, but also to low capacity organizations.

5.4 Promoting Equitable Outcomes with Big Data

The third dimension of social equity relates to the promotion of equitable outcomes using Big Data. This could be conceived in two ways. First, directing Big Data analysis to reduce disparities across various social dimensions (i.e. income, race, ethnicity, and gender) for different groups. Second, targeting disadvantaged or underserved groups by using Big Data to improve their quality of life. The city of

Louisville offers some innovative ways to address equitable outcomes using Big Data.

5.4.1 Sensor Technology in Louisville

Among government agencies, public health agencies have been leading the way in using Big Data to reduce health disparities. New technologies offer innovative ways to assist low-income individuals to manage their healthcare and improve their health. In 2010, the city of Louisville created an inhaler sensor called Asthmapolis, which also comes with a supplementary mobile application that allows asthma patients and their doctors to understand asthma's triggers and provides an effective way to control asthma, while simultaneously generating data for public health researchers (Propeller 2013). More than 500 sensors have been deployed to low-income residents suffering with asthma in Louisville. While the program is still in the early stages, the benefits for residents have been notable. Interviews with program participants highlight their increased confidence in their disease management due to the "smart inhalers." Furthermore, participants are happy to be part of the program because the inhaler sensor is provided free of charge through funding from philanthropic grants (Runyon 2013).

This program in Louisville is believed to be truly transformative in "breaking down data silos in the public sector...[and] is a model project for how the public sector and communities should start working with informatics" (RWJF Public Health Blog 2012). Utilizing this technology provides a benefit to the user, as their disease management improves. It is also useful to doctors and public health officials because the individual-level data, geo-tagged by location, generated can be utilized to inform future public health decisions. The City of Louisville has pushed to incorporate innovations that improve public health because of the belief that having a healthy population contributes to regional and economic competitiveness, which encourages businesses to locate in the city (RWJF Public Health Blog 2012). This mindset is consistent with Louisville's strategy to improve data literacy as a workforce development tool to increase the city's economic competitiveness. Thus, for smaller cities such as Louisville, innovations in technology and Big Data that promotes the image and reputation of the city as cutting edge and with a good quality of life can improve economic competitiveness.

5.4.2 New York City: Improving Social Service Delivery

In 2009, HHS-Connect was developed in New York to collect all data relating to social services in one digital repository in order to streamline the intake process for clients visiting different social service agencies. HHS Connect has transformed service delivery for social services into a client-centric model. The increased coordination between city agencies has improved case management processes and provided clients with one access point to self-screen for over 30 benefit programs.

These types of internal innovations can make the experience easier for clients while helping overburdened agencies detect fraud, improve service delivery, and reduce costs (Goldsmith 2014).

These programs emphasize the need to develop partnerships between social service providers to allow data sharing between agencies and streamline intake services. This creates organizational efficiencies and also makes receiving services for socially vulnerable populations easier and more efficient, thus saving individuals time and money. However, there are a variety of issues that limit the power of Big Data. On the federal level, statutes vary about what health records, educational transcripts, and data related to homelessness, child welfare, drug abuse and mental health can be collected, published, or shared (Goldsmith and Kingsley 2013). On the state and local level, many laws were written prior to the digital age and can create conflict and confusion, thereby slowing down the adoption of innovations in these fields.

6 Lessons Learned About Big Data, Governance, and Social Equity

This case study of five U.S. cities with Offices of Innovation sought to answer two primary research questions. First, is Big Data changing decision-making in city hall? Second, is Big Data being used to address social equity and how? To varying degrees, Big Data in all of our case study cities is altering the way in which decisions are made in local government by supplying more data sources, integrating cross agency data, and to use predictive rather than reactive analytics to make decisions. This has the potential to improve administrative efficiency and reduce man hours spent on tasks, thereby saving time, energy, and money. While this may be true, cities often do not calculate the costs associated with collecting, cleaning, managing, and updating Big Data. No study to date has examined the cost-effectiveness of these programs to determine the return on investment. Furthermore, local government's focus on tame problems using a rational framework that promotes efficiency in government systems, raises long-standing concerns about "problem definition" within government (Rittel and Webber 1973; Dery 1984; Baumgartner and Jones 1993; Rochefort and Cobb 1994). In particular, top down models of decision-making that use technologies accessible by groups that are already advantaged may exacerbate social inequalities and inhibit democratic processes.

While major cities, such as New York, have high capacity public agencies that can populate the data required in a centralized repository, smaller cities may not. Louisville's open data portal, for example, relies on one staff member populating the data using a data portal that was developed in-house. What happens if this staff member leaves his post? New York City's MODA provides the needed expertise and capacity to assist public agencies and departments to conduct predictive analytics. Their model of support is dependent on a separate government entity staffed with ICT experts dedicated solely to supporting other agencies. Other models of predictive analytics can be found in training agency staff members. Training police officers in predicting policing analytics is an example of Big Data analytics altering the operations within a single public agency.

While there is great promise that predictive analytics will become widely accessible and affordable (e.g. Chicago's Smart Data platform), caution should be taken to ensure that there are checks and balances when using Big Data. Big Data analytics are only useful if the data is accurate and if the analysis of the data is context relevant. Therefore, Big Data analytics alone should not be used to make decisions, but rather, Big Data coupled with public engagement and experiential knowledge should make predictive analytics and decision-making more effective.

The second question is how Big Data is being used to address issues related to social equity? We examine three primary ways that Big Data relates to social equity. First, making data available and accessible can promote more social equity and open data portals are the primary way cities are doing so. But open data portals alone do not lead to equitable outcomes. While open data portals provide data to the general public, having data available does not ensure that every resident within a city is able to access, analyze, or use it for their benefit. Truly innovative municipal governments should aim to promote widespread access and understanding of data and technologies among a broad cross-section of the population, especially groups that are traditionally disadvantaged and digitally disconnected. Boston's MONUM, for example, does not operate Boston's open data portal and instead focuses their resources on utilizing new technologies to improve citizen engagement experiences, education learning tools, and streetscapes.

None of the Offices of Innovation studied have programs that directly engage or teach the public how to maximize the potential of open data portals. The one organization that is attempting to teach disadvantaged populations about the benefits of open data is LISC Chicago, a non-profit community-based organization. Expanding these types of sessions through partnerships established between Offices of Innovation and local community groups would be one way of making the open data movement have greater impact and reach.

The second dimension of equity, digital access and literacy, has been an area of concern for four of the five cities in our study. Using federal funding aimed at digital access and literacy programs, each city has struggled, to varying degrees, to continue these efforts after the funding expired. Chicago's efforts at bridging the digital gap has continued due to the efforts of non-profits, such as LISC Chicago and local public investment. Chicago's work highlights the importance of federal funding, local planning, and effective collaborations with non-government organizations. The city of Boston also has innovative programs, such as Tech Goes Home, that addresses digital access and literacy. Furthermore, bringing city services to neighborhoods through City Hall to Go is reframing the relationship between Boston's City Hall and the public by providing direct access to municipal staff and services on the van. Residents are able to interact directly with decision-makers, and benefit from spending less time and effort to receive municipal services when the van arrives in their neighborhood.

Understanding how Big Data can be used to address issues of equity is complex, due to the various dimensions of equity that can be considered. Each of the cities studied have been focusing on some issues of equity, but none have taken a comprehensive, multi-faceted approach to social equity. Big Data and new technologies have the potential to address thornier wicked issues if different policy questions and priorities were raised and if there is political support for it. Our study suggests that cities using Big Data have opted to more frequently tackle questions focused on system optimization rather than on targeting social inequality.

Appendix: Case Study Cities

Boston Department of Innovation and Technology (DoIT)

The city of Boston has a unique structure for their office of innovation. Their office is called the Boston Department of Innovation and Technology or DoIT and is housed in Boston's City Hall. Their primary role is to collect, manage, and organize Big Data. DoIt is also the city's internal social media team and operates a coordinated, data-driven, strategy across all social media platforms, such as Facebook and Twitter, with the goal of curating daily engagement to help improve residents' quality of life ("Boston's Mayoral Transition", NextBoston). The city has a social media policy and an organizational strategy to support this work with a social media liaison positioned in each of the city's departments. Due to these efforts, Boston's social media strategy has received national recognition and has seen exponential growth in terms of engagement with the public. For example, in 2013, the City of Boston's Facebook page followers grew by 200 % and the page's reach grew 400 % between 2012 and 2013 ("Boston's Mayoral Transition", NextBoston).

DoIt collaborates very closely with the Mayor's Office for New Urban Mechanics or MONUM. According to the Co-Founder of MONUM, the department, "serves as a complementary force for city departments to innovate city services and we're there to support them...[and unlike DoIt], MONUM has a great deal of independence and the ability to be innovative while not being encumbered by maintaining and supporting the innovation" (personal communication, March 19, 2014). Because MONUM is not managing the Big Data, the department focuses on piloting innovative, and sometimes risky programs that if successful, will be scaled up within a city department or city-wide. Thus, they are provided the freedom and flexibility to be creative and innovative. In 2013, Boston was named the #1 Digital City in America by the Center for Digital Government's annual Digital Cities Survey. Between the efforts of DoIt and MONUM, and their social media strategy, the city of Boston is widely regarded as one of the leading Big Data innovators in municipal government.

Chicago Department of Innovation and Technology (DoIT)

In Chicago, the office of innovation is also known as the Department of Innovation and Technology (DoIT). This department takes a "comprehensive approach to data and analytics" and focuses their efforts on several key programs, including Chicago's Digital Excellence Initiative, the Smart Communities program, and implementing Chicago's Technology Plan, a comprehensive plan of five strategies and 28 initiatives to improve Chicago's efforts in innovation and technology (Personal Communication, March 21, 2014). Since Brenna Berman was promoted to Chief Information Officer in late 2013, the department's efforts have emphasized Ms. Berman's personal vision of "resident-centered technology and innovation," as well as internal innovations that foster data-driven decision making, such as predictive analytics programs, internal dashboards, and modernizing existing systems into user-friendly applications (Thorton, "Chicago Welcomes New CIO Brenna Berman").

Louisville Department of Economic Growth and Innovation

In early 2012, Louisville's Economic Development Department was restructured to become the Department of Economic Growth and Innovation. Ted Smith, previously the Director of Innovation, was appointed the Director of this new department. Louisville's Department of Economic Growth and Innovation is a separate but coordinated department of the Louisville Metro Government, a regional government entity. According to Smith, the Department has three primary goals: (1) civic innovation, such as new approaches to community engagement, (2) government innovation, particularly innovating existing government processes, and (3) service of performance improvement, including creating internal dashboards, establishing cultural methodologies, and improving outcomes from a "bottom-up" perspective (Personal communication, February 27, 2014). Louisville's Department takes a very unique approach to innovation, fusing economic growth and development principles and innovation. Much of the Department's efforts are focused on making the city more attractive to private businesses as well as developing digital platforms and infrastructure that can benefit both the city's residents and the private sector.

New York Mayor's Office of Data Analytics (MODA)

New York City's innovation office is known as the Mayor's Office of Data Analytics (MODA). MODA works in coordination with New York City's Department of Information Technology & Communications (DoITT). NYC DoITT is primarily responsible for managing and improving the city government's IT infrastructure and telecommunication services to enhance service delivery for New York's residents and businesses. MODA was officially established by an executive order from Mayor Bloomberg in April 2013, but the agency had been working informally within New York City government for several years previously under the name "Financial Crime Task Force" (Personal Communication, February 24, 2014). MODA manages the city's Open Data portal and works extensively on data management and analytics using their internally developed data platform, known as DataBridge. In order to establish DataBridge, MODA collaborated with DoITT to consolidate references for each building address in the city throughout all of the city's agencies into one database, so that when one searches by address, all of the information from every department is accessible in one place (Nicholas O'Brien, personal communication, February 24, 2014). MODA operates specific projects to improve processes or gather more information about the city's operations. MODA's projects typically fall into one of these four categories: (1) aiding disaster response and recovery through improved information, (2) assisting NYC agencies with data analysis and delivery of their services, (3) using analytics to deliver insights for economic development, and (4) encouraging transparency of data between the city's agencies, as well as to the general public.

Philadelphia Office of Innovation and Technology (OIT)

Previously known as the Division of Technology, Philadelphia's Office of Innovation and Technology (OIT) was established in 2011 (Wink, "Office of Innovation and Technology to replace Division of Technology at City of Philadelphia", 2013). Prior to reinventing and restructuring the office to include innovation, the Division was responsible for the city's day-to-day technological operations. Philadelphia OIT was created as the city began changing its culture and projects that were more innovative externally, as well as internally within the infrastructure of municipal government. While Philadelphia OIT is responsible for all major technology initiatives in the city, the department is broken into 11 units, one of which is Innovation Management. The Innovation Management unit's responsibilities fall under three internal categories: (1) open data, (2) civic technology, including mobile applications, and (3) innovation (Personal communication, February 28, 2014). The innovation category was established as the Philly KEYSPOT initiative was launched, a federally funded public-private partnership that established approximately 80 public computing centers in communities and provides residents with Internet access and training.

As mentioned earlier, the city of Boston's Mayor's Office of New Urban Mechanics has a satellite office in Philadelphia with the same name. According to Almirall et al. (2014), Philadelphia's MONUM is referred to as a civic accelerator, which is an organization that "match(es) cities with start-ups, private firms, and non-profit organizations interested in partnering with government to provide better services, bring modern technology to cities, or change the way citizens interact with city hall" (p. 4). MONUM's Philadelphia office is also located in city hall and its mission to transform city services and engage citizens and institutions throughout the city to participate in addressing the needs of city residents.

References

Almirall E, Lee M, Majchrzak A (2014) Open innovation requires integrated competition-community ecosystems: lessons learned from civic open innovation. Bus Horiz 57(3): 391–400. http://dx.doi.org/10.1016/j.bushor.2013.12.009. Accessed 5 Mar 2014

Appallicious (2014) Case study on Appallicious, the City of San Francisco, open data, and Gov 2.0—Appallicious. Appallicious. http://www.appallicious.com/case-study-appallicious-city-san-francisco-open-data-gov-2-0-2/. Accessed 13 July 2014

Araque J, Maiden R, Bravo N, Estrada I, Evans R, Hubchik K et al (2013) Computer usage and access in low-income urban communities. Comput Hum Behav 29(4):1393–1401, http://dx.doi.org/10.1016/j.chb.2013.01.032. Accessed 8 Mar 2014

Ash Center Mayors Challenge Research Team (2014) Chicago pioneers open source analytics platform. Government Technology. http://www.govtech.com/data/Chicago-Pioneers-Open-Source-Analytics-Platform.html. Accessed 6 Apr 2014

Azari R, Pick JB (2005) Technology and society: socioeconomic influences on technological sectors for United States counties. Int J Inf Manag 25(1):21–37, http://dx.doi.org/10.1016/j.ijinfomgt.2004.10.001. Accessed 5 Mar 2014

Bailey A, Ngwenyama O (2011) The challenge of e-participation in the digital city: exploring generational influences among community telecentre users. Telemat Inform 28(3):204–214, http://dx.doi.org/10.1016/j.tele.2010.09.004. Accessed 8 Mar 2014

Batty M (2012) Smart cities, big data. Environ Plan B Plan Des 39(2):191–193, http://dx.doi.org/10.1068/b3902ed. Accessed 2 Mar 2014

Batty M (2013) Big data, smart cities and city planning. Dialogues Hum Geogr 3(3):274–279, http://dx.doi.org/10.1177/2043820613513390. Accessed 1 Mar 2014

Batty M, Axhausen KW, Giannotti F, Pozdnoukhov A, Bazzani A, Wachowicz M, Ouzounis G, Portugali Y (2012) Smart cities of the future. Eur Phys J Spec Top 214(1):481–518, http://dx.doi.org/10.1140/epjst/e2012-01703-3. Accessed 28 Mar 2014

Baumgartner FR, Jones BD (1993) Agendas and instability in American politics. Illinois. University of Chicago Press, Chicago

Boston's Mayoral Transition: The Problem Properties Task Force (2013) NextBoston: Blog. http://next.cityofboston.gov/post/65455635934/the-problem-properties-task-force-established-by. Accessed 30 Apr 2014

Burstein R (2013) Most cities don't need innovation offices. Slate Magazine. http://www.slate.com/articles/technology/future_tense/2013/06/big_ideas_for_cities_don_t_always_come_from_innovation_offices.html. Accessed 6 Apr 2014

Chen CLP, Zhang CY (2014) Data-intensive applications, challenges, techniques and technologies: a survey on Big Data. Inform Sci 275:314–347, http://dx.doi.org/10.1016/j.ins.2014.01.015. Accessed 28Mar 2014

Chourabi H, Nam T, Walker S, Gil-Garcia JR, Mellouli S, Nahon K, Pardo T, Scholl HJ (2012) Understanding smart cities: an integrative framework. In: Proceedings of 45th Hawaii international conference on system sciences. http://www.enricoferro.com/paper/CEDEM13.pdf. Accessed 2 Mar 2014

City of Chicago Public Computing Centers (2014) National Telecommunications and Information Administration: annual performance progress report for Public Computing Centers. http://www2.ntia.doc.gov/files/grantees/36-42-b10567_apr2013.pdf. Accessed 18 June 2014

City of New York Public Computing Centers (2014) National Telecommunications and Information Administration: annual performance progress report for Public Computing Centers. http://www2.ntia.doc.gov/files/grantees/17-43-b10507_apr2013.pdf. Accessed 18 June 2014

Correa T (2010) The participation divide among "online experts": experience, skills and psychological factors as predictors of college students' web content creation. J Comput Mediat Commun 16(1):71–92, http://dx.doi.org/10.1111/j.1083-6101.2010.01532.x

Cumbley R, Church P (2013) Is "Big Data" creepy? Comput Law Secur Rev 29(5):601–609, http://dx.doi.org/10.1016/j.clsr.2013.07.007. Accessed 9 Mar 2014

Deronne J, Walek G (2010) Neighborhoods get smart about technology. Smart Communities. http://www.smartcommunitieschicago.org/news/2443. Accessed 9 Apr 2014

Dery D (1984) Problem definition in policy analysis. University Press of Kansas, Lawrence

DiMaggio P, Hargittai E, Celeste C, Shafer S (2004) Digital inequality: from unequal access to differentiated use. In: Neckerman K (ed) Social inequality. Russell Sage, New York, pp 355–400

Domingo A, Bellalta B, Palacin M, Oliver M, Almirall E (2013) Public open sensor data: revolutionizing smart cities. IEEE Technol Soc Mag 32(4):50–56, http://dx.doi.org/10.1109/MTS.2013.2286421. Accessed 4 Mar 2014

Eaves D (2010) Learning from libraries: the literacy challenge of open data. eavesca. http://eaves.ca/2010/06/10/learning-from-libraries-the-literacy-challenge-of-open-data/. Accessed 13 July 2014

Feuer A (2013) The Mayor's geek squad. The New York Times. http://www.nytimes.com/2013/03/24/nyregion/mayor-bloombergs-geek-squad.html?pagewanted=all. Accessed 8 Apr 2014

Fuentes-Bautista M (2013) Rethinking localism in the broadband era: a participatory community development approach. Gov Inf Q 31(1):65–77, http://dx.doi.org/10.1016/j.giq.2012.08.007. Accessed 20 Mar 2014

Gilbert M, Masucci M, Homko C, Bove A (2008) Theorizing the digital divide: information and communication technology use frameworks among poor women using a telemedicine system. Geoforum 39(2):912–925, http://dx.doi.org/10.1016/j.geoforum.2007.08.001. Accessed 4 Mar 2014

Gilmour JA (2007) Reducing disparities in the access and use of Internet health information. A discussion paper. Int J Nurs Stud 44(7):1270–1278, http://dx.doi.org/10.1016/j.ijnurstu.2006.05.007. Accessed 10 Mar 2014

Goldsmith S (2014) Unleashing a community of innovators|data-smart city solutions. Data-Smart City Solutions. http://datasmart.ash.harvard.edu/news/article/unleashing-a-community-of-innovators-399. Accessed 1 May 2014

Goldsmith S, Kingsley C (2013) Getting big data to the good guys|data-smart city solutions. Data-Smart City Solutions. http://datasmart.ash.harvard.edu/news/article/getting-big-data-to-the-good-guys-140. Accessed 1 May 2014

Goodyear S (2013) Why New York City's open data law is worth caring about. The Atlantic Cities. http://www.theatlanticcities.com/technology/2013/03/why-new-york-citys-open-data-law-worth-caring-about/4904/. Accessed 8 Apr 2014

Guillen MF, Suarez SL (2005) Explaining the global digital divide: economic, political and sociological drivers of cross-national internet use. Soc Forces 84(2):681–708

Hargittai E (2002) Second-level digital divide: differences in people's online skills. First Monday 7(4)

Hemerly J (2013) Public policy considerations for data-driven innovation. Computer 46(6):25–31, http://dx.doi.org/10.1109/MC.2013.186. Accessed 7 Mar 2014

Hilbert M (2011) The end justifies the definition: the manifold outlooks on the digital divide and their practical usefulness for policy-making. Telecomm Policy 35(8):715–736, http://dx.doi.org/10.1016/j.telpol.2011.06.012. Accessed 14 Mar 2014

Hollywood JS, Smith SC, Price C, McInnis B, Perry W (2012) Predictive policing: what it is, what it isn't, and where it can be useful. NLECTC Information and Geospatial Technologies Center of Excellence. Lecture conducted from RAND Corporation, Arlington, VA. http://www.theiacp.org/Portals/0/pdfs/LEIM/2012Presentations/OPS-PredictivePolicing.pdf

Howard A (2011) How data and open government are transforming NYC. O'Reilly Radar. http://radar.oreilly.com/2011/10/data-new-york-city.html. Accessed 10 Apr 2014

Howard A (2012) Predictive data analytics is saving lives and taxpayer dollars in New York City. O'Reilly Data. http://strata.oreilly.com/2012/06/predictive-data-analytics-big-data-nyc.html. Accessed 10 Apr 2014

Jenkins H, Purushotma R, Clinton K, Weigel M, Robison AJ (2009) Confronting the challenges of participatory culture media education for the 21st century. MIT Press, Cambridge

Joseph RC, Johnson NA (2013) Big data and transformational government. IT Prof 15(6):43–48, http://dx.doi.org/10.1109/MITP.2013.61. Accessed 11 Mar 2014

Kim G, Trimi S, Chung J (2014) Big-data applications in the government sector. Commun ACM 57(2). doi:10.1145/2500873. Accessed 20 Mar 2014

Kitchin R (2014) The real-time city? Big data and smart urbanism. GeoJournal 79(1):1–14, http://www.nuim.ie/progcity/wp-content/uploads/2014/02/GeoJournal-Real-time-city-2014.pdf. Accessed 8 Apr 2014

Kvasny L, Keil M (2006) The challenges of redressing the digital divide: a tale of two US cities. Inf Syst J 16:23–53

LaRose R, Gregg J, Strover S, Straubhaar J, Carpenter S (2007) Closing the rural broadband gap: promoting adoption of the internet in rural America. Telecomm Policy 31(6–7):359–373, http://dx.doi.org/10.1016/j.telpol.2007.04.004. Accessed 10 Mar 2014

Lee H, Park N, Hwang Y (2015) A new dimension of the digital divide: exploring the relationship between broadband connection, smartphone use and communication competence. Telemat Inform 32(1):45–56, http://www.sciencedirect.com/science/article/pii/S0736585314000161. Accessed 25 Mar 2014

Light J (2001) Rethinking the digital divide. Harvard Educ Rev 72(4), http://hepg.org/her/abstract/101. Accessed 2 Mar 2014

Looker ED, Thiessen V (2003) Beyond the digital divide in Canadian schools: from access to competency in the use of information technology. Soc Sci Comput Rev 21(4):475–490

McAuley D, Rahemtulla H, Goulding J, Souch C (2011) How open data, data literacy and linked data will revolutionise higher education. Pearson Blue Skies RSS. http://pearsonblueskies.com/2011/how-open-data-data-literacy-and-linked-data-will-revolutionise-higher-education/. Accessed 13 July 2014

Mossberger K, Tolbert CJ, Hamilton A (2012) Measuring digital citizenship: mobile access and broadband. Int J Commun 6:2492–2528, http://ijoc.org/index.php/ijoc/article/view/1777. Accessed 6 Mar 2014

Novotny T (2013) Civic tech and adaptive change: learnings from young leaders in Louisville. Storify. https://storify.com/tamirnovotny/civic-tech-and-adaptive-change-lessons-from-louisv. Accessed 9Apr 2014

NYC DoITT (n.d.) Technology & public service innovation—broadband access. DoITT—Technology & Public Service Innovation—Broadband Access. http://www.nyc.gov/html/doitt/html/open/broadband.shtml. Accessed 9 Apr 2014

NYC DoITT (2012) Local law 11 of 2012 publishing open data. DoITT—Open Government/Innovation—Open Data. http://www.nyc.gov/html/doitt/html/open/local_law_11_2012.shtml. Accessed 8 Apr 2014

Orzag P (2009) Open government directive. The White House. http://www.whitehouse.gov/open/documents/open-government-directive. Accessed 13 July 2014

Prieger JE (2013) The broadband digital divide and the economic benefits of mobile broadband for rural areas. Telecomm Policy 37(6–7):483–502, http://dx.doi.org/10.1016/j.telpol.2012.11.003. Accessed 8 Mar 2014

Prieger JE, Hu W (2008) The broadband digital divide and the nexus of race, competition, and quality. Inf Econ Policy 20(2):150–167, http://dx.doi.org/10.1016/j.infoecopol.2008.01.001. Accessed 20 Mar 2014

Propeller Health (2013) Wyckoff Heights Medical Center is first New York Hospital to offer Asthmapolis Mobile Asthma Management Program. Propeller Health. http://propellerhealth.com/press-releases/. Accessed 9 Apr 2014

Revenaugh M (2000) Beyond the digital divide: pathways to equity. Technol Learn 20(10), http://eric.ed.gov/?id=EJ615183. Accessed 2 Mar 2014

Reyes J (2014) "Smart policing" movement training Philly cops to be data scientists. Technically Philly Smart policing movement training Philly cops to be data scientists Comments. http://technical.ly/philly/2014/02/18/philadelphia-police-smart-policing-crime-scientists/. Accessed 30 Apr 2014

Rich S (2012) E-government. Chicago's data brain trust tells all. http://www.govtech.com/e-government/Chicagos-Data-Brain-Trust-Tells-All.html. Accessed 8 Apr 2014

Rittel HW, Webber MM (1973) Dilemmas in a general theory of planning. Policy Sci 4(2):155–169

Rochefort DA, Cobb RW (1994) Problem definition: an emerging perspective. In the politics of problem definition: shaping the policy agenda. University Press of Kansas, Lawrence

Runyon K (2013) Louisville premieres new program to fight pulmonary disease. The Huffington Post. http://www.huffingtonpost.com/keith-runyon/louisville-chooses-asthma_b_4086297.html. Accessed 9 Apr 2014

RWJF Public Health Blog (2012) Asthmapolis: public health data in action. Robert Wood Johnson Foundation. http://www.rwjf.org/en/blogs/new-public-health/2012/08/asthmapolis_public.html. Accessed 9 Apr 2014

Shueh J (2014a) Big data could bring governments big benefits. Government Technology. http://www.govtech.com/data/Big-Data-Could-Bring-Governments-Big-Benefits.html. Accessed 6 Apr 2014

Shueh J (2014b) 3 Reasons Chicago's data analytics could be coming to your city. Government Technology. http://www.govtech.com/data/3-Reasons-Chicagos-Analytics-Could-be-Coming-to-Your-City.html. Accessed 6 Apr 2014

Tapia AH, Kvasny L, Ortiz JA (2011) A critical discourse analysis of three US municipal wireless network initiatives for enhancing social inclusion. Telemat Inform 28(3):215–226, http://dx.doi.org/10.1016/j.tele.2010.07.002. Accessed 4 Mar 2014

Thornton S (2013a) How open data is transforming Chicago. Digital Communities. http://www.digitalcommunities.com/articles/How-Open-Data-is-Transforming-Chicago.html. Accessed 8 Apr 2014

Thornton S (2013b) How Chicago's data dictionary is enhancing open government. Government Technology. http://www.govtech.com/data/How-Chicagos-Data-Dictionary-is-Enhancing-Open-Government.html. Accessed 8 Apr 2014

Tolbert C, Mossberger K, Anderson C (2012) Measuring change in internet use and broadband adoption: comparing BTOP smart communities and other Chicago neighborhoods. http://www.lisc-chicago.org/uploads/lisc-chicago-clone/documents/measuring_change_in_internet_use_full_report.pdf. Accessed 8 Apr 2014

Velaga NR, Beecroft M, Nelson JD, Corsar D, Edwards P (2012) Transport poverty meets the digital divide: accessibility and connectivity in rural communities. J Transp Geogr 21:102–112, http://dx.doi.org/10.1016/j.jtrangeo.2011.12.005. Accessed 8 Mar 2014

Vilajosana I, Llosa J, Martinez B, Domingo-Prieto M, Angles A, Vilajosana X (2013) Bootstrapping smart cities through a self-sustainable model based on big data flows. IEEE Commun Mag 51(6):128–134, http://dx.doi.org/10.1109/MCOM.2013.6525605. Accessed 20 Mar 2014

Warren M (2007) The digital vicious cycle: links between social disadvantage and digital exclusion in rural areas. Telecomm Policy 31(6–7):374–388, http://dx.doi.org/10.1016/j.telpol.2007.04.001. Accessed 8 Mar 2014

Wigan MR, Clarke R (2013) Big data's big unintended consequences. Computer 46(6):46–53, http://dx.doi.org/10.1109/MC.2013.195. Accessed 8 Mar 2014

Williams R (2013a) NYC's plan to release all-ish of their data. Sunlight Foundation Blog. http://sunlightfoundation.com/blog/2013/10/11/nycs-plan-to-release-all-ish-of-their-data/. Accessed 8 Apr 2014

Williams R (2013b) New Louisville open data policy insists open by default is the future. Sunlight Foundation Blog. http://sunlightfoundation.com/blog/2013/10/21/new-louisville-open-data-policy-insists-open-by-default-is-the-future/. Accessed 8 Apr 2014

Williams R (2014) Boston: the tale of two open data policies. Sunlight Foundation Blog. http://sunlightfoundation.com/blog/2014/04/11/boston-the-tale-of-two-open-data-policies/. Accessed 13 July 2014

Wink C (2013) What happens to OpenDataPhilly now? Technically Philly what happens to OpenDataPhilly now Comments. http://technical.ly/philly/2013/09/18/what-happens-to-opendataphilly-now/. Accessed 13 July 2014

Big Data, Small Apps: Premises and Products of the Civic Hackathon

Sara Jensen Carr and Allison Lassiter

Abstract Connections and feedback among urban residents and the responsive city are critical to Urban Informatics. One of the main modes of interaction between the public and Big Data streams is the ever-expanding suite of urban-focused smartphone applications. Governments are joining the app trend by hosting civic hackathons focused on app development. For all the attention and effort spent on app production and hackathons, however, a closer examination reveals a glaring irony of the Big Data age: to date, the results have been remarkably small in both scope and users. In this paper, we critically analyze the structure of The White House Hackathon, New York City BigApps, and the National Day of Civic Hacking, which are three recent, high-publicity hackathons in the United States. We propose a taxonomy of civic apps, analyze hackathon models and results against the taxonomy, and evaluate how the hackathon structure influences the apps produced. In particular, we examine problem definitions embedded in the different models and the issue of sustaining apps past the hackathon. We question the effectiveness of apps as the interface between urban data and urban residents, asking who is represented by and participates in the solutions offered by apps. We determine that the transparency, collaboration and innovation that hackathons aspire to are not yet fully realized, leading to the question: can civic Big Data lead to big impacts?

Keywords App • Hackathon • Participation • Representation • Open governance

---

S.J. Carr (✉)
School of Architecture/Office of Public Health Studies, University of Hawaii Manoa, 2410 Campus Road, Honolulu, HI, USA
e-mail: saracarr@hawaii.edu

A. Lassiter
Department of Economics, Monash University, 20 Chancellors WalkMenzies Building, Room E. 970, Clayton, VIC 3800, Australia
e-mail: allison.lassiter@monash.edu

© Springer International Publishing Switzerland 2017
P. Thakuriah et al. (eds.), *Seeing Cities Through Big Data*, Springer Geography,
DOI 10.1007/978-3-319-40902-3_29

1 Introduction

In the age of Big Data, mobile technology is one of the most crucial sources of data exchange. Analysts are examining the preferences, behaviors and opinions of the public through status updates, tweets, photos, videos, GPS tracks and check-ins. In turn, urban residents are accessing the same data as they view restaurant reviews with Yelp, find a ride with Uber, and stream Instagram photos. Untethered devices such as smartphones and tablets are critical to real-time, on-the-go data uploading and access. On these mobile devices, the public is connecting to data through apps. Apps are becoming the primary interface between data and the public.

At present, apps are primarily created by private companies seeking to profit from granular knowledge of urban behaviors. Yet, the allure and potential of apps is increasingly recognized by non-profit and government organizations, with development encouraged from the federal government all the way down to local municipalities. In the private tech industry, the "hackathon," a short and intense period of collaborative brainstorming, development, and coding, is a standard model of app development, and now the public sector is following suit. So-called "civic hackathons" are rapidly proliferating. Many view civic hackathons and app development as an exciting indicator of a new era of collaborative, open governance and bottom-up engagement. In the words of the promoters behind the National Day of Civic Hacking,

> *Civic hackers are community members (engineers, software developers, designers, entrepreneurs, activists, concerned citizens) who collaborate with others, including government, to invent ways to improve quality of life in their communities...Participants will use technology, publicly available data, and entrepreneurial thinking to tackle some of our most pressing social challenges such as coordination of homeless shelters or access to fresh, local, affordable food.* (Hack for Change 2014a)

In private industry, the utility of the hackathon is usually clear: employees work to innovate new products that will keep the company on the cutting edge of the market, often with potential shared profits (Krueger 2012). The goals of civic hackathons are less so. Ostensibly, they consist of citizen developers and representatives donating their time to create apps that address community wants and needs. The structure of the events, however, heavily influences the kind of apps produced, their intended users, and their long-term sustainability. In this study, we examine three models of civic hacking, develop a taxonomy of civic apps based on their structure, and offer cautions and suggestions for future civic hackathons.

2 Premises of the Hackathon

Hackathons first gained popularity through the 2000s, as technology companies informally hosted hacking marathons. Hackathons were intended to promote exploratory coding, new idea generation, and prototyping in a low-risk environment. As a Facebook software engineer Pedram Keyani wrote, "hackathons are our time to take any idea—big or small, sane or crazy—and build it into something real for people to react to" (Keyani 2012). Facebook held its first official, in-house hackathon in 2007. Since then, "[e]very couple of months, a few hundred of our engineers unleash their talents in epic, all-night coding sessions," working alongside people in different departments with different skill sets. Facebook's hackathons were greeted with the intensity and enthusiasm common among young coders, where "...everyone keeps working until around 6:00 am or when they pass out—whichever comes first" (Keyani 2012). Yet, since the first all-night pizza- and caffeine-fueled marathons, hackathons have expanded and become more inclusive. Facebook now hosts daytime events, allowing a wider range of people to participate (Krueger 2012).

Not surprisingly, the low cost, low risk, and often innovative hackathon model is now spilling out beyond the tech industry. For governments, hackathons seem to offer answers to two major issues. Firstly, governments often have a wealth of data, yet lack the capacity to process and innovate with it. Hackathons not only bring in virtually free labor in the spirit of "volunteerism and civic duty," (Hack for Change 2014b) but often facilitate mobile app development. Supporting app development helps governments seem contemporary, innovative and efficient. Moreover by releasing the data used in app development, governments can make the claim to transparency in their operations.

Since participation and team collaboration are built into hackathons, they foster a spirit of open governance. Citizens define their own problems and solutions, as well as a voice for their own communities, often while working with government-supplied Open Data. Material for the National Day for Civic Hacking lauds that hackathons are "... representative of a movement that is underway to leverage the power of technology and engaged neighbors to minimize barriers between government and citizens. National Day of Civic Hacking is truly about citizens stepping up to their role in government from the local to the state to the federal level...NDoCH represents the movement toward a truly collaborative government/citizen relationship of the future" (Hack for Change 2014b). Whether the hackathons have lived up to this promise has yet to be seen.

Hackathons profess a more fluid democracy, a seeming rebuke to the top-down problem solving of the previous age of urban informatics (Greenfield 2013; Townsend 2013). As an article on the Open NASA blog states, "hackathons frequently show us insights and applications that we never would have imagined coming from our own work. These technology development events don't give us all the answers—but they engage the public in exploration of our data and our challenges in creative and compelling ways, sparking a flame that just might become something big and powerful. A hackathon isn't a product, it's an approach..." (Llewellyn 2012). Hackathons have potential as an approach that engages the public and harnesses the creativity of the crowd. This focus on the hackathon as an approach is not consistently at the forefront, however. Some contest administrators tout the hacakathon's potential to problem solve and tackle challenges (NYCEDC 2016; Hack for Change 2014a).

As Open and Big Data proliferate, there is seemingly little reason not to hold a hackathon. The stakes can be very low. However, to use a tech industry term, are hackathons "disrupting" models of governance by widening participation? Is app generation leading to more efficient problem solving?

3 Civic Hackathon Models

As a way to examine the premises of civic hackathons, we examine three high profile events in the United States: the 2013 White House Hackathon, the 2013 New York City BigApps contest, and the 2014 National Day of Civic Hacking. These three hackathons embody the different approaches and goals of civic hackathons. The scales, organizational principles, and predominant types of apps generated vary, but together they reveal issues common across hackathon models.

3.1 2013 White House Hackathon

Since the Obama Administration took office, they have supported open data and open governance. Their agenda was formalized through 2009s Open Government Directive. The directive required all 143 United States federal agencies to upload all non-confidential data sets to a newly created sharing platform, data.gov, by November 9, 2013 (The White House, Office of Management and Budget 2009). Disseminating newly disclosed data through a hackathon is attractive because of the transparency and participation associated with the model. Hackathon proponents assert that creating civic apps is critical to maximally leveraging Open Data.

The 2013 White House Hackathon focused on an Application Program Interface (API), a code library that gives immediate access to data as it is updated. The API, We the People, provides data on federal petitions, such as information concerning when the petition was created and how many people have signed it. From this API, civic hackers developed apps that made the data accessible in other formats, show the number of signatures in real time, and map spatial patterns of support (The White House We the People 2013). Given the limited data at hand, the fact that over 30 apps developed could be considered a success. However, of the 66,146 available datasets on data.gov, which range from environmental to budget information, the choice to make petition information the focus of the first hackathon is puzzling. Possibly, petition data were selected as the locus because petitions indicate an open, collaborative government, but creating better access to petition data does little to change the minimal impact of petitions in federal governance. The second White House hackathon held in 2014 (results still pending at the writing of this paper) continued to focus on petitions with Petitions Write API, which is intended to expand the platforms and sites through which people can submit petitions (Heyman 2014).

The 2013 White House Hackathon produced many prototypes, but few enduring projects. Since the competition, none of the apps have been institutionalized by the government, nor does it appear the apps have been updated or distributed. Video demonstrations of the apps are available, but few of the apps are directly accessible or downloadable. One downloadable program, a code library that extends analytic possibilities by porting petition data into the statistical program R, does not work with the current version of R.

While the initial tenets of Obama's Open Government Directive were "transparency, participation, and collaboration," (The White House, Office of Management and Budget 2009), the collaboration component has largely fallen by the wayside (Peled 2011). Spokespeople for the US government's Open Data Initiative now choose to highlight the potential of transparency and downplay the failure of federal agencies to use their data. The White House has made a very public push to tout transparency as a virtue in of its own sake. While the hackathon may widen the range of participants in governance, it falls short of deeply collaborative, open governance.

One of the tangible successes of the hackathon is that all the code produced was made available through the public code repository, GitHub. Everything that was finalized in the hackathon is now a public resource, so it could potentially be accessed and built upon in the future. The optimistic view is that "[w]ith each hackathon, some of the detritus—bits of code, training videos, documentation, the right people trading email addresses—becomes scaffolding for the attendees of later ones" (Judd 2011). It may be left to the developer community, however, and not the White House, to expand the results.

3.2 2013 New York City BigApps Competition

New York City's Big Apps contest is one of the longest running civic hackathons. It is also considered one of the most successful, in terms of apps sustained past the contest period. The competition originated with former Mayor Michael Bloomberg, who also pioneered some of the first Open Data legislation in the country and was the first to appoint a Chief Information and Innovation Officer, Rahul N. Merchant, in 2012 (New York City Office of the Mayor 2013). Both Bloomberg and Merchant brought considerable experience in the business, financial and technology sectors. They were able to secure major sponsors such as Facebook, eBay, Microsoft, and Google for the contest.

Some believe this kind of private-public partnership is key to garnering talent and funding apps that will sustain beyond the competition. In 2013, a panel of judges awarded $55,000 to the BigApps winner, and amounts ranging from $5000 to $25,000 to runners up in several categories. In all 3 years of the competition, BigApps has also always awarded an "Investor's Choice" prize, underscoring its focus on apps that offer a financial return.

Yet, BigApps does not explicitly require financial returns. Instead, it asks that participants explore ways to use technology to make New York City a better place to "live, work, learn, or play" (NYCEDC 2016). It also offers participants a chance to tackle known challenges supplied by 30 private and public entities. These range from improving health access, finding charging stations for electric cars, and helping network parents of students in the New York City school system. Past participants have occasionally taken on these challenges, but most winners of the contest come up with their own ideas, usually with the intent of monetization. The first place winner of the 2013 contest, HealthyOut, uses a Yelp API to help connect people with healthier food delivery options. HealthyOut has since raised $1.2 million in venture capital, beyond the $55,000 in prize money earned from BigApps (Lawler 2013).

The apps that have endured since the 2013 competition were able to secure substantial funding and develop a revenue stream. Hopscotch, an educational coding iPad app for children, also raised $1.2 million in seed funding (Lomas 2014) and is a top seller in the iPad store. The majority of BigApps winners with more community minded goals are no longer functional, since they were unable to create a sustainable financial model. HelpingHands, one of the prize winning apps that helped NYC residents enroll in social services, was not available in app stores at the time of this writing and the domain name was for sale.

It could be argued that BigApps encourages entrepreneurship and feeds money back into the city. The leaders of BigApps claim successful ideas will be rewarded with the resources to sustain them (Brustein 2012). However, its additional claims "that [it] empowers the sharpest minds in tech, design, and business to solve NYC's toughest challenges" (NYCEDC) rings hollow. The slogan itself recognizes that only a select group is empowered. Perhaps as a course correction to this issue and response to several public criticisms of the non-civic goals of the winners (Brustein 2012), the 2014 and 2015 contests were considerably more focused in scope and engaged tech and civic organizations for mentorship as well as offered contestants a chance to work directly with public agencies to solve their specific, pre-defined problems. Most notably, the 2014 context expanded the scope of contest deliverables past smartphone apps to device design and data tools, among others. The contest has sustained past the Bloomberg administration, and in 2015, the contest attempted to link to policy initiatives by responding directly to Mayor de Blasio's OneNYC plan by asking participants to specifically address issues of Affordable Housing, Zero Waste, Connected Cities, and Civic Engagement (NYCEDC 2016). However, the contest still stops at the early idea stage, and as of this writing most winners were still seeking further development and financial support.

3.3 2014 National Day of Civic Hacking

Unlike the White House Hackathon and BigApps, the National Day of Civic Hacking is not a government-driven initiative. The hackathon was organized by consultant Second Muse with non-profits Code for America, Innovation Endeavors and Random Hacks of Kindness. It is sponsored by Intel and the Knight Foundation, with support from the White House Office of Science and Technology Policy, several federal and state agencies, and other private companies. NDoCH aims to address hyperlocal issues by promoting a coordinated set of nation-wide hackathons hosted by individual localities on their own terms (Hack for Change 2014c; SecondMuse 2013).

NDoCH has few rules and the hackathon is interpreted broadly by participating cities and states. The full set of projects from NDoCH is messy, but successfully conveys the varying interests in participating areas. Most groups created apps or websites, such as Hack 4 Colorado's FloodForecast, which notifies users if their home address is in danger of flooding. Other groups worked on alternative technical projects. Maine Day of Civic Hacking, for example, focused on repairing a stop motion animation film in a local museum.

Secondary to its local focus, NDoCH foregrounds some national issues that participants can choose to address. In 2014, formal Challenges were advertised by federal agencies like the Consumer Financial Protection Bureau and the Federal Highway Administration. The Peace Corps, for example, requested "a fun, engaging and easy-to-use interface with the numerous and diverse Peace Corps volunteer opportunities that helps the user find the right opportunity for them" (Hack for Change 2014d), which was subsequently prototyped at the San Francisco Day of Civic Hacking (Hack for Change 2014e). NASA's Challenge to increase awareness of coastal inundation spurred several related projects.

Like the White House Hackathon, the route between hack and implementation is unclear. But, NDoCH's guiding principles are oriented toward the process of the hackathon, rather than the results. Stated goals include "Demonstrate a commitment to the principles of transparency, participation, and collaboration" and "Promote Science, Technology, Engineering and Mathematics (STEM) education by encouraging students to utilize open technology for solutions to real challenges" (SecondMuse 2013). Participants may not clearly understand, however, that most apps will not survive past the hackathon. The tension between process and product is exacerbated by reports from the NDoCH that tout the number of apps produced, not just process-related goals. Ensuring authentic, collaborative processes over app development remains a challenge.

4 A Taxonomy of Civic Apps

How do civic apps propose to solve problems? In order to better understand the products of the civic hackathon, we examined the results for the White House 2013 Hackathon, New York BigApps Contest 2013 and the National Day of Civic Hacking 2014. We evaluate their descriptions, demonstrations, and locate them in the Apple and Android stores, as applicable. For White House and BigApps, we evaluated information on the winning entries. For NDoCH, there were no selected winners and information on associated efforts was available, so we examined all entries.

We propose five categories that describe the apps produced across all events: spatial customization and personal services; spatial awareness and data communication; community building; educational; data gateways. These categories are described further in the following sections.

4.1 Spatial Customization and Personal Services

Many of our most used commercial apps such as Google Maps, Yelp, specialize in the spatial customization of individual daily routines. These apps, powered by ever expanding GPS technology, have advanced the "spatially enabled society," where citizens are better able to communicate with the world around them (Roche et al. 2012: 222). In spatially enabled society, "the question ceases to be simply 'Where am I?' and becomes: 'What is around me?' (as in services, people, and traffic), 'What can I expect?' and 'How do I get there?'" These apps center around easing mobility and in many cases consumption in the city—finding parking, giving real-time transit alerts, or customizing personal routes given a set of favorable inputs.

It is no surprise that many civic apps have also seized on expanding the suite of spatial customization apps, as mobility and movement in the city is a continuing challenge, while self-locating with GPS remains a relatively new possibility. The majority of the spatial customization apps provide real time transit alerts. In addition, services like HealthyOut tells a user which restaurants near them are best suited to their personal diet, while another app, called Poncho, tells what the weather will be at every location in your daily routine. These spatial customization apps are focused on the desires of the individual, harnessing open data to ease everyday life.

Some of these apps also crowdsource data from users, aggregate the data, and then provide users with continually updated, socially-derived urban data. Some of these apps are oriented toward typically underserved populations. Ability Anyware Assistive Technology Survey and Enabled City, built during the NDoCH, identifies and find accessible routes and buildings for people with disabilities.

We also include personal services in this category. Even though this information is sometimes aspatial, these services do help make urban life more efficient to the individual. Two apps built at NYC BigApps, ChildcareDesk and HiredinNY, intended to connect users to child care centers and jobs, respectively.

4.2 Spatial Awareness and Data Communication

The spatially enabled society is also at the heart of many apps that are not expressly built for individual easing, but rather to simply visualize otherwise invisible information to increase awareness among app users. Apps produced at the Virginia Beach hackathon under NDoCH mapped the effects of potential sea level rise. Others mapped child hunger statistics (Maine Child Hunger Cartogram Viewer) and SNAP benefits (SNAPshot) with the professed intent of spurring empathy.

In addition, this category of apps uses spatial information to encourage individuals to act in their own community, by bringing visibility to difficult-to-perceive issues. Freewheeling NC, an app built at a North Carolina hackathon during the National Day of Civic Hacking, crowdsources bike routes with the intent of influencing urban planning. Many of these apps use public data on underutilized vacant land in order to help community members to use them as gathering places, urban agriculture, or new development (Minimum Adaptable Viable Urban Space (MAVUS); [freespace] ATX; Abandoned STL).

Some apps similarly raise awareness through aspatial data communication. Often aimed at government transparency, such as several apps developed at the White House Hackathon with the We the People API, or campaign finance information and city council agendas. At times, data communication and spatial awareness come together, such as Flood Forecast's flood notifications.

4.3 Community Building

Many apps built during the National Day of Civic Hacking focused on community building through peer-to-peer communication. These apps help people in niche groups find each other, such as connecting pet owners and teens. Community building apps also help pair volunteers with nonprofit organizations such as Habitat for Humanity, help people find places to donate leftover food, and allow citizens direct access government officials.

4.4 Educational

The smallest group of apps are educational, which are often aimed at youth and incorporate a gaming component. The aforementioned Hopscotch at NYC BigApps teaches children to code, and two apps built under NDoCH aimed to teach users about watersheds and urban geography by using a Minecraft-like interface.

4.5 Data Gateways

Lastly, some apps are simply focused on making data accessible in a different machine-readable format or providing analytic environments for the data, but not for specific purposes. These are often interfaces geared towards developers to create even more apps. Almost a third of the apps from the White House Hackathon fall into this category; other data gateways were created for Peace Corps project data, hospital discharge costs, and even presidential inaugural addresses for textual analysis.

5 Models and Results

Table 1 presents a summary of the results from each hackathon against the app taxonomy. Because of the large variations in end product numbers—17 for the White House Hackathon, 7 in the BigApps contest, and 71 during the National Day of Civic Hacking—the percentage of the total apps is given in Table 1.

The structure, funding, and data released in each type of hackathon not only influenced the scale and number of apps, but the predominant type of apps produced. At the White House Hackathon, no winners were declared, but results and code were posted for only 17 out of 30 completed projects executed at the hackathon. Of the 17 apps, 12 communicated Spatial Awareness and Data Communication, while 5 focused on Data Access. This is no doubt related to the focus on only one dataset. It is also notable that the White House hackathon goals were not to solve any particular urban challenge, but rather simply to see what technological expertise could do with a set of open data.

Table 1 Results from the White House Hackathon 2013, New York City's BiggApps Contest 2013, and the National Day of Civic Hacking 2014

	Spatial customization and personal services	Spatial awareness and data communication	Data gateways	Community building	Educational	Other
White House Hackathon 2013 (17)	0 % (0)	71 % (12)	29 % (5)	0 % (0)	0 % (0)	0 % (0)
NYC BigApps 2013 (7)	86 % (6)	0 % (0)	0 % (0)	0 % (0)	14 % (1)	0 % (0)
National Day of Civic Hacking 2014 (71)	13 % (9)	34 % (24)	10 % (7)	28 % (20)	8 % (6)	7 % (5)

The total number of apps produced is given in parentheses

The total number of entries in New York City's BigApps is not available, but all seven winning entries earned prize money. It was the only hackathon of the three to offer prize money and the only to boast significant private business partners and potential for investors. Among the apps, 6 out of 7 are focused on personal mobility and personal services, with the remaining app was educational. While the contest claims that it is bringing together experts and developers to "solve New York's toughest challenges," its results thus far indicate that it is more focused on apps that ease individual consumption and mobility; instead of public data sets, the winning entry used the commercial API from Yelp.

While the White House Hackathon and BigApps each produced narrow results, the National Day of Civic Hacking generated many apps, crossing all five categories. Of the 71 products from the competition, some of the results are as technically sophisticated as those developed in BigApps and the White House Hackathon. Others entries are not apps at all—they are requests for apps or brainstorming sessions regarding the potential for apps. These "Other" results make up 7 % of all the entries into NDoCH. Notably, however, 28 % of NDoCH projects focused on Community Building, which was absent from the other two hackathons. Of Spatial Customization and Personal Services apps, several focused on finding accessible facilities for disabled citizens or improving public amenities such as bike lanes, trails, and parks. As the NDoCH organizers hoped, the apps associated with NDoCH reveal community-driven, locally-specific issues and civic innovation. However, they also reveal the technological limits of many localities. It is not surprising that the greatest number and most sophisticated apps come from technology hubs such as Palo Alto and Austin, where some of the more distant outposts did not have the expertise to even produce an app at the end of the event.

6 Future Hackathons

The landscape of civic data and civic apps is rapidly changing, corresponding with the rise of Open Data and Big Data, expanding mobile technology, and trends toward technocracy (Mattern 2013). The 2014 National Day of Civic Hacking, then in its second year, claimed a 30 % increase in events from its first hackathon (Llewellyn 2014). Promoting civic hackathons is a not only a low-risk, adaptable method of embracing contemporary problem solving amidst change, but previous ones have generated enough success to keep attempting them. Hackathons have fostered new types of civic participation, created enthusiasm among some community members, developed some new apps, and may broadly encourage more technological innovation in government. Yet, while there is evidence of successes, civic hackathons face unique challenges that must be addressed in order to deepen their impact.

The examples proffered here show that there are three common and interrelated issues with the hackathon model. Firstly, defining problems that are meaningful for the community. Secondly, the challenge of aligning the goals of market-ready apps with civic services. Thirdly, and most importantly, that the civic hackathon has the responsibility of addressing the needs of its full constituency, not simply the smartphone-owning, tech-literate public.

6.1 Defining Problems

Many experts have identified that lack of structure in hackathons, meant to encourage out of the box approaches, can lead to unfocused results. NASA's open data portal, for example, states that the key to implementing a successful hackathon "is to invest the effort to identify the right problem statements, provide the supporting data, and get a good mix of people in the room" (Llewellyn 2012). As data scientist Jake Porway warns, "They are not easy to get right... You need to have a clear problem definition, include people who understand the data not just data analysis, and be deeply sensitive with the data you're analyzing" (Porway 2013).

When any local government is faced with a challenge, deeply understanding the data being analyzed, the physical and social context, and possible opportunities is crucial. Developing the web of knowledge necessary to solve most issues takes time. While the public can offer needed fresh perspectives and local insight, it can be difficult for outsiders to come in and hit the ground running, which is necessary in a short-lived, intense hackathon. Many of the apps that come out of these contests help users navigate the city, rate local places, or plan itineraries—all services that are already well-covered and arguably better developed by large tech companies (Brustein 2012).

Yet, "problem solving" is often touted as the key tenet of many hackathons. BigApps seeks to "solve specific New York City challenges, known as BigIssues" (New York City Office of the Mayor 2012). BigApps identified four BigIssues for the 2013 competition: Jobs and Economic Mobility, Healthy Living, Lifelong Learning, and Cleanweb: Energy, Environment, and Resilience. Of course, it is nearly impossible to *solve* any sort of complex issue, like a BigIssue, in the context of a hackathon. Creating nuanced solutions requires both quantitative expertise and experience. If hackathons are going to meaningfully address difficult city problems, it likely is necessary to create hybrid teams of public participants and government employees, as the NYC BigApps contest has started to do. However, most crucially, the contests need to consider a structure that can commit to working through proposals beyond the short-lived timeframe of the hackathon.

6.2 Market-Ready Solutions

Because the hackathon is temporary by definition, perhaps it is not surprising that the results are commonly temporary as well. Hackathon products are often still in the brainstorming stage and are rarely taken to completion. The bulk of the apps that survive past a hackathon are market-ready and able to attract venture capital during or shortly after the hackathon.

At NDoCH, the organizers recognize that governments will not bear the responsibility for apps after their creation:

> *"Each new technology has a unique path to implementation. The key to the development of technologies that make their way out of the hackathon environment and into your community are public and private partnerships. One path to sustainability is that a group of volunteers develops a new app to connect low-income residents to the nearest free tax preparation site over the course of National Day of Civic Hacking. Following the event, the volunteers reach out to economic justice groups in their community so they can promote their services using the app, seek a sponsor to offset the cost of the text message usage, and work with government officials to promote the app as well as the availability of free tax prep in your city."* (Hack for Change 2014b)

Not only is the onus of innovation and development shifted to a narrow swath of the data-literate public, but the growth and sustainability is as well.

Sustainability is identified as the primary issue by many technologists. Code for America's Dan Melton writes, "...some of the biggest examples of disconnect and potential opportunity come out of app contests or hackathons. Policy makers/political leaders champion city or social contests, to which, developers respond with dozens or even hundreds of submissions. So far so good. When the app contest is over, often too is the partnership" (Melton 2011). O'Reilly Media editor Andy Oram adds, "...how could one expect a developer to put in the time to maintain an app, much less turn it into a robust, broadly useful tool for the general public?... The payoff for something in the public sphere just isn't there" (Oram 2011). Organizations like CivicApps (http://civicapps.org/), help to overcome sustainability issues by promoting apps for wider distribution, but nonetheless few are self-sustaining. Even the more civic-minded winning apps at NYC BigApps 2013 such as Helping Hands and HiredinNY are nowhere to be seen 1 year later, despite funding awards.

There is some implication that failure may be, in part, because of app quality. Refining existing ideas could help improve sustainability. Joshua Brustein of the *New York Times* (2012) says, "Inevitably, most of these projects will fail. Start-ups usually do. And considering the modest price of the program—BigApps costs the city about $100,000 a year, according to the city's Economic Development Corporation—the bar for success should be set low. But it seems that a better investment might be to spend more time working with a few developers on specific ideas, rather than continually soliciting new ones" (Brustein 2012). Tackling this issue will require hackathon organizers to turn an eye to building communities before building apps. Clay Johnson of the Sunlight Foundation, a nonprofit dedicated to government transparency, notes that they see their hackathons as only the beginning of their engagement with both developers and volunteers (Johnson 2010). There are some cases of longer term partnerships, such as the Federal Registry's partnership with the winners of the 2010 Apps for America civic hackathon to create a redesigned data distribution portal (Oram 2011), but these examples of government commitment are rare.

6.3 Participation and Representation

Revenue and consumption are the bases of most of the successful, sustaining hackathon propositions. This makes it challenging to address problems that are not profitable. Issues of the "unexotic underclass," such as veterans and welfare recipients, often go unaddressed (Nnaemeka 2013). This is evident despite Open Data's promise of egalitarianism and the participatory goals of hackathons.

One piece of this challenge is the demographic of participants that hackathons typically attract. The majority of hackathon developers are young, well-educated, and relatively affluent. Unsurprisingly, the majority of apps cater to this demographic, even in a civic context. As technologist Anthony Townsend (2013: 166) writes, "...should we be surprised when they solve their own problems first?...Not only do they not represent the full range of the city's people; often these hackers lack a sense that it's even their duty to help others...". Representation bias is also noted by Porway (2013), when he similarly writes about a New York City hackathon focused on greening the city, "...as a young affluent hacker, my problem isn't improving the city's recycling programs, it's finding kale on Saturdays."

Organizers should particularly look to increase the participation of women and low-income communities. The all-night structure and lack of code of conduct can be intimidating to women (Rinearson 2013). The National Center for Women and Information Technology notes the importance of specifically recruiting girls for events, not only as coders but as judges and mentors (NCWIT n.d.). One organization, Yes We Code, is responding by hosting their own hackathons that support ideas from low-income teens (Yes We Code 2013). Rather than hosting separate events, however, it is the responsibility of the city to ensure that such organizations and their constituents have a voice in the city-sponsored hackathon.

Representation bias also extends to the hardware. Only 56% of the U.S. population own smartphones and owners are primarily people under 35, well educated, and affluent (Smith 2013). Apps are, themselves, a limiting format. Though they continue to increase in popularity, many difficult to reach groups cannot be accessed with apps. In a world where mobile data contributes to visibility and voice, those that are not able to partake become invisible.

Recent language in the NYC BigApps contest and National Day of Civic Hacking acknowledges that apps can be exclusive. In 2014, BigApps will accept competition entries that use a wider array of technology products (NYCEDC 2016). Expanding the technology to gather data and input may have surprising outcomes. When New York City's non-emergency reporting system, 311, added a website and then app to their telephone reporting system, they expected call volumes to go down. Instead, they saw an overall increase in reporting, showing that different mediums helped access different sectors of the population. Moving away from the app interface could encourage broader engagement.

7 Conclusion

As of this writing, the White House has not held another hackathon, but NYC BigApps and the National Day of Civic Hacking persist into 2016. While these contests have been a reasonably convenient, low investment way for governments to appear to engage in innovative, technology-enabled problem solving, the existing hackathon models make it difficult to address truly complex, non-monetizeable issues. As governments grapple with what to do with their newly opened datasets and how to handle much of their Big Data, they are left with some difficult challenges. How should governments ensure that civic innovations are institutionalized and sustained without being dependent on private backers? How do they ensure everyone is being fairly represented—that those on the far side of the digital divide are not left out of the wake of technological progress?

The first step to improving the civic hackathon is to subject it to the same scrutiny as any other urban practice. This includes clarifying the goals of hackathons and developing associated metrics. If apps have the possibility of creating efficiency gains, results should be internalized and governments should commit personnel resources to hackathons, support scaling, and dedicate money for ongoing operation. Alternatively, are hackathons intended to kickstart for-profit app businesses? If so, the role of public money in this process should be made clear and the surface claims to solving complex urban issues should be eliminated. Or, are hackathons a method of signaling open governance? If so, this method of participation should be examined against existing models of collaborative governance. While the analytical literature surrounding hackathons is scarce, it is necessary to develop best practices for running a hackathon and building on the results. In doing so, the hackathon may have the opportunity to become everything it wants to be: transparent, collaborative, and innovative. For now, however, the lofty goals remain unmet.

References

Brustein J (2012) Contest whose winners may not succeed. New York Times. http://www.nytimes.com/2012/03/04/nyregion/new-yorks-bigapps-contest-has-mixed-results.html

Greenfield A (2013) Against the smart city (the city is here for you to use) [Kindle eBook ed.]. Amazon Digital Services

Hack for Change (2014a) Key highlights. http://hackforchange.org/about/key-highlights/

Hack for Change (2014b) FAQ. http://hackforchange.org/about/faq/

Hack for Change (2014c) About. http://hackforchange.org/about/

Hack for Change (2014d) Challenges. http://hackforchange.org/challenges/

Hack for Change (2014e) Peace corps peace. http://hackforchange.org/projects/peace-corps-peace-2/

Heyman L (2014) Announcing the White House's second annual civic hackathon [Blog post]. The White House Blog. http://www.whitehouse.gov/blog/2014/05/01/announcing-white-houses-second-annual-civic-hackathon

Johnson C (2010) Build communities not contests [Blog post]. The Information Diet. http://www.informationdiet.com/blog/read/build-communities-not-contests

Judd N (2011) Code for America's chief geek says civic hackers should fix hackathons next. TechPresident. http://techpresident.com/short-post/code-americas-chief-geek-says-civic-hackers-should-fix-hackathons-next

Keyani P (2012) Stay focused and keep hacking [Blog post]. Facebook Engineering Notes. https://www.facebook.com/notes/facebook-engineering/stay-focused-and-keep-hacking/10150842676418920

Krueger A (2012) Hackathons aren't just for hacking. Wired Magazine. http://www.wired.com/2012/06/hackathons-arent-just-for-hacking/

Lawler R (2013) HealthyOut is like a personal nutritionist for healthy food deliveries. TechCrunch. http://techcrunch.com/2013/04/30/healthyout/

Llewellyn A (2012) The power of hackathons in government [Blog post]. NASA Blog. http://open.nasa.gov/blog/2012/06/29/the-power-of-hackathons-in-government/

Llewellyn A (2014) National day by the numbers. Hack for Change Blog. http://hackforchange.org/national-day-by-the-numbers/

Lomas N (2014) Hopscotch, an iPad app that helps kids learn to code, raises $1.2M. TechCrunch. http://techcrunch.com/2014/05/08/hopscotch-seed/

Mattern S (2013) Methodolatry and the art of measure. Places: Design Observer. http://places.designobserver.com/feature/methodolatry-in-urban-data-science/38174/

Melton D (2011) Scaling our movement [Blog post]. Code for America Blog. http://www.codeforamerica.org/blog/2011/08/17/scaling-our-movement/

National Center for Women in Technology (n.d.) Top 10 ways to increase girls' participation in computing competitions. http://www.ncwit.org/resources/top-10-ways-increase-girls-participation-computing-competitions/top-10-ways-increase-girls

New York City Economic Development Corporation (NYCEDC) (2016) NYC BigApps past competitions. http://www.nycedc.com/services/nyc-bigapps/past-competitions

New York City Office of the Mayor (2012) Mayor Bloomberg appoints Rahul N. Merchant as the city's first chief information and innovation officer [Press Release]. News from the Blue Room. http://www.nyc.gov/cgi-bin/misc/pfprinter.cgi?action=print&sitename=OM&p=1405371522000

New York City Office of the Mayor (2013) Mayor Bloomberg announces winners of NYC BigApps, fourth annual competition to create apps using city data [Blog Post]. http://www1.nyc.gov/office-of-the-mayor/news/215-13/mayor-bloomberg-winners-nyc-bigapps-fourth-annual-competition-create-apps-using

Nnaemeka C (2013) The unexotic underclass. MIT Entrepreneurship Review. http://miter.mit.edu/the-unexotic-underclass/

Oram A (2011) App outreach and sustainability: lessons learned by Portland, Oregon. Radar Blog (O'Reilly Media). http://radar.oreilly.com/2011/07/app-outreach-and-sustainabilit.html

Peled A (2011) When transparency and collaboration collide: the USA open data program. J Am Soc Inf Sci Technol 62(11):2085–2094

Porway J (2013) You can't just hack your way to social change [Blog post]. Harvard Business Review Blog Network. http://blogs.hbr.org/2013/03/you-cant-just-hack-your-way-to/

Rinearson T (2013) Running an inclusive hackathon. Medium. https://medium.com/hackers-and-hacking/running-an-inclusive-hackathon-630f3f2e5e71

Roche S, Nabian N, Kloeckl K, Ratti C (2012) Are 'smart cities' smart enough. In: Global geospatial conference, Quebec, Canada, 14–17 May

SecondMuse (2013) National day of civic hacking. http://secondmuse.com/project3.html

Smith A (2013) Smartphone ownership—2013 update. Pew Internet & American Life Project. Washington, DC. http://www.pewinternet.org/~/media/Files/Reports/2013/PIP_Smartphone_adoption_2013.pdf

The White House, Office of Management and Budget (2009) Open government directive [Press release]. http://www.whitehouse.gov/open/documents/open-government-directive

The White House, We the People (2013) We the people API gallery. https://petitions.whitehouse.gov/how-why/api-gallery

Townsend A (2013) Smart cities: big data, civic hackers, and the quest for a new utopia. W.W. Norton, New York

Yes We Code (2013) Homepage. http://www.yeswecode.org/

ERRATUM TO

Planning for the Change: Mapping Sea Level Rise and Storm Inundation in Sherman Island Using 3Di Hydrodynamic Model and LiDAR

Yang Ju, Wei-Chen Hsu, John D. Radke, William Fourt,
Wei Lang, Olivier Hoes, Howard Foster, Gregory S. Biging,
Martine Schmidt-Poolman, Rosanna Neuhausler, Amna Alruheil,
and William Maier

© Springer International Publishing Switzerland 2017
P. Thakuriah et al. (eds.), *Seeing Cities Through Big Data*, Springer Geography,
DOI 10.1007/978-3-319-40902-3_18

DOI 10.1007/978-3-319-40902-3_18

This chapter was funded by the California Energy Commission – PIER, California Climate Change Impacts Program (UCB 500-11-016) and the grant was unfortunately not acknowledged in the original version.

The online version of the original chapter can be found at
DOI 10.1007/978-3-319-40902-3_18

© Springer International Publishing Switzerland 2017
P. Thakuriah et al. (eds.), *Seeing Cities Through Big Data*, Springer Geography,
DOI 10.1007/978-3-319-40902-3_30